Cell and Molecular Biology of Artemia Development

NATO ASI Series

Advanced Science Institutes Series

A series presenting the results of activities sponsored by the NATO Science Committee, which aims at the dissemination of advanced scientific and technological knowledge, with a view to strengthening links between scientific communities.

The series is published by an international board of publishers in conjunction with the NATO Scientific Affairs Division

A	**Life Sciences**	Plenum Publishing Corporation
B	**Physics**	New York and London
C	**Mathematical**	Kluwer Academic Publishers
	and Physical Sciences	Dordrecht, Boston, and London
D	**Behavioral and Social Sciences**	
E	**Applied Sciences**	
F	**Computer and Systems Sciences**	Springer-Verlag
G	**Ecological Sciences**	Berlin, Heidelberg, New York, London,
H	**Cell Biology**	Paris, and Tokyo

Recent Volumes in this Series

Series A: Life Sciences

Cell and Molecular Biology of Artemia Development

Edited by

Alden H. Warner

University of Windsor
Windsor, Ontario, Canada

Thomas H. MacRae

Dalhousie University
Halifax, Nova Scotia, Canada

and

Joseph C. Bagshaw

Worcester Polytechnic Institute
Worcester, Massachusetts

Plenum Press
New York and London
Published in cooperation with NATO Scientific Affairs Division

Proceedings of a NATO Advanced Research Workshop on
Cell and Molecular Biology of Artemia Development,
held August 11–13, 1988,
in Montreal, Quebec, Canada

Library of Congress Cataloging in Publication Data

NATO Advanced Research Workshop on Cell and Molecular Biology of Artemia
Development (1988: Montreal, Quebec)
    Cell and molecular biology of artemia development / edited by Alden H.
Warner, Thomas H. MacRae, and Joseph C. Bagshaw.
        p.   cm.—(NATO ASI series. Series A, Life sciences; vol. 174)
    "Published in cooperation with NATO Scientific Affairs Division."
    "Proceedings of a NATO Advanced Research Workshop on Cell and Molecular
Biology of Artemia Development, held August 11–13, 1988, in Montreal, Quebec,
Canada"—CIP t.p. verso.
    Includes bibliographical references and index.
    ISBN 978-1-4757-0006-0   ISBN 978-1-4757-0004-6 (e-BOOK)
    DOI 10.1007/978-1-4757-0004-6
    1. Artemia—Development—Congresses. 2. Artemia—Cytology—Congresses.
3. Molecular biology—Congresses. I. Warner, Alden H. (Alden Howard),
1937–   . II. MacRae, Thomas H. III. Bagshaw, Joseph C., 1943–   . IV. North
Atlantic Treaty Organization. Scientific Affairs Division. V. Title. VI. Series: NATO
ASI series. Series A, Life sciences; v. 174.
QL444.B815N37   1988   89-16165
595.3′2—dc20   CIP

© 1989 Plenum Press, New York
Softcopy reprint of the hardcopy 1st edition 1989
A Division of Plenum Publishing Corporation
233 Spring Street, New York, N.Y. 10013

PREFACE

The brine shrimp <u>Artemia</u> has become an important experimental system
for studies of the developmental process. In recent years the shrimp has
yielded considerable information on the pattern of development, biochemistry, and gene structure and expression of crustaceans. This book is
a compilation of research activity from twenty five of the most active research laboratories working with brine shrimp in the above areas. It also
represents the proceedings of a NATO Advanced Research Workshop held in
Montreal, Canada, August 11-13, 1988.

The book contains twenty nine full papers covering the major areas
discussed at the workshop. In addition, one page abstracts representing
seventeen poster presentations which were given at the workshop, and which
were deemed to be most relevant to the theme of the book, are included.
These are designated with an [a] in the Table of Contents following the
title of each paper. A considerable amount of discussion which took place
during the workshop has not been included in the book because of space
limitations. However, the editors will endeavour to make some of this information available at a later date through the <u>Artemia</u> <u>Newsletter.</u>

In addition to the high percentage of invited speakers who attended
and contributed to the workshop, the organizers would like to thank a
number of participants who made valuable contributions to the major discussion sessions. These include: John Freeman, Michael Horst, Herman
Slegers, Jack Vaughn, Frank Conte, Sandy McLennan, Clive Trotman and
Patrick Sorgeloos.

The editors would like to acknowledge the support of the Scientific
Affairs Division of NATO for sponsoring the Workshop and underwriting the
cost of publishing this book. We are also grateful to the Natural Sciences
and Engineering Research Council of Canada for their financial support,
supplemented further by support from Dalhousie University, the University
of Windsor, Worcester Polytechnic Institute and Sanders Brine Shrimp Company.

The editors thank Melanie Yelity and Gregory Safford and their staff
at the Plenum for their expert technical advice and the scaling of the
numerous prints and illustrations appearing in this book. Finally, we
wish to thank Mrs. Pamela Tehan for typing of the book and her patience in
dealing with the numerous editorial comments and changes that had to be
made in the process.

A. H. Warner, Windsor
T. H. MacRae, Halifax
J. C. Bagshaw, Worcester

March, 1989

CONTENTS

DORMANCY AND DEVELOPMENT

CELLULAR AND STRUCTURAL ASPECTS

*[a] - Abstract only

BIOCHEMICAL ASPECTS

MOLECULAR ASPECTS AND GENE STRUCTURE

TUBULINS AND HEMOGLOBINS

COMPUTER SEARCHES

ASPECTS OF THE ANAEROBIC METABOLISM OF ARTEMIA CYSTS

James S. Clegg and Susan A. Jackson

University of California
Bodega Marine Laboratory
Bodega Bay, CA 94923

INTRODUCTION

Among the many adaptations associated with the rigorous life history of the brine shrimp, Artemia is the striking resistance of the encysted embryo to anaerobiosis. In 1966 Dutrieu and Chrestia-Blanchine[1] reported that these cysts tolerated total anaerobiosis when incubated in sea water for over five months without a decrease in viability[2]. Subsequent work indicated that these anaerobic cysts did not carry out a conventional lactate-producing metabolism and that breakdown of trehalose, the disaccharide serving the energy metabolism of aerobic cysts, could not be detected over an 8 hour period of anoxia[3]. Comprehensive studies of the nucleotide pool by Stocco et al.[4] suggested that utilization of the unusual nucleotide, Gp_4G, might provide the free energy presumably required to support the maintenance of anaerobic cysts. Most recently, Hand and Gnaiger[5] used calorimetric methods to show that anaerobic energy metabolism is reduced to only about 2% of aerobic values. Those authors also calculated that the utilization of Gp_4G accounted for only about 2% of the heat dissipation taking place during anaerobiosis, and suggested that the very slow catabolism of trehalose might be a more likely explanation for their results. We will present some results of studies designed to test that suggestion.

While much progress has been made on the potential regulatory mechanisms involved in aerobic-anaerobic transitions (for reviews see Busa and Nuccitelli[6], Busa[7], and Hand and Gnaiger[4] very little is known about the details of the metabolism that takes place during anoxia. It has been assumed that anaerobic cysts must carry out, albeit at reduced rates, anabolic reactions associated with cell maintenance such as the synthesis of proteins and other macromolecules, and that these reactions provide the energetic need for the cyst to carry out an energy metabolism of any sort. Because the cyst is provided with a permeability barrier to inorganic ions and other non-volatile solutes, there appears to be no need for provision of energy for transport processes across cell surfaces.

The present study was carried out to gain further information on the nature of anaerobic cyst metabolism. Hampered greatly by the inability to utilize exogenous labelled presursors and other ordinary metabolites we resorted to the use of ^{14}C- labelled bicarbonate which, although limited in scope, allows us to probe to some extent this metabolism[8]. Also to be reported are initial results on the content of metabolites that would be likely end products of the anaerobic breakdown of endogenous carbohydrates, and that might be involved in CO_2- fixation.

MATERIALS AND METHODS

Unless stated otherwise the cysts used in this study were collected from the solar salt ponds near Hayward, California during the summer of 1987. They were processed by methods described elsewhere[8] which includes brief exposure to hypochlorite, and dry storage over $CaSO_4$ at -20°C. This population is 91% viable as judged from nauplius production in sea water during a 72 hour incubation period (25°C).

^{14}C-Bicarbonate Labelling

NaH$^{14}CO_3$ was obtained from Amersham (0.1 mCi/mmole) and added to incubation flasks immediately before sealing them. The incubation medium consisted of 0.2M NaCl in 0.05M Tricine buffer, pH 8.0(ST). Aerobic incubations were carried out by placing cysts (about 50mg dry weight) in 2 ml of medium in 50 ml Erlenmeyer flasks which were sealed after replacing the gas phase with 50% O_2: 50% air. Anaerobic studies were done using 10 ml flasks and 1 ml medium which had been saturated with 100% N_2. The flasks were shaken (40 rpm) at 25°C. Long term studies were carried out with the addition of penicillin (1000 units/ml) and streptomycin (0.1 mg/ml) and cysts were treated with full strength "clorox" (0°C, 2 min) to reduce the contributions of contaminating microorganisms. Great care was taken to insure that anoxic conditions existed[2].

After incubations cysts were collected and washed on cloth filters. Fractionation was usually carried out by homogenizing the cysts in 6% perchloric acid (PCA) at 0°C, and the supernatant obtained by centrifugation at 5700g for 10 min. at 4°C. For high performance liquid chromatography (HPLC) the supernatant was treated with washed charcoal which was then removed by centrifugation. The PCA-insoluble fraction was washed twice with 6% PCA and resuspended in 80% formic acid. Aliquots of the latter, and PCA-soluble fractions were added to PCS-scintillation fluid (Amersham) and counted in a scintillation spectrometer.

High Performance Liquid Chromatography

The method used is that described by Womersley et al.[9]. Aliquots of charcoal-treated PCA-soluble extracts were injected into an organic acid analysis column (BioRad's Aminex HPX-87H cation exchange) and the effluent monitored at 210 nm using a Kratos UV detector and Waters HPLC system run at 41°C. Flow rate was 0.6 ml/min and the mobile phase was 0.004N H_2SO_4. The extracts were assayed within 4 hours of fractionation. One-half ml fractions were collected directly into scintillation vials and counted for radioactivity using 5 ml of PCS scintillation fluid. Standards were made up in 0.004N H_2SO_4 and used to calibrate the column by recording their elution times. Tentative identity was determined from elution times of known standards and by "spiking" the PCA extracts of cysts. In some cases fractions isolated from cyst extracts were co-chromatographed with standards.

Polyribosome Analysis

These studies were carried out as described in detail previously[10]. Briefly, the cysts were dechorionated with NaOCl at 0°C and homogenized gently with a Dounce grinder in 0.05M Tris-HCl pH7.8 containing 0.01M $MgCl_2$ and 0.1M KCl (TMK). The homogenate was centrifuged at 14,000 g for 30 min. and the supernatant layered on a linear sucrose gradient (15-50%) made in TMK. After centrifugation for 80 min. at 29,000 rpm (Spinco SW 41 rotor, Beckman Model L-2) the gradients were analyzed with an Isco gradient fractionator with continuous recording of optical density at 254 nm.

Carbohydrate Analysis

Cysts were homogenized (4°C) in 75% ethanol (30 mg wet weight/ml) and the supernatant obtained by centrifugation. Trehalose and glycerol were isolated by HPLC using the same column and conditions described for organic acid analysis except that 0.01N H_2SO_4 was used as the mobile phase and a Knauer Differential Refractometer was used to detect and quantify trehalose (7.4 minutes retention time) and glycerol (13.1 minutes retention time). The ethanol-insoluble pellet was digested with 50% KOH at 95°C for 1.5 hours and the KOH-soluble supernatant was mixed with 2 volumes of absolute ethanol to precipitate glycogen which was dissolved with water and assayed by the colorimetric method of Dubois et al.[11]. Results are expressed as glucose equivalents.

RESULTS

[14]C-Bicarbonate Incorporation

Figure 1 compares incorporation of ^{14}C-bicarbonate into PCA soluble and insoluble fractions of cysts under aerobic and anaerobic conditions. Results for aerobic conditions are similar to those obtained previously[7] where it was shown that radioactivity in the insoluble fraction is located chiefly in protein with some labelling of RNA, the precursors for these also being labelled by [14]CO_2 (soluble fraction). While incorporation into the soluble fraction under anaerobic conditions proceeds nicely, and labels a similar array of compounds as judged by thin layer chromatography[12], the rate of incorporation into the insoluble fraction is greatly suppressed. Table 1 shows an unexpected result for longer term anaerobic incubations.

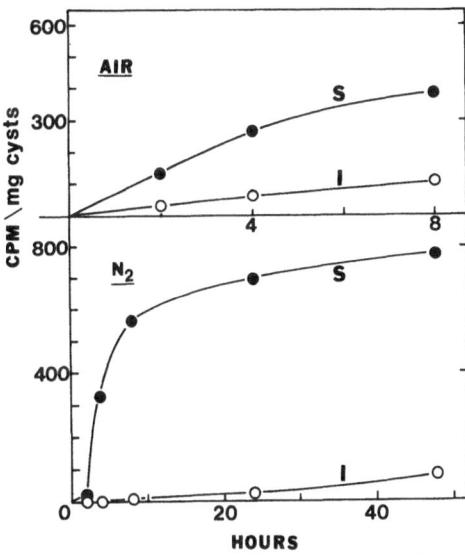

Fig. 1.   Incorporation of radioactivity from ^{14}C-bicarbonate
into the perchloric acid soluble (S) and insoluble
(I) fractions of cysts incubated aerobically (air)
and anaerobically (N_2). In both cases the cysts
were prehydrated at 0°C, and NaH[14]CO_3 was present
initially at 1μCi/ml.

3

Days Incubation	cpm/mg dry wt.		% Viable	% Mortality Per Day
	PCA Soluble	PCA Insoluble		
5	1,583	56	85	1.3
15	1,827	30	78	1.0
30	1,741	164	59	1.2

*100 mg dry cysts in 1.0 ml ST buffer, containing 1 μCi NaH^{14}CO$_3$/ml, in sealed 10 ml flasks; % viable refers to nauplius production in sea water which was 91% for unincubated cysts, and provides the basis to calculate % mortality; all data are averages of two experiments.

Instead of a constant increase in PCA insoluble radioactivity with time, a decrease appears to take place, followed by a significant increase. A possible explanation for this trend will be offered later. Data in Table 1 also show that considerable mortality occurs during anoxia, amounting to a death rate of about 1% per day.

The possibility arises that the ^{14}C-incorporation observed may be partially due to contaminating microorganisms embedded in the shells. That possibility has not been ruled out; however, a comparison of whole cysts with those decapsulated by hypochlorite, both after incubation with ^{14}C-bicarbonate, for the same time periods revealed no differences in incorporation (data not shown).

The distribution of ^{14}C in the PCA-insoluble fraction has been examined in preliminary fashion. Current results on incorporation into RNA have not been repeatable and will not be considered here. However, we have evidence that protein synthesis is occurring and we will concentrate on those observations.

The insoluble fraction from cysts (30 days of anaerobic incubation, Table 1) was washed twice with 10 volumes of methanol-ether (3:1) and heated for 2 hours in 5% trichloroacetic acid (TCA). The insoluble fraction was washed with TCA and suspended in 88% formic acid. Protein was pre-cipitated in aliquots from the formic acid soluble fraction with 20 volumes of 6% PCA. Some of these aliquots were dissolved with 0.5N NaOH and radio-activity and protein were measured, producing a mean value of 803 cpm/mg protein. Other aliquots were suspended in 8N HCl, sealed in thick walled glass tubes and hydrolyzed at 90°C for 24 hours. These solutions were evaporated to dryness, the residue resuspended in 40% ethanol and centri-fuged to remove insolubles. Thin layer chromatography (cellulose) was carried out, and the results are shown in Figure 2B. Most of the ^{14}C was localized in strips 7-10 of the hydrolyzate, and these were eluted from companion chromatograms and re-run with known amino acid standards. On this basis most of the radioactivity appeared to be in aspartate, glutamate, and alanine. The chromatogram shown in Fig. 2A was obtained similarly, but for the PCA-soluble fraction from anaerobic cysts (30 days, Table 1). Most of this radioactivity is in the previously mentioned free amino acids, pyrimidine nucleotides, and several organic acids.

From the foregoing results we conclude that anaerobic cysts incorporate ^{14}C from bicarbonate into a variety of low molecular weight metabolites, including amino acids, and that the latter are incorporated into proteins,

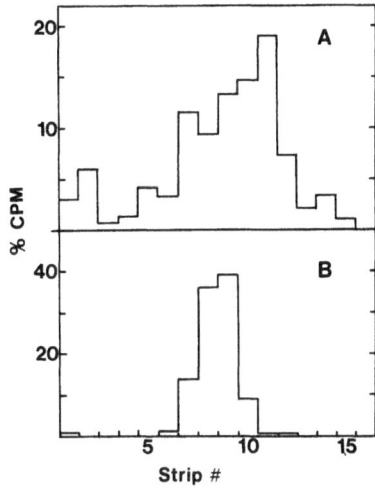

Fig. 2. Distribution of radioactivity in the PCA soluble
fraction (A) and acid hydrolyzate of the protein
fraction (B) from anaerobic cysts as judged by thin
layer chromatography on cellulose sheets. These
preparations were from the 30 day sample described
in Table 1. The developing solvent for chromatography was iso-propanol (7): formic acid (1):
water (2), volume/volume. The base line is strip
1. "% CPM" refers to the radioactivity in each strip
as a percentage of the total recovered.

all of these events taking place completely under anoxic conditions.

Polyribosomes in Anaerobic Cysts

One of us (J.S.C.) carried out these studies some time ago with Dr.
A.L. Golub at the University of Miami. This work has not previously been
published, but seemed relevant to the present study. Figure 3A illustrates
profiles from pre-hydrated cysts incubated aerobically for 2 hours. The
cysts are rich in ribosomes, as is well known, and the polyribosomes formed
aerobically are shown in better detail in Fig. 3B. Cysts incubated for an
additional 2 hours under anoxia have a profile that suggests ribosome runoff. That trend becomes more evident in profiles from this same group of
cysts incubated anaerobically for 4, 6 and 12 days (Figs. 4A,B,C). When
anaerobic cysts are returned to air, larger polyribosomes are formed,
suggesting active resumption of initiation (Figs. 4B' and C').

Although the dried cyst does not contain a substantial polyribosome
complement, some are present[10]. Figure 5A shows this profile for cysts
pre-hydrated at 0°C, and then incubated anaerobically (Fig. 5B) or
aerobically (Fig. 5C) for 1 hour. Little change can be discerned in the
anaerobic cyst profile; perhaps a very slight decrease in the polyribosome
region might take place. As expected, aerobic cysts are active in polyribosome formation. Unfortunately, this earlier work did not include the
study of cysts incubated under anoxia from the beginning for long periods
of time, nor was incorporation of $^{14}CO_2$ examined.

Are Carbohydrates Utilized during Anaerobiosis?

To examine this question we compared the contents of trehalose,

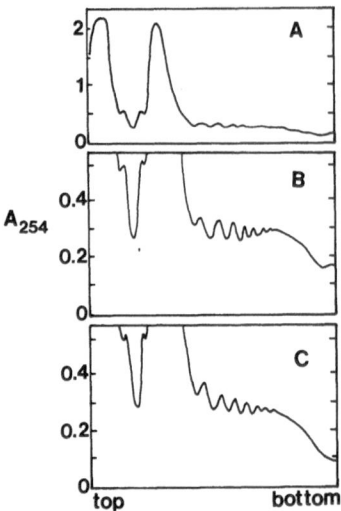

Fig. 3. Ribosome and polyribosome profiles from
cysts incubated aerobically for 2 hours
(A and B) and then for an additional 2
hours anaerobically (C). In all cases
10 optical density units (at 260 nm) of
the 14,000 g supernatant were layered on
the sucrose gradients.

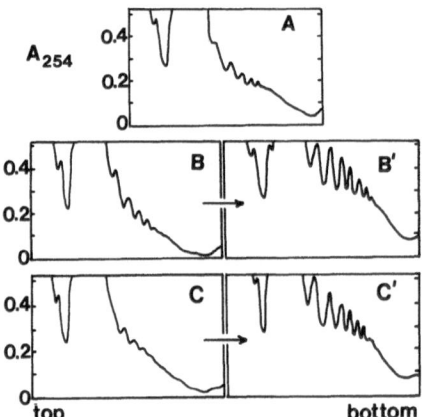

Fig. 4. Profiles are from cysts incubated for 4(A),
6(B), and 12(C) days under anaerobic conditions. B' and C' represent cysts that were
returned to aerobic conditions for 2 hours
before assay. This is a continuation of the
experiment described in Figure 3.

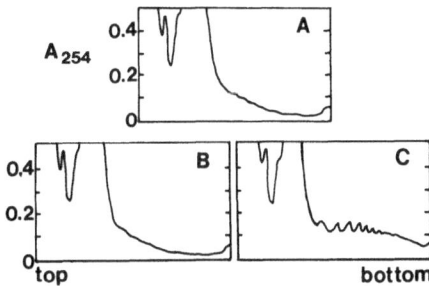

Fig. 5. Profiles of sucrose gradients in unincub-
         ated cysts pre-hydrated at 0°C (A), and
         in cysts incubated at 25°C for 1 hour
         under anaerobic (B) and aerobic (C) con-
         ditions.

glycerol and glycogen in unincubated cysts (pre-hydrated at 0°C) to cysts
that had undergone 95 days of anoxia (Table 2). The outcome indicates
that only glycogen decreased under anaerobic conditions; analysis of the
data by t-test gives a significant difference at the 1% level. These data
support the possibility that glycogen maybe mobilized very slowly during
anoxia, but provide no indication for trehalose or glycerol utilization.
The decrease in glycogen is not due to simple hydrolysis since glucose
does not accummulate, being present below the limits of detection (0.1μg/
mg dry cysts). Although anoxic cyst populations exhibit a decrease in
viability, the extent of this decrease is much less when seawater is used
as the incubating medium.

Organic Acids

     It has been known for some time that lactic acid is not produced
during cyst anoxia[2] and we have noted earlier in this paper that we know
virtually nothing about the nature of anaerobic intermediary metabolism.
If glycogen is being metabolized anaerobically, as it seems to be, its
utilization must result in the formation of one or more end-products. As
a preliminary step toward understanding this metabolism, and to gain further
information about the nature of CO_2-fixation, we have initiated a study
of organic acids likely to be involved in these processes.

     Figure 6 (center panel) shows the HPLC elution profile of 210 nm-
absorbing material present in the charcoal treated 6% PCA-soluble fraction
from unincubated cysts prehydrated at 0°C. The other panels in Figure 6
show results for cysts incubated for 8 hours under aerobic and anaerobic
conditions. The wealth of information prevents detailed discussion here.
However, certain features are worth noting. Peak 13 increases, while peak
11 decreases, both markedly in aerobic cysts compared to 0 time (controls)
and anoxic cysts. The profiles of anoxic cysts are surprisingly similar
to those of unincubated cysts, with the exception of peak 1. As expected,
lactic acid is not detectable.

     We have identified, tentatively, several of the peaks in these extracts
(Table 3) and have quantified changes in their concentration during incubation but we will not undertake that analysis here. However, it should
be noted that interpretation of compounds with short retention is complicated by co-elution. Although trehalose has very low absorbance at 210 nm

7

Table 2. Trehalose, Glycerol and Glycogen in Unincubated Cysts and in
Cysts Incubated for 95 days under Anoxic Conditions.

| Conditions | n | μg/mg dry wt. ± sd | | | % nauplii |
		Trehalose	Glycerol	Glycogen	
Unincubated	7	193 ± 11	28 ± 5	22 ± 5	90 ± 4
Anoxic in seawater	3	202 ± 7	25 ± 2	13 ± 3	56 ± 8
Anoxic in tricine-NaCl	4	201 ± 17	27 ± 4	11 ± 1	37 ± 7

n = number of measurements

its presence in large amounts contributes to most (all?) of peak 2. That
consideration, plus the fact that oxaloacetic acid is expected to be present
in extremely low concentrations, leads us to suspect that peak 2 is
trehalose and peak 4 is pyruvic acid. A useful simple way to compare these
changes during incubation is to express them as a ratio to the levels in
unincubated cysts. That has been done in Figure 7 using a computer program
to integrate the areas making up the peaks. These results show clearly
that most of these compounds stay the same or decrease during anaerobic
incubation; in fact, even the small increases may not be real since the
error in the integration process for small peaks is about 10%. Thus,
there is no indication whatsoever for the anaerobic accumulation of any
substance detected by this procedure.

In order to gain additional information on anaerobic intermediary
metabolism we have examined these metabolites for radioactivity using [14]C-
bicarbonate as the precursor. Here we consider only long-term anaerobic
incubation (Fig. 8). First it should be stressed that the amount of
extract applied to the column in this case was four times larger than that

Table 3. Tentative Identification and "Extinction Coefficients"
for the HPLC Peaks numbered in Figures 6 and 8.

Peak No.	Tentative ID	Peak Area/μg x 10^{-3}
2	oxaloacetic	934
	trehalose	0.3
3	citric	403
4	oxaloacetic	934
	pyruvic	1,170
5	malic	72
7	succinic	45
8d	fumaric	1,677
10	acetic	82

Peak areas are in units of microvolt sec, proportional to absorbance at 210 nm. These computer derived areas are divided by the
μg of known amounts of acids injected to give the "extinction coefficients" in the third column. Oxaloacetic acid produces two
peaks under these HPLC conditions, one of which coincides with
trehalose and the other with pyruvate.

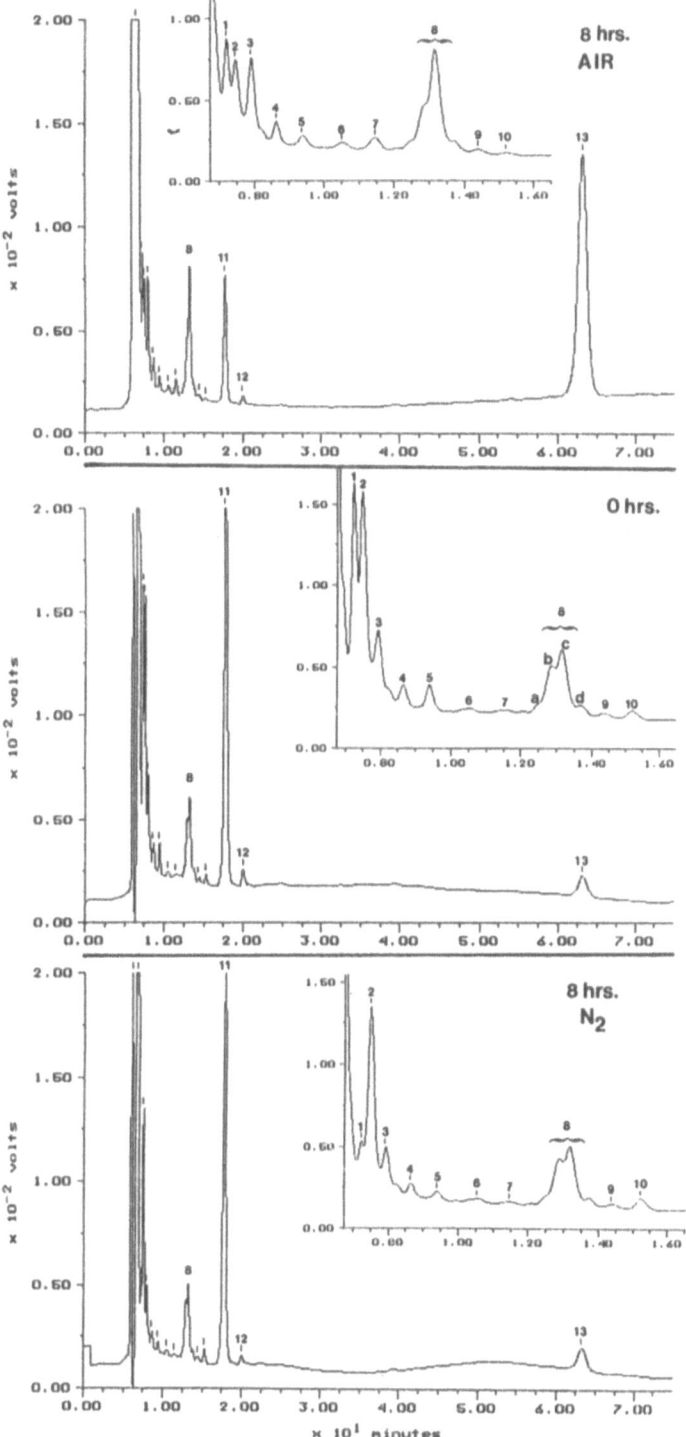

Fig. 6. (legend on next page)

Fig. 6.   Elution patterns of charcoal-treated PCA-soluble frac-
           tions obtained by HPLC. The ordinate is in units of
           absorbance at 210 nm, and the abscissa gives the re-
           tention times. Tentative identifications for some of
           the numbered peaks are given in Table 3. 0 Hrs. refers
           to cysts prehydrated at 0°C. The insets in each panel
           illustrate fine detail for retention times between 0-
           16 minutes.

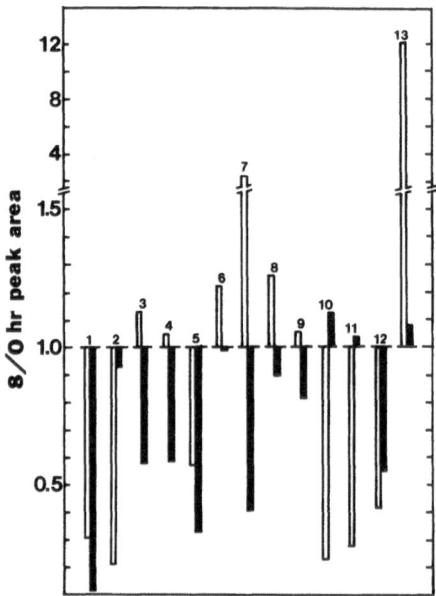

Fig. 7.   Ratios of peak areas obtained after 8 hours of
           incubation under aerobic conditions (open bars)
           and anaerobic conditions (black bars) compared
           to those present in unincubated cysts. Data are
           from Figure 6.

used for the analysis in Figure 6. Secondly, and also in comparison to
Fig. 6, no radioactivity was observed beyond peak 11 and up to a retention
time of 75 minutes; hence, only the shorter retention times are given in
Fig. 8. Interestingly, little ^{14}C was found associated with oxaloacetate
(peaks 2 and 4), an expected CO_2- fixation product. Also, fumarate (peak
8d) appears to contain little if any radioactivity. Instead, most of the
isotope was located in the areas coinciding with succinate (peak 7), malate
(peak 5) and peak 11 which is not identified but may prove to be propionate
(more on that later). An important outcome of these studies not shown in
Fig. 8 concerns recovery of radioactivity applied to the column: 12,200
cpm were applied, but only 1920 cpm were recovered in compounds with retention times less than 75 minutes. Thus, about 85% of the fixed $^{14}CO_2$ is
not accounted for, a point to which we will return.

     Finally, we note a few comparisons with the corresponding analysis
for nauplii (data not given). Peak 11 (propionate?) is very much reduced
and malate becomes, by far, the major acid of those identified; lactate,
not detectable in anaerobic cysts, appears in the profiles but is present
at low levels.

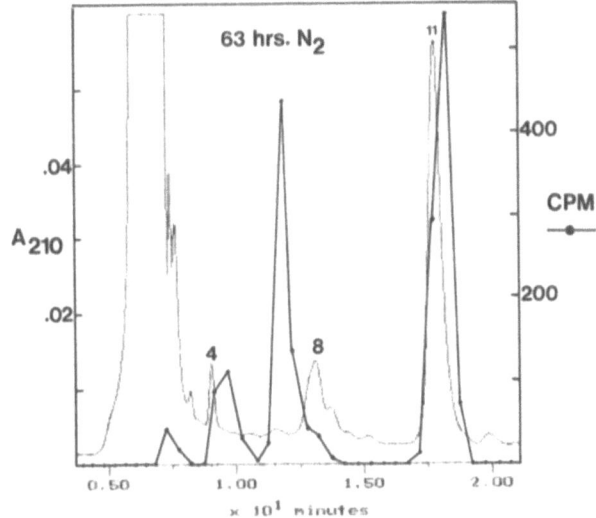

Fig. 8. HPLC profiles of charcoal-treated, PCA-soluble extracts
from cysts incubated anaerobically with ^{14}C-bicarbonate
for 63 hours. Incubation conditions were the same as
those described in Table 1.

DISCUSSION

It has generally been assumed that the anaerobic cyst must synthesize
proteins; however, to our knowledge there are no published data on the
subject. Evidence presented here (Figs. 1 and 2; Table 1) indicates that
protein synthesis is indeed taking place; however, because of the permeability problem it is extremely difficult to evaluate the rate or extent
because we lack detailed information about the labelled amino acid pool.
An order of magnitude approach, perhaps a bit naive, can be taken (for
short incubation times) from the data present in Figure 1. It has previously been shown that incorporated $^{14}CO_2$ in the PCA insoluble fraction
is chiefly in protein[12,13]. For aerobic cysts this amounts to about
14 cpm/hr/mg cysts and for anaerobic ones roughly 1.5 cpm/hr/mg cysts, indicating a factor of 10 difference between the two cases. While this
estimate is obviously crude it is perhaps partly justified by the observation that the array of small metabolites labelled with $^{14}CO_2$ is similar
under aerobic and anaerobic conditions[12]. Although we believe it
unlikely, we accept the caveat that some or even all of the observed incorporation into protein may be contributed by contaminating microorganisms.

This situation is confounded further by an apparent decrease in the
extent of labelling in the PCA-insoluble fraction of cysts during extended
anoxia (up to day 15) followed by an increase as anoxia continues (Table 1).
Inspection of polyribosome profiles provides information on this unexpected
result. The polyribosome content of cysts transferred to anaerobic conditions after aerobic incubation continues to decrease as anaerobiosis
continues over the 12 day period examined (Fig. 3 and 4). These profiles
strongly suggest that the runoff of pre-existing polyribosomes is taking

place. These data provide no insight into the question of the formation of new polyribosomes (ie, initiation) and that issue remains unresolved. Comparison of these polyribosome profiles to the long term anaerobic incubation data (Table 1) provides one possible scenario: up to about 15 days of anaerobiosis the system may be dominated by polyribosome runoff, after which new polyribosome formation may take place. Nevertheless, the decrease in radioactivity associated with the protein fraction that occurs between 48 hrs. (Fig. 1) and 15 days of anoxia strongly indicates that protein hydrolysis is also taking place, presumably by proteolytic enzymes. These sketchy and incomplete results provide only a hint at what these anaerobic cysts are actually doing with their protein synthesizing system, and the turnover of their proteins. It will be rewarding to examine the array of proteins made during anoxia and compare these to those of the aerobic, developing embryo.

It appears that the energy metabolism of anaerobic cysts may involve the slow mobilization of glycogen (Table 2). If we assume that the rate is linear with incubation time then glycogen will be used at $0.10\mu g/mg$ dry wt/day. At this rate all the glycogen would be gone after 225 days of anoxia. Our estimates of anaerobic mortality (Tables 1 and 2) indicate a substantial difference between cysts incubated in seawater compared to those in NaCl-Tricine. However, even in the former case the decline in viability during anaerobiosis is surprising in view of previous results showing that negligible death occurred during 5 months[1] or 4 months[3] of anaerobiosis. We can think of no obvious way to account for these differences from the results on hand.

Hand and Gnaiger[4] in a calorimetric study of anaerobic cyst metabolism hypothesized that mobilization of trehalose was involved in fueling anaerobic energy metabolism; our data provide no evidence for that idea (Table 2), but do indicate slow use of glycogen. We can use their data and ours to make some interesting calculations. Part of their work involved the activation of anaerobic cyst metabolism by an increase in intracyst pH caused by addition of NH_4Cl ("activated anoxibiosis"). They observed that about 0.11 μmole glucose equivalents were used per hour per mg cyst protein during this process, being equivalent to a theoretical heat production of about 1.4 mW/g of dried cysts (using some reasonable assumptions about the end product[s] of carbohydrate breakdown) and close to their actual measurements of heat production during activated anoxibiosis (0.9 - 1.0 mW/g). Converting their data to our units (using protein to be 40% of the cyst dried weight) we obtain a carbohydrate utilization rate of $8\mu g/mg$ dry wt/hr during activated anoxibiosis. If we take that rate to be equivalent to their estimate of heat production of 1.4 mW/g dried cysts during activated anoxibiosis, then we can use their actual measurement of heat production during anaerobiosis (lowest value = 0.12 mW/g) to estimate potential trehalose utilization rates under those conditions. Thus, ($8\mu g$/mg dry cyst/hr)/1.4 mW/g = (x)/(0.12 mW/g), where x is the calculated rate of trehalose utilization under complete anaerobiosis: $0.7\mu g/mg$ dry wt/hr. At that rate, all cyst trehalose would be depleted in about 10 days whereas we observe no use at all (Table 2). We now compare this to our estimate of glycogen utilization under these conditions: $0.1\mu g/mg$ dry wt/day, roughly 170 fold lower than the estimate made from the data of Hand and Gnaiger[4].

A number of assumptions have been made in the preceeding exercise which cast doubt on any conclusion to be drawn from the comparison. Nevertheless, the very large differences obtained suggest that a more detailed analysis is warranted. It could be that the mobilization of compounds other than carbohydrates might be involved. Protein seems a likely candidate in this regard since several amino acids have been implicated as important substrates for anaerobic metabolism in several invertebrates[14].

We wish to add one more comment about the excellent measurements of Hand and Gnaiger[4]. Inspection of Figure 2 of their paper indicates that values for heat dissipation during anaerobiosis (100% N_2) do not plateau to a constant value, but were continuing to fall when the study was terminated. Thus, heat dissipation may be reduced to much lower levels during prolonged anaerobiosis such as those examined in our study (Table 2). Perhaps this could account for part of the discrepancy noted above. The heat production data also suggest that carbohydrate utilization is not linear as we have assumed. Finally, the potential importance of Gp_4G as an anaerobic energy source becomes more likely under such conditions[3]. Be that as it may, the anaerobic utilization of substrates, whatever these might be, must result in the formation of end products and we turn our attention to that question next.

Our attempts to detect the accumulation of such end products have not yet met with much success; not only do none of the organic acids examined show significant increases during anaerobiosis, but most of them decrease (Fig. 7). Studies of end products of anaerobic metabolism in other animals have shown a diverse array of end products, including acetate, succinate and propionate (for reviews see books by Hochachka[15] and Hochachka and Guppy[14]. We are not yet convinced that propionate is present in Artemia cysts although the retention time of the prominent peak number 11 is very close to that of propionic acid. In any event, peak 11 shows no significant increase during anaerobiosis (Fig. 7). A common series of anaerobic end products involves enzymatic reactions between pyruvate and a variety of amino acids, and we intend to look for these in anaerobic Artemia cysts. Also worth considering is the possibility that a volatile end product is produced which escapes from the cysts.

The few studies carried out thus far on the incorporation of ^{14}C-bicarbonate into organic acids of anaerobic cysts can only provide some hints about the flow of carbon, and our interpretation of these data (Fig. 8) is obviously speculative. The three prominent areas of incorporated radioactivity in HPLC profiles coincide with malate (peak 5), succinate (peak 7) and peak 11, possibly propionate or a derivative. Although some ^{14}C may be present in oxaloacetate (peaks 2 and 4) that is highly uncertain, as is the possibility that CO_2 incorporation involves initial fixation into oxaloacetate, followed by conversion of the latter into malate, succinate, fumarate, and (perhaps) propionate. Variations on this pathway have long been known to exist in many animals (see Krebs[16], Wood[17], Ochoa[18], Hochachka[15], and Hochachka and Guppy[14]). The most puzzling aspect of the Artemia data is the apparent lack of labelled fumarate, and we have no explanation for that observation at present.

Although malate, succinate and (perhaps) propionate do become labelled with ^{14}C-bicarbonate, these either stay the same or decrease in chemical concentrations during anaerobiosis. Thus, although these acids may be on the catabolic route of trehalose and/or glycogen they do not seem to be end-products. We have noted that these acids, collectively, make up only 16% of the charcoal-treated PCA-soluble radioactivity applied to the HPLC column. We currently are tracking down this illusive radioactivity, hoping it will shed light on the flow of carbon and the end product(s) of carbohydrate catabolism during anoxia.

Finally, we return to the surprising and marked difference in cyst mortality due to anoxia in seawater and NaCl-Tricine noted in the present study (Tables 1 and 2) and also compared to the work of others[1,3]. Because the cyst is impermeable to non-volatile solutes some sort of direct ionic effect seems unlikely. The pH of seawater and NaCl-Tricine is similar, as is the osmotic pressure; these are not likely to be involved in any case. We believe the most likely difference is the presence of high levels of CO_2 in seawater and its near absence in the NaCl-Tricine incubation medium. While reviewing the work on animal anaerobes we noted that CO_2 is a vital requirement for the anaerobic metabolism of parasitic intestinal helminths[14,15]. It could well be that a similar situation exists in anaerobic <u>Artemia</u> cysts and we have begun to evaluate that possibility; CO_2 may be a crucial metabolite in their biochemical adaptation to long term anoxia. That suspicion is strengthened by the very recent analysis of CO_2 involvement in brine shrimp metabolism by Conte and Geddes [19].

ACKNOWLEDGMENTS

Supported by a grant from the National Science Foundation (DCB 8696048). We appreciate the skilled assistance of Victoria Milam in manuscript preparation and are grateful to Dr. Laurie Drinkwater for collecting and processing the cysts used in most of the work presented here.

REFERENCES

1. J. Dutrieu and D. Chrestia-Blanchine, Resistance des oeufs durables hydrates d'<u>Artemia</u> <u>salina</u> a l'anoxie. <u>C. R. Acad. Sci. Paris Serie D.</u> 263:998 (1966).
2. R. D. Ewing and J. S. Clegg, Lactate dehydrogenase and anaerobic metabolism during embryonic development in <u>Artemia salina</u>, <u>Comp. Biochem. Physiol</u> 31:297 (1969).
3. D. M. Stocco, P. C. Beers and A. H. Warner, Effects of anoxia on nucleotide metabolism in encysted embryos of the brine shrimp, <u>Devel. Biol.</u> 24:479 (1972).
4. S. C. Hand and E. Gnaiger, Anaerobic dormancy quantified in <u>Artemia</u> embryos: a calorimetric test of the control mechanism, <u>Science</u> 239:1425 (1988).
5. W. B. Busa and R. Niccitelli, Metabolic regulation via intracellular pH, <u>Am. J. Physiol.</u> 246:R409 (1984).
6. W. B. Busa, How to succeed at anaerobiosis without really dying, <u>Mol. Physiol.</u> 8:351 (1985).
7. J. S. Clegg, Protein synthesis in the absence of cell division during the development of <u>Artemia salina</u> embryos, <u>Nature, Lond.</u> 212:517 (1966).
8. J. S. Clegg, <u>Artemia</u> cysts as a model for the study of water in biological systems, <u>Meth. Enzymol.</u> 127:230 (1986).
9. C. Womersley, L. Drinkwater and J. H. Crowe, Separation of tricarboxylic acid cycle acids and other related organic acids in insect hemolymph by high performance liquid chromatography, <u>J. Chromatog.</u> 318:112 (1985).
10. J. S. Clegg and A. L. Golub, Protein synthesis in <u>Artemia salina</u> embryos, II. Resumption of RNA and protein synthesis upon cessation of dormancy in the encysted gastrula, <u>Devel. Biol.</u> 19:178 (1969).
11. M. Dubois, K. A. Gilles, J. K. Hamilton and F. Smith, Colorimetric method for determining sugars and related substances, <u>Analyt. Chem.</u> 28:350 (1956).
12. J. S. Clegg, Interrelationships between water and cell metabolism in <u>Artemia</u> cysts, IX. Evidence for the organization of soluble enzymes, <u>Cold Spring Harbor Symp. Quant. Biol.</u> 46:23 (1982).
13. J. S. Clegg, Interrelationships between water and cell metabolism in <u>Artemia</u> cysts, V. $^{14}CO_2$ incorporation, <u>J. Cell. Physiol.</u> 89:369 (1976).
14. P. W. Hochachka and W. Guppy, "Metabolic Arrest and the Control of Biological Time", Harvard University Press, Cambridge (1987).

15. P. W. Hochachka, "Living Without Oxygen", Harvard University Press, Cambridge (1980).

16. H. A. Krebs, Carbon dioxide fixation in animal tissues, _Symp. Soc. Exp. Biol._ 5:1 (1951).

17. H. G. Wood, A consideration of some reactions involving carbon dioxide, _Symp. Soc. Exp. Biol._ 5:9 (1951).

18. S. Ochoa, Biosynthesis of dicarboxylic and tricarboxylic acids by carbon dioxide fixation, _Symp. Soc. Exp. Biol._ N5:29 (1951).

19. F. P. Conte and M. C. Geddes, Acid brine shrimp: metabolic strategies in osmotic and ionic adaptation, _Hydrobiologia_ 158:191 (1988).

DEVELOPMENTAL ABNORMALITIES RELATED TO BICARBONATE ION STATUS DURING

EMERGENCE OF ARTEMIA

C. N. A. Trotman, S.P. Gieseg, R. S. Pirie and W. P. Tate

Department of Biochemistry
University of Otago
Dunedin, New Zealand

INTRODUCTION

Emergence of the Artemia prenauplius from its survival cyst after rehydration and the cessation of dormancy is far from being a simple process
of passive rupture and it requires metabolic and osmotic activity[1]. An
energy requiring mechanism is essential in the events leading up to successful emergence and inhibitors of oxidative metabolism can lead to unsuc-
cessful emergence[2]. Emergence evidently depends on a complete and relatively rapid rupture of all membranes and components of the cyst wall
external to the hatching membrane. Partial failure of this sequence is, in
effect, total failure of successful emergence. We describe below how
further exploitation of this biological assay system, namely incomplete
emergence, points to bicarbonate ion as being an essential component in
generation of the osmotic potential that energises emergence.

The emergence sequence features two recognized stages of the organism.
The first is E1, where the cyst wall develops a small opening to reveal the
embryo protruding slightly while restrained within one or more membranes.
The second is the E2 or prenauplius stage in which the embryo is enveloped
only by the hatching membrane although this may remain incidentally attached
to the empty cyst wall.

We find that the E1 stage is often prominent throughout the emergence
phase when cysts are hatched in artificial media under laboratory conditions.
Under optimal conditions, however, the E1 stage is transient and it is not
normal for excessive numbers to remain in the E1 stage. The abnormality of
a prolonged E1 stage, and other related abnormalities described below, are
interpreted in terms of a deficiency in environmental bicarbonate.

METHODS AND MATERIALS

Desiccated Artemia cysts were San Francisco Bay Brand or were collected
from Lake Grassmere, New Zealand; the species is the same. Natural Pacific
seawater (NSW) was obtained from coastal New Zealand, stored at 5°C and
filtered before use; its pH was 8.0 at 20°C. Artificial seawater (ASW)
contained 600 mM NaCl, 10 mM KCl and 10 mM $MgCl_2$[3]. Herbst artificial
seawater was made from 510 mM NaCl, 10.7 mM KCl, 55 mM $MgSO_4$, 11.7 mM $CaCl_2$
and 5.9 mM $NaHCO_3$[4]. NaCl medium was 600 mM. Buffers, drugs and

additional constituents of media were added as described with the results. *Artemia* were hatched by incubation at 20-25°C at various cyst concentrations up to 40 mg/ml. In experiments at low cyst concentrations, hatching was in dialysis tubing incubated in a larger volume of medium in order to confine the prenauplii at a higher density for observation. The dialysis tubing contained 20 mg of cysts and 5 ml of medium. Dechorionation was done by dissolving the cyst tertiary envelope in hypochlorite solution (10-14% available chlorine) for 5-10 min followed by washing with 600 mM NaCl.

Scanning electron microscopy was performed in a Cambridge Steroscan 360 instrument on specimens preserved in 2.5% glutaraldehyde and critical-point dehydrated. Light micrographs were taken on color transparency film under dark field illumination.

RESULTS AND DISCUSSION

Normal and Abnormal Emergence of the Prenauplius

In natural seawater (NSW) the El was a transient stage. Regardless of how low the concentration of cysts, they hatched efficiently and the proportion of well developed El visible at any time was typically fewer than 2%.

Abnormal emergence in artificial media consisted of excessive numbers of individuals failing to proceed beyond the El stage. El prenauplii of superficially normal appearance were present but their high incidence was abnormal, reaching in excess of 60% of total cysts under some circumstances. Furthermore, most failed to progress to E2 and remained arrested in an advanced stage of El (Fig. 1). The embryo was usually oriented with the eyespot in the part protruding through the cyst wall (Fig. 1) but we prefer to believe that this reflects a tendency for the embryo and hatching membrane to slip within the membranous envelope, rather than rupture happening adjacent to the eyespot.

A superficially different abnormal form, an oval prenauplius, frequently accompanied El arrest (Fig. 2). The oval prenauplius contained a live and moving embryo. If differed from the normal prenauplius however in having a shape closer to a prolate spheroid, but slightly wider at the eyespot end, and in the absence of the tuft of membrane attached to the tail end of a normal hatching membrane (Fig. 3). The nauplius rarely escaped from the oval stage, consistent with the possibility of an additional enclosing membrane.

Abnormal cultures contained varying proportions of arrested advanced El and ovals, together with cysts arrested with a small rupture. The term 'abnormal' used below refers to any of these manifestations of arrested development.

Isolation of the Lesion in Abnormal Emergence

The most characteristic feature of the oval is that it was never seen with an attached tuft of ruptured membrane at the posterior end, in contrast to the normal prenauplius which is not usually seen without what is considered to be this remnant inner cuticular membrane. Also not evident in the oval is the characteristic pear shape of the prenauplius, coinciding with the contour of the hatching membrane which exhibits a similar tapered shape even if not completely filled with embryo (Fig. 3). These two differences suggest that the abnormal may be constrained to its oval shape by retention of one or more additional membranes. The membranes of the developing cyst are well documented[5] and the most plausible candidates for a

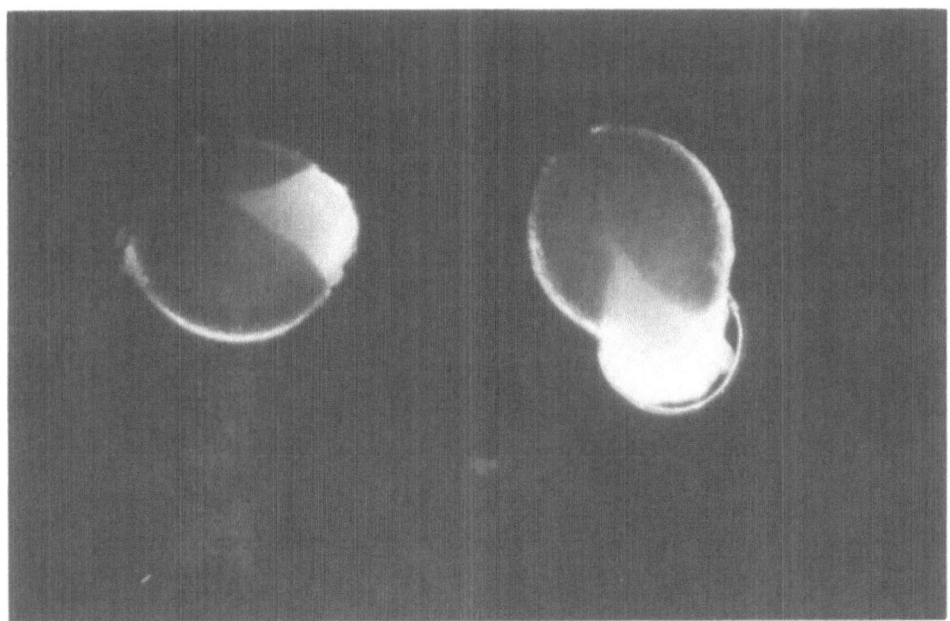

Fig. 1.   A light micrograph of arrested E1 stages found in incomplete
seawater media.

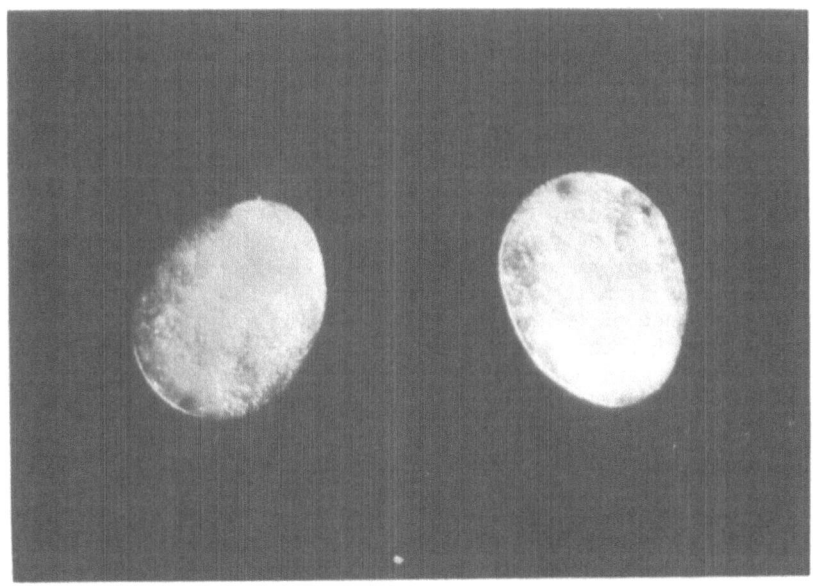

Fig. 2.   A light micrograph of oval prenauplii.

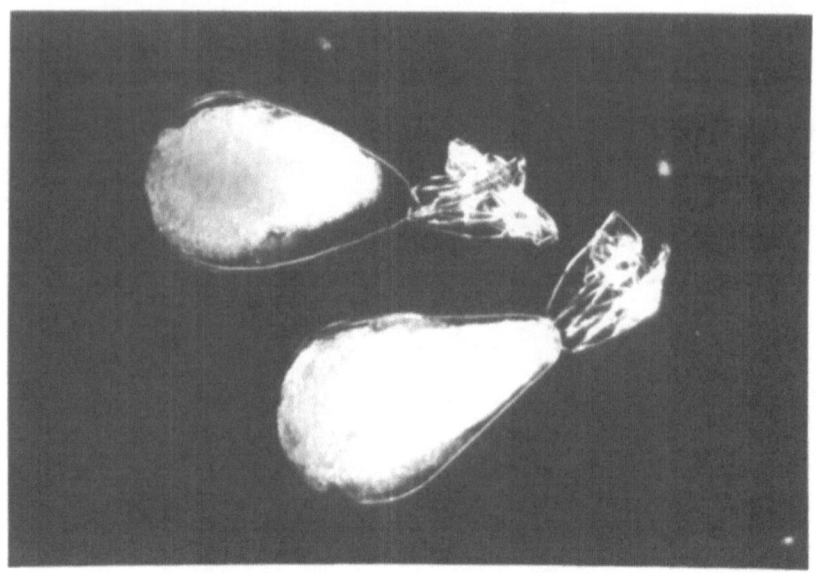

Fig. 3. A light micrograph of prenauplii emerged from dechorionated cysts showing a membranous attachment to the hatching membrane.

retained membrane external to the hatching membrane would appear to be in
the cuticular region, comprising inner and outer cuticular membranes and
intervening fibrous region.

Cysts stripped of their alveolar layer with hypochlorite reveal a
geodesic network of structural elements within the cuticular region[5].
These elements are prominent in scanning electron micrographs (Fig. 4) and
their centres provide the lines of preferential rupture in dechorionated
cysts emerging normally[6]. In sections (Fig. 5) the structural elements
are found to occupy the cuticular fibrous region but, significantly, not
the inner cuticular membrane, thus providing a distinguishing marker for
the fibrous region. The geodesic marker remains in the structure of the
abandoned cyst of normal prenauplii (Fig. 6) while the inner cuticular
membrane delaminates from it and is partly extracted by the vacating prenauplius. The absence of the marker from the membranes of oval prenauplii
(Fig. 7) indicates that they are not constrained by the fibrous cuticle or
more distal membranes and points to the inner cuticular membrane as the one
that has failed to be shed.

Biochemical and Environmental Factors in Abnormal Emergence

An understanding of the factors involved in defective emergence will
help us to construct a model of the normal process. A combination of two
factors was required to provoke the abnormality: the use of defined media
as seawater substitutes, and a cyst concentration less than about 1mg/ml
(Fig. 8). Abnormality did not occur in NSW or at high cyst concentrations
in defined media.

The two possible explanations explored were an essential constituent
lacking from defined media, or a substance, such as carbon dioxide, released

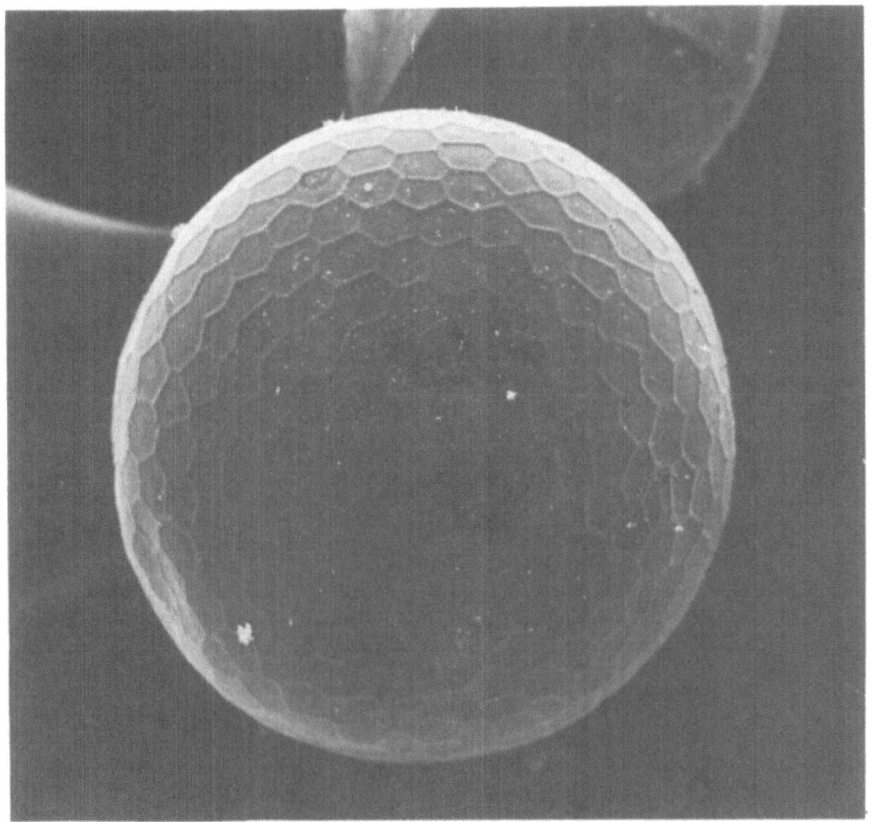

Fig. 4. A scanning electron micrograph of a dechorionated cyst revealing the irregular hexagonal network of structural
elements in the fibrous cuticular region. Cyst diameter,
180 μm.

by emerging embryos. Incubation of cysts in dialysis tubing confirmed
that any factor in the latter category was small and diffusible, thus not
in the nature of an enzyme active against the cyst wall, since the incidence
of abnormalities was related to the total incubation volume.

Abnormal emergence was largely overcome by the addition to defined
media of 5-10 mM HCO_3^-, this being the chemical fate of almost all CO_2 in
seawater at pH 8. The seawater was buffered to pH 8 with 10 mM Tris. This
finding was compatible with the twin factors identified with the abnormality, since CO_2 from respiring cysts would replenish HCO_3^-, normally present
in millimolar range in surface natural seawater[7]. The result was not
conclusive since it did not distinguish between the chemical and buffering
effects of HCO_3^-, which forms the major buffer in seawater through the
calcium and magnesium carbonate-bicarbonate system. However, buffering
with 10 mM Tris, phosphate, or borate buffers at pH 8 in the absence of
HCO_3^- was ineffective in preventing abnormal emergence. The principal
components of surface seawater[7] were also tested in 600 mM NaCl at and
above their environmental concentrations in case they showed the same effect
as HCO_3^- (Table 1). The cysts were incubated at 0.03 mg/ml by dialysis.

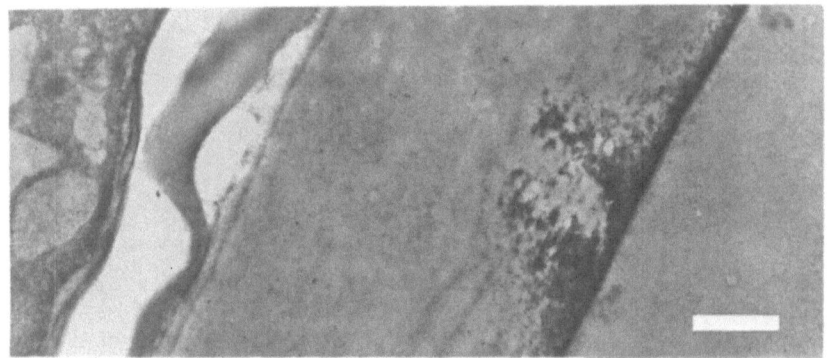

Fig. 5. An electron micrograph of the fibrous cuticular region in
         section. The dark stained area (right) is a structural
         element. The nauplius is at the left, with the hatching
         membrane separating from the inner cuticular membrane.
         Stained with uranyl acetate and lead citrate. Bar, 1 μm.

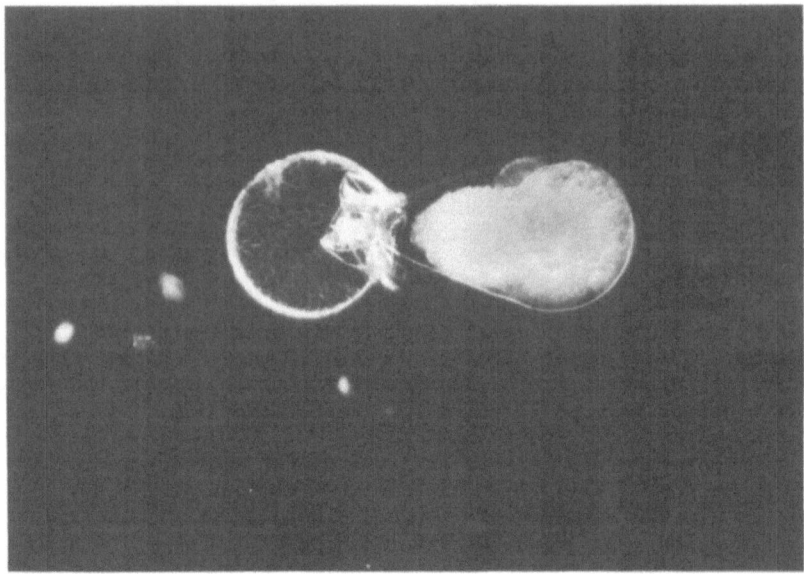

Fig. 6. Light micrograph of a vacated cyst of normal prenauplius
         hatched after dechorionation. The geodesic network is
         visible, showing that the fibrous cuticular region remains
         in the vacated cyst.

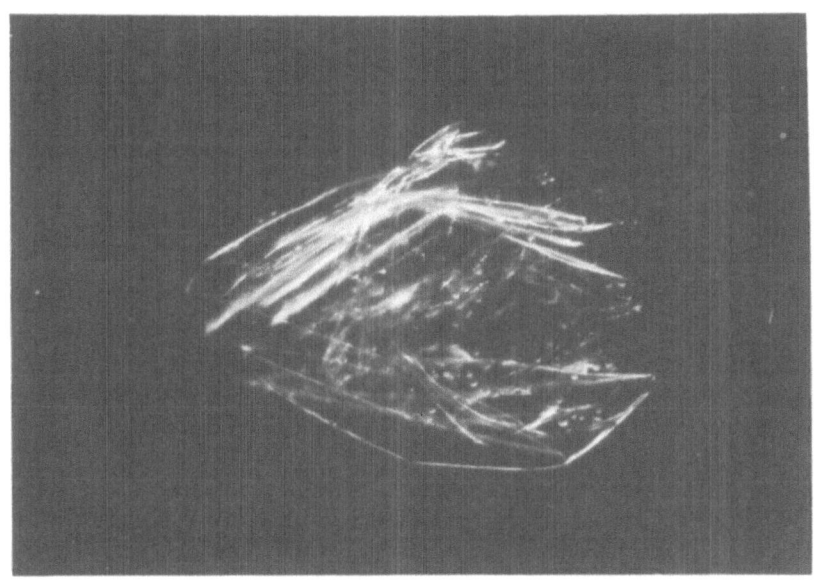

Fig. 7. A light micrograph of membranes from which an oval pre-
         nauplius has been artificially expressed. The geodesic
         marker is absent.

The conclusion of a specific requirement for HCO_3^- conflicted with
Sato's report of a requirement for Ca^{2+} in emergence[8]. Our experiments
showed that Ca^{2+} did not prevent abnormal emergence when prepared from the
stable analytical grade salt, $CaCl_2 \cdot 2H_2O$, but did in varying degrees when
the anhydrous or $.6H_2O$ salts, which generally contain absorbed carbonate,
were used. Tris buffer, which absorbs atmospheric CO_2 to form a carbonate,
restored normal emergence if present at unreallistically high concentrations
(e.g., above 0.1 M).

Supplying the HCO_3^- at different times up to 12 h after cyst rehydration showed the requirement for the ion to be localized to a period shortly

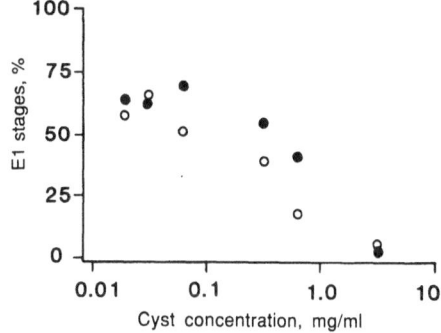

Fig. 8. Dependence of emergence on cyst
         concentration in NaCl medium;
         ○ at 17h; ● at 40h.

Table 1. Ions Found Ineffective in Restoring Normal Emergence in 600 mM
          NaCl

Ion	Form supplied	Maximum concentration, mM
Mg^{2+}	$MgCl_2$	66
Ca^{2+}	$CaCl_2.2H_2O$	66
K^+	KCl	10
NH_4^+	NH_4Cl	2.8
SO_4^{2-}	Na_2SO_4	55
phosphate	di,tri-Na	0.15

before emergence. In this period between dormancy and emergence, aerobic
metabolism is required and the intra-cyst pH is alkaline[9]. Therefore
metabolic CO_2 should exist chiefly as HCO_3^- which, unlike dissolved CO_2,
would not diffuse out of the cyst. Since rupture of the impermeable outer
cuticular membrane does occur in the abnormal emergence, it is possible
that the critical defect immediately follows this rupture and that a loss
of internal HCO_3^- into deficient medium arrests further progress. Being
closely associated with the emergence event, this effect of HCO_3^- is in
contrast to the reversible effect of CO_2 on dormancy itself, originally
described by Busa et al.[9,10].

A Model of Normal Emergence

     The pathology of the abnormality and its apparent chemical basis
suggest that HCO_3^- is involved in preparation of the embryo for efficient
escape from the cyst. The layers of the cyst wall external to the hatching
membrane rupture as an influx of water inflates the hatching membrane. Insufficient influx induces insufficient expansion to expel the prenauplius
or, from the present results, to rupture the inner cuticular membrane.

     An obvious mechanism of water influx would be osmotic, in response to
a high concentration of solutes. A space outside the embryo but inside the
cyst is a more logical location for an osmotic potential than within the
embryo, since the hatching membrane sac and not the young larva, needs to
inflate. The osmotic potential could exist internal to the hatching membrane or, since this is partially permeable, internal to the inner cuticular
membrane, the permeability of which is not certain. However, little
physical space is evident beneath either membrane in the developing cyst.
The outer cuticular membrane is impermeable[11] and if it provided the chief
permeability barrier for containment of the osmotic potential, the latter
could occupy the fibrous region in addition to an inner location. By scanning electron microscopy the fractured structural elements of the cuticle
appear to have a porous structure.

     HCO_3^- could contribute directly to the osmotic potential but the de-
pendence on oxidative metabolism for successful emergence is suggestive of
an energetic source, such as a pumped extrusion of ions or solutes from the
embryo into a location within the cyst. A requirement for HCO_3^- is readily
accommodated in a counter-ion role in models of membrane extrusion pumping.

Owing to the inaccessibility of the encysted embryo such a model is necessarily speculative but classically an import of HCO_3^- in exchange for Cl^- might be expected to complement an ATPase exporting 3 Na^+ in exchange for 2 K^+. A Na,K-ATPase certainly is present, in relatively low but rapidly increasing amount, in emerging prenauplii[12]. Conte[13,14] has proposed a Na,K-ATPase linked exchange of HCO_3^- for organic anions across the salt gland, an organ that could be a candidate for the osmotic generator.

Where might the embryo find HCO_3^- to import, since it is ionically oblivious to the environment outside the cyst? Its own metabolic CO_2 is a potential source if it diffuses across the embryo surface and is then converted to HCO_3^-, thus inhibiting loss from the cyst while making it available for Cl^- exchange. A cycle is proposed with metabolic CO_2 diffusing out as far as the cuticular region and HCO_3^- returning in exchange for Cl^- (Fig. 9). At rupture of the outer cuticular membrane the HCO_3^- reservoir is lost unless reflected from the medium, as observed, except that at high cyst concentrations the embryos would themselves contribute to the pool of environmental HCO_3^-.

The equilibrium ratio of CO_2 to HCO_3^- both inside and outside the embryo would depend on local pH but carbonate dehydratases, if located on either the inside or outside surfaces of the embryo, would increase the rate of transport by facilitating a gradient. Carbonate dehydratase has not been demonstrated beyond miniscule activity in Artemia, either by Conte[14] or ourselves, yet known inhibitors of the enzyme have deleterious effects on the organism. Acetazolamide and sulphanilamide, which react at the zinc-

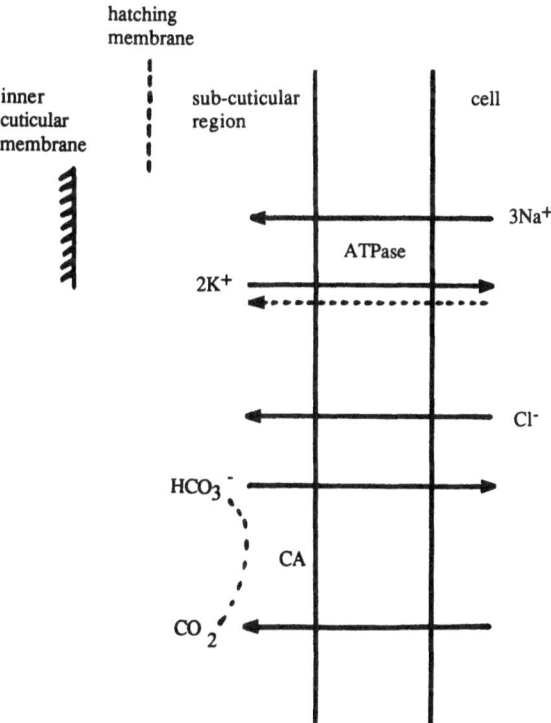

Fig. 9. Outline model of a pump mechanism for the generation of an osmotic potential within the cyst. CA, carbonic anhydrase.

associated HCO_3^--binding sites of carbonate dehydratases[15,16], were both inhibitory to nauplii (Figs. 10,11). These drugs were especially toxic to Artemia in high salinity, with a marked swelling of the nauplii compatible with an inhibition of osmoregulation. Since normal hatching was required, Herbst seawater, which contains bicarbonate, was used. Observations were confined to the post-emergence period owing to the impermeability of the cyst but they suggest that Artemia possesses either a carbonate dehydratase, despite the inability to confirm one directly, or another bicarbonate binding site of comparable structure. Other inhibitors of bicarbonate binding sites (thiocyanate) or of ATPases (ouabain, vanadate) inhibit Artemia development at the E2 to nauplius transition, insufficiently early to provide confirmation of the existence of such sites prior to the E1 stage.

Rafiee et al.[17] have recently reported the induction of similar abnormalities to those described above by cadmium ions in the micromolar range, invoking tubulin as a suggested site of the toxic action. Cd^{2+} forms an insoluble $CdCO_3$, but its effect in depleting the medium of carbonate family ions would be negligible at its toxic concentrations. We suggest, as an alternative, the possible precipitation of an inhibitory compound at surface bicarbonate binding sites.

CONCLUDING REMARKS

Successful emergence of Artemia from its remarkable state of cryptobiosis involves a delicately balanced sequence of events which must all be executed perfectly. The exquisite sensitivity of this mechanism to either positive or negative chemical disturbances is providing us with an excellent experimental system in which to study the interaction of an embryo with its immediate environment during both normal and abnormal early development.

ACKNOWLEDGEMENTS

We thank Dr. Philippa Wiggins for valuable discussions and Mr. Mark Gould for assistance with the SEM.

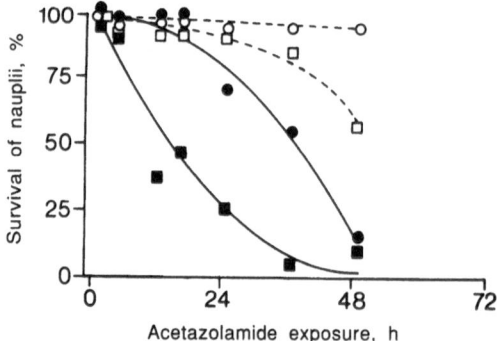

Fig. 10.   Response of nauplii to 1.0 mM acetazolamide added after 18 h. Nauplii were raised in Herbst medium (circles) or Herbst medium with 2.5 M total NaCl (squares) with drug (solid lines) or without drug (broken lines).

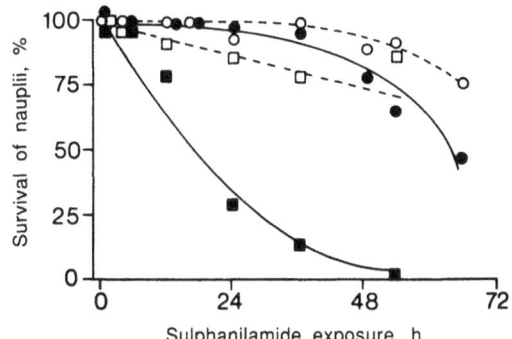

Fig. 11. Response of nauplii to 1.0 mM sulphanilamide; experimental details as in Fig. 10.

REFERENCES

1. J. S. Clegg, The control of emergence and metabolism by external osmotic pressure and the role of free glycerol in the developing cysts of Artemia salina, J. Exp. Biol. 41:879 (1964).
2. C. N. A. Trotman, B. C. Mansfield and W. P. Tate, Inhibition of emergence, hatching, and protein biosynthesis in embryonic Artemia salina, Devel. Biol. 80:167 (1980).
3. L. Moens and M. Kondo, The structure of Artemia salina hemoglobins. A comparative characterization of four naupliar and adult hemoglobins, Eur. J. Biochem. 67:397 (1976).
4. C. Herbst, Uber die zur Entwickelung der Seeigellarven Nothwendigen Anorganischen Stoffe, ihre Rolle und ihre Vertretbarkeit. Tl 3, die Rolle der Nothwendigen Anorganischen Stoffe, Arch. f. Entwicklungsmechanik der Organismen 17:306 (1904).
5. J. E. Morris and B. A. Afzelius, The structure of the shell and outer membranes in encysted Artemia salina embryos during cryptobiosis and development, J. Ultrastruct. Res. 20:244 (1967).
6. C. N. A. Trotman, S. P. Gieseg, R. S. Pirie and W. P. Tate, Abnormal development in Artemia: defective emergence of the prenauplius with bicarbonate deficiency, J. Exp. Zoo. 243:225 (1987).
7. K. Grasshoff, "Methods of Seawater Analysis," Verlag Press, Weinheim (1976).
8. N. L. Sato, Excystment of the egg of Artemia salina in artificial seawater of various condition, Gunma J. Med. Sci. 15:102 (1967).
9. W. B. Busa, J. H. Crowe and G. B. Matson, Intracellular pH and the metabolic status of dormant and developing Artemia embryos. Arch. Biochem. Biophys. 216:711 (1982).
10. W. B. Busa and J. H. Crowe, Intracellular pH regulates transitions between dormancy and development of brine shrimp, Artemia salina, embryos. Science 221:366 (1983).
11. D. De Chaffoy, G. De Maeyer-Criel and M. Kondo, On the permeability and formation of the embryonic cuticle during development in vivo and in vitro of Artemia salina embryos, Differentiation 12:99 (1978).
12. G. L. Peterson, R. D. Ewing and F. P. Conte, Membrane differentiation and de novo synthesis of the (Na$^+$ + K$^+$)-activated adenosine triphosphatase during development of Artemia salina nauplii, Develop. Biol. 67:90 (1978).

13. F. P. Conte, Role of C-4 pathway in crustacean chloride cell function, Am. J. Physiol. 238:R269 (1980).
14. F. P. Conte, Structure and function of the crustacean larval salt gland, Int. Rev. Cytol. 91:45 (1982).
15. Y. Pocker and S. Sarkanen, Carbonic anhydrase, structure, catalytic versatility, and inhibition, Advan. Enzymol. 47:149 (1978).
16. I. Bertini, C. Luchinat and.A. Scozzafava, Carbonic anhydrase, an insight into the zinc binding site and into the active cavity through metal substitution. Struct. Bond. 48:45 (1982).
17. P. Rafiee, C. O. Matthews, J. C. Bagshaw and T. H. MacRae, Reversible arrest of Artemia development by cadmium, Can. J. Zool. 64:1633 (1986).

LIGHT-INDUCED DEVELOPMENT OF ARTEMIA CYSTS*

A. Van der Linden[1], R. Blust[1], R. Dommisse[2], G. Criel[3] and W. Decleir[1]

1) Laboratory of Biochemistry and General Zoology, Groenenborgerlaan, 171, 2020 Antwerp, University of Antwerp
   RUCA, Belgium
2) Laboratory of Organic Chemistry, Groenenborgerlaan, 171
   2020 Antwerp, University of Antwerp, RUCA, Belgium
3) Laboratory of Anatomy, Ledganckstraat, 35, 9000 Ghent
   State University of Ghent, Belgium

INTRODUCTION

The role of light in the initiation of development of Artemia embryos has been noticed before[1-5]. Sorgeloos and his colleagues[6,7] also demonstrated that the light trigger is effective immediately after hydration and only under aerobic circumstances. When we started this study we focussed on the photoreceptor and the primary photoreaction mediating light induced hatching. The results of these studies revealed that the photoreceptor had absorption characteristics of a haempigment and that an oxidation-reduction reaction was involved in the primary photoreaction[8-10].

Several recent papers have discussed the regulatory mechanism of quiescence-development transitions in Artemia embryos; i.e. metabolic changes and the importance of intracellular pH changes as well as pH sensitive control points[11-18]. However, only a few have addressed the regulation of diapause in Artemia embryos[19]. However, the above cited papers do not explain why some cysts fail to hatch in hydrated and aerobic conditions as long as they are deprived from light. In this regard, the study of metabolic and intracellular changes occurring upon illumination of these cysts may yield some answers. The present work reports on respiration, intracellular pH (^{31}P-NMR), organelle organization, trehalose, glycerol and ATP levels of light deprived and illuminated cysts. Finally, light activation is discussed in the context of known mechanisms of diapause and quiescence activation.

MATERIAL AND METHODS

Materials

Dried Artemia franciscana cysts from the Great Salt Lakes in Utah, (USA) were purchased from San Francisco Bay Brand, Inc. USA. Cysts con-

---

*Parts of this work were published previously (see Ref. 20).

tained about 0.056mg protein /mg wet weight and the wet weight to dry weight ratio was (2.97 ± 0.12). Chemicals were reagent grade, and Hepes is 4-(2-hydroxyethyl)-1-piperazine-ethanesulfonic acid.

Isolation of Light-Requiring Cysts

A certain amount of viable cysts do not require light for hatching. In the batch used, 40% of the cysts hatched within 5 days at 25°C and in complete darkness. To study light-induced and light-deprived metabolism, we needed to remove the dark developing cysts. This was done following the procedure of Van der Linden et al.[20]. After 5 days of light deprived incubation, we obtained a stock of cysts containing 45% light requiring (27% of the original batch) and 55% non-hatching cysts (33% of original batch). The light-induced hatching percentage of these cysts remained constant (45%) over the dark incubation period examined (20 days).

Isolation of Non-Hatching Cysts

Non-hatching cysts present in the commercial batch were isolated after hydration of the cysts for 5 days at 25°C in aerobic and saturated light conditions. After this period no more hatching occurred. The hatchlings were removed daily by decantation. The minority of the remaining cysts (10%) were in diapause as could be demonstrated by applying the H_2O_2 activation method of Lavens et al.[21].

EXPERIMENTAL PROCEDURE

Light-Deprived Cysts

Light requiring cysts, isolated as described above, were kept for up to 21 days aerated at 25°C and in complete darkness in a 1% salt solution (Wimex seasalt) containing 10mM HEPES, pH 8, 1000 units/ml penicillin and 100μg/ml streptomycin sulphate. The medium was renewed frequently and the pH and pO_2 were monitored. Cysts were always manipulated under a red safety light[8]. At certain time intervals, samples were taken, rinsed and used for percent hatching determination, respiration, ^{31}P-NMR and EM studies or frozen immediately and stored in liquid nitrogen until further processing for trehalose, glycerol and ATP determinations.

Illuminated Cysts

Light-requiring cysts, sampled from the stock 5 days after the start of hydration, were divided into two treatment groups. One was kept as a control, while the other group was illuminated ($3.5mW/cm^2$) for 30 min with two fluorescent tubes (OSRAM 30 W-25-2 universal white). Thereafter, the illuminated and control cysts were maintained under the same dark conditions. After 12 hours (T_{12}), the first signs of emergence could be observed. Also, starting from zero hours (T_0) samples of both treatment groups were taken every 2 hours up to 12 hours. The samples were either frozen immediately or used directly for respiration measurements, ^{31}P-NMR and EM studies.

Respiration Measurements

Oxygen consumption was measured using a Clark type oxygen electrode attached to a YSI Biological Oxygen Monitor (model 53). Respiration measurements were done with samples of ± 60mg wet weight of cysts incubated in the earlier described buffered salt solution containing antibiotics[20].

Determination of ATP

Cyst homogenizations (Braun microdismembrator) and ATP extractions were done as described in Van der Linden et al.[20]. The ATP measurements were done with luciferase according to Strehler[21].

Determination of Trehalose and Glycerol

Sample preparation (Braun microdismembrator) for trehalose and glycerol extraction and their quantification (HPLC-refractometer) was done as described by Van der Linden et al.[20].

^{31}P-Nuclear Magnetic Resonance Spectroscopy (^{31}P-NMR)

^{31}P-NMR was used to determine the pHi and the phosphorus containing compounds of light deprived, illuminated and non-hatching Artemia cysts[23, 24]. For in vivo spectroscopy, a superfusion apparatus was used comparable with the one described by Busa et al.[12]. The packed cyst column (10mm in diameter and 20mm high) was retained by two perforated teflon disks of about 2mm covered with a nylon plancton sieve (125µ). The superfusion buffer (10mm HEPES, 1% Wimex salt, pH 8) was pumped from an aerated reservoir through the cysts column from bottom to top at a rate of about 10ml/min (Desega STA-Vielfach Schlauchpumpe 131900). This flow rate appeared to be sufficient as the cysts were able to emerge in the superfusion apparatus. Sample preparation and treatment was always carried out under a red safety light.

In vivo ^{31}P-NMR spectra of Artemia cysts were obtained at room temperature on an Oxford Research System Biospec 1.9 T operating in the Fourier Transform mode at 32 MHz. Typically, 1000 FID interferograms were accumulated with a 45° flip angle and 0.7 seconds repetitions and then Fourier transformed using a 10-Hz line broadening factor. Usually, at least two spectra taken under identical conditions were added. The spectral width was 2000 to 3000 Hz. Peak positions were obtained by referencing to external 85% H_3PO_4. The pHi was estimated from the chemical shift of Pi using the previously published calibration curve and equation for activated Utah cysts, with X equal to the chemical shift and pH equal to 6.7 + log (X-0.57/3.08-X)[12]. Due to the rather low strength of the magnetic field (1.9 T) and the absence of proper RF (Radio Frequency) shielding, some sensitivity was lost. The obtained Pi resonance values should be considered within an error margin of ± 0.1 ppm or ± 0.2 pH unit.

Electron Microscopy (EM)

Three groups of cysts were prepared for EM investigations. These were: 1) light requiring cysts, 2) light requiring cysts, illuminated with saturated light energy doses and incubated for another 4 hours prior to fixation, and 3) non-hatching cysts. The cysts were fixed and perforated in a solution containing 3% glutaraldehyde, 0.1M Na cacodylate buffer, pH 7.4, and 12% sucrose, according to Hayat[25]. After post fixation with OsO_4 in the same buffer, they were treated with uranyl acetate according to Farquhar and Palade[26], dehydrated in ahcohol and embedded in Epon. Ultrathin sections, stained with uranyl acetate and lead acetate, were examined with a Siemens Elmiskop 1A supplied with Kodak Electron Image Film.

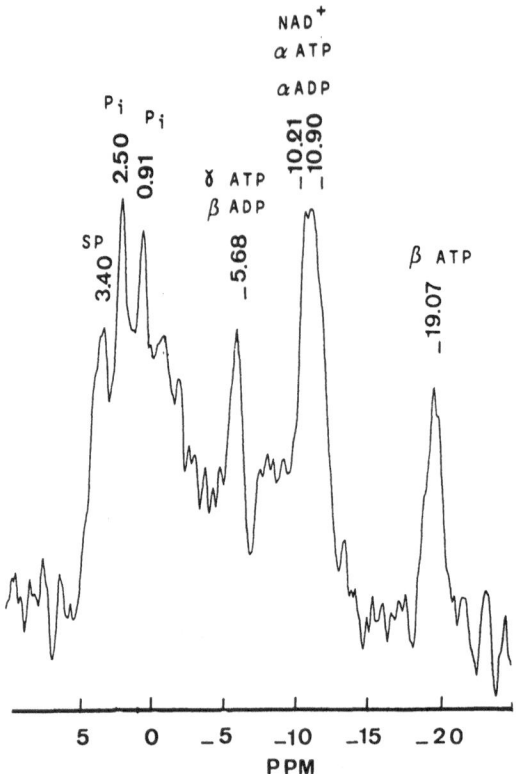

Fig. 1. ^{31}P-NMR spectrum of <u>Artemia</u> cysts which were
hydrated and illuminated during 4 hours (1000
FIDs, 0.7 s).

RESULTS

^{31}P-NMR Studies of Illuminated Cysts

The ^{31}P-NMR spectrum of 4 hours hydrated and illuminated cysts shown
in Fig. 1 displays two inorganic orthopohosphate resonances, one at 0.91
and the other at 2.50 ppm to which an intracellular pH of 5.89 and 7.22,
respectively could be assigned within ±0.2 pH units. The presence of a
sugar phosphate resonance (3.4 ppm) and a high amount of nucleosides tri-
and di- phosphates) (-5.68; -10.90; -19.07 ppm) reflects the active
metabolism of developing cysts. This spectrum is very similar to what was
found by Busa et al. for aerobically developing cysts, but they obtained
higher pH values (6.4 and 7.9)[12]. The reason two P_i peaks were found
could be due either to the involvement of two intracellular compartments
with different pH's or two different groups of cysts. We favor the last
proposition with developing and non-developing cysts as two distinct groups.
This view is supported by the spectrum of non-hatching cysts (Fig. 2.) and
knowing that these cysts represent 32 % of the original batch.

^{31}P-NMR Studies of Non-Hatching Cysts

Results in Fig. 2 demonstrate that only one major P_i resonance at

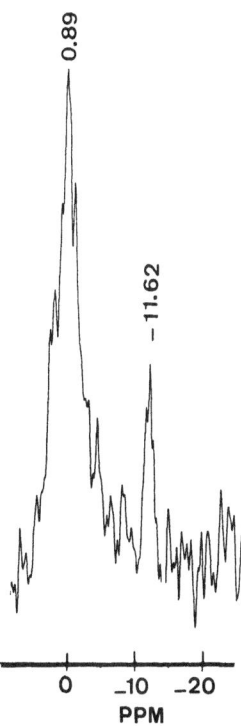

Fig. 2. ^{31}P-NMR spectrum of non-hatching <u>Artemia</u>
cysts which were isolated from a batch
after 5 days incubation in light, (1000
FIDs, 0.7s).

0.80 ppm is visible, compared to the one at 0.91 ppm in the previous
spectrum (Fig. 1) indicating a pH of about 5.86. Furthermore no NTP/NDP
resonances are detectable and the resonance at -11.62 ppm indicates the
presence of dinucleotides such as NAD$^+$[24].

^{31}P-NMR Studies of Light-Deprived Cysts

The spectrum for the 5 days light deprived cysts (Fig. 3.) shows great
similarity with the spectrum of non-hatching cysts. Sugar phosphates and
ATP resonances are absent and only the dinucleotide resonance remains. The
pH_i values (±0.2 pH units) of both groups in this sample must coincide
since only one P_i resonance (0.98 ppm; pH 5.99) is present. However, upon
illumination a shift of the P_i resonance can be attributed to the light
deprived cysts, while the P_i resonance of the non-hatching cysts remained
unchanged (0.87 ppm; pH 5.83) (see Fig. 4.). The latter was tested separately (results not shown). This doubling of the P_i resonance was also
registered by Busa et al.[12] upon resumption of aerobic development.
Concomitant with the shift in intracellular pH from 5.99 to 6.68 (0.98 ppm
to 1.80 ppm), there is the appearance of sugar phosphates and NTP/NDP
resonances. This is in agreement with the ATP measurements. Here the pH
shift is still in progress as can be seen by comparing this spectrum with
the spectrum of illuminated cysts shown in Fig. 1 (pH 7.22).

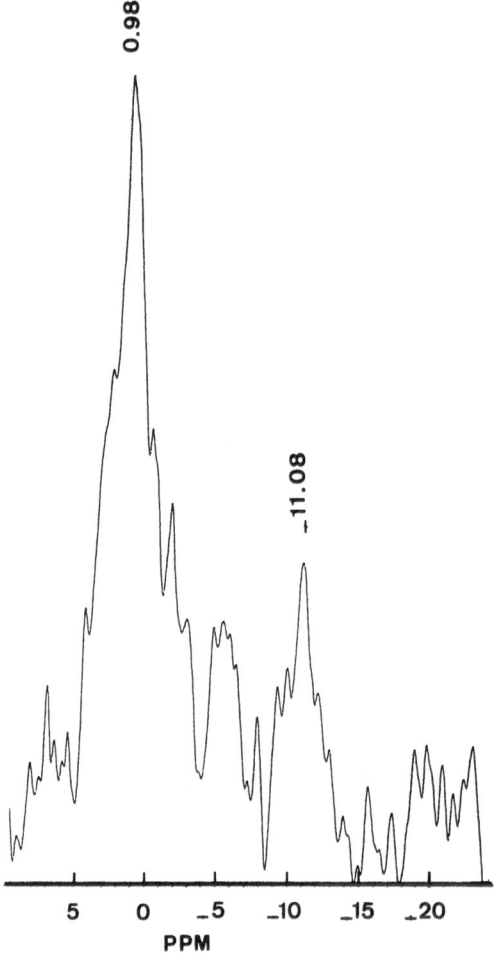

Fig. 3. ^{31}P-NMR spectrum of <u>Artemia</u> cysts which were incubated for 5 days under light-deprived conditions.
(1000 FIDs, 0.7s).

        Since it was already determined that developing cysts had an alkaline
pH[19], it was not surprising that the pH increased upon illumination. It
is, however, too early to make conclusions on the pH change as a function
of time after illumination. The low field strength of the magnet, the
absence of RF shielding, the uncontrolled temperature (19°C) retarding
development and only one-half of the cysts responding to illumination (45%)
made it necessary to obtain a high amount of FIDs (at least 1000) and made
measurement of immediate and small pH changes impossible.

Respiration

        The oxygen consumption of light-deprived cysts was found to be very
low (0.014 O_2 μl/h/mg wet weight). Although the light-induced hatching
percentage of these cysts remains constant over the whole incubation period,
the oxygen uptake stabilizes only after 14 days of incubation (see Fig. 5
upper panel).

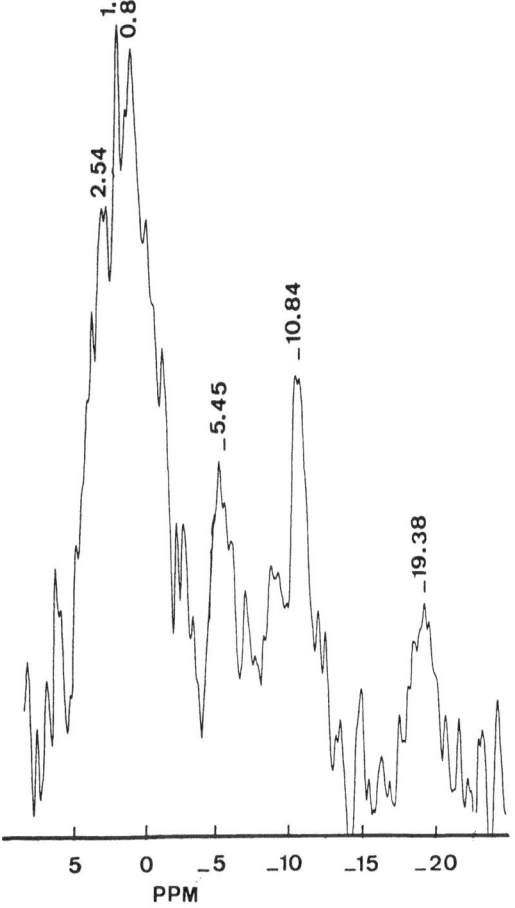

Fig. 4. ^{31}P-NMR spectrum selected from a sequence obtained
upon illumination of 5 days light-deprived <u>Artemia</u>
cysts. (1000 FIDs, 0.7s).

Upon illumination of a 5 days light-deprived treatment group of cysts,
the respiration increases rapidly and finally reaches values 10 times
higher than those obtained in darkness (0.03 to 0.3 µl/h/mg wet weight;
see Fig. 6, upper panel).

ATP Concentrations

The ATP concentration and the respiration of dark dormant cysts follow
a similar pattern (see Fig. 5). In darkness, the ATP content drops to a
constant level, which is less than 20% of that found during development.
The low ATP level is in agreement with the ^{31}P-NMR data of light-deprived
cysts (see Fig. 3).

Upon illumination, and concomitant with the increase in respiration
and the breakdown of trehalose, a 60% increase in ATP content is observed,

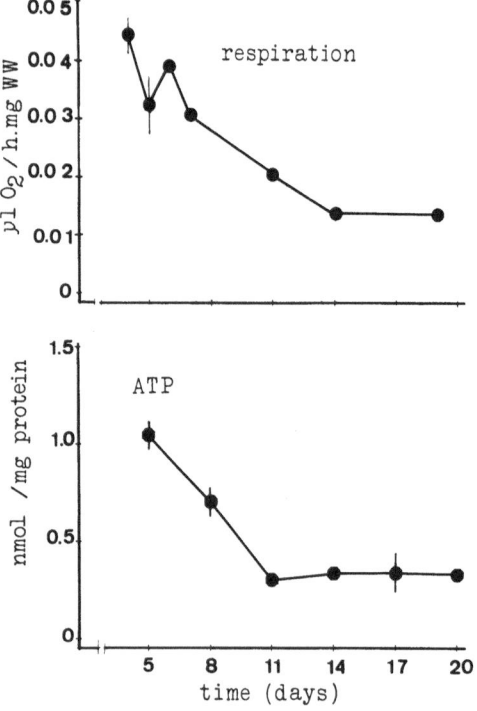

Fig. 5. Respiration (µl O_2/h/mg wet weight) and ATP concentrations (nmol/mg protein) of light requiring Artemia cysts during prolonged incubation in darkness. Each point is the mean (± SD) of at least 3 independent samples. No vertical bars means that the SD is smaller than the size of the symbol.

reaching a constant level after 4 hours (Fig. 6).

Trehalose and Glycerol Concentrations

The results of Fig. 7 show that the concentrations of trehalose do not vary significantly (Kruskal-Wallis; H=1.8005; n=5; P > 0.5) during light-deprived incubation. However the glycerol concentrations show a slight decrease (K.-W.; H=18.57; n=5; P< 0.005). This means that the cysts remain viable when deprived from light without consuming detectable amounts of their endogenous trehalose reserves. Even supposing that trehalose is the primary fuel used by cysts during aerobic incubation in darkness[27-31], the amount of trehalose breakdown that can account for the measured oxygen consumption is too small to be detected.

The results in Fig. 8 demonstrate that trehalose breakdown and glycerol synthesis start upon exposure to light while in darkness no changes were observed. Trehalose concentrations decrease from 2.2 to 1.4 µmol/mg protein in 12 hours. This means a utilization of 36% of the trehalose reserves. On the other hand, there is a doubling of the glycerol content between 6 and 12 hours after illumination while there is no change in the control cysts.

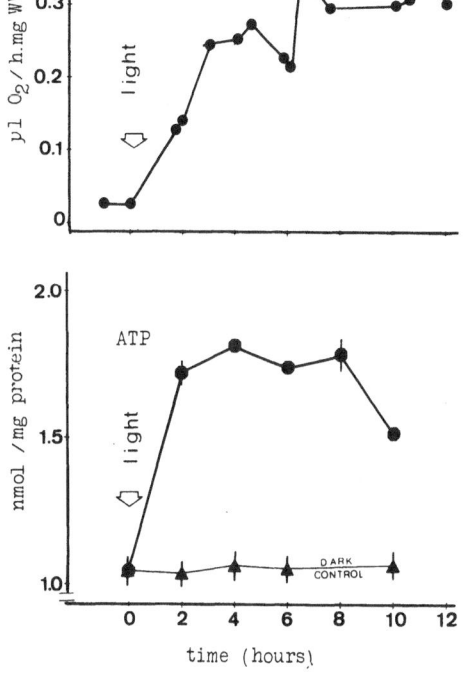

Fig. 6. Respiration (μl O_2/h/mg wet weight) and ATP concentration (nmol/mg protein) of light requiring
<u>Artemia</u> cysts as a function of time after 30 min-
utes light exposure (-●-). Control cysts (-▲-)
were maintained in darkness. Each point represents the mean ± SD of 4 independent samples. No
vertical bars means that the SD is smaller than
the size of the symbol.

Studies of Different Cysts by Electron Microscopy

The results of electron microscopic (EM) level research of various
cysts illustrate once again the suppressed development of light-deprived
cysts (Plate 1). Only spherical and ellipsoidal electron dense mitochondria
are found and there is a total absence of polyribosomes. Upon illumination
of these cysts the mitochondria loose their circular outline and become
elongated and electron lucent (Plate 2). The appearance of rough endoplasmic reticula reflects the development of polyribosomes.

Since the isolation procedure to obtain light-requiring cysts doesn't
exclude non-hatching cysts, we isolated non-hatching cysts and studied
their intracellular appearance. Plate 3 illustrates that the cell and
organelle membranes are disrupted in these cysts and that no organelles
other than yolk platelets could be identified.

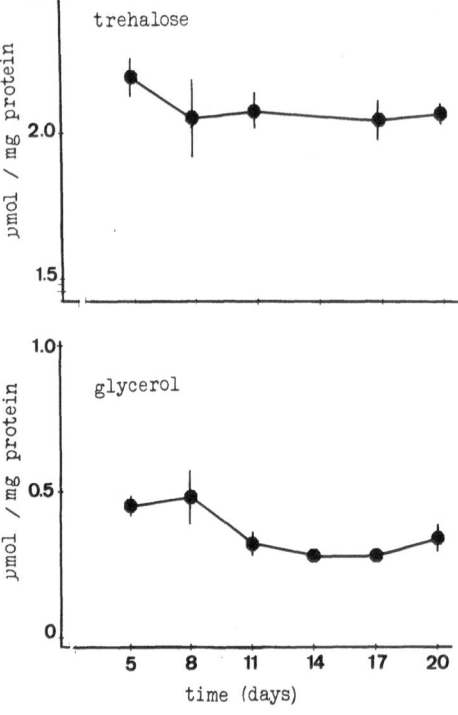

Fig. 7. Concentrations of trehalose and glycerol (μmol/
mg protein) during prolonged incubation of
light-requiring <u>Artemia</u> cysts in darkness. Each
point is the mean (± SD) of 4 independent
samples. No vertical bars means that the SD is
smaller than the size of the symbol.

DISCUSSION

The results of this study demonstrate that a large percentage (27%)
of the commercial <u>Artemia</u> cysts used in our study are metabolically arrested
when deprived of light. These cysts have a very low respiration rate when
incubated for long periods in dark, hydrated and aerated conditions and
the trehalose and glycerol levels remain relatively constant. However,
the light-induced hatchability of these cysts was constant at least over
the period investigated (20 days).

In view of, 1) the observation of an acid intracellular pH in these
cysts, 2) the information on the regulatory role of the pH_i in metabolic
transitions in <u>Artemia</u> cysts[12-14,16,18], and 3) the demonstration that
trehalase is the major pH sensitive control point in this regulation[15,17],
we suggest that the absence of trehalose consumption and the suppression
of metabolism in light-deprived cysts may be caused by the acid pH. This
suppressed metabolism, and especially the very low respiration rate of
light-deprived cysts, is consistent with the presence of non-functional
mitochondria. These findings suggest two possible reasons for a depressed
metabolism in light-deprived cysts: the presence of non-functional

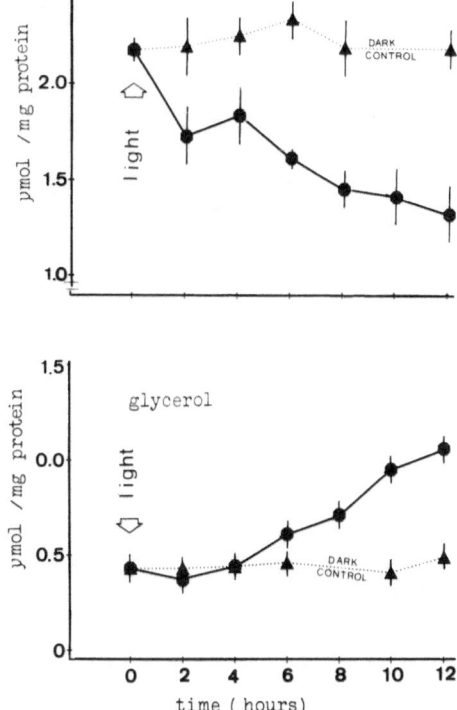

Fig. 8. Changes in the concentrations of trehalose and
glycerol (μmol/mg protein) in light-requiring
Artemia cysts after a 30 minutes exposure to
light (-●-). Control cysts were maintained in
darkness during the whole period (12 hours)
(-▲-). Each point is the mean ± SD of 4 independent samples. No vertical bars means that
the SD is smaller than the size of the symbol.

mitochondria and the acid intracellular pH. However, it is not possible
to conclude which of these two mechanisms is the initial cause for the
metabolic arrest since the absence of functional mitochondria can, by
themselves, elicit suppression of metabolism creating a negative energy
balance and a consequent drop in pH_i [14].

In addition to the fact that light activates the development of
Artemia cysts, we have demonstrated that carbohydrate metabolism of light-
deprived cysts also can be activated by light. This is manifested by an
enhancement of respiration accompanied by a decrease in trehalose concentration following illumination. These light-induced changes are visible
within the first two hours after illumination. Also, trehalose which is
metabolized during development is thought to be the fuel for the synthesis
of glycerol[30]. Glycerol concentrations were observed to increase 6 hours
after the onset of illumination. This light-induced resumption of
metabolism is also associated with the light-induced pH_i increase and the
change in mitochondrial and ribosomal organisation. The sequence of these

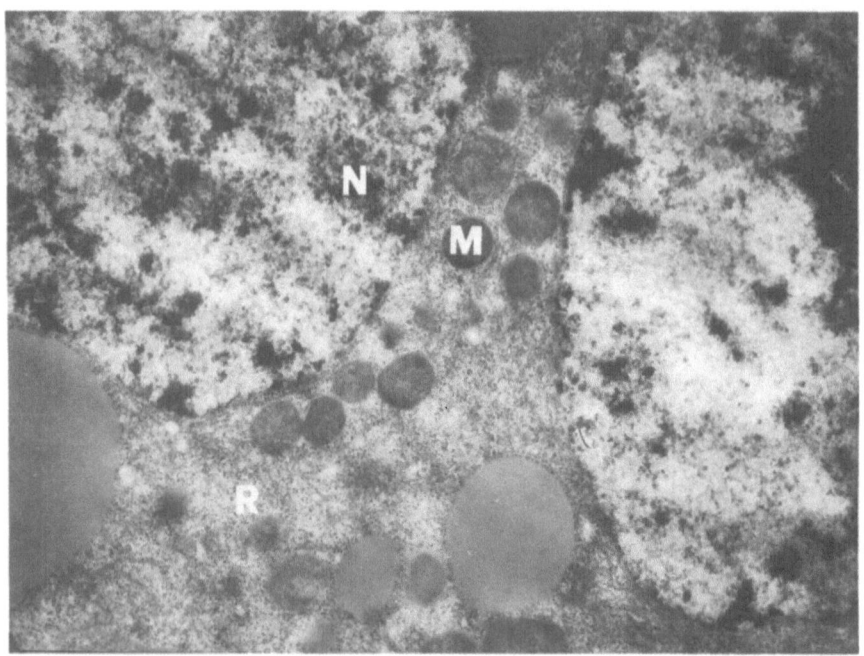

Plate 1.   EM view of a cross-section of light requiring <u>Artemia</u> cyst,
incubated for 5 days in darkness. (ribosomes, R, mitochodria,
M, and the nucleus, N)

light-induced events, however, remains unknown. Is the acid pH_i the
initial cause of the metabolic arrest of light-deprived cysts which light
could then induce an increase in pH_i causing resumption of metabolism?
Unfortunately we were not able to measure rapid pH_i changes. Therefore,
we have no information on whether light induces an increase in pH_i with a
consequent resumption of development or whether light induces, in some
other way, metabolism accompanied by an increase in pH_i [14].

The appearance of electron dense mitochondria in dormant cysts with
a different shape than in developing cells is well studied in dormant
fungi[32,33]. In <u>Artemia</u> a morphometric study on mitochondrial maturation
upon hydration of the cryptobiotic cysts was done by Jardel and Nayudu[34].
They demonstrated that upon hydration there is a change in mitochondrial
shape from circular to elongated. This is parallel to what we find upon
illumination of hydrated cysts. These authors also demonstrated a correlation between the rise in respiration rate and the increase in cristae
during hydration. This supports their hypothesis that the increase in
respiration is related to an increase in inner membrance surface area of
the mitochondria rather than an activation of specific mitochondrial sites
of oxidative phosphorylation. Although we did not study the mitochondrial
changes as a function of time after illumination, our results agree with
this hypothesis. We also observed a change in the mitochondrial size
associated with an increase in respiration rate.

Since the literature on <u>Artemia</u> development often deals with quiescence
or diapause in <u>Artemia</u> embryos, the question arose as to which of these
two forms (stages) represents metabolic arrest of light-deprived activated

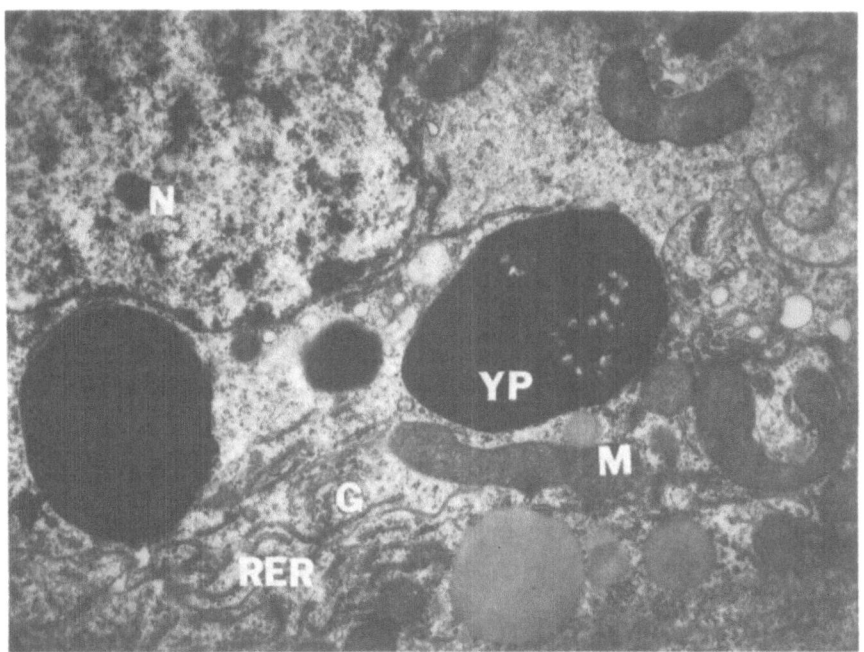

Plate 2. EM view of a cross-section of light requiring <u>Artemia</u>
cyst 4 hours after light activation. Polyribosomes are
formed and a rough endoplasmic reticulum (RER) becomes
visible. (mitochondria, M, Golgi apparatus, G, nucleus,
N and yolk platelets, YP)

cysts. The terms diapause, dormancy and quiescence which are often used
under different meanings are explained and illustrated in relation to
<u>Artemia</u> in the review of Lavens and Sorgeloos[35]. Quiescence is charact-
erised by a temporary arrest in development caused by unfavourable environmental conditions, e.g. anhydrobiosis, cryobiosis, anoxibosis and
osmobiosis. Upon restoration of favourable circumstances, development
resumes. The environment in which dormancy and development can be investigated are quite different. In <u>Artemia</u> a lot of work has been done on
quiescence-development transtions and the regulatory role of the intracellular pH[11-18]. Diapause on the other hand is due to an innate property
of the cysts. Metabolic and developmental arrest are under endogenous
control. Development resumes, naturally only in favourable circumstances,
upon activation by a transient physical or chemical treatment. The diapause
and the activated form can be studied under identical environmental conditions. The only studied and recognized form of diapause in <u>Artemia</u> is
the developmental arrest of cysts newly released from the female ovisac[16,
21, 35]. This diapause can be broken (activation) by different treatments,
e.g. dehydration-hydration cycles, storage at low temperature, CO_2 and
H_2O_2 treatment before bringing them back to favourable hatching conditions
[19, 21, 35].

What about the cysts which remain developmentally arrested as long as
they are deprived of light? We hypothesize that because of the following
observations these cysts are in post-diapause and are not quiescent. The
suppressed metabolism of light-deprived cysts is caused by internal

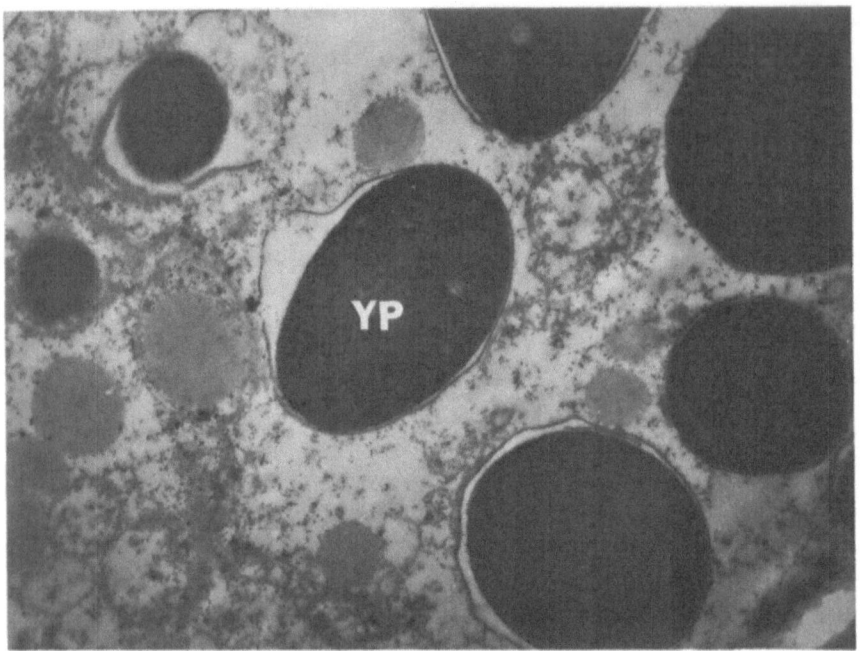

Plate 3. EM view of a cross-section of non-hatching <u>Artemia</u> cyst.
The membranes are ruptured and only yolk platelets (YP)
can be observed.

regulation; the intracellular pH and the mitochondrial structure are
possible candidates. Light is not indispensable for the growth of
<u>Artemia</u>[3] and hence darkness is not at all an unfavourable circumstance
comparable to anoxia. Development of light-deprived cysts is resumed upon
a transient illumination treatment. Bringing these illuminated cysts back
in darkness does not revert the resumption of development which would
indeed be the case if we were dealing with quiescence-development transitions. Hence, the diapause and the activated form can be both studied in
the same dark circumstances which is typical for diapause-development
transitions.

Based on our observations we have been able to show that by artific-
ially inserting a light deprived period upon hydration of commercial cysts,
we were able to observe that the final activation mechanism of diapause
(at least for some cysts, 27% of batch used) can be subdivided in a treatment to make the cysts light sensitive and a consequent illumination
treatment. A classic diapause activation treatment in <u>Artemia</u> requires a
change in the environmental circumstances which the cysts must undergo to
obtain intracellular changes inherent with breaking of diapause and with
growing light sensitivity. In doing so the similarity with the diapause
activation mechanism in <u>Daphnia</u> becomes apparent. In this cladoceran,
light might play a central role in the termination of diapause[36-40]. The
diapaused embryo of <u>Daphnia</u> contains a minimum of two phases, a photo-
refractory phase followed by a photo-sensitive phase. The photo-refractory
phase in embryos collected from the wild is completed by exposure to low
temperatures or to elevated CO_2 concentrations in the medium. In most
cases, dependent on the strain, light is required during the photo-sensitive

phase for the termination of the embryonic diapause. <u>Daphnia</u> embryos held in darkness retain their viability but are photo-refractory. The process underlying activation is basically photoperiodic. The photo-refractory phase and its completion to enter the photo-sensitive phase in <u>Daphnia</u> shows many similarities with the diapause of <u>Artemia</u> cysts newly released from the ovisac and the classic diapause activation turning the cysts light sensitive.

The artificial insertion of a light deprived period upon hydration of cysts has another advantage. It shows that unlike newly released diapause cysts and developing cysts which have an alkaline pH, but more like quiescent cysts, activated cysts that are light deprived have an acid pH_i. In view of the proposed diapause activation mechanism, where illumination plays the role of a post-diapause activator, a possible role for the pH_i in regulating diapause-development transitions could not be ruled out. However we do not know when this drop in pH was accomplished, since the light deprived cysts we have been working with were always incubated for at least 5 days, and this incubation accompanied by the demonstrated depressed metabolism could have caused a drop in pH_i. But some arguments favor a drop in pH_i during classic diapause activation, e.g. the known activation treatments create a negative energy balance and hence a drop in pH_i. This could mean that a drop in pH might be an essential intracellular event accompanying completion of diapause and becoming light sensitive. However, this remains speculative as long as we have not tested the changes in pH_i of newly released cysts upon diapause activation and upon consequent illumination. Working with commercial cysts offered us a nice opportunity to study light-induced development, but it offers no possibility to draw conclusions on the classic diapause activation.

REFERENCES

1. P. Sorgeloos, First report on the triggering effect of light on the hatching mechanism of <u>Artemia</u> <u>salina</u> dry cysts, <u>Mar. Biol</u>. 22:75 (1973).
2. P. Sorgeloos, De invloed van abiotische en biotische faktoren op de levenscyclus van het pekelkreeftje <u>Artemia</u> <u>salina.</u> Ph. D. thesis, State University Ghent, Belgium (1975).
3. J. P. Royan, Effect of light on the hatching and growth of <u>Artemia</u> <u>salina</u>. Mahasagar-Bull. Natn. Inst. Oceanogr., Vol. 9:83 (1976).
4. L. V. Spektorova and A. M. Syomik, The influence of incubation conditions upon <u>Artemia</u> hatching efficiency in three strain models, in: "Book of Abstracts. International Symposium on the Brine Shrimp, <u>Artemia</u> <u>salina</u>", Artemia Reference Centre, Ghent (1979).
5. P. Vanhaecke, A. Cooreman and P. Sorgeloos, International study on <u>Artemia</u>. XV: Effect of light intensity on hatching rate of <u>Artemia</u> cysts from different geographical origin, <u>Mar. Ecol. Prog.</u> Ser. 5:111 (1981).
6. P. Sorgeloos and G. Persoone, Technological improvements for the cultivation of invertebrates as food for fishes and crustaceans. II. Hatching and culturing of the brine shrimp, <u>Artemia salina</u>, <u>Aquaculture</u> 6:303 (1975).
7. P. Sorgeloos, M. Eaeza-Mesa and F. Benijts and G. Persoone, Current research on the culturing of the brine shrimp <u>Artemia</u> <u>salina</u> at the State University of Ghent, Belgium in: "Proceedings 10th European Symposium on Marine Biology", Vol 1, G. Persoone, and E. Jasper, eds., Universa Press, Wetteren (1976).
8. A. Van der Linden, R. Blust and W. Decleir, The influence of light on the hatching of <u>Artemia</u> cysts (Anostraca: Branchiopoda: Crustacea), <u>J. Exp. Mar. Biol. Ecol.</u> 92:207 (1985).
9. A. Van der Linden, I Vankerckhoven, R. Caubergs and W. Decleir, Action

spectroscopy of light-induced hatching of Artemia cysts
(Branchiopoda, Crustacea), Mar. Biol. 91:239 (1986).

10. A. Van der Linden, R. Blust, K. Cuypers, C. Thoeye and F. Bernaerts, An action spectrum for light-induced hatching of Artemia cysts, in: "Artemia Research and its Applications," Vol. 2, W. Decleir, L. Moens, H. Slegers, E. Jaspers and P. Sorgeloos, eds. Universa Press, Wetteren (1987).

11. W. B. Busa, Cellular dormancy and the scope of pH_i-mediated metabolic regulation, in, "Intracellular pH: Its Measurement, Regulation, and Utilization in Cellular Functions." R. Nuccitelli and D. W. Deamer, eds., A. R. Liss, Inc. New York (1982).

12. W. B. Busa, J. H. Crowe and G. B. Matson, Intracellular pH and the metabolic status of dormant and developing Artemia embryos, Arch. Biochem. Biophys. 216:711 (1982).

13. W. B. Busa and J. H. Crowe, Intracellular pH regulates transitions between dormancy and development of brine shrimp (Artemia salina) embryos, Science 221:366 (1983).

14. W. B. Busa and R. Nuccitelli, Metabolic regulation via intracellular pH, Amer. J. Phys. 246:R409 (1984).

15. J. F. Carpenter and S. C. Hand, Arrestment of carbohydrate metabolism during anaerobic dormancy and aerobic acidosis in Artemia embryos: determination of pH sensitive control points, J. Comp. Physiol. B 156:451 (1986).

16. J. H. Crowe, L. M. Crowe, L. Drinkwater and W. B. Busa, Intracellular pH and anhydrobiosis in Artemia cysts, in "Artemia Research and its Applications," Vol. 2, W. Decleir, L. Moens, H. Slegers, E. Jaspers and P. Sorgeloos. eds., Universa Press, Wetteren, (1987).

17. S. C. Hand and J. F. Carpenter, pH - induced metabolic transitions in Artemia embryos mediated by a novel hysteretic trehalase, Science 232:1535 (1986).

18. S. C. Hand and E. Gnaiger, Anaerobic dormancy quantified in Artemia embryos: A calorimetric test of the control mechanism, Science 239:1425 (1988).

19. L. E. Drinkwater and J. H. Crowe, Regulation of embryonic diapause in Artemia: environmental and physiological signals, J. Exp. Zool. 241(3):297 (1987).

20. A. Van der Linden, R. Blust, A. J. Van Laere and W. Decleir, Light induced release of Artemia dried embryos from diapause; analysis of metabolic status, J. Exp. Zool. 247:131 (1988).

21. P. Lavens, W. Tackaert and P. Sorgeloos, International study on Artemia XLI. Influence of culture conditions and specific diapause deactivation methods on the hatchability of Artemia cysts produced in a standard culture system. Mar. Ecol. Prog. ser. 31:179 (1986).

22. B. L. Strehler, Adenosine 5' triphosphate and creatine phosphate. Determination with luciferase. In: "Methods of Enzymatic Analysis," H. V. Bergmeyer, ed., Academic Press, New York (1965).

23. R. B. Moon and J. H. Richards, Determination of intracellular pH by ^{31}P magnetic resonance, J. Biol. Chem. 248:7276 (1973).

24. R. J. Gillies, R. Alger, J. A. den Hollander and R. G. Shulman, Intracellular pH measured by NMR: Methods and Results, in: "Intracellular pH: Its measurement, Regulation and Utilization in Cellular Functions," R. Nuccitelli and D. W. Deamer, eds., Alan R. Liss, Inc., New York (1982).

25. M. A. Hayat, "Basic Techniques for Transmission Electron Microscopy", Academic Press, Inc., New York (1986).

26. M. Farquhar and G. Palade, Cell junctions in the amphibian skin, J. Cell Biol. 26:263 (1965).

27. J. Dutrieu, Observations biochimiques et physiologiques sur le developpement d'Artemia salina Leach, Arch. Zool. Exp. Gen. 99:1 (1960).

28. S. Muramatsu, Studies on the physiology of Artemia embryos. I. Respiration and its main substrate during early development, Embryologia, 5:95 (1960).
29. D. N. Emerson, The metabolism of hatching embryos of the brine shrimp Artemia salina. Proc. S. Dakota Acad. Sci. 42:131 (1963).
30. J. S. Clegg, The control of emergence and metabolism by external osmotic pressure and the role of free glycerol in developing cysts of Artemia salina, J. Exp. Biol. 41:879 (1964).
31. J. S. Clegg and F. P. Conte, A review of the cellular and developmental biology of Artemia. in: The Brine Shrimp Artemia, Physiology, Biochemistry, Molecular Biology, Vol. 2., G. Persoone, P. Sorgeloos, O. Roels and E. Jaspers, eds., Universa Press, Wetteren (1980).
32. R. Brambl, Respiration and mitochondrial biogenesis during fungal spore germination, in: "The Fungal Spore, Morphogenetic Controls," G. Turian and H. R. Hohl, eds., Academic Press, New York (1981).
33. L. E. Hawker, The dormant spore, in: "The Fungal Spore," D. J. Weber and W. H. Hess, eds., J. Wiley, New York (1976).
34. J. P. Jardel and P. R. V. Nayudu, Mitochondrial maturation of Artemia salina. Micron 13:365 (1982).
35. P. Lavens and P. Sorgeloos, The cryptobiotic state of Artemia cysts, its diapause deactivation and hatching: a review, in: "Artemia Research and its Application," Vol. 3, P. Sorgeloos, D. A. Bengtson, W. Decleir, and E. Jaspers, eds., Universa Press, Wetteren (1987).
36. R. G. Stross, Light and temperature requirements for diapause release and development in Daphnia, Ecology 47:368 (1966).
37. R. G. Stross and J. C. Hill, Photoperiod control of winter diapause in freshwater crustacean, Daphnia, Biol. Bull. 134:176 (1968).
38. R. G. Stross, Photoperiod control of diapause in Daphnia III, Biol. Bull. 137:359 (1969).
39. R. G. Stross, Photoperiod control of diapause in Daphnia. IV, Light and CO_2 sensitive phases within the cycle of activation, Biol. Bull. 140:137 (1971).
40. J. R. Pancella and R. G. Stross, Light induced hatching of Daphnia resting eggs, Chesapeake Science 4, n°3:135-140.

ADAPTATION TO HYPOXIA IN ARTEMIA

W. Decleir, G. Wolf and B. De Wachter

Laboratory of Biochemistry and General Zoology
University of Antwerp (R.U.C.A.)
Groenenborgerlaan 171, B-2020 Antwerp, Belgium

INTRODUCTION

The genus Artemia has diversified to inhabit ecological niches characterized by high salinity and high temperatures. Both of these abiotic environmental factors contribute to a third ecological characteristic of Artemia habitats, i.e. low oxygen content. Unfortunately precise data of partial oxygen pressures during the day and night and/or seasonal cycles of the different natural Artemia habitats are extremely scarce. One important reason for this is the technical problems of measuring the O_2 - content of high salinity waters.

This paper describes some experiments which illustrate our approach towards understanding the adaptation of different Artemia strains to hypoxia. Although this experimental approach does not represent the exact natural conditions, we have studied the adaptation of Artemia to these effects by submitting the animals under laboratory conditions.

MATERIAL AND METHODS

Cultures of Artemia

Cysts of different Artemia strains were supplied by the Artemia reference center (A.R.C., Ghent) and subsequently hatched and cultured at 25°C in 35% artificial seawater (HW-Wimex salt) in 40 L air-water lift operated raceways[1-3]. In order to work in standard conditions we preferentially used (unless otherwise stated) female pre-adult individuals between 21 and 28 days old.

O_2-Acclimation

Oxygen acclimation was obtained by lowering of the oxygen content stepwise in 6L acclimation jars as described by Vos et al.[4]. The acclimation periods were 3 days in seawater, containing successively oxygen concentrations of 4.8, 2.4 and 1.0 ml/liter.

Respirometry

Oxygen consumption experiments were carried out in a Warburg constant volume respirometer (15ml flasks with 6 individuals and 3ml of seawater in each flask flushed throughly with O_2 -N_2 gas mixtures to obtain the desired specific oxygen concentration).

Hemoglobin Characterization

Hemolymph was collected from 2 to 3 animals. After puncturing the animals with a micro-dissection device at the base of the thoracopods, a glass capillary of 2µl was filled. The samples were immediately introduced into the diffusion chamber at 25°C. Thereafter oxygen binding curves were determined with the diffusion chamber technique as described by Sick and Gersonde[5] and modified by Vanpachtebeke et al.[6] and Hens et al.(in preparation).

Electrophoresis

The hemoglobins in the hemolymph (Hb1,2 and 3) were separated on cellulose acetate strips in 200 mM glycine-Tris, pH 8.58. Each Hb was cut out from the strip and introduced directly into the diffusion chamber, whereafter the oxygen binding curves, the p50- and the n-values were determined. The influence of GTP on oxygen binding was studied after dipping the surface of a Hb containing strip in a freshly prepared GTP solution (6.2 mg/ml).

RESULTS

The results in Fig. 1 show the respiratory rate (as a measure for total activity) of adult <u>Artemia</u> (San Francisco Bay) as a function of the environmental partial oxygen pressure for 4 different lots of animals acclimated to seawater containing oxygen concentrations of 4.8, 2.4, 1.5 and 1.0 ml/liter, respectively.

Submission of preadult female individuals from two <u>Artemia</u> strains (Great Salt Lake and Kazakhstan) to mild hypoxia (1ml O_2/liter) results in the increase of the Hb production, which is about three times higher for Kazakhstan as compared to Great Salt Lake. The concentration of the three Hb types in the hemolymph as calculated after electrophoresis is presented in Fig. 2.

As can be seen in Table 1 the oxygen affinity (p50) of total hemolymph of both strains are rather similar, but cooperativity (n50) of Great Salt Lake shrimp is about 1.75 times higher compared to shrimp from Kazakhstan. It is striking that similar p50 values of the total hemolymph of both strains result from differing properties of the composing hemoglobin types. The p50 values of the separated Hb1,2 and 3 of both strains differ considerably, as shown in Table 1, but in both strains the Hb3 shows the highest affinty for oxygen (lowest p50) and Hb1 the lowest (highest p50). This is also illustrated by Fig. 3. The p50 values for Hb3 are respectively 2.63 and 3.82mm Hg for Great Salt Lake and Kazakhstan shrimps. This means that the Hb3 of the Great Salt Lake strain is adapted to lower environmental oxygen concentrations than the Hb3 of the Kazakhstan strain. The p50 range between Hb1 and Hb3 is 4.63mm Hg for the Great Salt Lake strain and only 2.45mm Hg for the Kazakhstan strain. This means that the pO_2 range between which the three Hbs of the Great Salt Lake strain are active is wider.

Addition of GTP to Hb1 and Hb2 of both strains (after electrophoresis) shows that the binding of this organic phosphate to the pigment results in a shift to the right of the oxygen dissociation curve (Fig. 4). This,

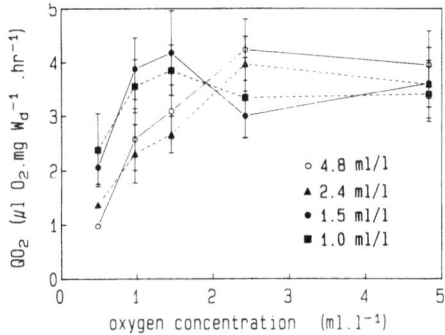

Fig. 1. Relation between the respiration rate QO_2 (μl O_2. mg W_d^{-1}.h^{-1}) and the environmental oxygen concentration ($ml.l^{-1}$) for four differently acclimated lots of the San Francisco Bay Artemia strain, to 4.8, 2.4, 1.5 and 1.0 ml O_2L^{-1}, respectively.

however, is not yet proof that this organic phosphate plays a role in the control of oxygen binding in the living organisms.

When the culture from Kazakhstan Artemia is older than 6 weeks the colour of the animals, adapted to hypoxia, changes from bright red to brown. The extinction spectrum of the hemolymph (1µl in 1ml 50mM glycine-Tris, pH 8.58) of such brown-coloured animals shows two typical peaks for artemocyanin at 625 and 680nm adjacent to the Hb-peaks at 415, 540 and 580nm, respectively. There results are shown in Fig. 5. The artemocyanin was separated electrophoretically from the Hbs, and submitted to oxygen binding analysis in the diffusion chamber. We found no reversible oxygen binding associated with this compound confirming that artemocyanin is not a respiratory pigment. Nevertheless the pigment can be dehydrogenated with sodium dithionite and this results in a slight decrease in absorbance of the 625 and 680 nm peaks. This observation may be related to the electron

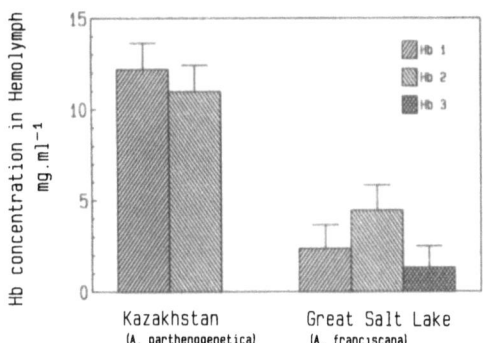

Fig. 2. Hemoglobin concentration of Hb1, 2 and 3 of pre-adult females of two Artemia strains (Kazakhstan and Great Salt Lake).

Table 1. Oxygen Binding Properties of Hemolymph and of Individual Hemoglobins after Electrophoresis on Cellulose Acetate strips[a]

Sample	Kazakhstan	n	Great Salt Lake	n
p50 hemolymph	5.86 ±0.24	(10)	5.81 ±0.17	(10)
Hb1	6.27 ±0.40	(7)	7.26 ±0.52	(4)
Hb2	4.40 ±0.43	(10)	5.07 ±0.32	(12)
Hb3	3.82 ±0.22	(7)	2.62 ±0.20	(6)
n50 hemolymph	2.44 ±0.17	(10)	4.27 ±0.57	(10)
Hb1	2.39 ±0.39	(7)	3.15 ±0.38	(4)
Hb2	1.56 ±0.21	(10)	3.27 ±0.52	(12)
Hb3	1.59 ±0.23	(7)	1.41 ±0.25	(6)

[a]Oxygen affinity (p50) is the partial oxygen pressure (in mm Hg) at which the respiratory pigment is 50% loaded with oxygen. Cooperativity (n50) is an expression of the Hill coefficient at 50% saturation. All analyses were performed at 25°C and Hb1, 2 and 3 were measured at pH 8.58. N is the number of analyses performed using 2-3 animals per assay. Only pre-adult females between 21 and 28 days old were used.

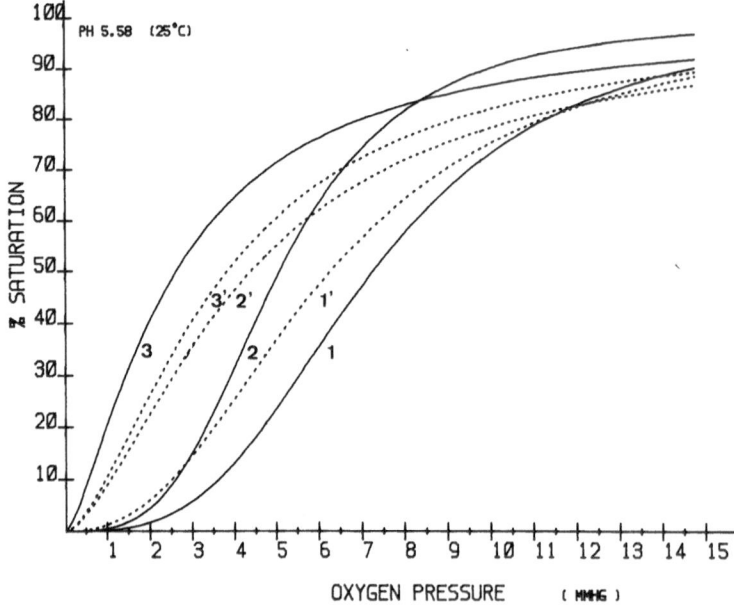

Fig. 3.   Oxygen binding properties of electrophoretically separated hemoglobins of two Artemia strains, i.e. Great Salt Lake (Hb 1, 2 and 3 ——) and Kazakhstan (Hb1', 2' and 3' ····)

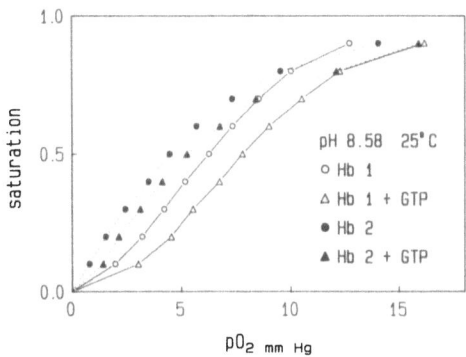

Fig. 4. Effect of GTP on oxygen binding of electrophoretically separated hemoglobins 1 and 2 of the
Kazakhstan *Artemia* strain.

charge of the copper in the molecule, but it is not clear what this has to
do with artemocyanin I and/or II[7].

DISCUSSION

   Artemia appears to be a respiratory regulator above a critical oxygen
tension of 1 to 2 ml O_2/l. At lower oxygen tensions we find a significant
decrease in activity, which is less pronounced in animals that were previously acclimated to lower oxygen tensions. As has been demonstrated in
previous studies[4,8,9], there are two underlying mechanisms of this adaptation to low environmental oxygen tensions.
   1. In animals acclimated to high partial oxygen pressures, a low
oxygen environment induces the production of lactic acid as the end-product
of anaerobic metabolism. Its concentration rises in a few hours and may
attain lethal levels at very low partial oxygen pressures. This can be
considered as a short term- or stress-response to hypoxia. Under our experimental conditions (Fig. 1) this resulted in a high mortality (92%) of
the animals acclimated at 100% oxygen and kept in the Warburg vessels at
0.5 ml O_2/l. This anaerobic pathway of adult *Artemia* (production of lactic
acid) is also found in many other crustaceans and differs from the anaerobic
pathways found by Clegg and Conte[10] and Conte et al.[11] in cysts and
larvae of *Artemia*, which represents a remarkable specialization of the
early life stages of brine shrimp.
   2. In animals acclimated to low oxygen pressures, a low oxygen
environment is accompanied by a much slower decrease in aerobic metabolism
and a very low level of lactic acid production. Under our experimental
conditions (Fig. 1) this resulted in 97% survival of the animals acclimated
at lower oxygen tensions and kept in the Warburg vessels at 0.5 ml O_2/l.
This can be considered as a long term adaptation to hypoxia and the main
mechanism of it is the production of more respiratory pigments and the
change of the fractional composition of three types of hemoglobin molecules,
each adapted to a different environmental oxygen concentration.

   In the context of a study of the production and characterization of
the different hemoglobins produced in hypoxia, in the different *Artemia*
strains, we present in this paper the first results obtained with a new
technique that allows us to work with freshly taken hemolymph (instead of
total extracts) from a few (and even from one single) animals. As a

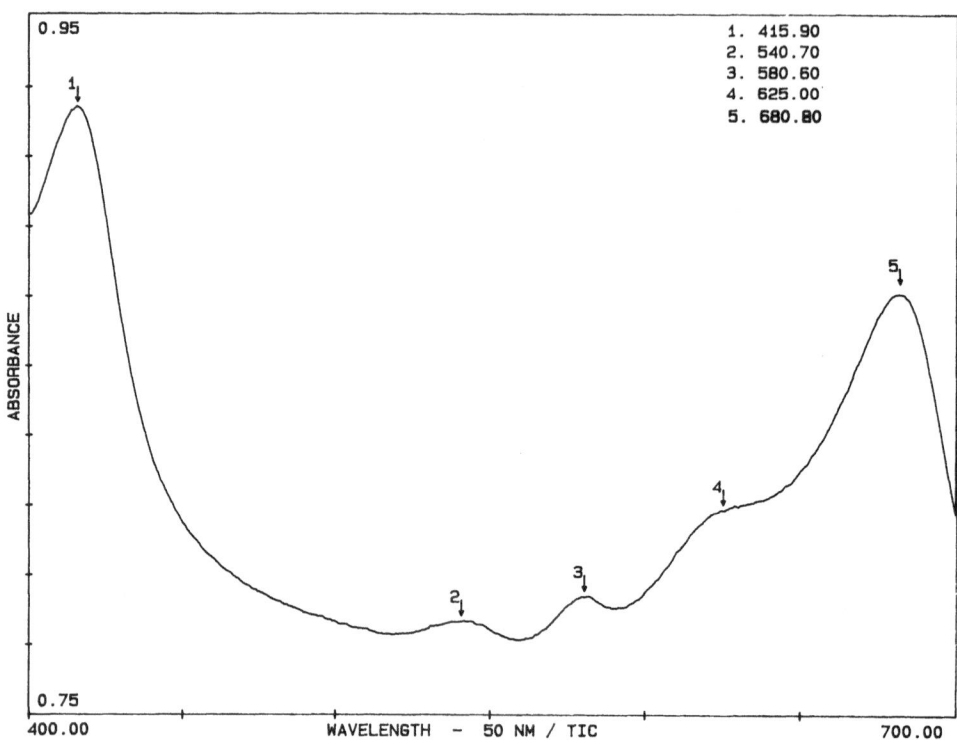

Fig. 5. Extinction spectrum of the hemolymph of ageing females from
the Kazakhstan Artemia strain (50 mM glycine-Tris, pH 8.58).

standard approach the hemolymph was always taken from preadult female
animals. In this condition the blood showed sufficient stability to be
analysed by electrophoresis or in the diffusion chamber without rapid change
of the components. This is not the case with blood from older animals,
which seemed to undergo some enzymatic deterioration after the puncturing,
and which always showed a more complex composition. The latter is due, at
least in part, to the presence of hemoglobin degradation products and to
the presence of artemocyanin II[12].

The presence of the copper containing hemoprotein artemocyanin has
been demonstrated by Krissansen et al.[7,13-16] and Trotman et al.[12].
Artemocyanin II has proteolytic activity and is responsible for Artemia Hb
degradation[12]. Our results (Fig.5) definitely indicate that this
molecule is not a respiratory pigment. Another probable role of
artemocyanin is in metal transport in the blood.

The existence of three different types of hemoglobins (Hb1, Hb2 and
Hb3) was shown for the first time by Bowen et al.[17]. Vos et al.[4] showed
that the fractional composition of these 3 hemoglobins changed in response
to adaptation to hypoxia. The oxygen binding curves of these 3 hemoglobin
types in total extracts were studied by Wolf et al.[18]. The properties
of the hemoglobins in native hemolymph, with or without prior electrophoresis, has been studied in two strains of Artemia and presented here
(see Fig. 3 and Table 1). The first strain is the Great Salt Lake one
(A. franciscana) and the second is the Kazakhstan strain (A. parteno-

genetica). As these two strains show an important genetic distance[19,20], we may also expect to find differences in their adaptation to hypoxia. Our data (see Fig. 2) show that under our experimental conditions the response to hypoxia is different in the two strains. The Kazakhstan strain synthesizes more hemoglobin with predominantly Hb1 and Hb2 (54.2 and 46.7%, respectively). Thus the oxygen carrying capacity of the hemolymph is increased by making more respiratory pigment. The Great Salt Lake strain synthesizes three times less hemoglobin but it also synthesizes Hb3 (16.2%), which is much better adapted to low oxygen pressures. It increases the oxygen carrying capacity of the hemolymph by making a pigment that is more specialized to bind oxygen in low oxygen environments.

The influence of GTP on the different hemoglobins was also observed in both strains (see Fig. 4). As the hemoglobins of Artemia are relatively high molecular weight proteins (M.W. = 260,000) in the hemolymph, it is unlikely that active organic phosphates such as GTP play an important role in the control of hemoglobin properties as in the intracellular vertebrate hemoglobins. Nevertheless, our results indicate that GTP probably acts by binding to the hemoglobin molecule, increasing slightly the P50, and this seems to be a general property of most hemoglobins.

In summary, we conclude that the increase in hemoglobin content and the change in the fractional composition of the three hemoglobins in Artemia, as a long term adaptation to hypoxia, depends on the environmental oxygen pressure and on the genetic properties of each Artemia strain. Experiments have also shown that both total hemoglobin and hemoglobin composition are slightly altered by age, sex and the physiological condition of the animal. This will be a subject of further research .

ACKNOWLEDGEMENT

This work is part of the project 2.0012.82 of the F.K.F.O.

REFERENCES

1. P. Sorgeloos, De invloed van abiotische en biotische factoren op de levenscyclus van het pekelkreeftje, Artemia salina, Ph.D. Thesis, State University Ghent, Belgium (1975).
2. P. Sorgeloos, Het gebruik van het pekelkreeftje Artemia spec. in de aquacultuur, Thesis of higher Aggregate, State University Ghent, Belgium (1979).
3. D. Versichele and P. Sorgeloos, Controlled production of Artemia cysts in batch cultures, in "The Brine Shrimp Artemia", Vol. 3, G. Persoone, P. Sorgeloos, O. Roels and E. Jaspers, eds., Universa Press, Wetteren (1980).
4. J. Vos, F. Bernaerts, I. Gabriels and W. Decleir, Aerobic and anaerobic respiration of adult Artemia salina acclimated to different oxygen concentrations, Comp. Biochem. Physiol. 62A:545 (1979).
5. H. Sick, H. and K. Gersonde, Method for the continuous registration of oxygen binding curves of haemoproteins by means of the diffusion chamber, Anal. Biochem. 32:362 (1969).
6. M. Vanpachtebeke, G. Wolf and W. Decleir, Mathematical approach of the measuring conditions for the determination of oxygen dissociation curves of hemocyanin from Sepia officinalis with the diffusion chamber technique, Arch. Int. Physiol. Biochem. 90:81 (1982).
7. G. Krissansen, C. Trotman and W. Tate, Identification of the blue green chromophore of an abundant biliprotein from the haemolymph of Artemia, Comp. Biochem. Physiol. 77B:249 (1984).
8. W. Decleir, J. Vos, F. Bernaerts and C. Van den Branden, The respira-

tory physiology of *Artemia* sp, in "The Brine Shrimp *Artemia*," Vol 2, G. Persoone, P. Sorgeloos, O. Roels and E. Jaspers, Universa Press, Wetteren (1980).

9. F. Bernaerts, C. Van den Branden and W. Decleir, The aerobic metabolism in *Artemia* sp., in "Exogenous and Endogenous Influences on Metabolism and Neural Control," A.D.F. Addink, ed., Pergamon Press, New York (1981).

10. J. Clegg and F. Conte, A review of the cellular and developmental biology of *Artemia*, in "The Brine Shrimp *Artemia*," Vol 2, G. Persoone, P. Sorgeloos, O. Roels and E. Jaspers eds., Universa Press, Wetteren (1980).

11. F. Conte, J. Lowry, J. Carpenter, A. Edwards, R. Smith and R. Ewing, Aerobic and anaerobic metabolism of *Artemia* nauplii as a function of salinity, in "The Brine Shrimp *Artemia*," Vol 2, G. Persoone, P. Sorgeloos, O. Roels and E. Jaspers, eds., Universa Press, Wetteren (1980).

12. C. Trotman, G. Krissansen, B. Kastner and W. Tate, Occurence and properties of the glycoprotein artemocyanin, in "*Artemia* Research and its Application," Vol. 2, W. Decleir, L. Moens, H. Slegers, E. Jaspers and P. Sorgeloos, eds., Universa Press, Wetteren (1987).

13. G. Krissansen, C. Trotman and W. Tate, A novel protease may explain widely differing models for the structure of *Artemia salina* haemoglobin, *Biochim. Biophys. Acta*, 671:104 (1981).

14. G. Krissansen, C. Trotman and W. Tate, Purification of Artemocyanin, a haemolymph protein from the brine shrimp *Artemia, Biochem. Internat.* 7:369 (1983a).

15. G. Krissansen, C. Trotman and W. Tate, Physical and biochemical characterization of Artemocyanin, an abundant glycoprotein complex with latent proteolyric activity from the brine shrimp *Artemia, Biochim. Biophys. Acta* 747:151 (1983b).

16. G. Krissansen, C. Trotman and W. Tate, Two forms of the glycoprotein Artemocyanin from the brine shrimp *Artemia*. A relationship between latent proteolytic activity and structural modifications of the protein, 747: 159-164(1983c).

17. S. Bowen, H. Lebherr, M. Poon, V. Chow and T. Grigliatti, The haemoglobins of *Artemia salina*. I. Determination of phenotype by genotype and environment, *Comp. Biochem. Physiol.* 31:733 (1969).

18. G. Wolf, M. Vanpachtebeke, F. Hens, W. Decleir, L. Moens and M. Van Hauwaert, Study of the oxygen binding properties of *Artemia* hemoglobins, in "*Artemia* Research and its Applications," Vol. 2, W. Decleir, L. Moens, H. Slegers, E. Jaspers and P. Sorgeloos, eds., Universa Press, Wetteren (1987).

19. F. Abreu-Grobois and J. Beardmore, International study on *Artemia* II, Genetic characterization of *Artemia* populations-an electrophoretic approach, in: "The Brine Shrimp *Artemia*," Vol. 1, G. Persoone, P Sorgeloos, O. Roels and E. Jaspers, eds., Universa Press, Wetteren (1980).

20. F. Abreu-Grobois, A review of the genetics of *Artemia*, in: "*Artemia* Research and its Applications," Vol. 1, W. Decleir, L. Moens, H. Slegers, E. Jaspers and P. Sorgeloos, eds., Universa Press, Wetteren (1987).

OXYGEN BINDING IN ARTEMIA HEMOLYMPH

G. H. Wolf, B. De Wachter, F. Hens and A. Van der Linden

Laboratory of Biochemistry and General Zoology (RUCA)
Groenenborgerlaan, 171 B-2020
Antwerpen, Belgium

ABSTRACT

Two Artemia species, i.e. Great Salt Lake (A. franciscana) and Kazakh-stan (A. parthenogenetica) were submitted to a stepwise lowering of the partial oxygen pressure until 1 ml $O_2.L^{-1}$ was reached in the saline medium (35%). The increase of hypoxia resulted in the increase of the hemoglobin concentration of the hemolymph. Oxygen binding was determined in small freshly collected hemolymph samples (2 to 4 ul) of preadult females. The oxygen binding curves show an undulating biphasic course. This phenomenon only occurs with Artemia hemolymph and was never noticed when measuring Artemia hemoglobin extracts. Similar curves have been reported [1,2] for the vertebrate sickle cell Hbs, which are sensitive to aggregation, gelation and crystallization effects resulting in biphasic oxygen binding curves. As far as we know, this is the first time that biphasic oxygen binding curves have been registrated for blood solutions containing invertebrate Hbs. Our results reveal that both Artemia species show biphasic oxygen binding curves for hemolymph with and without buffer (pH 8.58) and/or added GTP (3.1 mg.mL^{-1}). Oxygen affinity (p50) and cooperativity (n_{50}) were also determined (25°C).

ACKNOWLEDGEMENTS

This work is part of the program n° 2.0007.88 of the FKFO. B.D.W. is under contract of the IWONL, and F.H. of the FKFO.

REFERENCES

1. A. Colosimo, M. Brunori and J. Wymon, Polysteric linkage, J. Mol. Biol. 100:47 (1976).
2. J. Wyman, Linked functions and reciprocal effects in hemoglobin: a second look, Adv. Protein Chem. 19:223 (1964).

DISRUPTION OF ARTEMIA DEVELOPMENT BY METALS

A. S. Pandey, J. E. Breckenridge and T. H. MacRae

Department of Biology
Dalhousie University
Halifax, Nova Scotia B3H 4J1, Canada

ABSTRACT

Development of the brine shrimp is influenced by low concentration of several metals with the extent of disruption, determined by hatching inhibition after seventy two hours of metal exposure, depending on the type of metal and its concentration. Mercury toxicity is generally enhanced by organic derivation, however larger mercury compounds such as diphenylmercury exert less effect on Artemia development. The decline of toxicity in relation to increased size of organic mercury compounds may reflect reduced movement of the toxic compounds through cell membranes. All metals examined enter the developing organisms upon rupture of the cyst shell and may, if in sufficient concentration, delay or prevent emergence and hatching. In some cases, incompletely emerged organisms hatch while still enclosed in the cyst shell, demonstrating that completion of early developmental processes is not a prerequisite for later events. The delayed reversion of metal poisoning suggests brine shrimp are able to sequester intracellular metals, perhaps by induced synthesis of metal-binding proteins such as the metallothioneins. We have tested cloned metallothionein genes from mouse, Neurospora and Drosophila for their ability to hybridize to Southern blots of restriction-digested Artemia DNA. Our preliminary results reveal that the Drosophila metallothionein gene, cDM51, binds to Artemia DNA under moderately stringent conditions. We are just beginning to clone the Artemia genes for metallothionein or other metal-binding proteins as one approach to develop a bioassay for metals in the marine environment.

MOLECULAR AND CELLULAR ASPECTS OF CHITIN SYNTHESIS IN LARVAL ARTEMIA

Michael N. Horst

Division of Basic Science/Biochemistry
Mercer University, School of Medicine
Macon, GA 31207 USA

INTRODUCTION

The cuticle of Artemia and other crustaceans contains chitin which is covalently attached to protein[1-3]. While the role of such proteins in structural support of the cuticle seems clear, the possible functions of the proteins in the biosynthesis of the cuticle are not obvious. Recent work in my laboratory has indicated that crustaceans synthesize a chitin oligosaccharide via a dolichol-linked intermediate[4]. Subsequent transfer of the oligosaccharides to either endogenous polypeptides or synthetic peptide acceptors[5] yields a chitoprotein which may serve as a substrate for chitin synthetase, the enzyme responsible for polymerization of macromolecular chitin[6]. However, the pathway of synthesis for this material within the crustacean epithelial cell is not clear; one notion is that the chitoprotein is synthesized in the rough endoplasmic reticulum (RER), moves to the Golgi apparatus (where chitin synthetase is presumably located) and serves as a primer molecule for chitin synthetase, yielding a chitin-protein complex. The mechanism whereby this product is exported to the outside of the apical membrane and incorporated into the growing cuticle is not known.

The present study involves an investigation of several of the unknown aspects of chitin synthesis in larval Artemia. First, chitin synthetase has been solubilized and partially purified; some of the physical and catalytic properties of the enzyme have been determined. Secondly, the chitoprotein from Artemia has been isolated, purified and partially characterized. Finally, a variety of autoradiographic and immunochemical methods have been utilized at the light and electron microscope level to study the pathway of chitin synthesis within the epithelial cells.

MATERIALS AND METHODS

Radiochemicals

UDP-6-^{3}H GlcNAc (6.6 Ci/mmol) was from New England Nuclear. ^{35}S-

59

Methionine (1064 Ci/mmol), 1, 6-^{3}H GlcNAc (33 Ci/mmol) and 6-^{3}H GlcN (40 Ci/mmol) were from ICN Biomedicals, Inc.

Chemicals

UDP-hexanolamine agarose, UDP, PMSF (phenylmethane sulfonyl fluoride), soybean trypsin inhibitor (STI), monensin, chitosan, Streptomyces chitinase, tomato lectin and cibacron blue agarose were from Sigma. Trition X-100 (TX-100) was obtained from Pierce, octyl-D-glucopyranoside (OG) was from Calbiochem; 4-chloro-1-naphthol was from BioRad. Goat anti-rabbit IgG (GAR Ig) and GAR-Ig-peroxidase were from Jackson Labs. All electrophoresis reagents were BDH grade from Gallard Schlesinger. Nitrocellulose was from Schleicher and Schull. Reacetylated chitosan (RC) was prepared as described[7]. Granular chitin (GC) was prepared according to Berger and Reynolds[8]; chitin oligosaccharides were prepared by digestion of GC with wheat germ endochitinase[9]. Polyclonal antibodies to chitin oligosaccharides[10] were obtained from Dr. Byron Anderson, Northwestern University.

Purification of Chitin Synthetase

Larval Artemia were grown for 2-3 days, homogenized and crude microsomes were prepared as described[4,6]. Microsomes were solubilized with 2% TX-100 in Buffer A: 50 mM HEPES, pH 7.1 containing 30 mM MgCl$_2$, 0.2 mM PMSF, and 10 µg/ml STI. After Dounce homogenization and bath sonication (30 sec/RT), samples were centrifuged (100,000 x g/1h) and the supernatant removed. The pellet was resuspended in Buffer A containing 2% octyl-D-glucoside; after Dounce homogenization, sonication and centrifugation, the OG soluble and insoluble fractions plus the TX-100 soluble material were assayed for chitin synthetase activity using UDP-^{3}H GlcNAc (150,000 dpm) as substrate[6]. Total protein was measured by a fluorescamine procedure using BSA as standard[6].

Affinity Chromatography

Samples (50 ml) of TX-100 or OG soluble material were mixed with RC (20 ml) at 4°C for 2-15 h. The slurry was packed into a column (2 x 20 cm) and washed with buffer B: 20 mM HEPES, pH 7.1 containing 30 mM MgCl$_2$. When the absorbance at 280 nm reached a steady baseline, the bound material was eluted with 8 M urea. The peak of eluted protein was pooled and dialyzed against water and then Buffer B prior to mixing with UDP-agarose (10 ml) for 4-15 h at 4°C. The slurry was packed into a column (2 x 10 cm) and washed with buffer B. Bound material was eluted with either 3 M KCl or 5 mM UDP in buffer B. The peak of eluted protein was pooled dialyzed and assayed.

Samples of buffer and urea soluble chitoproteins were purified by affinity chromatography on wheat germ agglutinin-agarose as described [10]; 0.1 M GlcNAc was used to elute bound material. Antichitin antibody-agarose was prepared by coupling purified antichitin antibody to agarose[11]; samples were mixed with the matrix in Tris-buffered saline and bound material was eluted with 0.2 M glycine, pH 2.8.

Electrophoresis Procedures

Samples were analyzed either by 10% SDS-PAGE [12] or by two dimensional electrophoresis [13]. Samples were electroblotted from gels to nitrocellulose paper in Tris-glycine-20% methanol buffer [14]. The paper was subsequently blocked with 0.5% Tween 20 in TBS: 20 mM Tris, pH 7.4, containing 0.15 M NaCl [15]. Blots were immunostained with specific primary antibody followed by GAR Ig-peroxidase and a 4-chloro-1-naphthol [16]; thereafter, blots were stained for total protein with colloidal gold [15].

Preparation of Polyclonal Antibodies

Female New Zealand white rabbits were immunized with either RC or glycopeptides prepared from <u>Artemia</u> cuticle by partial acid hydrolysis (4 M HCl/ 4h/37°C). After weekly injection of antigen (0.8 ml PBS containing 0.1 to 0.6 mg antigen) for 4 weeks, animals were ear bled using a vacuum cuff. After clotting, the antiserum was removed, titered by dot blot assay, and purified by affinity chromatography on either Protein A agarose or GAR-Ig-agarose[16]. Bound antibody was eluted with 0.2 M glycine buffer, pH 2.9, neutralized, dialyzed and stored at -20°C until needed.

In vivo Labeling and Extraction of Artemia Proteins

<u>Artemia</u> larvae (2-5 ml packed volume in 10 to 20 ml (seawater) were labeled for 2 to 4 h with either ^{35}S methionine (0.05 Ci/ml), [^{3}H]-GlcN (0.1 mCi/ml) or [^{3}H]-GlcNAc (0.1 mCi/ml). Afterward, larvae were washed with 50 % methanol and homogenized in chloroform:methanol (2:1). The insoluble residue was extracted with acetone, dried and homogenized in buffer A (50 ml); after centrifugation (5500 x g/15 min), the buffer soluble fraction was removed and dialyzed. The residue was homogenized in 9.5 M urea, pH 10.3, and centrifuged as before. The urea insoluble pellet was boiled in 0.1 M borate buffer, pH 9.0 containing 2 % SDS (30 ml). After centrifugation, the urea and SDS soluble fractions were dialyzed. For preparation of <u>in vitro</u> labeled proteins, crude microsomes were incubated with UDP-[^{3}H]-GlcNAc as described[4,6]; following incubation, the samples were extracted as described above for <u>in vivo</u> labeled proteins.

Light and Electron Microscopy Procedures

Larvae were fixed in 0.2 M collidine buffer, pH 7.5, containing 2.5 % glutaraldehyde and 5 % paraformaldehyde: after microwave irradiation (7-10 sec/high power), samples were fixed at room temperature for 2 h. Samples were dehydrated and embedded in paraffin and either LR White[17] or Lowicryl K4M[18] resin. Gold or silver sections were cut from plastic blocks and immunostained by standard procedures[19]. Colloidal gold and gold labeled probes were prepared as described[20,21]; proteins were labeled at the following pH values: chitinase, 7; GAR-Ig, WGA and tomato lectin, 9; Protein A, 6.2. Poststaining of thin sections was carried out with uranyl acetate, lead citrate or osmium tetroxide[22]; alternatively, sections were poststained with permanganate[23]. Samples were examined using a Hitachi HU-11E-2 transmission electron microscope operated at 75 kV. Paraffin sections (4 um) were deparaffinized and immunostained by standard procedures using the PAP procedure[16]; aqueous fast green was used as counterstain. Samples for autoradiography were prepared by labeling larvae at 30°C with [^{3}H]-GlcN as described above in the presence or absence of 10 mM monensin for 2 to 4 h; chase periods of from 2 to 24 h were carried out in non-radioactive seawater. After fixation, samples were embedded in LR white resin and thick sections (0.5 µm) were prepared for light microscope autoradiography using Ilford K5D emulsion[24]; after exposure for 2 to 6 weeks, slides were developed and counterstained with Polychrome stain. Samples for electron microscope autoradiography were prepared as described[25]. Gold sections were covered with Ilford L-2 emulsion via the loop technique and allowed to expose for 2 to 6 months prior to development and examination.

RESULTS

Purification of Chitin Synthetase

When brine shrimp larvae were grown for either two or three days, they could be used for isolation of satisfactory quantities of chitin synthetase activity. After a series of preliminary experiments using a variety of nonionic detergents to solubilize the enzyme, the two step extraction procedure described under Methods was selected. As shown in Table 1, Triton X-100 extracts about 75% of the total protein, but little of the chitin synthetase activity. Subsequent extraction with octyl glucoside solubilized most of the remaining protein and the majority of the chitin synthetase activity (Table 1). At this point, the enzyme had been purified 12-fold. During chromatography on reacetylated chitosan, elution of enzyme was achieved with 8 M urea; other salts such as 3 M NaCl or KCl were not as effective in removing bound enzyme. Following chitin affinity chromatography, the enzyme had been purified about 350-fold. Finally, the enzyme was bound to UDP-agarose; elution of activity was achieved using either 3M KCl or 5 mM UDP in the buffer. The specific activity of the KCl eluted enzyme from the UDP-agarose affinity column was 193 pmol GlcNAc incorporated/h/mg protein, representing an overall purification of 653-fold. Since the solubilized enzyme was extremely labile to storage, it was necessary to proceed through the purification scheme as quickly as possible. Preliminary experiments indicate that Cibacron Blue-agarose may be used as an alternate affinity absorbent.

Table 1. Purification of Artemia Chitin Synthetase

Sample	Protein mg	Volume ml	Activity cpm/mg	S.A.	Purification -fold
Microsomes	204	40	3367	0.30	-
2% TX-100 Soluble	147	30	541	0.05	-
2% TX-100 Insoluble	13.2	12	8234	0.72	2.4
2% OG Soluble	3.15	10.5	41787	3.66	12.4
2% OG Insoluble	1.41	3.0	4983	0.44	-
RC: Urea eluted	0.007	28	1.16×10^6	102	346
UDP-agarose KCl eluted	0.074*	35	2.2×10^6	193	653

Larvae were grown for 72 h, homogenized and crude microsomes were prepared. The enzyme was solubilized and purified as described in the text. Specific activity (S.A.) is pmol GlcNAc incorporated/h/mg protein. The reacetylated chitosan (RC) preparation used above was a Triton X-100 (TX-100) soluble sample. Total protein contained in the sample was estimated by absorbance at 280 nm; for specific activity, all samples were assayed by the fluorescamine procedure.

Properties of the Partially Purified Chitin Synthetase

When the UDP-agarose purified enzyme was chromatographed on Sepharose CL-6B in HEPES/MgCl$_2$, a single peak of protein and activity was observed indicating that the solubilized enzyme was not present as aggregates of different size (data not shown). Analysis of the purified enzyme by 2D-PAGE showed that the preparation contained several major protein spots (Fig. 1). Unfortunately, chitin synthetase cannot be positively identified in such a gel, since the enzyme is denatured with sodium dodecyl sulfate (SDS) prior to electrophoresis in the second dimension. If one assumes that the enzyme represents one of the major spots in Fig. 1, then the apparent molecular weight of the solubilized enzyme would be 60-80,000 d; the approximate pI would be between 4.6 and 6.7. It is possible that the enzyme exists in different isozyme forms.

When the purified enzyme was incubated with [^{35}S]-labeled urea soluble proteins in the standard assay plus 5 mM (nonradioactive) UDP-GlcNAc and then analyzed by 2-D PAGE, autoradiograms of the resultant gels showed that several new polypeptides were present (Fig. 2). Presumably, these proteins were glycoslated by the enzyme and moved to a different position in the 2-D map because of added carbohydrate which alters the size and pI.

Finally, the purified chitin synthetase was analyzed for the presence of bound carbohydrate. In this experiment, partially purified enzyme was subjected to SDS-PAGE and the resultant gel was blotted to nitrocellulose paper. The blot was probed with polyclonal antibody to chitin; binding was detected using GAR-Ig-peroxidase followed by staining with 4-CN. As shown in Fig. 3, the partially purified enzyme was positive with this immunostaining procedure, indicating the enzyme contains bound chitin oligosaccharides. The results suggest that Artemia chitin synthetase may carry its own primer for chitin synthesis. A similar situation has been suggested for the purified enzyme from Saccaromyces cerevisciae (Cabib, E., personal communication).

Future Trends

Once chitin synthetase has been purified to homogeneity, we plan to sequence the amino terminus, prepare a cDNA probe and identify the gene for chitin synthetase in an Artemia cDNA library. We hope to compare the sequence of the Artemia enzyme with that of Saccharomyces cerevisceae [26, 27].

Isolation and Characterization of Artemia Chitoproteins

Following extraction of radiolabeled Artemia larvae with buffer, urea and SDS, samples were analyzed by SDS-PAGE. As shown in Fig. 4, many radiolabeled proteins were detected by autoradiography of the gel. The labeled components represent those larval proteins which were being synthesized during the time period when the larvae were exposed to ^{35}S-methionine.

In order to identify chitoproteins in the complex mixture of proteins contained in the buffer and urea soluble fractions, samples were purified by WGA agarose chromatography (see Methods). The resultant preparations were analyzed by SDS-PAGE and immuno-blotting; in most cases the primary antibody used as probe was polyclonal anti-chitin antibody. As shown in Fig 5A, a number of different WGA-agarose purified preparations exhibit immunoreactive bands, ie. polypeptides which contain chitin oligosaccharides. The most dominant chitoprotein in the urea soluble samples has a molecular weight of approximately 25,000 daltons. Other chitoproteins, present at lower levels, have molecular weights of 22, 38 and 50,000 daltons. A slightly smaller chitoprotein was detected in the buffer

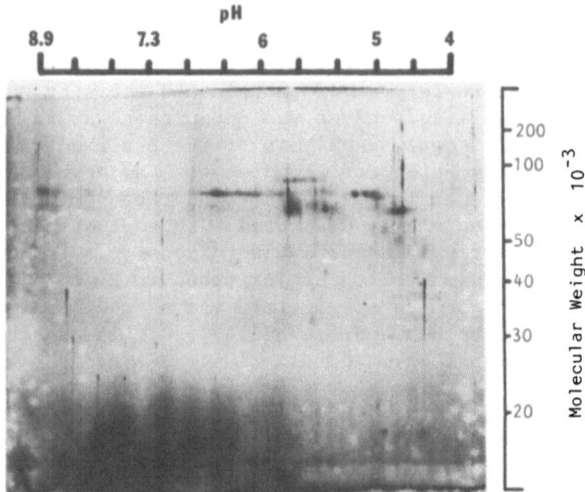

Fig. 1. 2-D Electrophoresis of UDP-agarose affin-
         ity purified chitin synthetase. Scales
         for the isoelectric focusing (horizontal)
         and SDS-PAGE (vertical) dimensions are
         indicated. The gel was silver stained.

soluble sample. Also, when urea soluble chitoproteins were further purified
by immunoaffinity chromatography on antichitin-agarose, the same bands
were detected. As a control experiment, both urea soluble chitoproteins
and standard glycoproteins were blotted and immunostained with and without
primary antichitin antibody. As shown in Fig. 5B, one band was observed
in the standard, but none in the chitoprotein sample in the absence of
primary antibody. The band in the standard is due to endogenous peroxidase
activity in the cytochrome C and not to nonspecific antibody binding. The
polyclonal antibody reacts with the chitobiosyl core of ovalbumin which
contains an N-linked high mannose oligosaccharide [28]. The antibody

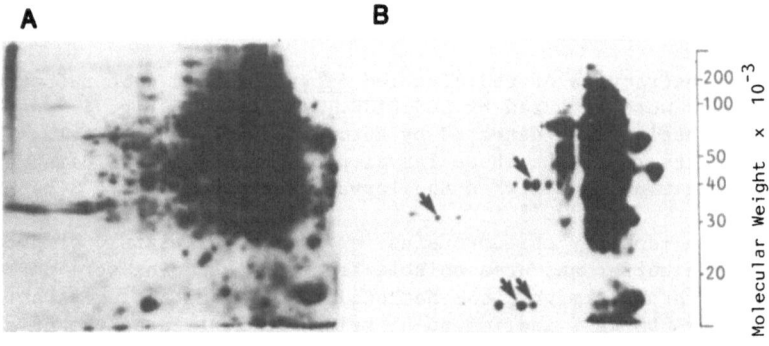

Fig. 2. Autoradiography of 2-D gel analysis of
         ^{35}S-labeled urea soluble proteins from
         Artemia before (a) and after (b) incub-
         ation with purified chitin synthetase.
         Position of new polypeptides is indicated
         (arrows).

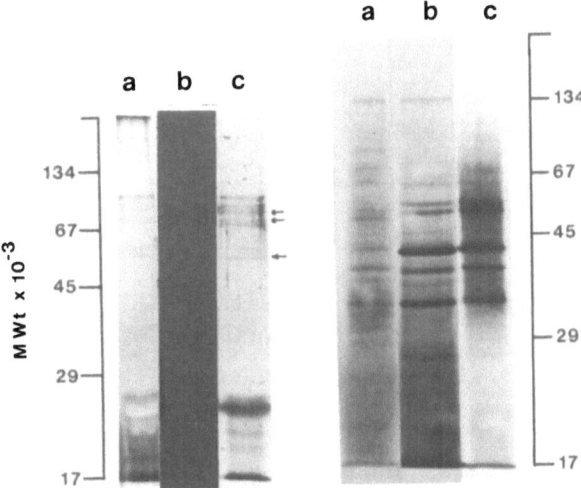

Fig. 3.   (Left) SDS-PAGE analysis of affinity purified
chitin synthetase. The UDP purified enzyme
was stained with silver (a), or electroblotted
to nitrocellulose and immunostained with
antichitin antibody (b); afterward, the blot
was stained for total protein with colloidal
gold (c). The position of immunostained bands
is indicated by arrows.

Fig. 4.   (Right) SDS-PAGE analysis of ^{35}S-labeled
polypeptides following in vivo labeling of
Artemia larvae. Samples analyzed were: a,
buffer soluble b, urea soluble; c, SDS soluble.
After electrophoresis, the gel was autoradiographed.

also reacts with ovomucoid which contains high amounts of GlcNAc in complex
N-linked oligosaccharides [28]. When the polyclonal antibody was partially
purified by chromatography on ovalbumin-agarose, the unbound portion of
the antibody still retained reactivity toward chitoproteins, indicating
that these polypeptides contain larger oligosaccharides. In order to
study this possibility further, the urea soluble chitoproteins were subjected to a dot-blot analysis using a series of polyclonal antibody preparations against various sizes of chitin oligosaccharides; thus, one
antiserum was directed against N-acetylchitobiose, another against N-
acetylchitotriose and so on up to an antiserum against a hexasaccharide
of chitin [11]. All antisera tested positive against the Artemia chitoprotein in the dot-blot assay (data not shown), suggesting that the chitoproteins contain chitin oligosaccharides ranging from at least N=2 up to
N=6.

Immunostaining of Cuticular Chitin

When paraffin sections of larval Artemia were immunostained using
antichitin antibody and the GAR-Ig/4-CN procedure, intense labeling of the
cuticle was observed (Fig. 6A). Generally, labeling of gut cuticle was
also observed as was a strong reaction with the gut contents. Since larval
Artemia are cannibalistic, this material likely represents cuticle residue

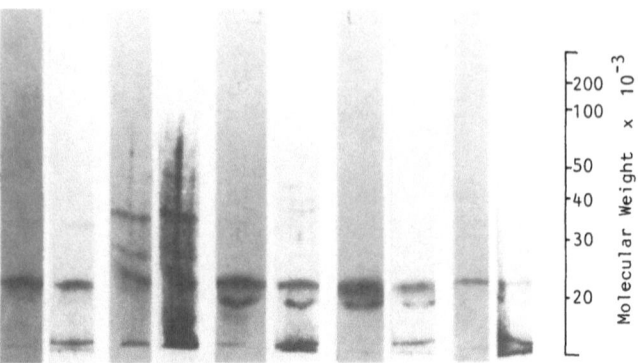

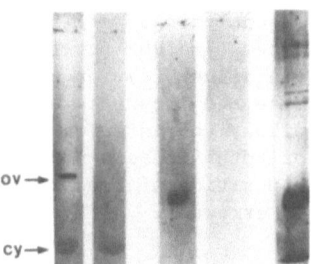

Fig. 5A. (Top) Immunostaining of chitoproteins. All samples were purified on WGA-agarose, separated by SDS-PAGE, electoblotted and immunostained with antichitin antibody (a lanes); blots were stained for protein with colloidal gold (b lanes). Samples were; 1, 2, in vitro (UDP) ^{3}H-GlcNAc labeled urea soluble (US); 3, in vivo ^{35}S-met labeled US; 4, in vivo ^{3}H-GlcN labeled US after antichitin-agarose; 5, in vivo ^{3}H-GlcN labeled buffer soluble.

Fig. 5B. (Bottom) Immunostaining control experiments. Samples were separated on SDS gels, blotted to paper and stained with (1,3) and without (2,4) primary antibody. Samples were: 1 and 2, standard cytochrome c (cy) and ovalbumin (ov); 3,4 and 5, WGA purified chitoproteins. In lane 5, blot from lane 4 was stained for total protein with colloidal gold.

of partially digested brethren. In control experiments, omission of primary antibody followed by GAR-Ig/4-CN yielded no staining of the gut contents or cuticle (Fig. 6B). Similar results were obtained when the antichitin antibody was partially purified by chromatography on ovalbumin-agarose; the unbound material gave a strong positive reaction with cuticle like that

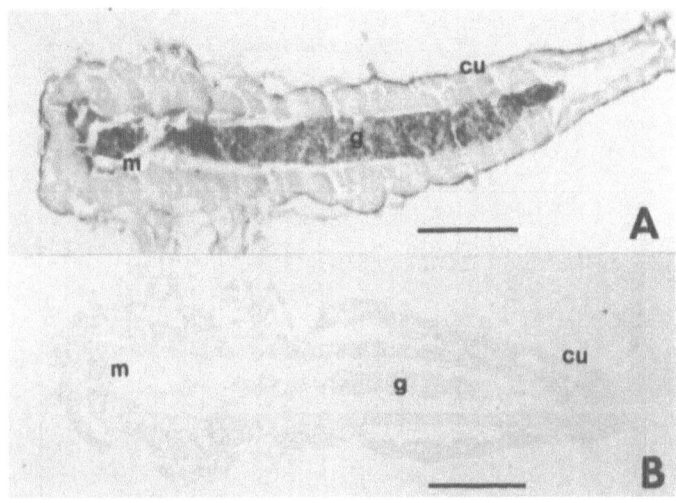

Fig. 6. Immunostaining of <u>Artemia</u> larvae with antichitin
antibody/PAP. Paraffin sections were examined
with (A) and without (B) primary antibody included
in the staining protocol. The antibody stains
cuticle (cu), gut contents (g) and adjacent peri-
trophic membranes (m). Bar, 10 μm.

seen in Fig. 6A. Finally, larval cuticle was immunostained with antiglycopeptide antibody; since this preparation contained antibodies to both
the carbohydrate and peptide portions of the immunizing molecule, we chose
to purify the peptide binding antibody and examine its properties. Thus,
crude antiserum was partially purified by affinity chromatography on chitin;
the unbound antibody, presumably directed against the peptide portion of
the molecule, was used in the immunostaining reaction with results identical to those shown in Fig. 6A. Thus, it would appear possible to stain
both the carbohydrate and peptide portions of the <u>Artemia</u> cuticle using
immunochemical methods.

Since the resolution obtained by light microscopy is limited, we next
investigated immunolabeling of the larval cuticle at the electron microscope level. In a series of preliminary experiments, immunolabeling of
Lowicryl embedded material was found to be far superior in intensity and
low background to LR White embedded material. For all of the immunolabeling
work described here, Lowicryl K4M was used; UV polymerization was carried
out at 4°C with good retention of antigenic properties. As shown in Fig.
7, immunolabeling of thin sections with anti-chitin antibody/Protein A-
gold was positive in the cuticle of larvae. Additionally, some labeling
at the apical surface of the epithelial cells was detected, often in the
perinuclear region. On the other hand, labeling inside the epithelial
cells was not intense; this may be due to sensitivity of the antigen to
fixation and/or embedding. Labeling of thin sections with the anti-peptide
antibody was confined to the area beneath the cuticle and at the surface
of the epithelial cells (Fig. 7b). One possible reason for the lack of
staining in the cuticle is that the antibody may not be able to bind to a
crosslinked antigen. Another possibility is that the antibody cannot
penetrate into the section to bind.

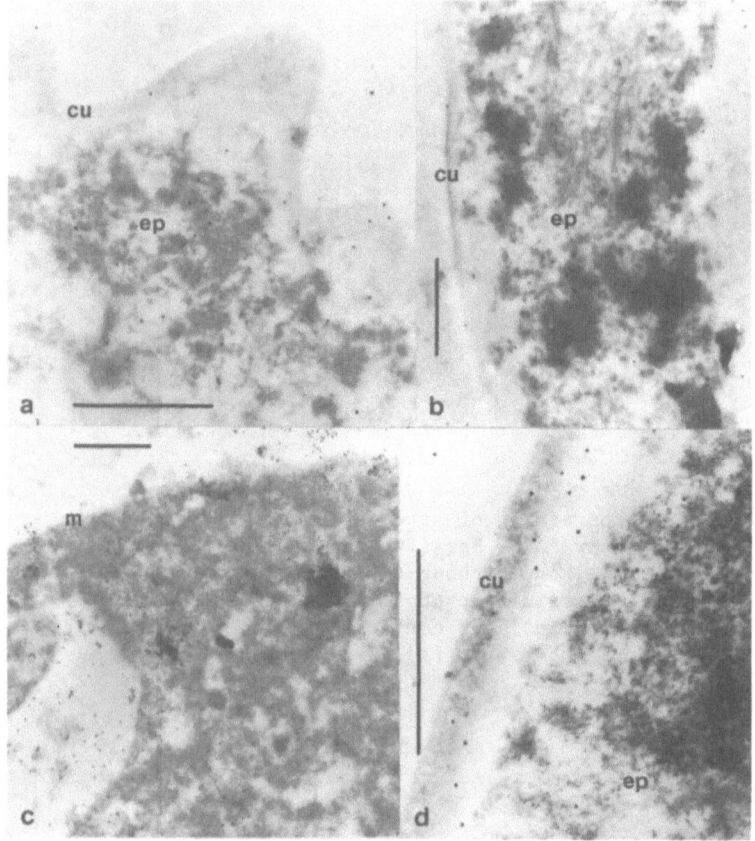

Fig. 7.   Immunostaining at the electron microscope level. Sections in a and b were stained with antiglycopeptide
antibody/Protein A-gold. Label is found in the cuticle
(cu) and epithelial cells (ep). (No post-stain was
used. For comparison see Fig. 10). Sections in c and
d were stained with antichitin antibody/Protein A-gold.
Dense labeling is seen inside epithelial cells and at
the apical membrane (m) in c; cuticle (cu) also binds
antibody in d. Bars, 1 μm.

Gold-Labeled Probes for Chitin at the Electron Microscope Level

In order to provide comparative data to the immunolabeling studies,
several gold labeled lectins which bind to chitin oligosaccharides were
prepared and tested on thin sections. As shown in Fig. 8, wheat germ
agglutinin-gold (WGA-gold) gave an intense labeling of the larval cuticle.
In addition, some labeling of vesicles near the apical membrane of the
epithelial cells was observed. Finally, some labeling in the perinuclear
area was detected. Labeling of cuticle with tomato lectin-gold was similar
to that seen with WGA-gold. As shown in Fig. 9, tomato lectin-gold gave
intense labeling of the cuticle and apical membrane. In some cases, dense
concentrations of label were observed within the cell and in the perinuclear
area. Finally, chitinase from Streptomyces [8] was labeled with gold and
used as a probe for cuticular chitin. As shown in Fig. 9d, the majority

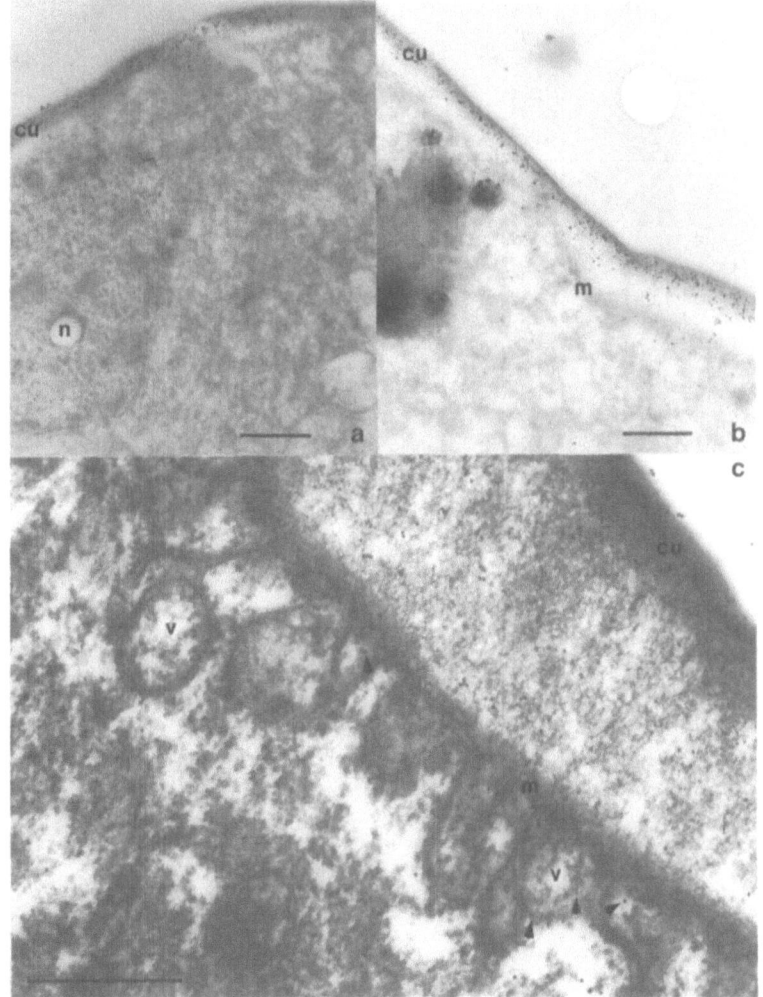

Fig. 8. Binding of WGA-gold to Artemia cuticle (cu) and epith-
         elial cells. Labeling is observed adjacent to the
         nucleus (n) as in a. In b, details of WGA-gold binding
         to the cuticle and material between the cuticle and
         apical membrane (m) are shown. In c, WGA-gold binding
         to cuticle and vesicles (v) within epithelial cells
         (arrowheads) is seen. Uranyl acetate counterstain;
         bar, 1 μm.

of the label was bound to the cuticle. Little binding was observed within
the epithelial cells.

    Taken together, the light and electron microscope antibody and gold
labeled probe data suggest chitin can be detected in the cuticle and at
the apical surface of the epithelial cells. Binding in the perinuclear
area may suggest a possible role for the Golgi apparatus, but proof of this
notion must await further study. In addition to the perinuclear area, the
Golgi apparatus in larval Artemia is often located at the border between

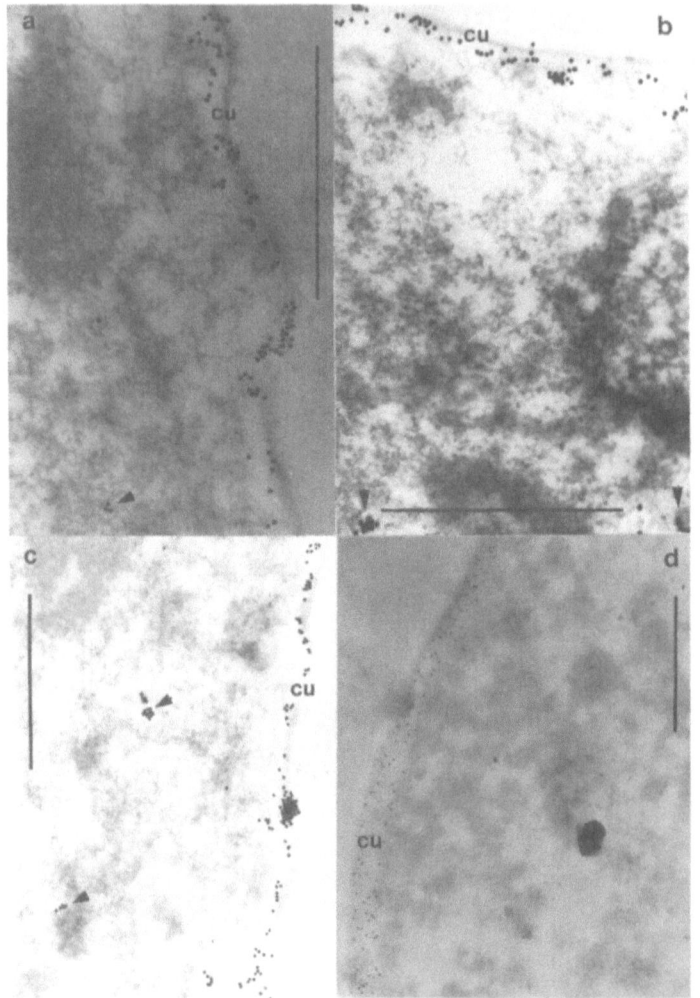

Fig. 9. Binding of gold labeled probes to <u>Artemia</u> cuticle
and epithelial cells. Tomato lectin-gold, a-c;
chitinase-gold, d. Tomato lectin-gold is bound
at localized sites (arrowheads) within the epithelial cells and gives intense labeling of the
cuticle. Labeling with chitinase-gold occurs
primarily in the cuticle. Bar, 1 μm.

two adjacent epithelial cells (Fig. 10a,b). We have used osmium tetroxide
post-fixation/post-staining to detect the Golgi (Fig. 10c).

Autoradiographic Studies of Artemia Chitin Synthesis

In order to examine the pathway of chitin synthesis in larval <u>Artemia</u>
in greater detail, autoradiographic studies were conducted using ^{3}H-GlcN
at both the light and electron microscope level. Using the procedures
described under Methods, clear labeling of the larval cuticle was observed
after incubation of larvae with ^{3}H-GlcN for 1 or 2 h. Thereafter, chase

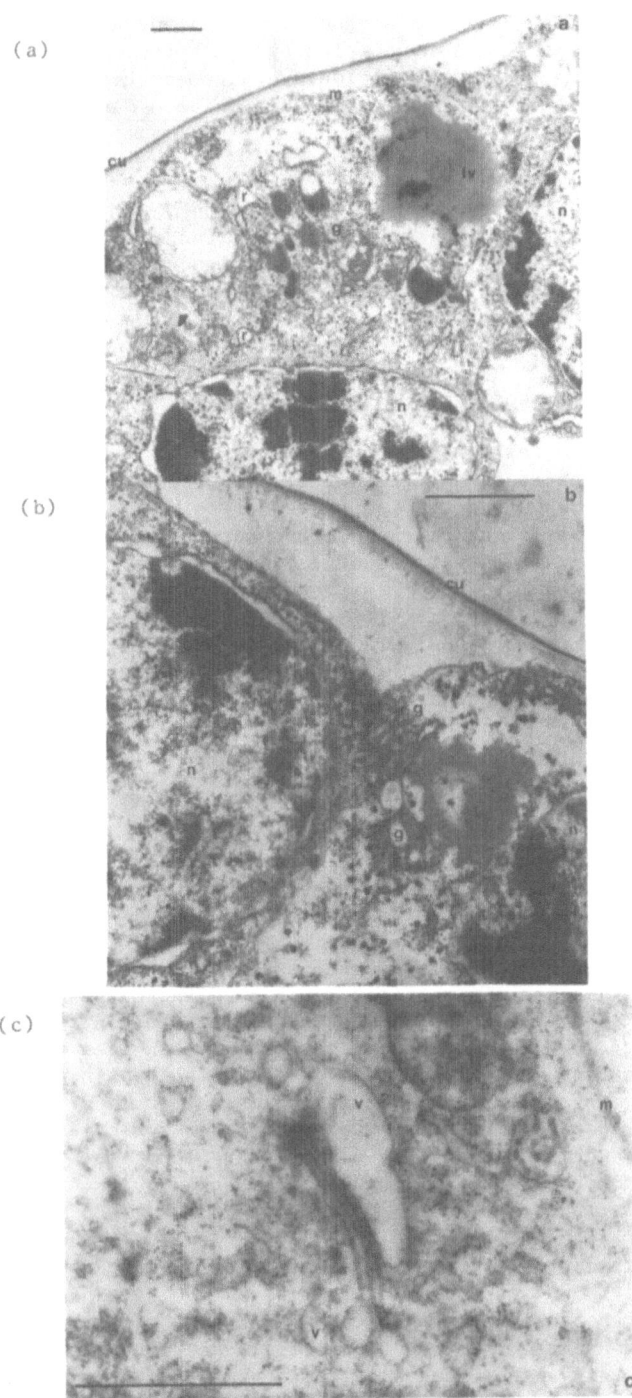

(a)

(b)

(c)

Fig. 10. Ultrastructure of larval <u>Artemia</u> showing cuticle (cu)
and several nuclei of epithelial cells (n). The Golgi
apparatus (g) appears between adjacent nuclei in both
a and b; apical membrane (m), rough endoplasmic

(Continued)

Fig. 10
(Continued)
reticulum (r) and lipovitellin (lv) are indicated.
The section was poststained with uranyl acetate and
lead citrate. Details of Golgi apparatus structure
in _Artemia_ larvae are shown in c; cisternae appear as
flat leaflets with vesicles (v) attached or nearby.
Large vesicles to one side are a common feature; c
was poststained with osmium tetroxide. Bar, 1 μm.

periods in non-radioactive seawater were used to follow the course of the
isotope into the larval cuticle. Labeling was often observed near the
nuclei of cells at the light level (Fig. 11). Electron microscope autoradiograms revealed incorporation near the apical membrane of the epithelial
cells (Fig. 12). As yet, no conclusive evidence of Golgi labeling has
been observed at the electron microscope level.

Effect of Monensin on Artemia Chitin Synthesis

Since we wished to test the hypothesis that _Artemia_ chitin synthesis
involves the Golgi apparatus, a series of light level autoradiographic
studies were carried out using the ionophore monensin, which is known to
block transit of materials out of the Golgi apparatus in eukaryotic cells
[29]. As shown in Fig. 11, treatment of larvae with monensin blocks the
incorporation of ^{3}H-GlcN into the larval cuticle. Analysis of this experiment at the electron microscope level is now underway. The tentative
conclusion is that chitin synthesis in _Artemia_ appears to require transit
of secreted material through the Golgi apparatus. Whether this is a chitoprotein primer or a macromolecular chitin-protein complex remains to be
determined.

DISCUSSION

During the present study, purification of chitin synthetase from
larval _Artemia_ was greatly facilitated by the use of affinity chromatography
on reacetylated chitosan. This matrix is stable, easy to prepare and
affords a substantial purification of the enzymatic activity. The UDP-
agarose matrix is much less stable and represented the most expensive step
in the procedure. Alternative matrices, such as Cibacron Blue-agarose,
appear promising. The _Artemia_ chitin synthetase has been purified 653-
fold; by comparison, the enzyme from _S. cereviscise_ has been purified 20-
fold [30]. The purified _Artemia_ enzyme (or some other polypeptide which co-
purifies with it) appears to contain chitin oligosaccharides. Future
studies are required to determine if the enzyme itself becomes part of the
macromolecular product, ie. does chitin synthetase serve as its own primer?
Similar results have been observed in fungal systems [30].

Chitoproteins from larval _Artemia_ have been isolated and partially
purified by affinity chromatography. These glycoproteins appear to have
a variety of molecular weights ranging from 22–50,000 daltons. The size
of the bound oligosaccharides appears to range from 2 to at least 6 GlcNAc
residues. Studies are now underway to precisely characterize these oligosaccharides before and after incubation with purified chitin synthetase.

Immunolabeling studies have revealed chitin in the larval cuticle but
have been somewhat less successful in localizing chitin inside the epithelial cells which synthesize the product. Future studies will explore
alternative fixation and embedding procedures which may allow improved

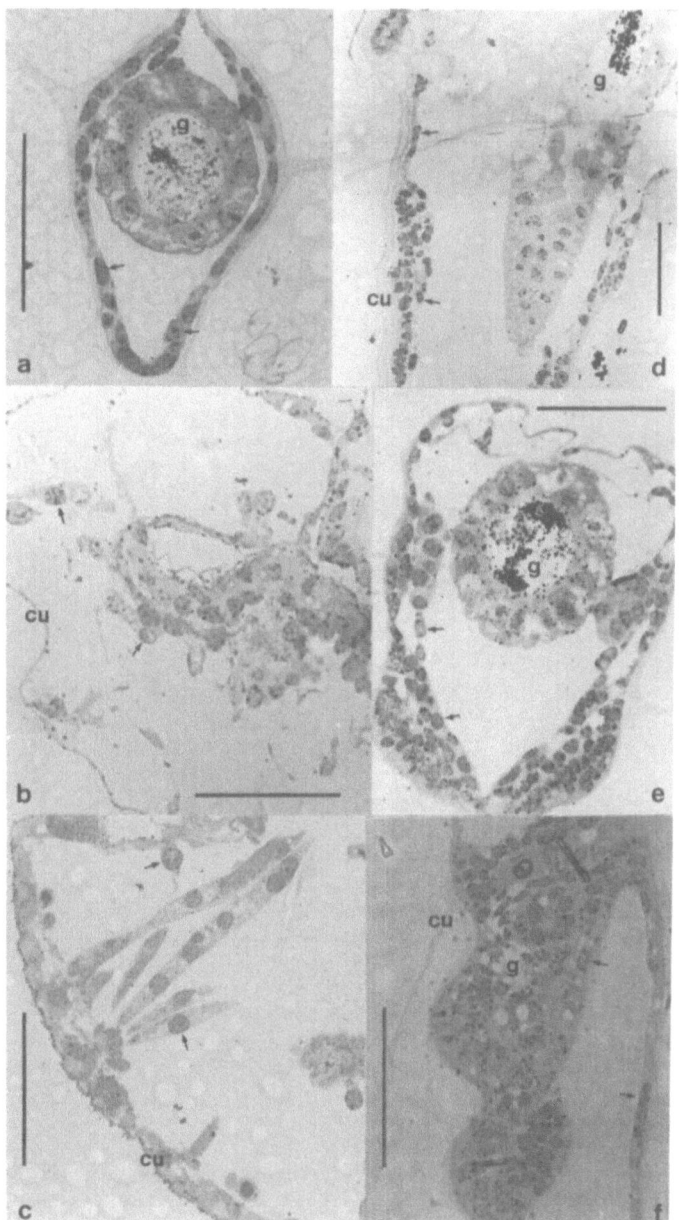

Fig. 11. Light microscope autoradiography of larval brine shrimp labeled
with ³H-GlcN for 1 h and chased for 30 min (a,d), 60 min (b,e)
or 4 h (c,f). The photomicrographs show cross-sections through
the thorax of larvae. Normal controls are shown in d,e and f.
In controls, the majority of radiolabel appears in the gut (g)
at 30 min; there is some label in cuticle epithelial cells at
60 min. Synthesis of new cuticle is clearly underway at 4 h
(c). Treatment of larvae with monensin blocks movement of
radiolabeled GlcN into the cuticle (e,f: right side), while
most remains in the gut (center in e,f). Nuclei are indicated
by arrowheads. Bar, 10 μm.

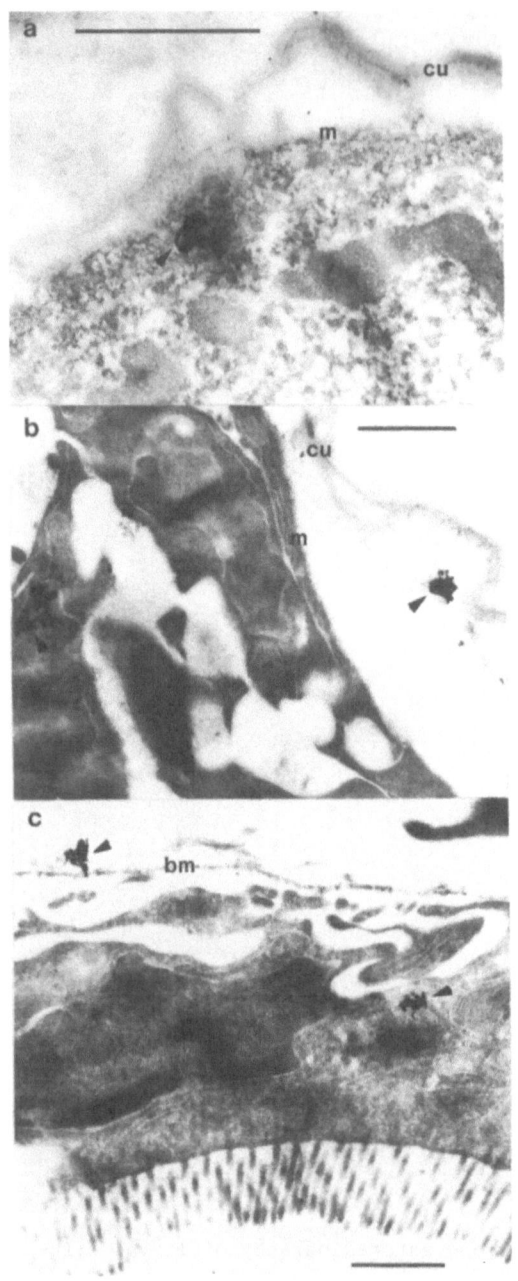

Fig. 12.   High resolution autoradiography of ^{3}H-GlcN labeled <u>Artemia</u>
larvae. Emulsion coated sections were exposed for 60
days; grids were poststained with uranyl acetate and lead
citrate after gelatin removal. In (a), label (arrowsheads)
is found just under the apical membrane (m) near the
cuticle (cu); (b), radiolabel is observed underneath the
cuticle near the apical membrane; (c), label appears in
gut epithelial cells and near the basement membrane (upper
left). Bar, 1 μm.

resolution of chitin inside the secretory cells. To date, most of the lectin and antibody studies suggest chitin synthesis takes place at or near the apical membrane of the epithelial cells. However, studies with monensin indicate that a primer molecule or oligosaccharide must pass through the Golgi in order for chitin deposition to occur. Studies on the effect of monensin at the electron microscope level are now in progress.

ACKNOWLEDGEMENTS

This work was supported by grant GM-30952 from the National Institutes of Health. I would like to thank Ms. Ely Klar, J. W. Hightower and Ms. Sharon Werner for excellent technical assistance. Some of the electron micrographs were taken in the laboratory of Dr. Henry C. Aldrich, University of Florida. The advise and support of Dr. Aldrich, Dr. Bill Buhi and Dr. Greg Erdos is greatly appreciated. I thank Ms. Ginger Sanders for help in typing the manuscript.

REFERENCES

1. K. Rudall, The chitin/protein complexes of insect cuticles, _Advances in Insect Physiol._ 1:257 (1963).
2. S. Hunt, "Protein Polysaccharide Complexes in Invertebrates," Academic Press, New York (1970).
3. P. R. Austin, C. Brine, J. Castle and J. Zikakis, Chitin. New facets of research, _Science_ 212:749 (1981).
4. M. N. Horst, The biosynthesis of crustacean chitin. Isolation and characterization of polyprenol-linked intermediates from brine shrimp microsomes, _Arch. Biochem. Biophys._ 223:254 (1981).
5. M. N. Horst, Glycosylation of exogenous peptide acceptors by larval brine shrimp microsomes, in: "UCLA Symposium on Synthetic Peptides: Approaches to Biological Problems," Plenum Press, New York (in press).
6. M. N. Horst, The biosynthesis of crustacean chitin by a microsomal enzyme from larval brine shrimp, _J. Biol. Chem._ 256:1412 (1980).
7. S. Hirano, Y. Ohe and H. Ono, Selective N-acetylation of chitosan, _Carbohydrate Res._ 47:315 (1976).
8. L. Berger and D. Reynolds, The chitinase system of a strain of _Streptomyces griseus, Biochem. Biophys. Acta_ 29:522 (1958).
9. J. Molano, I. Polacheck, A. Duran and E. Cabib, An endochitinase from wheat germ. Activity on nascent and preformed chitin, _J. Biol. Chem._ 254:4901 (1979).
10. S. Makhlouf, L. Davis, P. Deepika and B. Anderson, Immuno-chemical characterization of antisera reactivities to N-acetyl-D-glucosamine oligosaccharides with the B(1-4)-glycosidic linkage, _Glycoconjugate J._ 3:351 (1986).
11. D. Livingston, Immunoaffinity chromatography of proteins, _Methods in Enzymology_ 34:723 (1974).
12. U. Laemmli, Cleavage of structural proteins during the assembly of the head of the bacteriophage T4, _Nature_ 227:680 (1970).
13. M. N. Horst and R. M. Roberts, Solubilization, electrofocusing and two-dimensional electrophoresis of plasma membrane polypeptides from Chinese hamster ovary cells, in: "Electrofocus/78" Elsevier, New York (1979).
14. H. Towbin, T. Staehelin and J. Gordon, Electrophoretic transfer of proteins from polyacrylamide gels to nitrocellulose sheets: procedure and some applications, _Proc. Natl. Acad. Sci._ USA 76:4350 (1979).
15. M. Moeremans, G. Daneels and J. DeMey, Sensitive colloidal metal (gold or silver) staining of protein blots on nitrocellulose membrane, _Anal. Biochem._ 145:315 (1985).

16. A. Johnstone and R. Thorpe, "Immunochemistry in Practice," Blackwell, Boston (1982).
17. G. Newman, B. Jasani and E. Williams, A simple postembedding system for the rapid demonstration of tissue antigens under the electron microscope, Histochem. J. 15:543 (1983).
18. E. Carlemalm, M. Garavito and W. Villiger, Resin development for electron microscopy and analysis of embedding at low temperature, J. Micros. 126:123 (1982).
19. J. Polak and I. Varndell, "Immunolabeling for Electron Microscopy," Elsevier, New York (1984).
20. J. Roth, The preparation of protein A-gold complexes with 3 and 15 nm gold particles and their use in labeling multiple antigens on ultra-thin sections, Histochem. J. 14:791 (1982).
21. J. DeMey, Colloidal gold as a marker and tracer in light and electron microscopy, EMSA Bulletin 14:54 (1984).
22. P. Lewis and D. Knight, Staining methods for sectioned material, in: "Practical Methods for Electron Microscopy," A. Glauert, ed., Elsevier, New York (1982).
23. H. Hoch, Use of permanganate to increase electron opacity of fungal walls, Mycologia 59:1209 (1977).
24. M. Williams, Autoradiography and immunocytochemistry in: "Practical Methods in Electron Microscopy," A. Glauert, ed., Elsevier, New York (1977).
25. A. Stevens, High resolution autoradiography, Methods in Cell Biology 2:255 (1966).
26. C. Bulawa, M. Slater, E. Cabib, J. AuYoung, A. Sburlati, L. Adair and P. W. Robbins, The S. cerevisciae structural gene for chitin syn-thetase is not required for chitin synthesis in vivo, Cell 46: 213 (1986).
27. A. Sburlati and E. Cabib, Chitin synthetase 2, a presumptive partici-pant in septum formation in Saccharomyces cerevisciae, J. Biol. Chem. 261:15147 (1986).
28. R. Kornfeld and S. Kornfeld, Comparative aspects of glycoprotein structure, Ann. Rev. Biochem. 45:217 (1976).
29. P. Ledger and M. Tanzer, Monensin: a perturbant of cellular physiology, Trends in Biochem. Sci. 9:313 (1984).
30. A. Duran and E. Cabib, Solubilization and partial purification of yeast chitin synthetase, J. Biol. Chem. 253:4419 (1978).

SEGMENT MORPHOGENESIS IN ARTEMIA LARVAE

John A. Freeman

Department of Biology
University of South Alabama
Mobile, Alabama   36688

INTRODUCTION

Development in brine shrimp nauplii includes the sequential formation
of segments in the growing thorax and abdomen beginning at the anterior
end[1,3]. After nineteen instars the segments are complete. In each segment two ventro-lateral thoracopods, segmental and intersegmental muscles,
segmental ganglia, and peripheral nerves are formed over a period of three
to four molt cycles[2,4,5]. This type of development is typical of primitive crustaceans and may mimic the embryonic development of the higher
crustaceans[6].

The cellular and molecular mechanisms guiding segment formation have
received little attention, although the early larval stages of Artemia are
well suited for studying tissue morphogenesis. Evagination of the thoracopod limb bud is similar to the epithelial folding and evagination observed
during early development in many invertebrates and vertebrates[7,8]. The
ectodermal epithelium, which exists as a cone-shaped undifferentiated
monolayer at hatching, develops to a limb bud stage during the second
instar and completes segment morphogenesis by the fifth instar. Evagination of this epithelium to the limb bud of the thoracopod takes place at a
point in development which is essentially post-gastrular[3,5] and the
epidermal cells undergo evagination in a very short period of time[9].
The epithelium is clear and not surrounded by other tissues, making microscopic examination of living animals possible. Our research endeavors
have focused on the cellular mechanisms active in epithelial evagination
during segment morphogenesis and the means by which the events are regulated.

SEQUENCE OF EVENTS IN THE VENTRAL THORAX DURING THE SECOND INSTAR

Cell Division

Cell division during embryogenesis leads to an increase in mass and
contributes to form and shape of the tissue. Freeman[9] demonstrated that
the thorax epidermal cells of Artemia undergo replication in a spatial and
temporal manner. Cell division in the longitudinal (anterior-posterior)
direction leads to an increase in the number of cells in each longitudinal
file. Increase in length of the molt is, in part, a result of the increase

in cell number in these files. At ecdysis to the second instar (hour 1, instar II) five to six longitudinal files lie on either side of the mid-line [9] and there are no transverse files in the anterior region (Fig. 1,A).

During the second instar, cell division in the antero-ventral thorax changed orientation by 90° resulting in the formation of transverse files. The first transverse file became evident in the anterior ventral thorax 2-3 hours after ecdysis to instar II (Fig. 1,B,C). This file formed in what will become the central-most region of the first thoracopod bud (Fig. 1,A-D). The size of the file and the number of files in this region con-tinued to increase for the next five hours until 6 files with 15 cells in each were established (Fig. 1D). Although our observations indicate that the transverse files are a product of, and are maintained by directed cell division, we can not, at this time, eliminate the possibility that some cell neighbor rearrangement [7,10,11] is taking place early on in the pro-cess. During the first four larval instars transverse file formation is more pronounced in the first segment than in the second segment (Fig. 1,D).

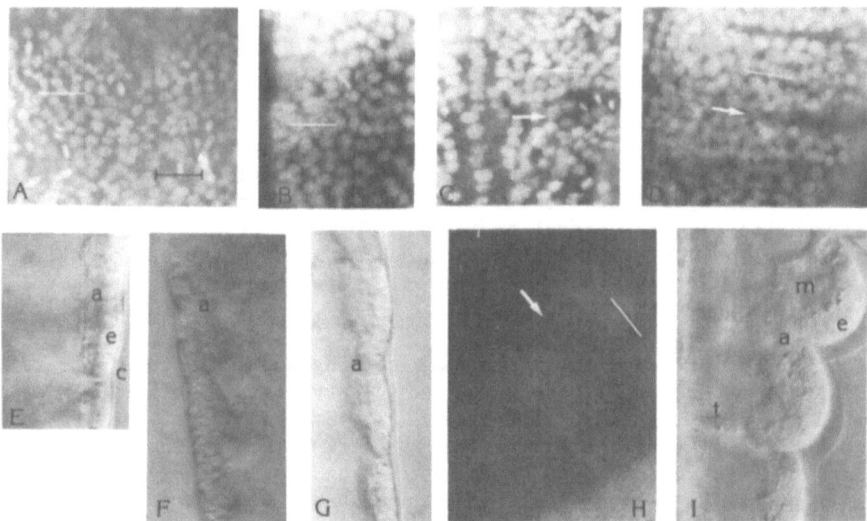

Fig. 1. Appearance of the epidermal cells in the anterior-
ventral thorax during segment morphogenesis in instar II.
A-D, bisbenzimide fluorescence, nuclei stained. H, NBD
fluorescence. E-G, I, differential interference contrast. A. Ventral view, hour 2, transverse files not
yet formed. B. Hour 3, transverse file beginning to
form. C. Hour 4, several transverse files formed in
segment 1, AM forming immediately behind it. D. Hour
7, transverse files well formed, AM has invaginated. E.
Hour 3, AM cells beginning to become squamous, cuticle
(c) still attached to epidermis. F. Early apolysis in
anterior region, AM region beginning to invaginate. G.
Hour 5, complete apolysis, evagination continuing. H.
Epifluorescence of tubulin staining by NBD-colcemid in
thoracopod bud cells. AM region contains little staining. I. Segments in instar III showing evaginated limb
bud epidermis (e), completely invaginated AM (a), tendon
cells (t), and differentiating mesoderm cells (m).
White bars indicate central region where first transverse
files form. Arrow indicates region where AM forms.
Black bar in A=25 μm; in B-I = 20 μm.

The thin cuticle in the articulating regions of jointed appendages in arthropods is known as the arthrodial membrane, a structure that acts as a "joint" and permits the appendages to move [12]. The arthrodial membrane (AM) forms between the regions of transverse files (Fig. 1,2). By the end of the second instar the borders of the first segment are morphologically defined by AM at the anterior and posterior borders and a region of squamous

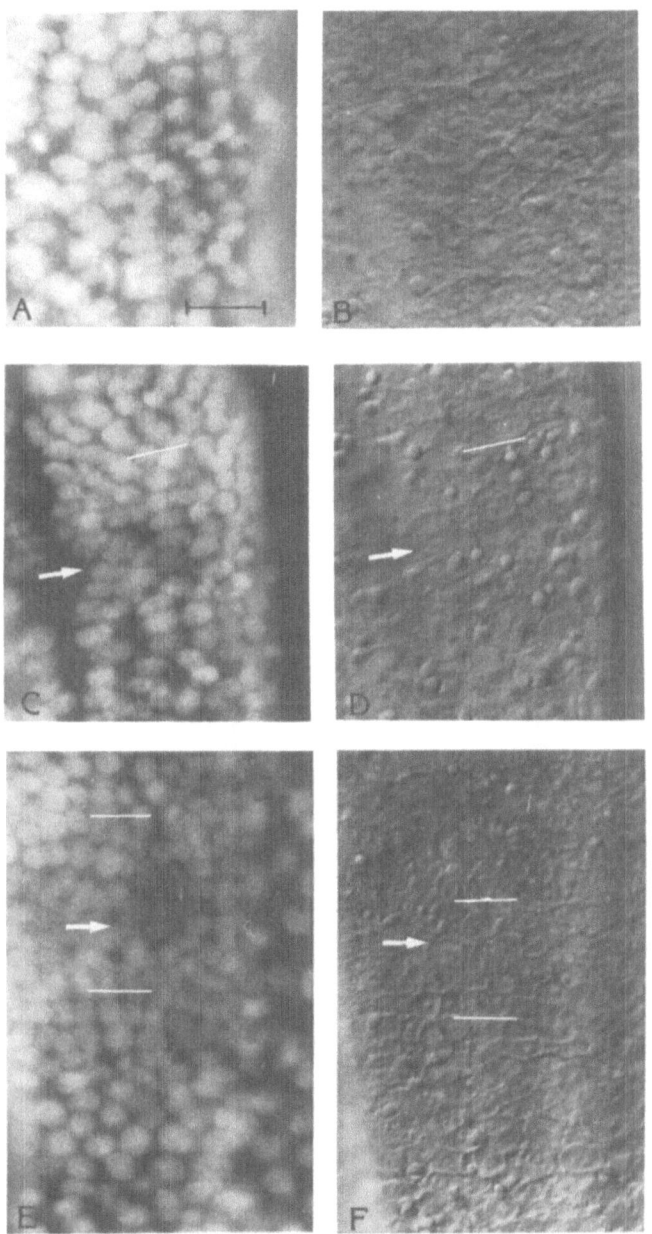

Fig. 2. Appearance of nuclei (A,C,E, bisbenzimide fluorescence) and whole cells (B,D,F, differential interference contrast of A,C,E, respectively) viewed from the ventral surface at hours 2 (A,B), 3 (C,D) and 4 (E,F) showing AM cell shape change to the spindle form. White lines and arrows as in Fig. 1. Bar = 25 μm.

cells along the ventral midline (Fig. 1). Cell replication in this region was less than in the bud region and differences in cell density in the AM and thoracopod bud regions were evident by hour 4-5. The change in cell density was measured using a rectangular overlay with an ImageAction@ video image analysis system. A rectangle was alternately placed over the thoracopod bud and AM region and the nuclei counted. Early in instar II the cell density was similar in both regions (Fig. 3). As cell proliferation continued in the thoracopod bud the difference in cell density between the bud and AM regions increased until, by hour 5-6, the cell density in the AM region was only 60% of that in the bud region.

The dorsal roof of the thorax consists of a thin epithelium that changed little during instar I. By the end of instar II, however, these cells had changed from a pattern of longitudinal files to a clump of cells organized in a segmental array with nuclear free cytoplasm at the anterior, posterior, and medial borders [13]. These cells eventually formed attachment sites for tendon cells growing up from the ventral surface (Fig. 1,I). Tendon cells were present by the end of the second instar and were clearly observed by the beginning of instar III (Fig. 1,I).

Differentiation of the Arthrodial Membrane Cells

Attainment of the squamous form marks the differentiated condition of the AM cells. Within three hours of molting to instar II the presumptive AM cells assumed a squamous, spindle shape. The neighboring limb bud cells remained cuboidal (Figs. 1,E,F,G, 2). This is evident when observed in profile (Fig. 1,F,G) or from a frontal view (Fig. 2). A direct comparison of the pattern of nuclei and the AM cells hour 2 to hour 4 can be seen

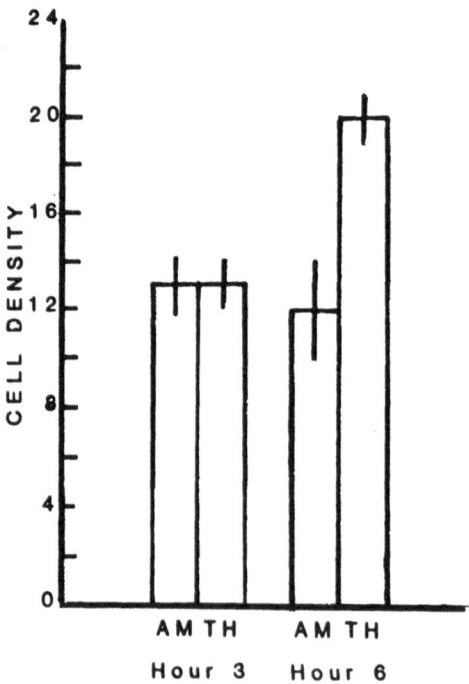

Fig. 3. Comparison of cell density in AM and thoracopod bud cells (TH) in instar II larvae at hour 2-3 (Hour 3) and hour 5-6 (hour 6). The density rises in the limb bud region while it does not change in the AM region.

in Fig. 2 and a schematic comparison of the time course of the developmental changes in the thorax epidermal cells is shown in Figure 4.

In addition to the relationship of the first transverse file to the first limb bud another morphological trait of the presumptive thoracopod limb bud cells is the proximity to the bilateral clumps of mesodermal cells which lie directly beneath the epidermis. Although the AM cells appear to have some contact with the mesoderm cells the AM region always arises in the space between the mesodermal clumps (Fig. 1,G,I).

Mechanism of Cell Shape Change in AM Cells

Epithelial evagination (or invagination) generates tissue form in embryonic tissues in many organisms. This has been clearly demonstrated in vertebrate nerve cord, thyroid gland invagination, lung bud branching, salivary gland branching, retina formation, and sea urchin gastrulation [7,8]. It has also been documented in arthopods [14] and, more specifically, in _Artemia_ [1,5,9]. Among the proposed cellular mechanisms involved in invagination are cell adhesion [15,16], change in cell shape through apical constriction [17-21] and cell neighbor rearrangement [10,11]. Microfilaments have been postulated to be the active factors on the grounds that apical constriction in a few cells could initiate change in an entire epithelium [20]. Sanger et al. [22] discussed the role of stress fibers in morphogenesis. Ettensohn [23], however, has pointed out that although these processes have been observed in many developing tissues, little evidence has been forthcoming to substantiate any one process as generating

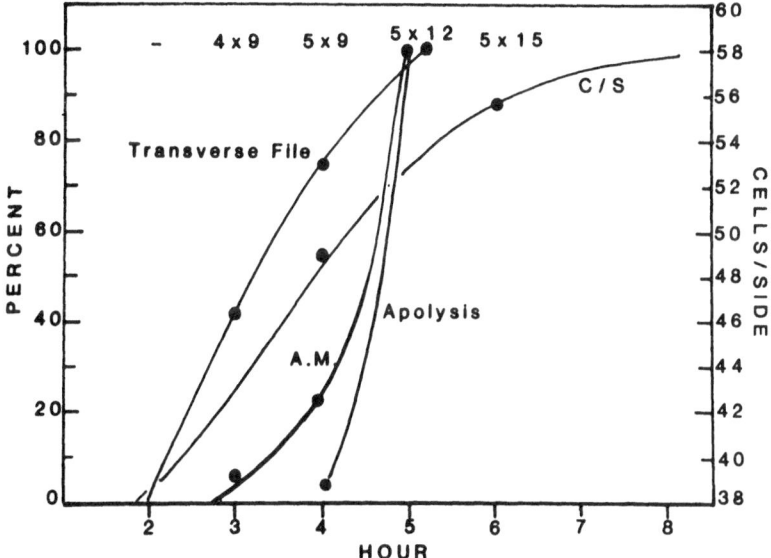

Fig. 4. Comparison of the sequence and timing of cellular events occurring in the ventral thorax during instar II. Ordinate, percent of larvae (n=40) demonstrating formation of transverse files (at least one file of bisbenzimide-stained nuclei), spindle shape arthrodial membrane cells (AM), and apolysis. The size of the cell pattern (number files by number of nuclei in central file) in the thoracopod bud at hours 3,4,6,7 is indicated at the top. For comparison to general growth the number of cells per side (C/S, see Freeman [9] for the entire thorax is also indicated.

the force for epithelial shape change. It is likely that several, or all, of the processes are acting simultaneously or sequentially in an orchestrated manner.

Cell proliferation increases after gastrulation. The fact that epithelial cell replication takes place within a limited amount of space (or mitotic pressure) has led to the hypothesis that mitotic pressure could provide the motive force for tissue shape change [15,18,24-28]. The morphogenetic potential is present in epithelial tissues during evagination, tube formation, and tube branching [29], although, as Kolega [8] points out, most of the tests used to demonstrate mitotic pressure have not sufficiently eliminated other possible forces, such as cytoskeletal elements.

It is possible that evagination of the thoracopod bud during segmentation in Artemia involves several of the primary mechanisms for epithelial morphogenesis. Our initial interests have centered on the interaction of mitotic pressure and cell density with cytoskeletal elements in the epidermis. Cell shape change from cuboidal to spindle shape could involve the action of microtubules. Moreover, changes in the microtubule pattern in the AM and limb bud cells may be essential to evagination. We first analyzed the temporal nature of microtubule presence by staining fixed larvae with NBD-colcemid [30]. A transient increase in the staining of the AM cells, as compared to other regions of the thorax, was evident by hour 3-4. By hour 5-6, however, the staining pattern had reversed. Staining in the thoracopod bud region had markedly increased and little staining could be seen in the AM region (Fig. 1,H). Similar results have been obtained in a separate study employing a polyclonal antibody to tubulin in dissected specimens [31]. With both methods of microtubule detection the staining pattern was abolished when the larvae were treated with colchicine prior to hour 3.

To further examine the importance of microtubules, synchronized instar II larvae were exposed to colchicine [32] for two hours beginning at hour 1. These larvae demonstrated no cell shape change and the thin AM region never formed at hour 3-4, when AM cell differentiation normally took place (Fig. 5, A-C). These findings support the contention that microtubules play an active role in the cell shape change and epithelial evagination. At this time, however, we have no evidence to suggest that the motive force in the shape change is based on microtubules. It is possible that the microtubules form a scaffold upon which other cytoskeletal elements act, as has been suggested for several developmental systems [8]. Cheshire [31] demonstrated the presence of actin and myosin in the AM cells and further showed that cytochalasin B blocked evagination. Thus, another possible activity of the microtubules is to interact with microfilaments during cell shape change. Following AM cell shape change the microtubules may be necessary for maintaining the spindle shape in the AM cells, although this contention is not supported by the weak staining pattern after hour 5. The strong staining in the limb bud region after apolysis suggests microtubules may be active in the evagination process and in the maintenance of the cuboidal shape of the cells as mitosis continues and the epithelium is pushed out against the old cuticle.

Interaction of AM Morphogenesis and Molt Cycle Events in Limb Bud Evagination

The crustacean cuticle essentially forms the external shape of the organism. Due to the adhesion of the epidermal cells to the restrictive cuticle any morphogenetic change in the epithelium will not be evident until apolysis (release of the old cuticle) occurs, permitting the epidermis to assume a new form. In instar II brine shrimp, early apolysis occurs in the region of the first segment at hour 4-5 (Fig. 4), shortly before it occurs in the more posterior regions of the presumptive thorax and abdomen.

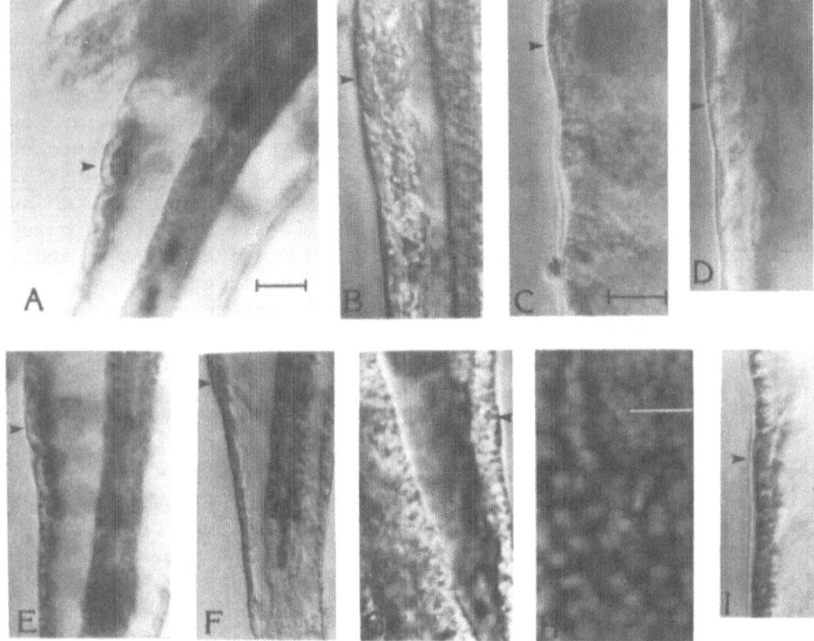

Fig. 5. Appearance of instar II larvae after exposure to inhibitors. A. Control larvae, hour 6, instar II. B. Colchicine treated larva at hour 3. C. Colchicine treated larva at hour 6. D. Larva exposed to 50 µg/ml 20-HE for two hours showing precocious apolysis. The epidermis forms a ripple throughout the length of the thorax. E. Control larva from BrdUrd and F3rdT experiments, hour 6, instar II. F. and G. Larvae exposed to BrdUrd at 100 µg/ml, hour 3 and 6, respectively. H. Bisbenzimide fluorescence of nuclei in BrdUrd-exposed larva. Transverse files have not formed. Compare region with white bar in H to Figures 1,D and 2,E. I. Larva exposed to 10 µg/ml F3rdT. The arrowhead indicates the equivalent region of the thorax in each of the light microscopic photographs. The bar in A = 25 µm in A, E and F; the bar in C = 20 µm in C, D, G, H and I and 24 µm in B.

The onset of early apolysis in the anterior region overlapped with AM cell shape change.

The close temporal relationship between AM formation and early apolysis forms the basis for a model of the mechanisms of epithelial deformation that may explain limb bud evagination and AM invagination. The potential for deformation would be established prior to apolysis by the combination of unequal mitotic (lateral) pressure within the epithelium and AM cell shape change. Since the AM cells become thinner and thus less able to resist the pressure, these cells would be more likely to deform at apolysis. Thoracopod bud epidermis, which remains plate-like and thus better able to resist deformation, would be pushed out to the old cuticle by hydrostatic pressure. Continued mitotic activity in the limb bud region [9] following apolysis would result in continued expansion of the bud outwardly and the

AM region inwardly until the new cuticle becomes rigid.

To examine the hypothesis that limb bud evagination results from epithelial deformation upon apolysis, larvae were treated with molting hormone (20-hydroxyecdysone, 20-HE, 50 µg/ml seawater) beginning at hour 1 of instar II and examined at hourly intervals for the next three hours. Molting hormone stimulates apolysis in larval and adult crustaceans [33,34]. The exposure period covers the time when cell shape change in the AM cells is occurring. If early apolysis is in fact critical for dissipation of localized mitotic pressure, then the model would predict that, if apolysis occurred prior to cell shape change, ie. in reverse order, deformation would not take place. Larvae exposed to 20-HE demonstrated region-specific or total apolysis by hour 3 (Fig. 6). Only 8% of these larvae showed any AM cell differentiation and less than 10% demonstrated limb bud evagination. The presence of low levels of evagination in both control and treated larvae was probably due to slight asynchrony within the population.

Although most of the larvae exhibited no change in the shape of the thorax wall, a few showed a ripple effect in the epidermis which extended the entire length of the thorax (Fig. 5,D). This would be expected if deformation was not restricted to the anterior region and the lateral force generated by the mitotic pressure was equally distributed along the length of the thorax.

We can not rule out the possibility that the evagination mechanism does not involve the interaction of shape change and early apolysis but, instead, is initiated at hour 4-5 by some other mechanism and, therefore, would not have begun at hour 3. However, the fact that most of the hormone-treated larvae failed to evaginate even by hour 7 does not support this possibility.

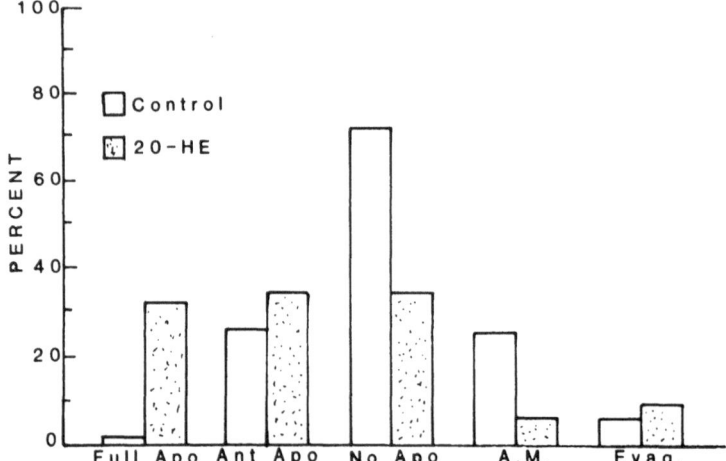

Fig. 6. Effect of 20-HE (50 µg/ml) on early apolysis (Apo). AM cell formation (spindle shape, AM), and evagination of the limb bud (Evag) in instar II larvae. Degree of apolysis is divided into complete apolysis along the thorax (Full Apo), apolysis limited to anterior ventral thorax (Ant Apo) and no apolysis (No Apo). Ordinate, percent of 40 synchronized larvae at hour 3-4.

The epidermis of crustacean larvae is capable of synthesizing all the components of the complex cuticle and, thus, could be considered as differentiated. However, specialized cuticle formation takes place during development as the epidermal cells synthesize specific integumental structures in the juvenile. Thus, the early larval integument contains regions of epidermal cells which are determined along a somewhat restricted developmental pathway. They are capable of producing cuticle but have not completely differentiated to the adult form.

Formation of a segment involves the differentiation of several cell types in the epidermis, each type identified by specific cellular morphology. Manifestation of the differentiation process in the segments includes cuticle shape (setal cells), spindle cell shape (AM), axopodial extension (tendon cell), and change in cell adhesion (neuroblasts) [2,5,9]. The fact that these activities must occur in a coordinated manner for successful segment development suggests that these cells become determined at the same time and that differentiation of the diverse cell types is controlled by similar mechanisms. Due to the unusual development in brine shrimp, differentiation could begin at cleavage, gastrulation, postgastrular encystment, rehydration, emergence, the first larval instar, or at any one of the later larval instars.

To study the regulation of the onset and timing of differentiation, we employed bromodeoxyuridine (BrdUrd), a drug known to block differentiation in a wide variety of cells [35], and trifluorothymidine (F3rdT), shown to block differentiation in insect tissues [36]. Larvae were exposed to the compounds beginning at hatching and observed for cell differentiation during instar II. Both bromodeoxyuridine and trifluorothymidine inhibited segmentation in a dose-dependent manner (Fig. 7). As previously found [13], all the cellular events of segment formation were inhibited by the compounds in an all-or-none manner. Differentiation of AM cells was completely blocked in 92% of the BrdUrd-exposed larvae at a concentration of 200 µg/ml and all the larvae exposed to 10 µg/ml F3rdT (Figs. 5,E-I; 7). The squamous spindle shape was never attained by these cells (Fig. 5, G, I) and, after apolysis, there was no evagination of the epithelium (Fig. 5,F,G,I). Addition of equal concentrations of thymidine and BrdUrd significantly reduced the inhibitory effect of BrdUrd (Fig. 7), probably by competing with BrdUrd for incorporation into DNA. The difference in the dose response curves in Figure 7 may be due to differences in rate of uptake or incorporation of the compounds. Alternatively, F3rdT may more effectively disturb gene regulation after incorporation into the DNA, as has been found in Drosophila [36]. Neither compound caused death when present during the first 24hr of the larval period. Moreover, neither compound inhibited molting or growth of the larvae. Thus, we feel justified in stating that, at least at the level tested, only cell differentiation was affected by the drugs. These findings are in keeping with comparable studies with other systems [35,37, 38].

For BrdUrd to block differentiation it must be incorporated into the DNA prior to the molecular event leading to change in gene expression. Tanaka et al. [38] estimated that BrdUrd was effective in inhibiting scale differentiation ('cell shape, polarity, pattern and alignment) in avian epidermis if more than 50% of the cells incorporated the compound. The results presented here and those of Freeman [13] show that the most sensitive time for incorporation was the first 10 hours of instar I. Addition of BrdUrd or F3rdT after this period had little inhibitory affect on development through instar II. Freeman and Chronister (unpublished findings) demonstrated by cell cycle analysis that during this period the cells in the anterior ventral region of the developing thorax progressed from G2-M

through Gl and S and re-entered G2. Although the cell cycle transit did
not appear to be totally synchronous, all of the cells were cycling. Thus,
the 8-10 hr period was long enough for incorporation of BrdUrd and F3rdT
in these cells.

The cells in the anterior ventral thorax continue to cycle and undergo
a second cell cycle transit prior to AM formation (Freeman and Chronister,
unpublished findings). Although exposure to the drugs during this period
(the last 8-10 hours of instar I and hours 1-3 of instar II) did not affect
segmentation [13], the second cell cycle would be the first in which BrdUrd
inhibition could take place. Conducting similar experiments with embryos

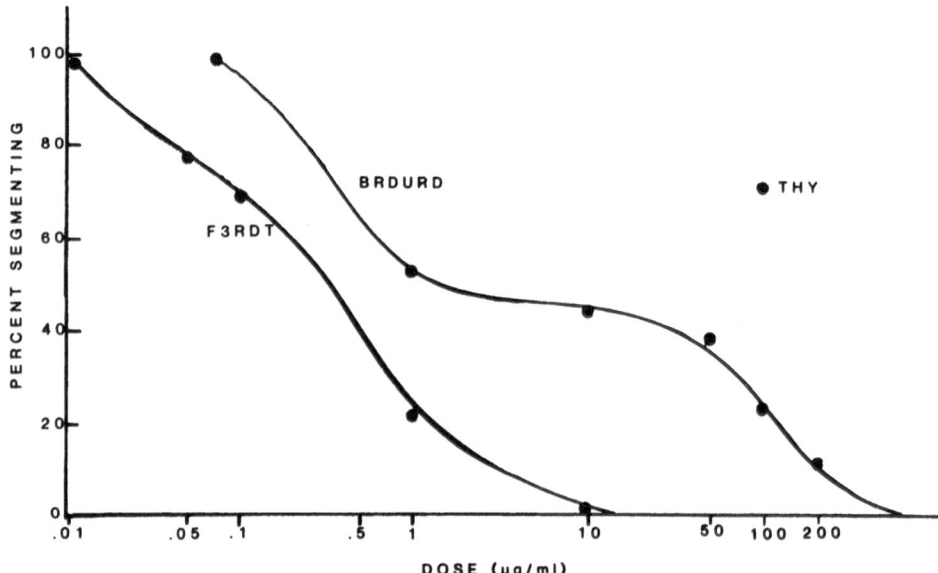

Fig. 7. Dose response curve for bromodeoxyuridine (BRDURD) and
trifluorothymidine (F3RDT) beginning at hatching and
observed for segmentation at the end of instar II.
Each point represents percent of 75 to 250 larvae. THY,
Larvae exposed to equal concentrations (100 µg/ml) of
BrdUrd and thymidine. All morphological traits of segmentation (AM, transverse file, evagination, tendon
cell) were considered.

prior to encystment, during rehydration, or during emergence has not been
done and, considering the difficulty in uptake of compounds during these
periods, it may not be possible to carry out similar experiments with these
stages. Thus, at this time we can not eliminate these periods in development as the point when cell differentiation begins. If, however, differentiation does begin during one of these periods then BrdUrd and F3rdT
must be acting at a level which: 1) is later in the sequence of regulatory
events and, 2) encompasses, in some manner, a diverse set of regulatory
switches in many different genes. Future research will be aimed at determining the molecular action of these interesting compounds in the inhibition
of cell differentiation in brine shrimp larvae.

DISCUSSION

The findings reported here suggest that segment morphogenesis involves a wide range of cellular activities (Fig. 4). The earliest detectable change was patterned cell division giving rise to transverse files of epidermal cells. At some point, a cue is received by the presumptive AM cells which directs them along a developmental path which is different from that of the thoracopod bud cells. The switch elicits nuclear translocation and change in cell shape to the spindle form as the cells become thinner than the adjacent thoracopod cells. The mechanism of this cue is not understood. One possibility is a gradient of some morphogen which is recognized by the individual cells [39]. Another possible mechanism, similar to epithelio-mesenchymal interactions in vertebrate development, is induction by the underlying mesodermal cells. The spatial location, established at gastrulation [5], interacting with some temporal trigger (or clock) may be sufficient to induce change in the adjacent epidermis. Exactly which group of cells in the anterior epidermis would be affected by the mesoderm is difficult to determine since direct contact would primarily affect the thoracopod bud cells rather than the AM cells. It is possible that the result of mesoderm stimulation is cell proliferation in the bud region. A similar inductive mechanism has been demonstrated in mesenchyme-induced lung development [28].

Differential tubulin staining and tests with colchicine treatment showed that cell shape change in AM cells and evagination is dependent upon microtubule function. Future studies will determine whether microtubules have a structural support function or are an active motive force in the shape change. Actin and myosin form the molecular basis for cell shape change and movement in many developing tissues. They may be active in epithelial morphogenesis in Artemia as well [31]. The observation that actin immunofluorescence staining increased after AM cell formation supports the hypothesis that invagination of the AM region may involve microfilament-dependent stretching of the cells to the spindle shape, possibly through interaction with myosin in a manner similar to a stress fiber [22], followed by constriction of the microfilament band during invagination.

The model for epithelial evagination/invagination developed here has, as its basis, an inequality in lateral pressure arising from differences in cell density (mitotic pressure) and epithelial thickness (AM morphogenesis). The potential for deformation of the epithelium is established prior to apolysis and is dissipated after apolysis, resulting in invagination of the AM region. Many different, but not necessarily mutually exclusive mechanisms have been advanced to explain epithelial morphogenesis [8]. Considering the fact that two of the mechanisms have been shown to function in brine shrimp segmentation, it is likely that segment morphogenesis involves the interaction of each of the mechanisms in a specific (and possibly cell-specific) temporal sequence.

Hormonal induction of apolysis during instar II demonstrates the interdependency of morphogenesis and molting in crustacean larvae. Our results are the first to show the importance of the timing of premolt events in limb bud evagination and, in addition, to demonstrate hormonal control of molting in these larvae. The control of the molt cycle of instar I and II larvae may be through a Y-organ-based molt controlling system, such as exists in adult crustaceans. It is also possible that stored supplies of 20-HE in yolk granules provide the means of controlling the onset of apolysis in the developing larvae prior to Y-organ development.

Brine shrimp larvae exhibit an unusual developmental state in that the cephalon contains differentiated limbs and epidermal structures while the thorax consists of undifferentiated tissues. It is difficult to precisely define the stage of differentiation of the thorax epidermal cells. One of the classical characteristics of the state of terminal differentiation, namely enhanced synthesis of a small number of "unique" proteins which function specifically in the differentiated tissue, is already present at hatching. For epidermal cells, these proteins would include all those involved in forming the cuticle. Yet, the segmental structures and appendages still exist in a rudimentary form. Differentiation of the thorax epidermis may involve an initial phase, in which the ability to synthesize components of the cuticle is acquired, followed by a longer secondary phase in which cuticle synthesis and cuticle shape become specific to a region of the integument, such as the AM and setae. If such a sequence exists in Artemia, then the process of differentiation may be regulated at more than one step and , possibly, over a relatively long period of time. Our experiments with BrdUrd suggest that it is the second phase, epithelial morphogenesis, that is affected by this compound. Cells in the segmenting region underwent changes in proliferation and shape change without any obvious alteration in cuticle synthesis. The epidermal cell thus appears to be unique in its rate of developmental path selection. Moreover, the path is not totally rigid or limited for all epidermal cells. Presumptive neural cells along the ventral midline, for example, function as normal epidermal cells for several instars and then leave the epidermal epithelium, stop cuticle synthesis, and begin neural morphogenesis [2,5]. Thus far, we have not examined the effect of BrdUrd on later stages of epidermal differentiation in older larvae. Such experiments may further elicidate the regulatory steps in differentiation of the epidermal cells of Artemia.

ACKNOWLEDGEMENTS

I wish to thank Ms. Kathleen E. Mignone for excellent technical assistance. This research was supported by a grant from the University of South Alabama Research and Public Service Council and NATO grant no. 0865/87.

REFERENCES

1. P. B. Weisz, The space-time pattern of segment formation in Artemia salina, Biol. Bull. 91:119 (1946).
2. P. B. Weisz, The histochemical pattern of metameric development in Artemia salina, J. Morph. 81:45 (1947).
3. J. S. Clegg and F. P. Conte, A review of the cellular and developmental biology of Artemia, in: "The Brine Shrimp Artemia," Vol. 2, G. Persoone, O. Sorgeloos, O. Roels and E. Jaspers, eds., Universa Press, Wetteren (1980).
4. D. T. Anderson, Larval development and segment formation in the branchiopod crustaceans Limnadia stanlejana (Conchostraca and Artemia salina, Aus. J. Zool. 15:47 (1967).
5. R. Benesch, Zur ontogenie und morphologie von Artemia salina, Zool. Jahrb. Abt. fur Anat. Ontog. Tiere 86:307 (1969).
6. D. T. Anderson, "Embryology and Phylogeny in Annelids and Arthropods," Pergammon Press, New York (1973).
7. C. A. Ettensohn, Gastrulation in the sea urchin embryo is accompanied by the rearrangement of invaginating epithelial cells, Develop. Biol. 112:383 (1985).
8. J. Kolega, The cellular basis of epithelial morphogenesis, in: "Developmental Biology, A Comprehensive Synthesis," Vol. 2, L. W. Browder, ed., Plenum Press, New York (1986).
9. J. A. Freeman, Epidermal cell proliferation during thoracic development in larvae of Artemia, J. Crust. Biol. 6:37 (1986).

10. D. Fristrom, The mechanism of evagination of imaginal discs of
    Drosophila melanogaster. III. Evidence for cell rearrangement,
    Develop. Biol. 54:163 (1976).
11. J. D. Hardin and L. Y. Cheng, The mechanism and mechanics of archenteron
    elongation during sea urchin gastrulation, Develop. Biol. 115:490
    (1986).
12. R. Dennell, Integument and Exoskeleton, in: "The Physiology of Crusta-
    cea," Vol. 2. T. Waterman, ed., Academic Press, New York (1960).
13. J. A. Freeman, The integument of *Artemia* during early development, in:
    "Biochemistry and Cell Biology of Brine Shrimp," T. MacRae, J.
    Bagshaw and A. Warner, eds., CRC Press, Boca Raton (1988).
14. C. A. Poodry, Imaginal discs: morphology and development, in: "The
    Genetics and Biology of *Drosophila*," Vol. 2d, M. Ashburner and T.
    R. F. Wright, eds., Academic Press, New York (1980).
15. T. Gustafson and L. Wolpert, Cellular movement and contact in sea urchin
    morphogenesis, Biol. Rev. Camb. Philos. Soc. 42:442 (1967).
16. J. B. Nardi and I. Reynolds, Bidirectional folding of an insect epith-
    elial monolayer, J. Exp. Zool. 237:209 (1986).
17. B. Burnside, Microtubules and microfilaments in amphibiam neurulation,
    Amer. Zool. 13:989 (1973).
18. S. R. Hilfer, Extracellular and intracellular correlates of organ
    initiation in the embryonic chick thyroid, Amer. Zool. 13:1023
    (1973).
19. B. S. Spooner, Microfilaments, cell shape change, and morphogenesis
    of salivary epithelium, Amer. Zool. 13:1007 (1973).
20. G. M. Odell, G. Oster, P. Alberch and B. Burnside, The mechanical
    basis of morphogenesis. I. Epithelial folding and invagination,
    Develop. Biol. 85:446 (1981).
21. G. C. Schoenwolf and M. V. Franks, Quantitative analysis of changes
    in cell shapes during bending of the avian neural plate, Develop.
    Biol. 105:257 (1984).
22. J. M. Sanger, B. Mittal, M. Pochapin and J. W. Sanger, Observations of
    microfilament bundles in living cells microinjected with fluores-
    cently labelled contractile proteins, J. Cell Sci. Suppl. 5:17
    (1986).
23. C. E. Ettensohn, Mechanisms of epithelial invagination. Quart. Rev.
    Biol. 60:289 (1985).
24. R. L. Pictet, W. R. Clark, R. H. Williams and W. J. Rutter, An ultra-
    structural analysis of the developing embryonic pancreas, Develop.
    Biol. 29:436 (1972).
25. J. Zwaan and R. W. Hendrix, Changes in cell and organ shape during
    early development of the ocular lens, Amer. Zool. 13:1039 (1973).
26. D. P. Richman, R. M. Stewart, J. W. Hutchinson and V. S. Cavinoss,
    Mechanical model of brain convolution development, Science 189:18
    (1975).
27. M. S. Smuts, S. R. Hilfer and R. L. Searls, Patterns of cellular
    proliferation during thyroid organogenesis, J. Embryol. Exp. Morphol.
    48:269 (1978).
28. G. V. Goldin and N. K. Wessels, Mammalian lung development: the possible
    role of cell proliferation in the formation of supernumerary tracheal
    buds and in branching morphogenesis, J. Exp. Zool. 208:337 (1979).
29. S. R. Hilfer and R. L. Searls, Cytoskeletal dynamics in animal morpho-
    genesis, in: "Developmental Biology. A Comprehensive Synthesis,"
    Vol. 2, L. W. Browder, ed., Plenum Press, New York (1986).
30. T. Hiratsuka and T. Kato, A fluorescent analog of colcemid, N-(7-
    nitrobenz-2-oxa-1,3-diazol-4-yl)-colcemid, as a probe for the
    colcemid-binding sites of tubulin and microtubules, J. Biol. Chem.
    262:6318 (1987).

31. L. B. Cheshire, The role of microfilaments, microtubules, and inter-
    mediate filaments in epithelial evagination in the brine shrimp
    Artemia, M. S. Thesis, University of South Alabama (1987).
32. M. M. Mareel and M. de Mets, Effect of microtubule inhibitors on in-
    vasion and on related activities of tumor cells, Int. Rev. Cytol.
    90:125 (1984).
33. J. A. Freeman and J. D. Costlow, The cyprid molt cycle and its hormonal
    control in the barnacle Balanus amphitrite. J. Crust. Biol. 3:173
    (1983).
34. D. M. Skinner, Molting and regeneration, in: "The Biology of Crustacea,"
    Vol. 9, D. B. Bliss and L. H. Mantel, eds., Academic Press, New York
    (1985).
35. H. Holtzer, H. Weintraub, R. Mayne and B. Mochan, The cell cycle, cell
    lineages, and cell differentiation, in: "Current Topics in Develop-
    mental Biology," Vol. 7, A. Moscona and A. Monroy, eds., Academic
    Press, New York (1972).
36. T. M. Rizki and R. M. Rizki, Developmental modifications induced by
    DNA base analogs, Amer. Zool. 17:649 (1977).
37. W. E. Wright, BUdR, probability and cell variants: towards a molecular
    understanding of the decision to differentiate. BioEssays 3:245
    (1985).
38. S. Tanaka, H. Sugihara-Yamamoto and Y. Kato, Epigenesis in developing
    avian scales, I. Stage-specific alterations of the developmental
    program caused by 5-bromodeoxyuridine, Develop. Biol. 121:467 (1987).
39. H. Meinhardt, Heirarchical inductions of cell states: a model for seg-
    mentation in Drosophila, J. Cell Sci. Suppl. 4:357 (1986).

HEAT SHOCK INDUCES DEVELOPMENTAL ANOMALIES IN ARTEMIA*

A. Hernandorena

Laboratoire du Museum National d'Histoire Naturelle
Plateau de l'Atalaye
64202 Biarritz Cedex, France

INTRODUCTION

The morphogenesis of supernumerary appendages on otherwise apodous abdominal segments can be induced in Artemia by specific nutritional conditions imposed on young feeding larvae[1-6]. This manipulation, qualified as startling[7], was difficult to integrate into the context of our knowledge on morphogenesis. Artemia larvae start feeding after having performed three molts at the expense of vitelline reserves. During early postembryonic development, that is after hatching and before feeding, they are not sensitive to nutritional conditions and were therefore submitted to abnormal thermal conditions. Heat shocks delivered to newly hatched nauplii have long term effects on subsequent development. Now, it becomes easier to conceive that nutritional conditions imposed on older feeding larval stages may have morphogenetic effects.

MATERIAL AND METHODS

The Utah strain of Artemia was used. Resting eggs (cysts) were disinfected according to a method adapted from Provasoli and Shiraishi[8]. Disinfected cysts were incubated in a set of tubes each containing 10ml of sterile sea water. Nauplii hatched 24 h after hydration were transferred to a second set of tubes containing 10ml of sterile sea water. Larvae were transferred 24 h after hatching, in groups of five, to a third set of culture tubes containing 10ml of autoclaved nutritive medium. Transfers were performed aseptically using a transfer hood equipped with a germicidal lamp[9]. Hatching and culturing of control animals took place at 25±0.5°C with a 10 h light-14 h dark photoperiod. Adults were obtained in 10 days.

Heat shocks of 40°C were delivered to embryos during preemergence development or to newly hatched nauplii by transferring tubes from the water bath at 25±0.5°C to a bath at 40°±0.5°C. Durations of the shocks are given in the results section. Heat shocks to encysted embryos were delivered by transferring tubes, after 1, 6 or 18 h of incubation, to 40°C. Heat shocks to nauplii were delivered by transferring tubes to 40°C at 0, 4, 8 or 24 h after hatching. A known number of larvae surviving after the

---

*Preliminary results of this study were presented at the X Reunion des Carcinologistes de langue francaise at Concarneau (France) in June 1987.

shock were cultured and morphogenesis was checked by a microscopical examination of grown individuals anesthesized with chloroform. Normal appendicular morphogenesis is diagrammed in Fig. 1.

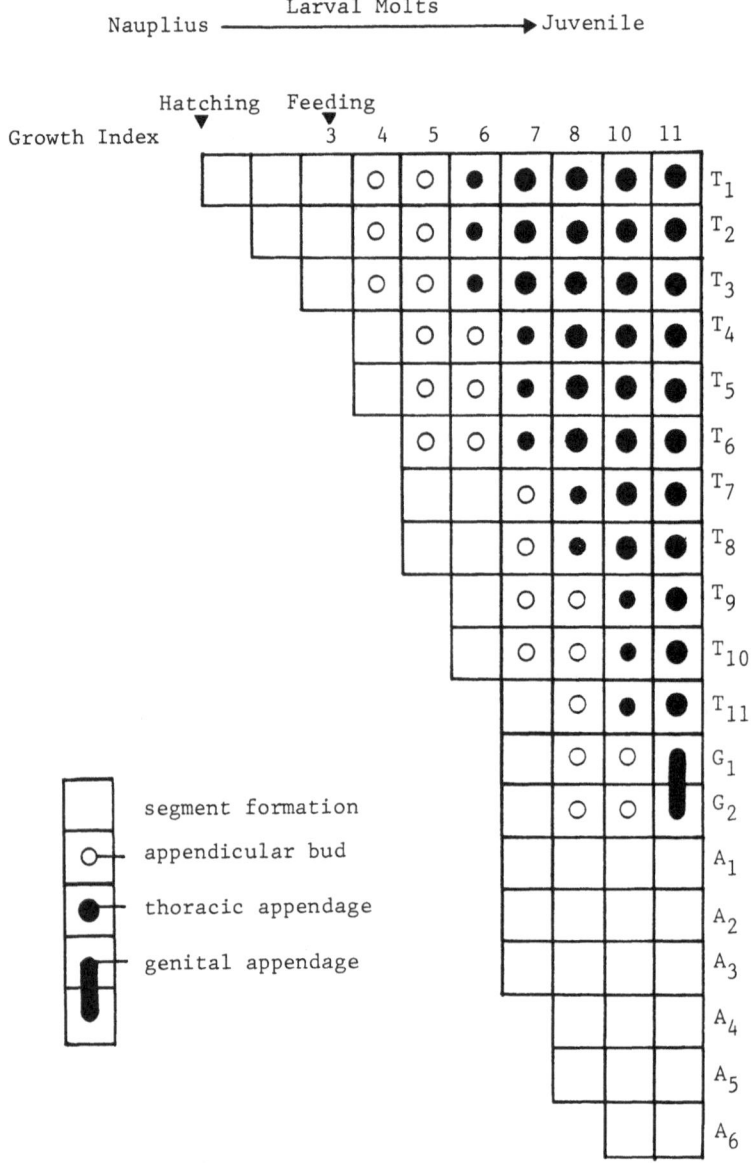

Fig. 1. Diagrammatic representation of _Artemia_ postembryonic development from new-born nauplius to juvenile through larval molts. Segments are added successively with appendages gradually coming into service. T_1-T_{11}, thoracic segments; G_1-G_2, genital segments; and A_1-A_6, abdominal segments. The growth index defined by Provasoli and d'Agostino [10] and Hernandorena [1] approximates number of molts.

Table 1. Effect of a 2.5 h Heat Shock at 40°C Delivered to Newly Hatched (0), 4 and 8 h Old Nauplii

Time after hatching	Number of nauplii shocked	Survival at 24 h post-shock	Number of shocked larvae cultured	Shocked larvae completing development	Abnormal animals
h		%		%	%
0	421	68	100	64	30
4	409	33	125	43	30
8	389	18	70	33	26

RESULTS

In the first set of experiments (Table 1), about 400 nauplii were submitted at 0, 4 or 8 h after hatching to a heat shock lasting 150 min. The thermosensitivity of nauplii increased sharply from 0 to 8 h as shown by the decreased percentage of larvae surviving 24 h after the shock. The percentage of shocked individuals capable of completing development decreased when the shock was postponed to 4 or 8 h after hatching but the percentage of abnormal individuals did not increase.

In the second set of experiments (Table 2), the shock was delivered to newly hatched nauplii. The number of nauplii submitted to the shock was not counted. The percentage of shocked individuals capable of completing development decreased and the percentage of surviving individuals showing morphological anomalies increased when the duration of the shock increased from 60 to 240 min.

The abnormal individuals recorded in Table 1 and 2 had the same type of anomalies as those recorded in Table 3. In this table, defects obtained in 117 individuals after a 4 h shock delivered to newly hatched nauplii are detailed. Of these individuals, 3 presented an abnormal spiraled abdominal segmentation (Fig. 2). This anomaly corresponds to an abnormal positioning of the genital appendage which is lateral instead of ventral

Table 2. Effect of a 1-4 h Heat Shock at 40°C Delivered to Newly Hatched Nauplii

Duration of heat shock	Number of shocked larvae cultured	Shocked larvae completing development	Abnormal individuals
h		%	%
1.0	100	89	0
2.0	102	79	16
2.5	100	64	30
3.5	75	60	31
4.0	715	44	37

Table 3. Abnormal Individuals Obtained After a 4 h Heat Shock at 40°C
Delivered to Newly Hatched Nauplii

	Fused thoracic appendages	Abnormal genital appendages	Abnormal antennae	Spiraled abdomen	Total number of organisms
Number of organisms	74	34	6	3	117
Percentage of organisms	63.2	29.1	5.1	2.6	100

(Fig. 3). Of the 117 abnormal individuals, 114 manifested anomalies, which
were most often unilateral, in appendicular morphogenesis. Of these 114
individuals 74 had fused adjacent thoracic appendages (Fig. 4). This
anomaly is identical to that resulting from a known mutation, "swimmerette",
described by Squire and Grosh [11]. The fusion of adjacent appendages,
which is abnormal in the case of thoracic segments, is normal in the case
of genital segments. The unilateral fusion of appendicular buds of the
two genital segments leads in the males to the formation of two latero-
ventral gonopodes (Fig. 1). Figure 5 illustrates an abnormal genital
appendage. The anterior part, rising from the first genital segment looks
like a thoracic appendage and the posterior part arising from the second

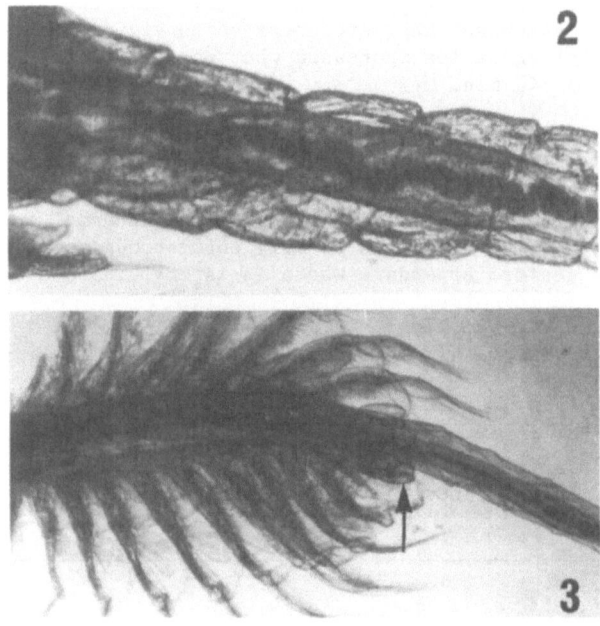

Fig. 2. Spiraled abdominal segmentation resulting
from a 4 h heat shock to newly hatched
nauplii.

Fig. 3. Lateral positioning of the genital appendage (arrow) resulting from a 4 h heat shock
to newly hatched nauplii.

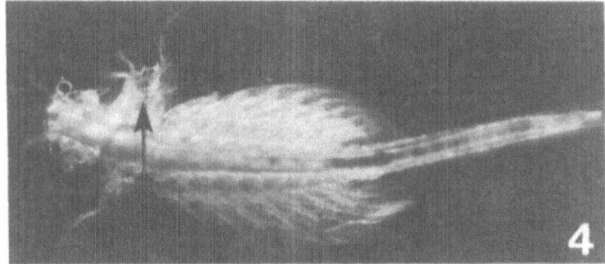

Fig. 4. Fused thoracic appendages (arrow) resulting
from a 4 h heat shock to newly hatched nauplii.

genital segment looks like a genital appendage with a typical accessory
gland and a vas deferens. The abnormal gonopode has a lateral position
corresponding to that of a thoracic appendage compared to the ventral
position of the normal gonopode in the same individual (Fig. 6). The
genital appendicular buds which normally fuse unilaterally to form the
gonopode, may not fuse. In this case, a supernumerary appendicular bud
appeared on the first abdominal segment (Fig. 7). Of the 114 individuals
manifesting anomalies in appendicular morphogenesis, 6 females had truncated
antennae (not shown).

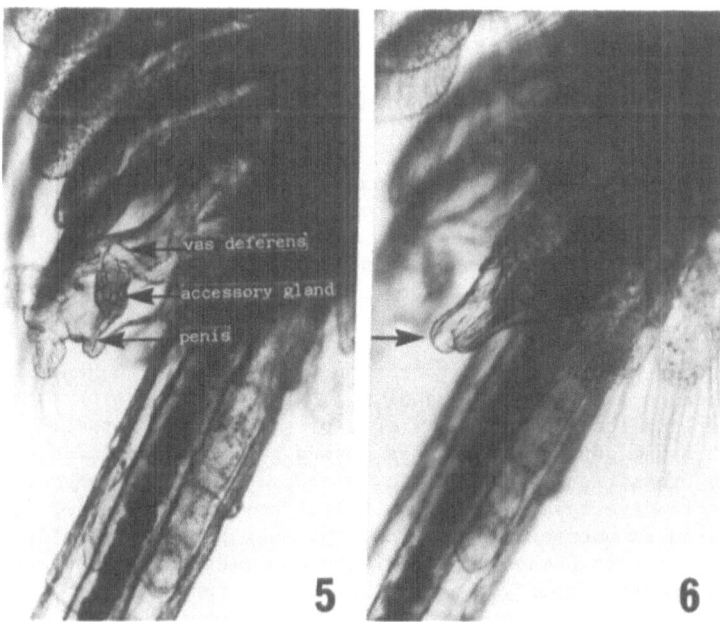

Fig. 5. Abnormal genital appendage resulting from
a 4 h heat shock to newly hatched nauplii.

Fig. 6. Normal genital appendage (arrow) assumes a
ventral position as opposed to the lateral
position of the abnormal genital appendage
in Fig. 5.

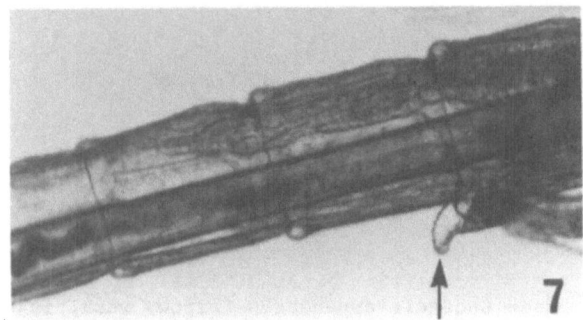

Fig. 7. Unilateral supernumerary appendicular (arrow)
bud on abdominal segment A_1 resulting from a
4 h heat shock to newly hatched nauplii.

In the last set of experiments, heat shocks at 40°C were delivered to
encysted embryos 1, 6 or 18 h after hydration of the cysts. Heat shocks
lasting 1, 2 or 4 h had no effect on subsequent development. Growth and
survival rates of animals born from shocked embryos were identical to those
of control animals. The microscopical examination of adults revealed no
morphological anomalies.

DISCUSSION

Data on the short-term effects at the molecular level of heat shocks
delivered to Artemia embryos during pre-emergence development and to 24 h
old larvae are available [12]. According to these data, the inhibition of
the synthesis of non-heat shock proteins is virtually complete after a 60
min shock at 40°C delivered to 24 h old larvae. During the present experiments, no larvae were capable of completing development after a 60 min
shock at 40°C, when the shock was postponed until 24 h after hatching
(result not shown). When delivered to newly hatched nauplii, a 60 min
shock at 40°C had no effect on subsequent development whereas heat shocks
lasting 2 to 4 h had long-term effects (Table 2). A total of 189 abnormal
individuals was observed. Although low, the number is sufficient to support
this discussion. Results indicate that during Artemia postembryonic development, decisions governing the transition from thoracic to genital segments are taken in newly hatched nauplii. Larvae had to go through 12
molts and complete the morphogenesis of eleven pairs of thoracic appendages
before the abnormal genital appendages formed. In Drosophila, decisions
governing the transition from thoracic to abdominal segments are taken
during embryonic development by the switch of homeotic genes [13] and heat
shocks delivered to embryos induce phenocopies of homeotic mutations [14,
15]. The existence of homeotic genes is not documented in Crustacea. Of
the 230 genes known to modify the development of Crustacea, 215 affect their
color and 15 affect other aspects of their development [16]. None of these
15 genes corresponds to a homeotic gene. The "swimmerette" mutation involves a fusion or alternatively a failure of separation of adjacent
appendages [11]. If the genital appendicular anomalies described in this
paper correspond to homeotic transformations, heat shocks delivered to new-
born nauplii induce phenocopies of homeotic mutations. Since homeotic
mutations are not yet discovered in Crustacea, the description of copies
of such mutations induced by heat shocks in Artemia may have preceded that
of the originals.

Knowing that thermal stress imposed upon young non-feeding larvae can induce long-term morphological anomalies, it becomes easier to conceive that nutritional stress imposed upon older feeding larvae may induce the morphogenesis of supernumerary appendages on otherwise apodous abdominal segments. The expression of genes coding for the morphogenesis of abdominal appendages which is normally repressed, may be derepressed, and as under-lined by Sang [7], this derepression corresponds to the production of homeoses. As with other homeoses, we have here an atavistic change. The problem of the morphogenesis of abdominal appendages raises, in original terms, the same problem as that raised by the induction of morphogenetic effects after heat shocks. That is, the problem of the molecular require-ments for gene activation.

The negative results obtained with heat shocks delivered to encysted embryos may be explained by a peculiarity of Artemia embryonic development. In resting cysts, embryonic development is arrested at the gastrula stage [17,18]. The pre-emergence period between hydration of the resting cysts and hatching of the nauplii corresponds to that of an intense morphogenesis leading to the formation of the three naupliar segments: the antennular, antennal and mandibular. The period of pre-emergence development is char-acterized by the absence of cellular multiplication and DNA biosynthesis [19,20]. Heat shocks become efficient at hatching when DNA biosynthesis resumes and the observed increasing number of nuclei per animal from 0 to 24 h after hatching [20] parallels the increased thermosensitivity of nauplii. Thus the effects of heat shocks may depend on the presence of replicating DNA.

Artemia offers unique opportunities to determine what kinds of de-cisions governing postembryonic development are taken before gastrulation since females can be subjected to thermal and nutritional stress with the unique advantage of a clear view, through the transparent cuticle, of the process of oocyte maturation along the ovaries. By stressing individually reared parthenogenetic females at precise moments of oogenesis, important questions on Crustacea development may be answered.

ACKNOWLEDGEMENT

Mrs. Anne-Marie Escaffre (Unité Elevage Larvaire, INRA, St Pée-sur-Nivelle, France) is greatly acknowledged for the photographs.

REFERENCES

1. A. Hernandorena, Obtention de morphogèneses appendiculaires abortives et surnuméraires chez Artemia salina (L.) (Crustacé Branchiopode) par carences alimentaires de base pyrimidique et de nucléotide purique, C.R. Acad. Sc. Paris 271:1406 (1970).
2. A. Hernandorena, Besoin alimentaire en acide adénylique, croissance et morphogénèse chez Artemia salina (L.) (Crustacé Branchiopode), Ann. Nutr. Alim. 28(2):65 (1974).
3. A. Hernandorena, Action antagoniste de nucléotides purique et pyrimid-que alimentaire sur la morphogenese appendiculaire d'Artemia salina (L.) (Crustacé Branchiopode), C.R. Acad. Sciences, Paris 284:1337 (1977).
4. A. Hernandorena, Relationships between purine and pyrimidine dietary requirements and Artemia salina morphogenesis, Comp. Biochem. Physiol. 62B:7 (1979).
5. A. Hernandorena, Action de la methionine alimentaire sur la morphogénèse appendiculaire d'Artemia. Hypothèse sur l'évolution du phénotype de ce crustacé, Reprod. Nutr. Dévelop. 25:75 (1985a).

6. A. Hernandorena, Purine-mediated phosphate influence on _Artemia_ morpho-genesis, _Reprod. Nutr. Develop._ 25:883 (1985b).
7. J. H. Sang, "Genetics and Development", Longman, London, New York (1984).
8. L. Provasoli and K. Shiraishi, Axenic cultivation of the brine shrimp _Artemia salina, Biol. Bull._ 117:347 (1959).
9. L. Provasoli, K. Shiraishi and J. R. Lance, Nutritional idiosyncrasies of _Artemia_ and _Tigriopus_ in monoxenic culture, _Ann. N.Y. Acad. Sci._ 77:250 (1959).
10. L. Provasoli and A. d'Agostino, Development of artificial media for _Artemia salina, Biol. Bull._ 136:434 (1969).
11. R. D. Squire and D. S. Grosch, "Swimmerette" a new sex-linked recessive mutant in the brine shrimp _Artemia salina_ L., _Biol. Bull._ 133:487 (1967).
12. D. Miller and A. G. McLennan, Changes in intracellular levels of Ap3A and Ap4A in cysts and larvae of _Artemia_ do not correlate with changes in protein synthesis after heat shocks, _Nucleic Acid. Res._ 14:6031 (1986).
13. E. B. Lewis, A gene complex controlling segmentation in _Drosophila, Nature_ 276:565 (1978).
14. A. H. Maas, Uber die Auslöbarkeit von temperatur-modifi-cationen wärend der Embryonalent-wieklung von _Drosophila melanogaster, Wilhelm's Roux Arch._ 143:315 (1948).
15. P. Santamaria, Heat shock induced phenocopies of dominant mutants of the bithorax complex in _Drosophila melanogaster, Molec. Gen. Genet._ 172:161 (1979).
16. D. Hedgecock, M. L. Tracey and K. Nelson, Genetics, in: "The Biology of Crustacea", Vol. 2, D. E. Blics ed., Academic Press, New York and London (1982).
17. J. Dutrieu, Observations biochimiques et physiologiques sur le développement d'_Artemia salina_ L., _Arch. Zool. Exp. Gen._ 99:1 (1960).
18. R. Benesch, Zur ontogenie und morphologie von _Artemia salina_ L., _Zool. Jb. Anat._ 86:307 (1969).
19. Y. H. Nakanishi, J. Iwasaki, T. Okigaki and H. Kato, Cytological studies of _Artemia salina_. I. Embryonic development without cell multiplication after the blastula stage in encysted dry eggs, _Annot. Zool. Jap._ 35:223 (1962).
20. C. S. Olson and J. S. Clegg, Cell division during development of _Artemia salina, Wilhelm's Roux Arch._ 184:1 (1978).

MORPHOLOGICAL STUDY OF THE OVARY OF ARTEMIA

G.R.J. Criel

Laboratorium voor Anatomie, Rijksuniversiteit Gent
K. L. Ledeganckstraat, 35
B-9000 Gent, Belgium

INTRODUCTION

Clegg and Conte reviewed the present knowledge of the ovaries in
Artemia, rightly stating that some problems such as the differentiation
between oocytes and nurse cells, the fate of the nurse cells, the origin
of the yolk and the presence of cell types other than the germinal cell line
are but partly solved[1]. An account of some morphological aspects of the
development of the eggs was given by Criel[2] however, ultrastructural
studies on ovogenesis are still separate[2-13] and therefore fail to give
a clear insight into the entire course of events. Hoping to solve some of
these problems, we undertook a systematic study of the ovaries throughout
the vitellogenic cycle with the light and electron microscope, applying
cytochemical and other techniques as required.

MATERIAL AND METHODS

Some of the Artemia used in this study belonged to the San Francisco
strain and were reared in our laboratory in a permanent culture. Others
were reared in the Artemia Reference Center and belonged to the Great Salt
Lake strain. Both of these strains belong to the Artemia franciscana
sibling species and no differences were found between them.

For paraffin embedding whole Artemia were fixed for 2 hours in AFA
(ethanol: formaldehyde: acetic acid; 20:75:5, V/V/V). Four to five micron
sections were cut and then stained with Mallory's triple stain. For Epon
embedding uteruses were partly dissected or ovaries were prelevated and
fixed in a glutaraldehyde-paraformaldehyde mixture[14] diluted 3:1 with
0.2M cacodylate buffer. Postfixation was done in 2% OsO4, in the same buffer
used for morphological study. For cytochemical studies postfixation was
omitted. Tissues were dehydrated in ethanol and embedded. Two micron
sections were stained with toluidine blue for light microscopical investigation. For electron microscopy ultrathin sections were stained with uranyl
acetate and lead citrate.

Acid phosphatases were localized by the technique of Gomori adapted
for electron microscopy[15,16]. Incubation with horseradish peroxidase was
done to demonstrate pinocytosis. The peroxidatic activity was detected by
the technique of Graham and Karnovsky[17]. Impregnation with lanthanum was

99

used to demonstrate the permeability of ovarian tissue to electron dense tracers[18].

Paraffin sections were examined with a Leitz Ortholux microscope, two micron Epon sections with a Wild phase-contrast microscope, and ultrathin sections with a Siemens Elmiskop 1A.

RESULTS

Gross Morphology

The ovaries are paired tubular structures which extend from the 11th thoracic segment through the genital segments and end in the abdominal segments. Macroscopically they show cyclic changes which can be observed with the naked eye and reflect the cyclic event of vitellogenesis occuring in some of the germ cells. Adult female Artemia ovulate approximately every 140 hours depending on oviparity, ovoviviparity and rearing conditions. During previtellogenesis the ovaries are translucent. At the beginning of vitellogenesis small opaque dots appear here and there in the ovary. The dots become more and more numerous and enlarge as vitellogenesis proceeds. The ovary gradually acquires a beaded aspect which develops into a knotty rod.

Light Microscopy

At the light microscopic level two fundamentally different cell types can be discerned: the somatic cells named after the "special somatic cells" of Lochhead and Lochhead[19] and the germ cells which include the oogonia and the cells which later differentiate into oocytes and nurse cells. In longitudinal sections (Fig. 1) the somatic cells are seen to form an interrupted layer of clear cells facing the gut and separating the clusters of germ cells. Some somatic cells show irregular dense inclusions.

On transverse sections the ovaries are club-shaped. The thicker end faces the gut and consists mainly of somatic cells. In the thinner end we find the oogonia which are the most lateral elements of the ovary. In previtellogenic ovaries we find oogonia over the entire length and in vitellogenic ovaries, they form islands between the vitellogenic oocytes. Maturing

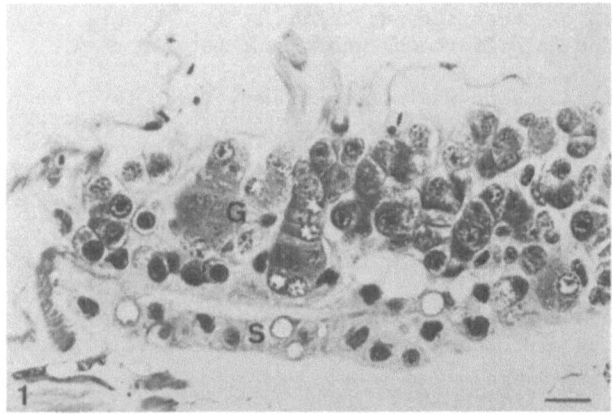

Fig. 1. Longitudinal section through paraffin-embedded previtellogenic ovary. S, somatic cells; G, germ cells. Bar, 20 μm.

germ cells lie in between the oogonia and the somatic cells and the more
mature they are the closer they lie to the gut. The development of the germ
cells can be divided into previtellogenesis and vitellogenesis phases.
Early previtellogenesis is marked by a generalized increase in cell size
and at the end of this stage the cluster of young oocytes transforms into
a ribbon. We presume each ribbon consists of a maximum of 32 oocytes as the
longest ribbon we were able to reconstruct consisted of 18 oocytes. Later
in previtellogenesis there is an increase in cytoplasmic volume and the
nuclei move from their central position to the convex side of the ribbon
which becomes curved. Now the differentiation between oocyte and nurse
cells begins (Fig. 2). One cell of each ribbon rounds up and soon protrudes
above its sister cells, its nucleus transforms into a germinal vesicle and
it becomes the oocyte. The other cells of the ribbon do not change markedly
and they are the nurse cells. In 2 μ Epon embedded sections a dense grain
is seen in the germinal vesicle (Fig. 13a) which is not visible in paraffin
embedded material.

The onset of vitellogenesis is marked by the appearance of a yolk
nucleus in the centre of the oocyte (Fig. 3) and the first yolk droplets
appear midway between this nucleus and the oocyte surface. Gradually the
whole oocyte fills with yolk. In paraffin sections it is difficult to
follow the evolution of the nurse cells as they are pressed against the
oocyte and their volume diminishes markedly. At the end of vitellogenesis
(Fig. 4) clear vacuoles which are accessory nuclei, appear on the oocyte
surface.

Ultrastructural Study of Yolk Products of Artemia

In mature eggs of Artemia four different yolk products are discerned:
electron dense protein yolk platelets, clear lipid droplets, vesicular yolk
bodies surrounded by an undulating membrane and covered with ribosomes and
intracisternal granules found in cisternae of the endoplasmic reticulum
(Fig. 5).

Electron Microscopy of Oogonia

These are the smallest cells of the germinal cell line (Fig. 6). The
oogonia are polygonal with a nucleus of 4 to 5 μ surrounded by a narrow
cytoplasmic rim. The diameter of the whole cell does not exceed 5 to 6 μ.
In the nucleus a barely ramified nucleolus lies in a nucleoplasm clouded
with small lumps of chromatin. The cytoplasm consists only of ribosomes
and some few mitochondria. Gonial mitoses most frequently occur at the end
of a vitellogenic cycle.

Electron Microscopy of Premeiotic Germ Cells

The mitoses by which the oocytes arise are incomplete with electron
dense rims arising at the level of the midbody of the mitotic spindle
(Fig. 7). All cells originating from the same oogonium are linked together
by a wavy membranous system which has evolved from the mitotic spindle and
is termed the fusome (Fig. 8). The germ cells are still polygonal at this
stage and their height does not exceed 12 μ. Their central, rather oval
nucleus has its largest axis no greater than 6 μ, a slightly ramified
nucleolus and small patches of heavily stained chromatin. The cytoplasm
consists of ribosomes and a few mitochondria.

Prophase of Meiosis: Formation of Synaptonemal Complexes

The onset of meiosis coincides with the disappearance of the fusomes
and the alignment of the oocytes in ribbons. During pachytene (Fig. 9a,b)
the characteristic appearance of the oocytes is that the whole group of germ

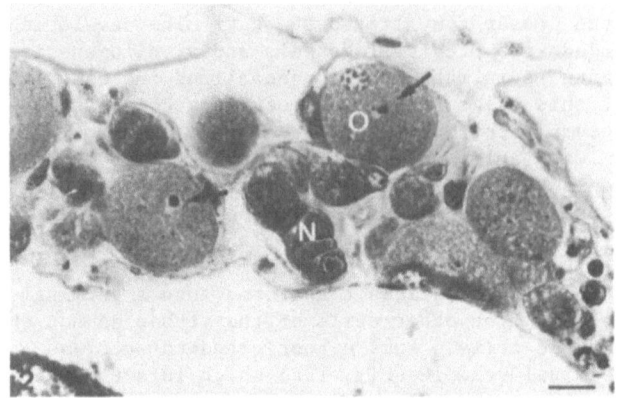

Fig. 2. Longitudinal section through a paraffin-embedded
ovary at the end of previtellogenesis: large
rounded oocytes (O) can be discerned from nurse
cells (N) constituting the rest of the germ cell
ribbon. The nucleus of the oocyte is clear in
comparison with that of the nurse cells. Arrows:
dense droplets typical for AFA-fixed material,
and probably representing clumped 'nuage' material.
Bar: 20 μm.

cells is less electron dense than neighbouring groups in other developmental
stages. The long axis of the oval nucleus reaches 10 μ and the nucleoplasm
is cleared by the disappearance of the chromatin. An electron dense
structure, the nucleolus or nucleolar organizer, adheres to the nuclear
membrane which also is the point of attachment of the synaptonemal complexes. In the cytoplasm, mitochondria are found in a condensed configura-
tion, a typical feature of this stage, and they are often dumb-bell shaped
or circular. Rarely, rough endoplasmic reticulum is found between the
ribosomes and polyribosomes which are the main cytoplasmic constituents.

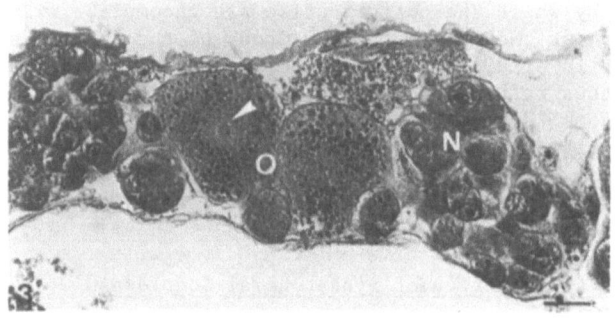

Fig. 3. Longitudinal section through paraffin-embedded
early vitellogenic ovary. O, oocyte; N, nurse
cell; arrowhead, yolk nucleus. Bar, 20 μm.

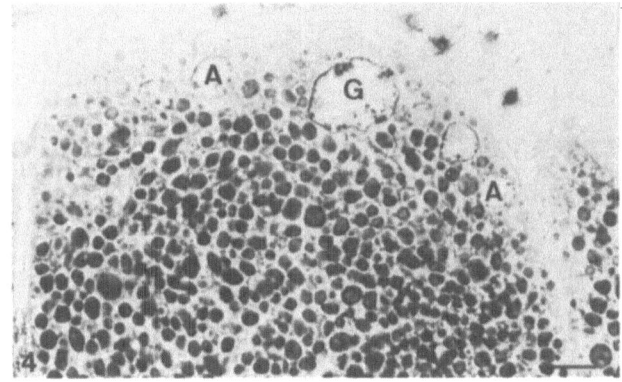

Fig. 4.   Section through paraffin-embedded late vitellogenic
          ovary. G, germinal vesicle; A, accessory nuclei.
          Bar, 10 μm.

Late Previtellogenesis: Increase in Volume of the Cytoplasm

During pachytene or immediately thereafter the fusomal material disappears and the oocytes align into a ribbon, linked together by cytoplasmic bridges delineated by electron dense rims and filled with cytoplasm. The nucleus enlarges to approximately 12 μ whereas the height of the cells increases to 25 μ. The shape of the oocytes changes from polygonal to a truncated cone (Fig. 10). In the nucleus the chomatin reappears in dense

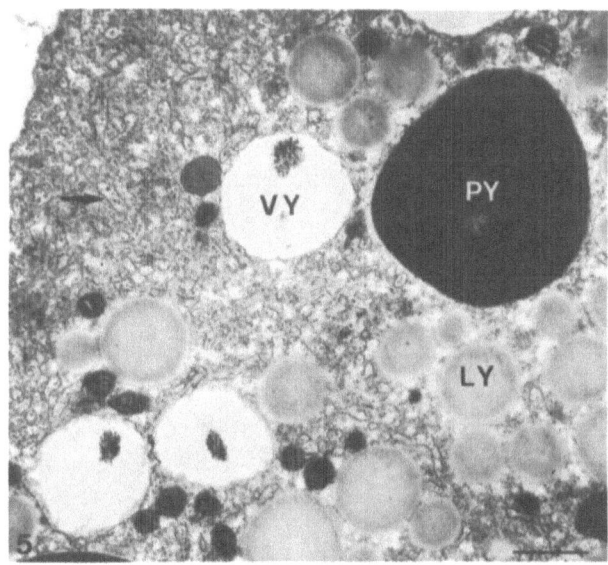

Fig. 5.   Yolk products of _Artemia_ shown in an unfertilized
          egg. PY, proteinaceous yolk platelets; LY, lipid
          yolk droplets; VY, vesicular yolk bodies; arrow,
          endoplasmic reticulum cisternae with granular
          material. Bar, 1 μm.

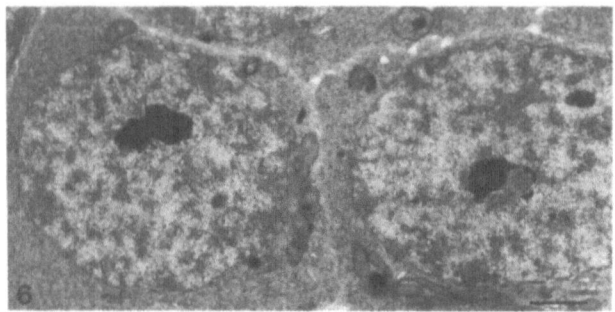

Fig. 6. Two oogonia each containing poorly ramified
         nucleolus in a nucleoplasm crowded with small
         chromatin blocks. Free ribosomes and scarce
         mitochondria are the only constituents of the
         cytoplasm. Bar, 1 µm.

fragments mostly attached to the nuclear membrane. The large central
nucleolus is highly ramified. An electron dense material is thought to
diffuse through the nuclear pores (Fig. 11) and is seen in the cytoplasm
as "nuage" material, often associated with annulate lamellae which develop
at this stage (Fig. 12). Ribosomes still constitute the mass of the cytoplasm, more and more mitochondria are found, cisternae of rough endoplasmic
reticulum develop and ·some intracisternal granules are formed; dictyosomes
are still scarce. Some small lipid and protein droplets appear together
with some vesicular yolk bodies. Their diameter never exceeds 2 µ.

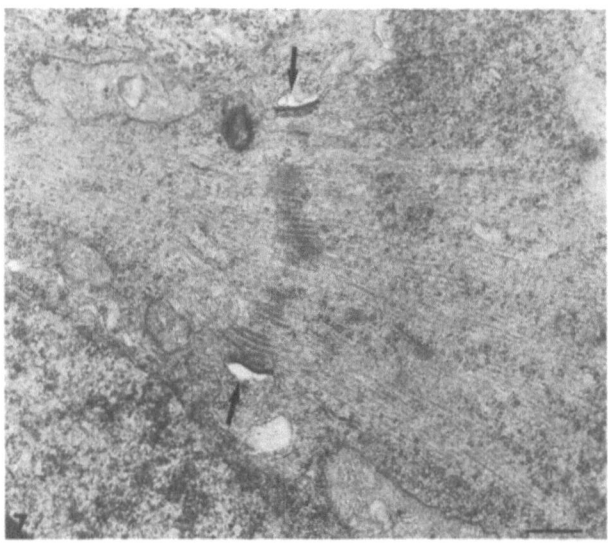

Fig. 7. Electron dense bridge rims (arrows) form at the
         midbody of the mitotic spindle. Bar, 0.5 µm.

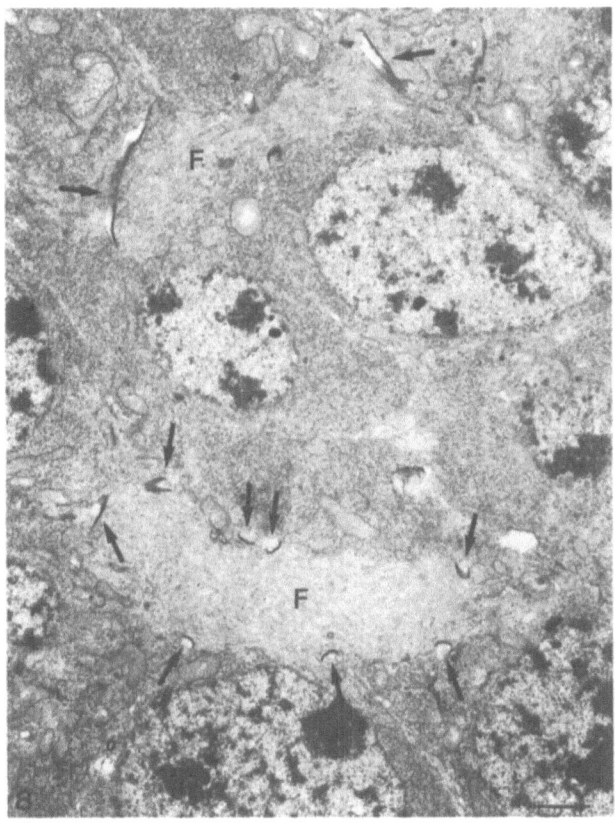

Fig. 8.   Two fusomes (F) between young oocytes.   Arrows,
          bridge rims.   Bar, 1 µm.

End of Previtellogenesis: Differentiation of the Oocyte and the Nurse Cells

The cell of the ribbon which is the oocyte rounds up, becomes covered
with short villi and soon protrudes above its sister cells.   Its nucleus
evolves into a germinal vesicle, the nuclear membrane becomes wavy and the
nucleoplasm clears completely except for some dense structures which were
unidentified until now (Fig. 13a,b,c).   Accessory nuclei develop between
the oocyte surface and the germinal vesicle.   Their surrounding membrane
shows nuclear pores and they often contain a small nucleoid (Fig. 14).
Cytoplasmic organelles develop further with small granules forming in some
endoplasmic reticulum cisternae which often concentrate in the neighbourhood
of dictyosomes (Fig. 15).   The lipid droplets and the clear vesicular yolk
bodies increase in size to approximately 10 µ, but the small protein droplets remain unchanged.   A yolk nucleus appears in the centre of the oocyte,
surrounded by "nuage" material and dictyosomes (Fig. 16).   It is composed
of small vesicles, multivesicular and dense bodies (Fig. 17).

The remaining cells of the ribbon are the nurse cells which become
lenticular as they seem to be pushed against the oocyte.   This gives an
impression of engulfing, but interdigitating plasma membranes are seen on
both sides of the cytoplasmic bridges (Fig. 18a,b).   No marked differences
between the cytoplasm of oocyte and nurse cells were found.   The changes

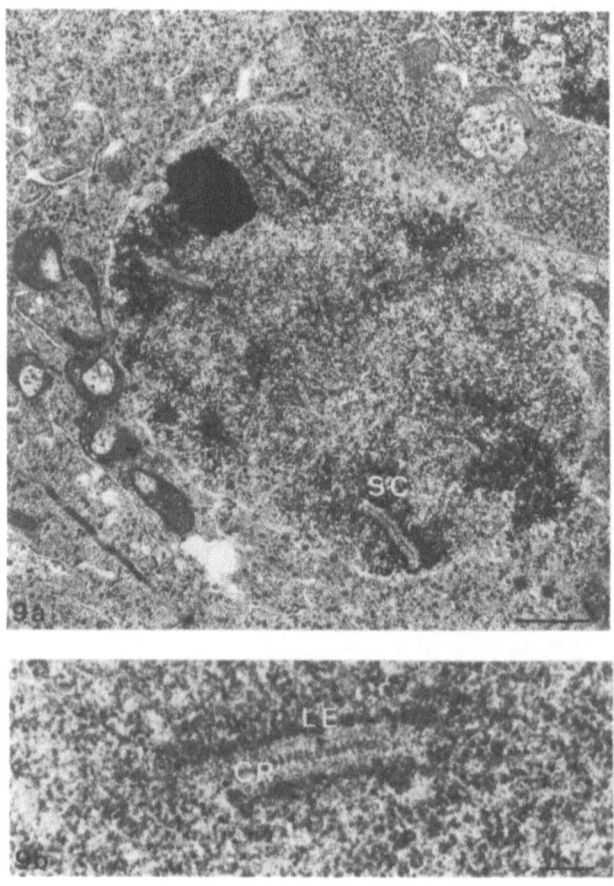

Fig. 9a. Nucleus of an oocyte in pachytene. Synaptonemal
          complexes (SC) surrounded by chromatin. Bar,
          1 μm
Fig. 9b. Enlargement of synaptonemal complex. LE,
          lateral element; CR, ladder structure of central
          region. Bar, 200 nm.

of the nurse cell nucleus and the relationship between the lenticular nurse
cells found at the onset of vitellogenesis and the rounded polyploid nurse
cells found at the end of vitellogenesis remain to be elucidated.

Early Vitellogenesis: Endogenous, Yolk Formation

     The first accumulation of yolk products is seen in a layer midway
between the yolk nucleus and the oocyte surface (Fig. 17). The layer is
composed of lipid yolk droplets which arise without any contact with other
organelles, clear vesicular yolk bodies and electrondense protein yolk
platelets with a vesiculated periphery. The mechanism involved in the increase in size of these yolk platelets is probably fusion, as is suggested
by the irregular contour of many immature granules (Fig. 19). As the yolk
granules mature they loose their vesiculated periphery and a small clear
rim separates the dense centre from the surrounding membrane. More and more

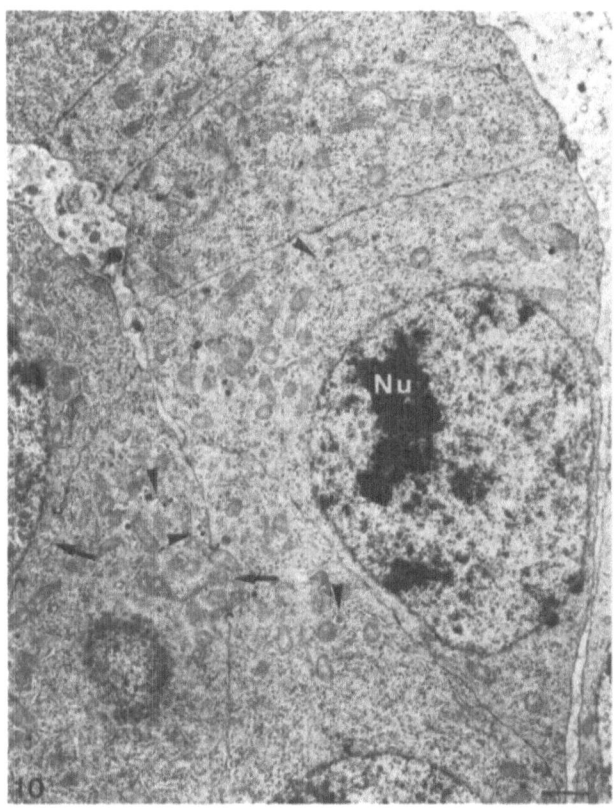

Fig. 10. Reorganization of the oocytes into ribbons after
pachytene. Nu, highly ramified nucleolus; arrowheads,
intracisternal granules of the endoplasmic reticulum;
arrows, cytoplasmic bridges. Bar, 1 μm.

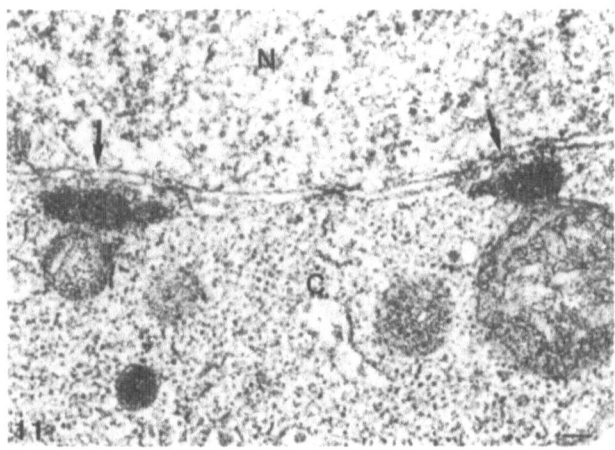

Fig. 11. Localization of "nuage" material near the nuclear pores
(arrows). Association here with mitochondria is fortuitous.
N, nucleus; C, cytoplasm. Bar, 100 nm.

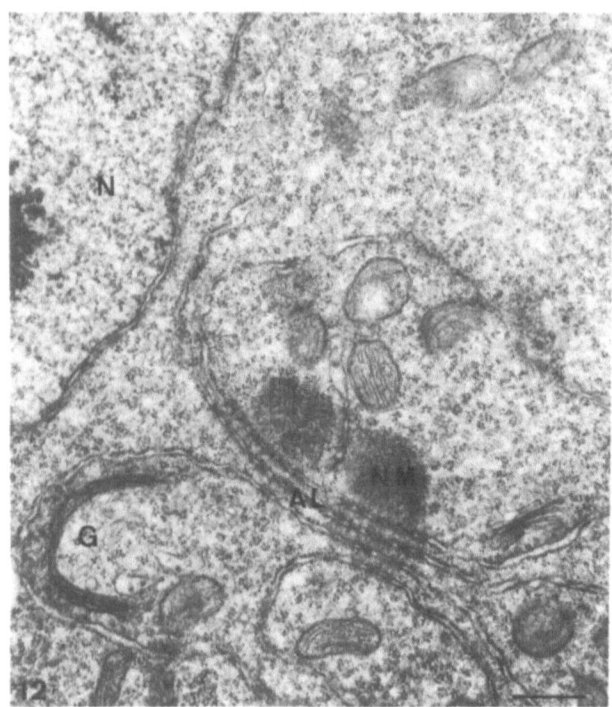

Fig. 12. "Nuage" material (NM) associated with
annulate lamellae (AL) in late previtellogenic oocyte. The cytoplasm is full of
polyribosomes. N, nucleus; G, Golgi
apparatus. Bar, 0.5 μm.

granules form within endoplasmic reticulum cisternae and concentrate around
the dictyosomes. Although the granules are well developed they do not show
an obvious secretory activity. Vesicles containing several intracisternal
granules pinch off from the ends of the cisternae and lose their ribosomes.
The relationship of the vesicles with the yolk platelets is not evident.

Late Vitellogenesis: Exogenous, Yolk Formation

Forty to sixty hours after the onset of a vitellogenic cycle an intense
pinocytotic activity is observed on the side of the oocyte directed to the
ovarial surface (Fig. 20). It is also in this region that most of the
accessory nuclei occur. Several stages in the uptake of material are seen.
The coated vesicles loose their coat and fuse to form larger and larger yolk
spheres while migrating centripetally. The yolk platelets are morphologically indistinguishable from those formed by endogenous synthesis. To
analyse the pinocytotic activity the ovaries were incubated with horseradish peroxidase. This study has shown that the peroxidase is actively
taken up even at incubation times as short as 5 minutes. It is not transported to the yolk platelets, however, but accumulates in clear vesicles
(Fig. 21). To exclude artifacts due to possible invaginations of the oocyte
surface we impregnated the ovaries with lanthanum which adheres to the
oocyte surface but never penetrates into it.

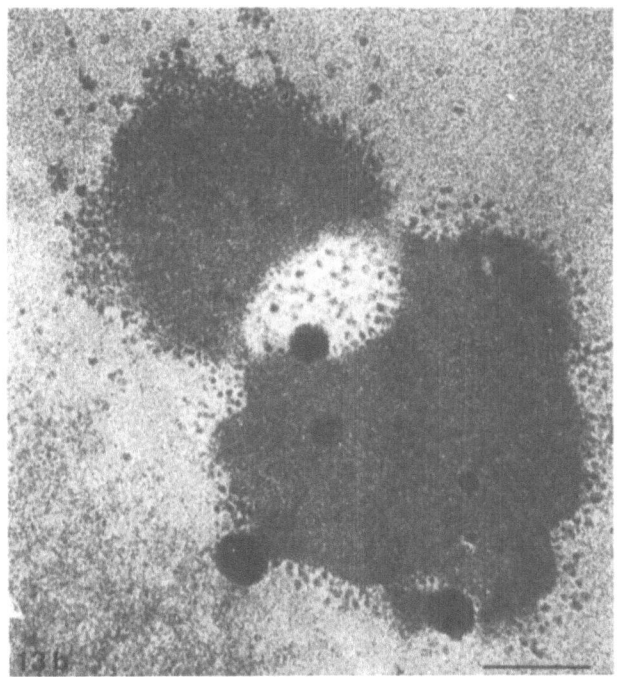

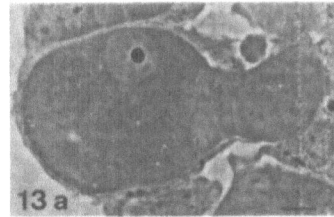 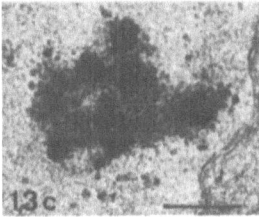

Fig. 13.   Differentiation between oocytes and nurse
           cells.   a, 2 μ epon-embedded section through an
           oocyte ribbon at the end of previtellogenesis.
           The nucleus of the oocytes evolves to a germ-
           inal vesicle containing a very dense round
           structure.   Bar, 10 μm.   b, electron micrograph
           of a section through such a dense structure.
           Morphology alone cannot identify this structure
           as a nucleolus or a karyosphere.   Bar, 0.5 μm.
           c, unidentified cluster against the wall of the
           germinal vesicle.   Bar, 0.5 μm.

    Important changes also occur in the nurse cells during vitellogenesis.
They enlarge and become almost circular acquiring a large nucleus with
several nucleoli and many small chromatin blocks (Fig. 22).   "Nuage" material may diffuse through the nuclear pores.   Their scarce cytoplasm contains
many mitochondria and lipid droplets often associated one with another.
The nurse cells are still linked with each other and with the oocyte by
cytoplasmic bridges (Fig. 23).   At the end of vitellogenesis the nurse cells
separate from the oocytes, are phagocytized by somatic cells and degenerate.
After ovulation large zones of somatic cells filled with degenerating

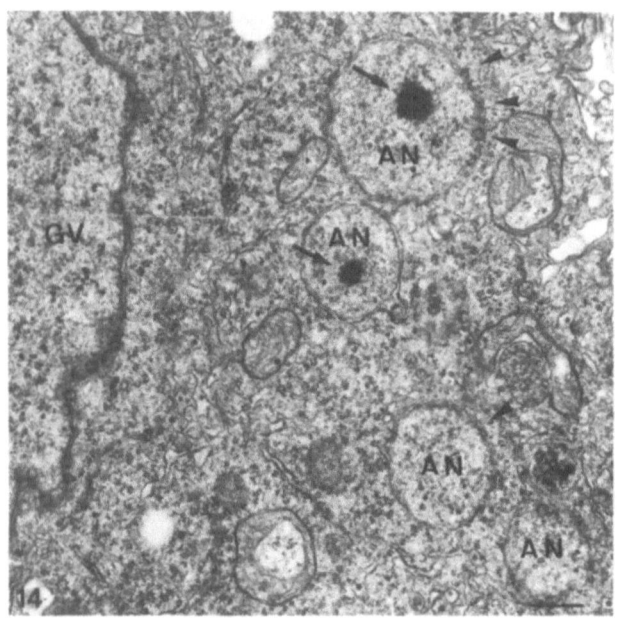

Fig. 14. Accessory nuclei (AN) originate between the germinal
vesicle and the oocyte surface at the onset of vitellogenesis. Arrows point to the nucleoids in the accessory nuclei; arrowheads point to their nuclear pores.
Bar, 0.5 μm.

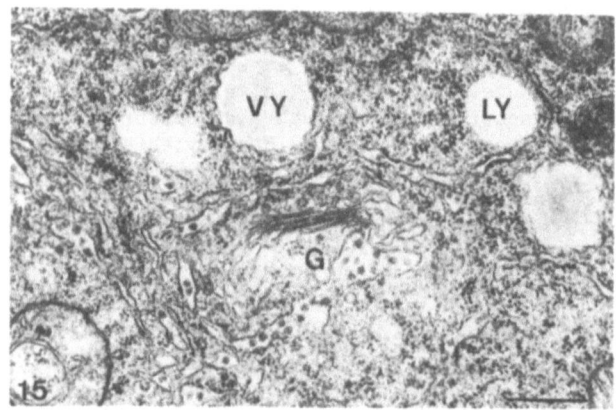

Fig. 15. Cytoplasm at the onset of vitellogenesis. Endoplasmic
reticulum cisternae containing intracisternal granules
in the neighbourhood of a Golgi apparatus (G). VY,
vesicular yolk body; LY, lipid yolk droplet. Bar,
0.5 μm.

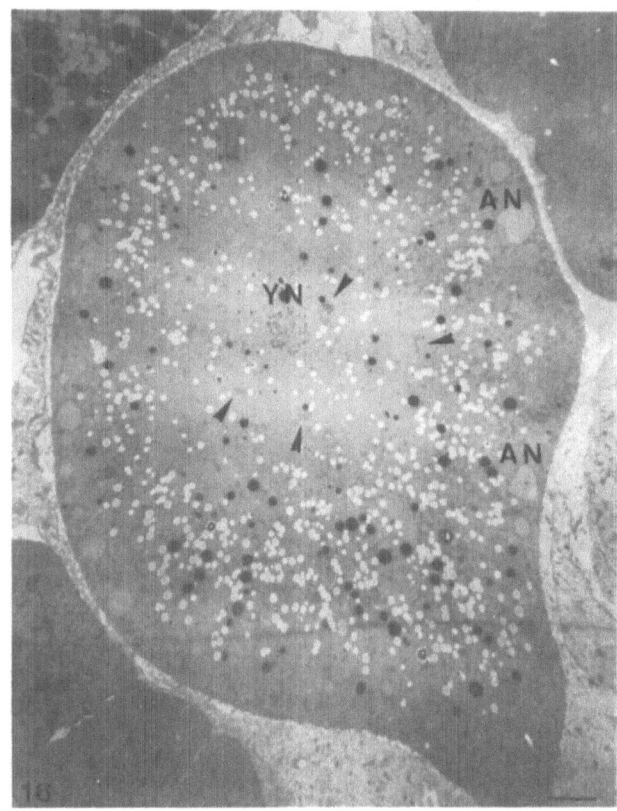

Fig. 16.   Very low magnification of an oocyte in early
            vitellogenesis. The yolk nucleus (YN) lies in
            the centre of the oocyte. The excentric germinal
            vesicle is not in the plane of this section. Many
            accessory nuclei (AN) lie at the periphery of the
            oocyte. Yolk products accumulate midway between
            the yolk nucleus and the oocyte surface. The
            clear granules are the lipid yolk droplets and
            the vesicular yolk bodies which cannot be discerned
            at this low magnification. The dense granules
            are the protein yolk platelets. Arrowheads point
            to the "nuage" material surrounding the yolk
            nucleus. Bar, 5 µm.

polyploid nurse cells are found in some regions of the ovaries (Fig. 24,25).

Electron Microscopy of the Somatic Cells

     The somatic cells form a continuous layer under the ovarian basement
membrane and separate the groups of germ cells and oocyte-nurse cell complexes. Their clear cytoplasm is in contrast with the electron dense germ
cells (Fig. 26). The nucleus is lobated with a clear nucleoplasm and scarce
chromatin and they show no secretory activity. A striking feature of the
somatic cells is an extensive smooth tubular system which is particularly

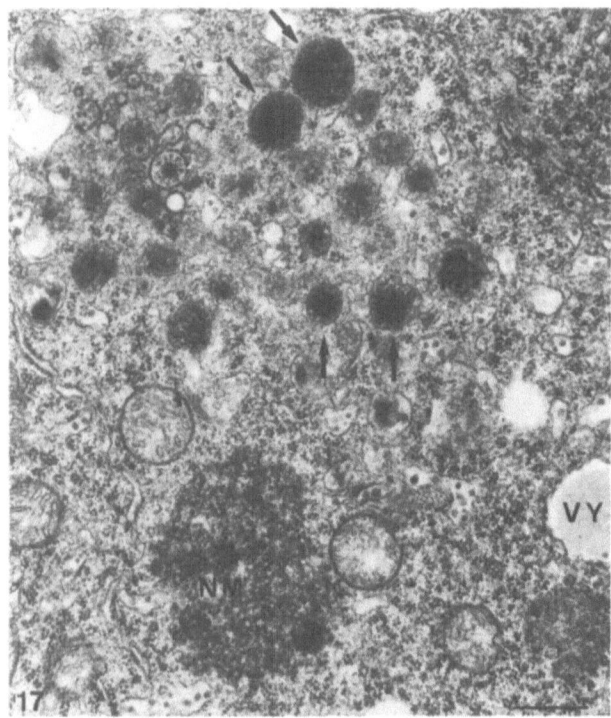

Fig. 17. A higher magnification of the yolk nucleus
shows its multivesicular bodies (arrows) and
other vesicles. NM, "nuage" material; VY,
vesicular yolk body. Bar, 0.5 μm.

developed in the most superficial cells (Fig. 27). The superficial cells
show a fine fibrillar network which seems to sustain the protuberances on
the surface of previtellogenic ovaries. Impregnation with lanthanum shows
that the tubular system is formed by deep infoldings of the plasma membrane
and horseradish peroxidase penetrates into the tubules. Both tracers are
found to a lesser extent in the somatic cells separating the germinal cells
(Fig. 28a,b). A broad capacity for phagocytosis is another characteristic
of the somatic cells as degenerating nurse cells are found embedded in
their cytoplasm. The degenerating cells become more and more electron dense
and are gradually broken down into smaller and smaller residual bodies
(Fig. 29). Acid phosphatases were detected in the small residual bodies
(Fig. 30). Somatic cells with large residual bodies were found in the
medial zone of the ovary, while the somatic cells with small residual bodies
occur in the lateral regions between the oogonia.

DISCUSSION

An accurate description of the sequence of macroscopical events of
oogenesis is given by Metalli and Ballardin[20]. Their description of the
cytological events however is incomplete as are most studies on oogenesis
in Artemia up to now. The first studies on the ovary of Artemia mainly
concerned the differentiation of oocytes and nurse cells[21-24]. Later

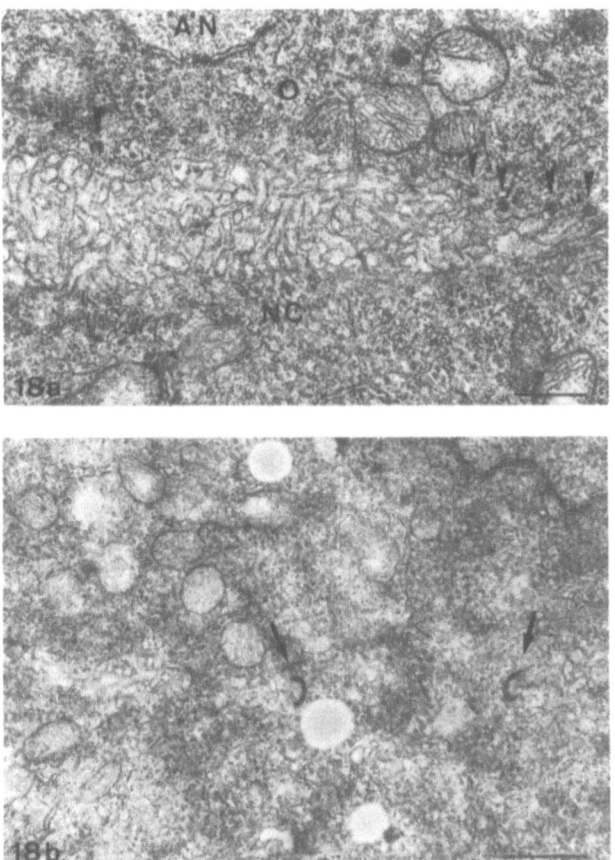

Fig. 18. Zone of contact between oocyte and nurse cells
in early vitellogenesis. a, highly inter-
digitated cell surfaces. Coated vesicles are
seen on the oocyte plasma membrane (arrowheads)
AN, accessory nucleus; O, oocyte; NC, nurse
cell. Bar, 0.5 μm. b, dense rims (arrows)
border the cytoplasmic bridge between oocyte
and nurse cell. Bar, 1 μm.

studies directed attention to the nuclear events of oogenesis[20,25-27] or
to some of its cytochemical aspects[28-36]. Ultrastructural studies hitherto only give fragmentary information[2-13]. For example, relevant facts
by Lochhead and Lochhead[19] were unfortunately limited to an abstract.
Linder[37] made a thorough study of ovogenesis in Chirocephalus, a species
closely related to Artemia while Munuswamy and Subramoniam[38,39] studied
oogenesis at the light and ultrastructural level in another related species,
Streptocephalus. In this discussion we try to associate phenomena observed
in Artemia with those in other animal species in the hope of obtaining a
better insight into the events of oogenesis. Oogonia are found over almost
the entire later-ventral length of previtellogenic ovaries as in Branchipus
[21]. Later in vitellogenesis groups of oogonia alternate with groups of
developing oocytes. It is possible that the oogonia arise from stem
cells[27] located at the posterior end of the ovary where Nitsche[23] and

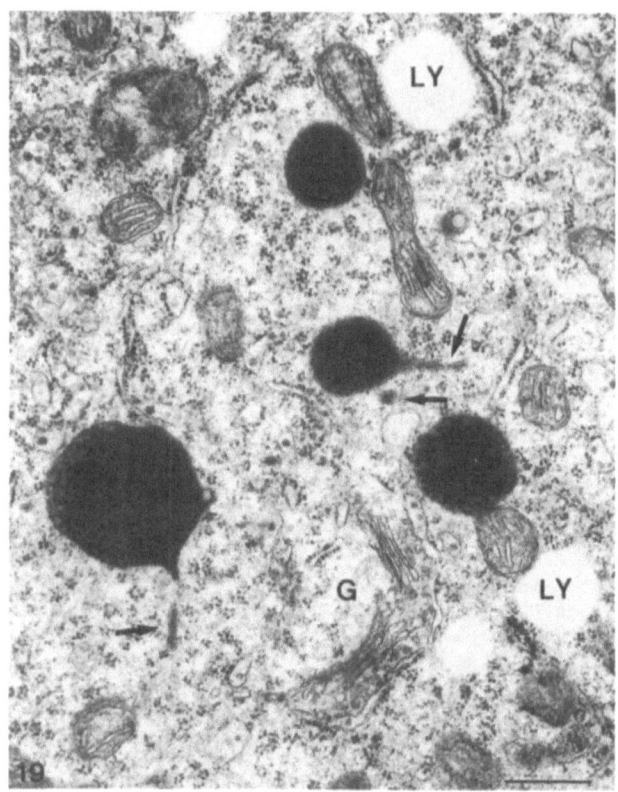

Fig. 19. Onset of endogenous yolk synthesis. Immature
yolk platelets show a vesiculated periphery.
Fusion with dense vesicles and tubules (arrows)
is seen. An endoplasmic reticulum cisterna
with intracisternal granules lays at the
forming face of a Golgi apparatus (G). LY,
lipid yolk droplet. Bar, 0.5 µm.

Zograf[40] describe "indifferent cells" in Branchipus and in Chirocephalus
respectively.

Oogenesis starts with the mitotic multiplication of oogonia. The cell
divisions are incomplete giving rise to the cytoplasmic bridges which were
first described in Artemia by Anteunis et al.[6-8]. Transitory cytoplasmic
bridges are described in vertebrate male and female germ cells[41-49] and
in Ascaris[50]. In insects with a polytrophic ovary[51] some new descriptions have appeared[52-56] as have some for Notostraca[57] and an annelid
[58]. The cytoplasmic bridges persist throughout vitellogenesis and are
thought to provide an intracellular transport route for nurse cell products
to the oocyte. As stated earlier the formation of oocyte-nurse cell complexes was one of the first known aspects of phyllopod and cladoceran
oogenesis[40].

The appearance of synaptonemal complexes is an obvious sign of the
onset of meiosis. In Artemia, synaptonemal complexes are found in all the
oocytes surrounding the fusome, which proves that unlike Notostraca[57] the

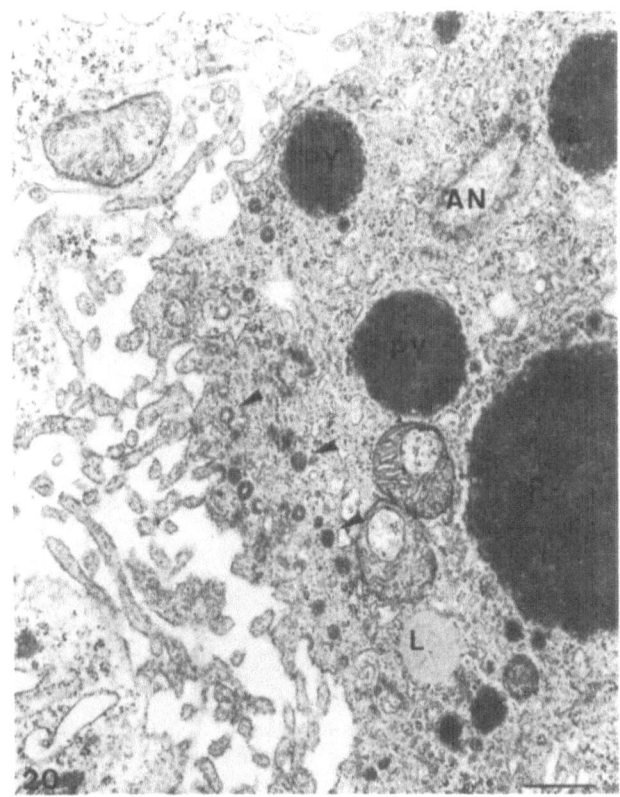

Fig. 20. Exogenous yolk synthesis. Coated vesicles (arrowheads)
          are pinched off between the villi and are transported
          to the protein yolk platelets (PY). AN, tangential
          section through an accessory nucleus; LY, lipid yolk
          droplets. Bar, 0.5 μm.

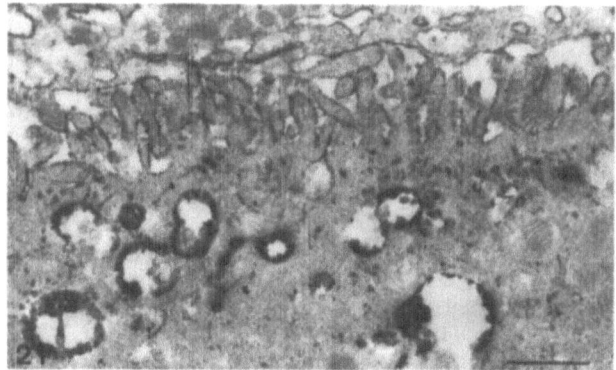

Fig. 21. An unstained section showing the uptake of horseradish
          peroxidase and its transport to clear vesicles. Bar,
          1 μm.

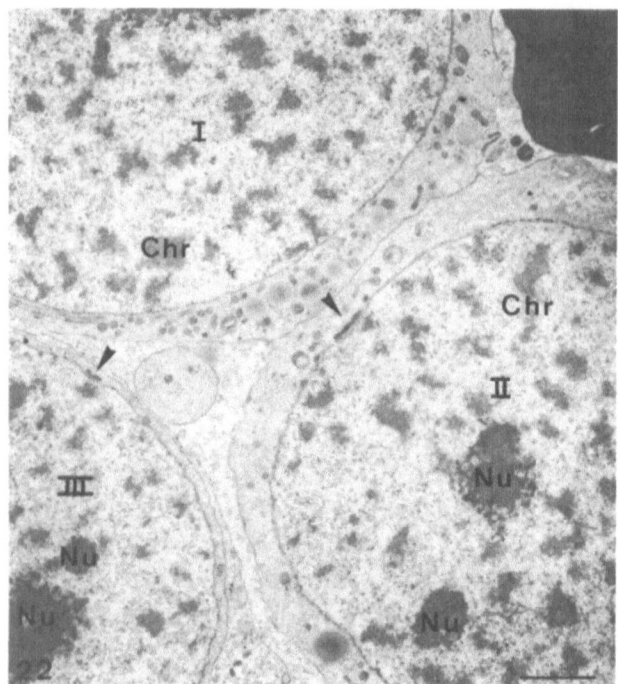

Fig. 22.   Parts of three nurse cells (I,II,III). Nurse
cell I and II belong to the same ribbon. The
large nucleus contains several nucleoli (Nu)
and many clumps of chromatin (Chr). Diffusion
of "nuage" material continues (arrowheads).
Note the many lipid yolk droplets and mitochondria. Bar, 2 μm.

differentiation between oocytes and nurse cells has not yet taken place at
that moment. The increase in volume of cytoplasm which starts after the
pachytene is common to oogenesis in most species. In Artemia it occurs in
the oocytes as well as in the nurse cells. Another regular phenomenon in
oogenesis occurring at this stage and in the whole germ cell ribbon is
the passage of nuclear material through the nuclear pores into the cytoplasm. In Artemia this was first described by Anteunis et al.[9]. In the
oocytes the diffusion ends with the formation of the germinal vesicle,
whereas in the nurse cells it lasts till the end of vitellogenesis. The
diffusing material is found throughout the cytoplasm as "nuage" material
and it is generally accepted to be of nucleolar origin. In Artemia this is
not proven by studies at the ultrastructural level but we presume that the
dense granules pointed out in Fig. 2 represent clumped "nuage" material and
correspond to the "corps enigmatiques" which were described and shown to
contain RNA by Fautrez-Firlefyn[30].

Annulate lamellae and the fibrogranular bodies, which often form in
association with "nuage" material, are other common features extensively
discussed in the review by Kessel[59]. Despite their widespread occurrence
only hypotheses have been proposed as to their possible function[60-64].
Anteunis[3] was the first to report annulate lamellae in Artemia.

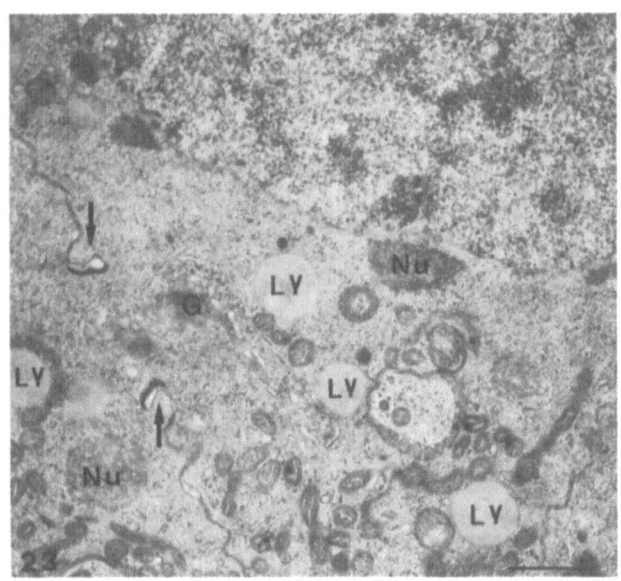

Fig. 23. Higher magnification of nurse cell cytoplasm. Cell
organelles pass through the cytoplasmic bridge
bordered by a dense bridge rim (arrows). G, Golgi
apparatus; LY, lipid yolk droplet; Nu, "nuage" material. Bar, 1 μm.

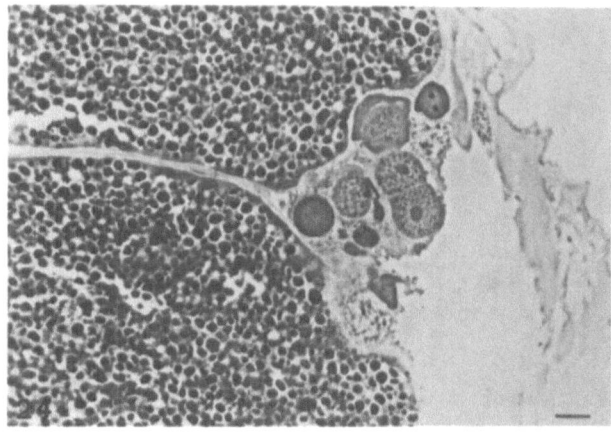

Fig. 24. Phase contrast micrograph of a 2 μ section through
oocytes at the end of vitellogenesis. The nurse
cells are phagocytized by somatic cells. Bar, 20 μm.

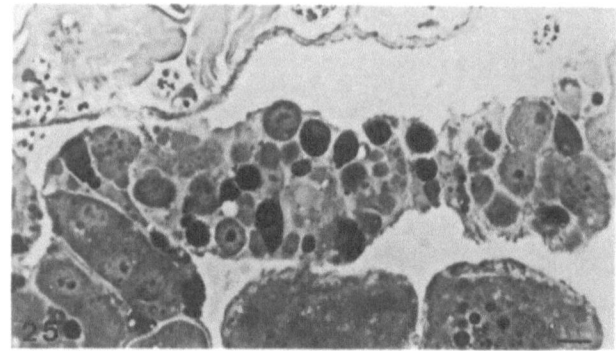

Fig. 25. Phase contrast micrograph of the central part of an ovary shortly after ovulation. Phagocytized nurse cells can still be recognized in the somatic cells. Bar, 20 μm.

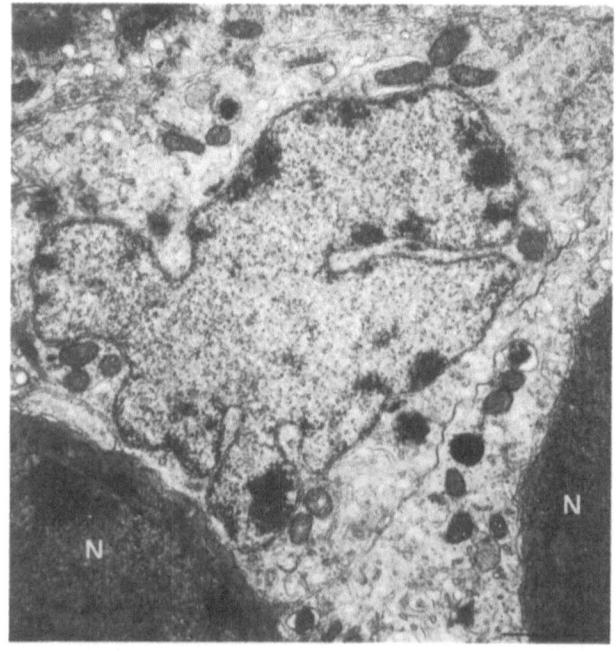

Fig. 26. Clear somatic cell between two nurse cells (NC). Note the clear, lobated nucleus and cytoplasm. Bar, 1 μm.

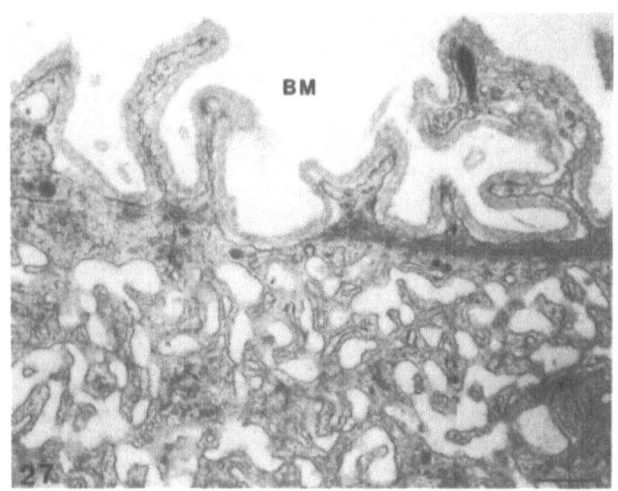

Fig. 27. Somatic cell at the surface of the ovary.
Note the extended tubular system and the sub-
surface filaments. BM, Basement Membrane.
Bar, 0.5 μm.

In most crustaceans, dense granules are formed in cisternae of the
rough endoplasmic reticulum in a way similar to that observed in Artemia
[65-81]. The significance of the contribution of these granules to the
yolk platelets is questioned however, as is the role of the Golgi apparatus
in some cases[67,72]. Dhainaut and Deleersnijder[82] and Zerbib[83] suggest
that the glycoprotein moiety of yolk is elaborated by the endoplasmic reticulum while lipoproteid yolk results from material incorporated into the
oocyte by micropinocytosis.

The end of previtellogenesis is marked by the transformation of the
oocyte nucleus into a germinal vesicle. As the nucleus of the nurse cells
apparently does not change, this event also marks the differentiation
between oocyte and nurse cells. This has been described in Artemia by
Iwasaki[26] and was found by Linder in Chirocephalus[37]. The dense nuclear
structure shown in Fig. 13 a and b may be a very dense nucleolus but could
also represent chromosomes condensed into a "karyosphere" as described by
Bier et al. in meroistic insect ovaries[84]. However, the electron micrographs of the karyosphere in Anopheles[85] and Chrysopa perla[86] show a
fibrous capsule around the chromosomes which differs completely from our
images. The disappearance of Feulgen positive material demonstrated by
Fautrez-Firlefyn[29] nonetheless supports the idea of the formation of a
karyosphere. Rue and Gontcharoff[87] describe micronucleoli under the
membrane of the germinal vesicle in Lineus ruber. These images too differ
from ours shown in Fig. 13c. A definite answer on the nature of the structures we describe in the germinal vesicle requires further investigations.

The yolk nucleus, which appears in the centre of the oocyte at the
start of endogenous yolk formation, is an assemblage of multivesicular
bodies, dense bodies and other vesicles[5]. Although yolk nuclei have been
described in various species[59], as far as we know no other types have
been reported since, except by Mothes-Wagner and Seitz[88] who call the previously described fibrogranular bodies in the spider mite yolk bodies.

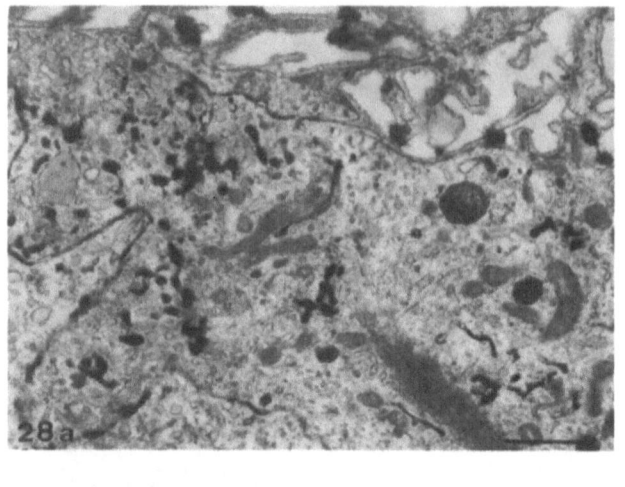

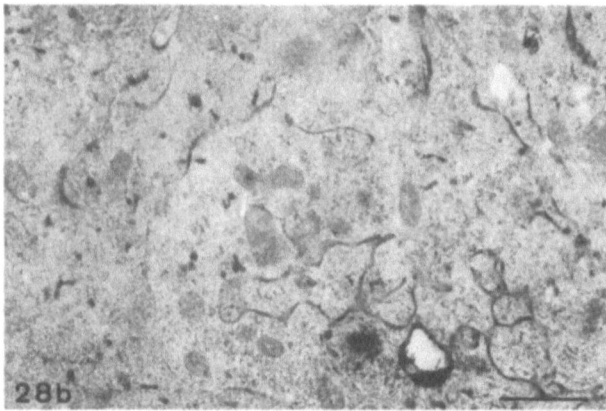

Fig. 28. Lanthanum impregnation (a) and horseradish
          peroxidase incubation (b) both pass through the
          tubular system before they reach the oocytes.
          Bars, 1 μm.

None of these nuclei can be compared to the one in Artemia. Although multivesicular bodies, are often seen to play a role in endogenous yolk formation
[58,89-96] this has never been mentioned in crustaceans nor in insects. As
the early yolk products of Artemia form a layer midway between the central
yolk nucleus and the oocyte surface, a major influence of the yolk nucleus
on endogenous yolk formation can be questioned. The contribution of the
granules contained in the endoplasmic reticulum cisternae, and of the
dictyosomes they surround, to the endogenous yolk is likewise questionable.
Some rough endoplasmic reticulum cisternae with intracisternal granules
persist till the fertilization of the mature egg and disappear with the
formation of the vitelline membrane in Streptocephalus[97] and Artemia[98].

    The exogenous yolk product that is taken up by pinocytosis is probably
lipovitellin. Its presence in the blood of vitellogenic females was demonstrated by Van Beek et al.[99] and with immunocytochemical techniques they
also showed that it was synthesized in the fat storing cells. By means of
an incubation with horseradish peroxidase we demonstrated the possibility

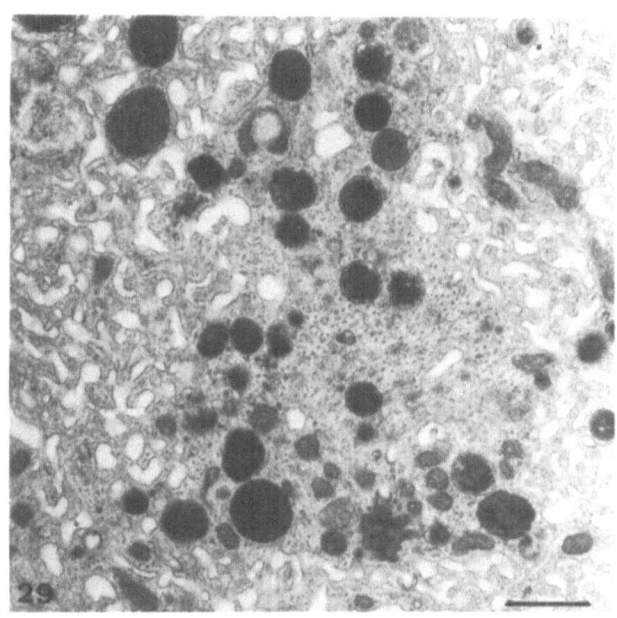

Fig. 29. Part of a somatic cell filled with dense granules
          which probably are remnants of phagocytized
          nurse cells. Bar, 1 μm.

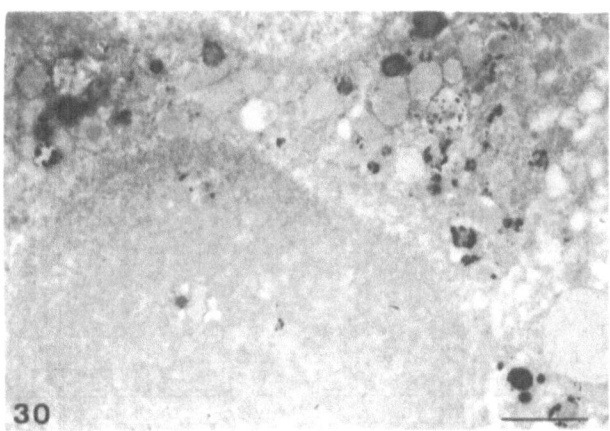

Fig. 30. Electron micrograph of an unstained section
          showing acid phosphatase in the dense granules
          of the somatic cells. Bar, 1 μm.

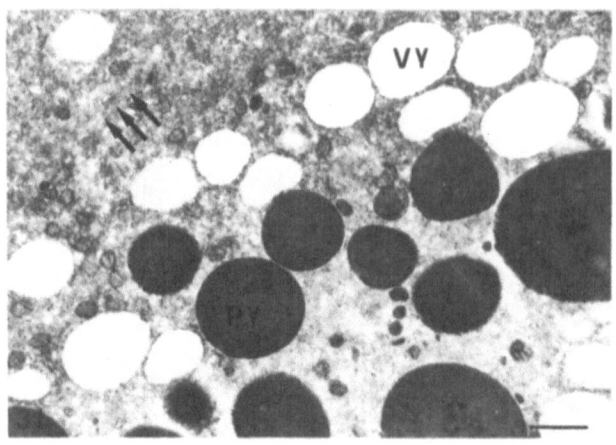

Fig. 31. Electron micrograph of a section through a
slightly centrifuged fertilized egg. Vesicular
yolk bodies (VY) form a layer at the top of the
protein yolk platelets (PY). Arrows point to
the centrifugal direction. Bar, 1 μm.

of protein uptake. However, as in crayfish[100] the peroxidase is transported to vacuoles and not to the yolk platelets. An explanation of this
phenomenon was proposed by Raikhel and Lea[101].

The rim of yolk products in between the yolk nucleus and the oocyte
surface seems to indicate that the mature yolk products arise there by the
confluence of endogenous material moving from the oocyte centre to the
periphery, and exogenous material moving in the opposite direction, which
indicates why no difference is found between endogenous and exogenous yolk
products.

The vesicular yolk bodies also arise within cisternae of the endoplasmic reticulum. In the mature egg, we find them near the surface but
they do not represent cortical granules, however, since we still find them
after the second egg division. In slightly centrifuged eggs we find the
vesicular yolk bodies under the lipid droplets and the cytoplasm on top of
the yolk platelets (Fig. 31). Their chemical composition and function are
unknown. They differ from the peroxidase containing bodies found under the
surface of mature oocytes[102,103].

The only germ cells where accessory nuclei are hitherto described
ultrastructurally are those of the hymenopterans Bombus terrestris[104],
Ophion luteus and Apanteles glomeratus[105], Habrobracon juglandis[106] and
Formica[53], the dipterans Phryne cincta and P. fenesralis[107] and Artemia
[8]. Their occurrence in Hymenoptera however is known since Blochmann[108].
In Artemia, Fautrez-Firlefyn[29] and Anteunis et al.[6-8] presumed the
accessary nuclei to be remnants of engulfed nurse cells. In our study
however it appeared that they arise in the oocyte early in vitellogenesis
and can thus be compared to the accessory nuclei of the Hymenoptera. All
authors agree that they are Feulgen negative, their granular inclusions
contain some RNA and they contain some protein. Although they are not sites
of intensive synthetic activity of nucleic acids[107] their occurrence in
Artemia in the same regions as where pinocytosis occurs is striking. As in

122

Habracon[106] the accessory nuclei of Artemia breakdown at the time the oocyte resumes meiosis and liberate their contents into the ooplasm.

At the onset of vitellogenesis the nearest nurse cells are pressed against the growing oocyte, and therefore in many light microscopical studies they were thought to be engulfed by the growing oocyte. At the ultrastructural level we never observed any engulfing, and neither did Cassidy and King in Habracon[106]. No yolk products are formed in the nurse cell, except for the presence of some lipid droplets, but these could also point to an incipient degeneration. In species such as Campodea[67] and Ophiotrocha[91] nurse cells synthesize yolk bodies which are transported to the oocyte. As in insects[51], the nurse cells of Artemia probably supply the oocyte mainly with cytoplasm and ribosomes.

Polyploidy of the nurse cells of Artemia was first described by Barigozzi[25] and later by Lison and Fautrez-Firlefyn[36] and Lochhead and Lochhead[19]. In Chirocephalus, Linder[37] finds large nuclei in the nurse cells but does not mention polyploidy and neither does Munuswamy[38] in his description of oogenesis in Streptocephalus. At the ultrastructural level we have not detected any signs of endoreduplication. Polyploid nurse cells are common in insects[51,109-116] and are thought to be related to the intensive RNA synthesis achieved by the nurse cells. Telfer[51] and Lutz and Huebner[115] stress the important phenomenon of asynchrony in the behavior of oocyte and nurse cells. The presence of intercellular bridges does not preclude asynchrony in nuclear differentiation and endomitotic divisions of nurse cell chromatin occur even though they are interconnected with the oocyte.

Zograf[40] described the degeneration of nurse cells in Streptocephalus. Their nuclear degeneration products are dispersed throughout the ovary and are presumed to be reabsorbed by the developing oocyte. In Chirocephalus Linder[37] also found dispersed chromatic droplets to which he ascribed the same origin as those found in Streptocephalus. In Artemia the degeneration of nurse cells is described[36] but they did not relate this to the end of their role in vitellogenesis and the fate of the degeneration products remained unsettled.

The somatic cells we describe are named after the "special somatic cells" of Lochhead and Lochhead[19] who described them as a strip along the full length of the ovary. Nitsche[23] describes "epithelial cells" on the dorsal side of the ovary of Branchipus, and in Chirocephalus Linder[37] finds "follicle cells". We opted for the name "somatic cells" because it does not point to a function or to an embryonic origin, but opposes them clearly to the germ cells. Our electron microscopic study enabled us to attribute the functions of phagocytosis and transport to the somatic cells. A phagocytotic function is suggested by the presence of more or less recognizable remnants of nurse cells after ovulation (Fig. 24), and of larger and smaller residual bodies during the rest of the reproductive cycle. The presence of acid phosphatases in the smaller residual bodies strenghtens this hypothesis. The occurrence of somatic cells containing small residual bodies in the oogonial region corroborates the opinion of Zograf[40] that degenerating nurse cells provide the developing oocytes with nutritive elements. A phagocytozing function was also ascribed to follicle cells in Gastropods[119] because of their lysosomes and residual bodies.

A tubular system similar to what we described in the somatic cells of Artemia was found by Arcier and Brehelin[118] in the follicle cells of Palaemon and by Jugan and Zerbib[119] and Zerbib and Jugan[120] in the follicle cells of Macrobrachium rosenbergii. In Macrobrachium, free passage of horseradish peroxidase through the tubular system was shown. In addition to the free passage of horseradish peroxidase we demonstrated the free passage of lanthanumnitrate. In Artemia the somatic cells completely surround the oocyte-nurse cell complexes. The tubular system thus probably enhances the diffusion of material of high molecular weight from the blood to the oocytes.

In conclusion we may say that our study although it has solved some problems of the oogenesis in Artemia has brought about many new questions. To our opinion the most interesting finding is the many similarities between oogenesis in Artemia and some insect groups. Since Warner and Bagshaw[121] also found DNA similarities between these different species findings may influence our insights in phyllogenesis.

ACKNOWLEDGEMENTS

We thank Dr. N. Goossens for critically reading the manuscript, Mrs. S. Van Hulle for the technical assistance, and Mr. F. De Wispelaere for the reproduction of the photographs.

REFERENCES

1. J. S. Clegg and F. P. Conte, A review of the cellular and developmental biology of Artemia, in: "The Brine Shrimp Artemia," Vol. 2, G. Persoone, P. Sorgeloos, O. Roels and E. Jaspers, eds., Universa Press, Wetteren (1980).
2. G. Criel, Morphology of the female genital apparatus of Artemia: a review, in: "The Brine Shrimp Artemia," Vol. 1, G. Persoone, P. Sorgeloos, O. Roels and E. Jaspers, eds., Universa Press, Wetteren (1980).
3. A. Anteunis, Les membranes annelées dans l'oeuf d'Artemia salina. Etude de microscopie électronique, Bull. Assoc. Anat. 49:168(1964).
4. A. Anteunis, N. Fautrez-Firlefyn and J. Fautrez, Ultrastructure du cortex et du plasme sous-cortical de l'oeuf d'Artemia salina, C. R. Soc. Biol. 155:1393(1961).
5. A. Anteunis, N. Fautrez-Firlefyn and J. Fautrez, L'ultrastructure du noyau vitellin de l'oeuf d'Artemia salina, Exp. Cell Res. 35:239 (1964).
6. A. Anteunis, N. Fautrez-Firlefyn and J. Fautrez, A Propos d'un complexe tubulo-mitochondrial ordonné dans le jeune oocyte d'Artemia salina, J. Ultrastr. Res. 15:122 (1966a).
7. A. Anteunis, N. Fautrez-Firlefyn and J. Fautrez, La structure des ponts intercellulaires "obturés" et "ouverts" entre les oogonies et les oocytes dans l'ovaire d'Artemia salina, Arch. Biol. 77:645 (1966b).
8. A. Anteunis, N. Fautrez-Firlefyn and J. Fautrez, L'incorporation de cellules nourricières par l'oocyte d'Artemia salina. Etude au microscope électronique, Arch. Biol. 77:665 (1966c).
9. A. Anteunis, N. Fautrez-Firlefyn and J. Fautrez, Ultrastructure du nucléole expulsé dans le cytoplasme de l'oocyte d'Artemia salina. C. R. Acad. Sc. Paris 266:1862 (1968).
10. J. D. Cassidy, Aspects of ultrastructure and cytochemistry during oogenesis in Artemia salina, Biol. Bull. 129:402 (1965).
11. D. Cassidy, Ultrastructural relationship between the developing oocyte and auxiliary cells in adult Artemia, Biol. Bull. 131:385 (1966).
12. J. D. Cassidy, C. F. Starmer and L. J. Beauregard, Quantitative radio-autography of vitellogenesis in Artemia salina, Biol. Bull. 135:412 (1968).
13. G. Criel, Electron microscopic study of autosynthetic and heterosynthetic yolk protein synthesis in Artemia, Third International Symposium of Invertebrate Reproduction, Tubingen (1983).
14. M. J. Karnovsky, A formaldehyde glutaraldehyde fixature of high molarity

for use in electron microscopy, J. Cell Biol. 27:137A (1965).

15. G. Gomori, "Microscopic Histochemistry. Principles and Practice," University of Chicago Press, Chicago (1952).

16. H. Sheldon, H. Zetterqvist and B. Randes, Histochemical reactions for electron microscopy: acid phosphatase, Exp. Cell Res. 9:592 (1955).

17. C. Graham and M. S. Karnovsky, The early stages of absorption of injected horseradish peroxidase in the proximal tubules of the mouse kidney: ultrastructural cyochemistry by a new techique, J. Histochem. Cytochem. 14:291 (1966).

18. J. Overton, Localized lanthanum staining of the intestinal brush border, J. Cell Biol. 38:258 (1968).

19. J. H. Lochhead and M. S. Lochhead, The development of oocytes in the brine shrimp, Artemia, Biol. Bull. 133:453 (1967).

20. P. Metalli and E. Ballardin, Radiobiology of Artemia: radiation effects and ploidy, Curr. Top. Rad. Res. Quart. 7:181 (1970).

21. A. Brauer, Uber das Ei von Branchipus grubii v.Dyb. von der Bildung bis zur Ablage. Abhandlungen der königlichen Akademie der Wissenschaften zu Berlin. Phys.Abh. nicht zur Akad.gehör. Gelehrter II :1 (1892).

22. C. Claus, Untersuchungen über die Entwickelung van Branchippus und Artemia nebst vergleichenden Bemerkungen über andere Phyllopoden, Arbeiten aus dem Zoologischen Institute Wien 6:267 (1886).

23. H. Nitsche, Ueber die Geschlechtsorgane von Branchipus grubei (von Dybowsky), Z. Wiss. Zool. Suppl. 25:281 (1875).

24. F. Spangenberg, Zur Kenntnis von Branchipus stagnalis, Z. Wiss. Zool. (Suppl.) 25:1 (1875).

25. C. Barigozzi, I fenomeni cromosomici nelle cellule somatiche di Artemia salina Leach, Chromosoma 2:251 (1941).

26. T. Iwasaki, Incorporation of 3H-thymidine during oogenesis in Artemia salina, Annotationes Zoologicae Japoneneses 43:132 (1970).

27. R. D. Squire, The effects of acute gamma irradiation on the brine shrimp Artemia. II. Female performance, Biol. Bull. 139:375 (1970).

28. N. Fautrez-Firlefyn, Etude cytochimique des acides nucléiques au cours de l'ovogénèse chez Artemia salina, Communications 3ièmes journées Cyto-Embryologyques Belgo-Néerlandaises Gand (1949).

29. N. Fautrez-Firlefyn, Etude cytochimique des acides nucléiques au cours de la gamétogénèse et des premiers stades du développement embryonnaire chez Artemia salina L, Arch. Biol. 62:391 (1951).

30. N. Fautrez-Firlefyn, Protéines, lipides et glucides dans l'oeuf d' Artemia salina, Arch. Biol. 68:249 (1957).

31. N. Fautrez-Firlefyn and J. Fautrez, Répulsion d'acides thymonucléiques hors du noyau de certaines cellules de l'ovaire d'Artemia salina L, C. R. Soc. Biol. 144:1127 (1950).

32. N. Fautrez-Firlefyn and J. Fautrez, Intranuclear proteins of the oocytes in Artemia salina L, Nature 172:169 (1953).

33. J. Fautrez and N. Fautrez-Firlefyn, La teneur en acide désoxyribonucléique des cellules de l'ovaire d'Artemia salina L, Biol. Jaarb. 20:127 (1953).

34. J. Fautrez and N. Fautrez-Firlefyn, Sur la présence de phospholipines dans les nucléoles de l'oocyte d'Artemia salina, R. Ass. Anat. 42: 506 (1955).

35. J. Fautrez and N. Fautrez-Firlefyn, Sur la présence et la persistance d'un noyau vitellin atypique dans l'oeuf d'Artemia salina, Develop. Biol. 9:81 (1964).

36. L. Lison and N. Fautrez-Firlefyn, Deoxyribonucleic acid content in ovary cells in Artemia salina, Nature 166:610 (1950).

37. H. J. Linder, Studies on the fresh water fairy shrimp Chirocephalus bundyi (Forbes) I. Structure and histochemistry of the ovary and accessory reproductive tissues, J. Morphol. 104:1 (1959).

38. N. Munuswamy and T. Subramoniam, Oogenesis and shell gland activity in a fresh water fairy shrimp Streptocephalus dichotomus Baird

(Crustacea: Anostraca), Cytobios 44:137 (1985).

39. N. Munuswamy and T. Subramoniam, Histochemical studies on the vitello-genesis in a fairy shrimp Streptocephalus dochotomus Baird (Crustacea : Anostraca), Proc. Indian Acad. Sci. 95:171 (1986).

40. N. Zograf, Phyllopodenstudien, Z. Wiss. Zool. 86:446 (1907).

41. P. Andreucetti, C. Taddei and S. Filosa, Intercellular bridges between follicle cells and oocytes during the differentiation of follicular epithelium in Lacerta sicula Raf, J. Cell Science 33:341 (1978).

42. D. W. Fawcett, "A textbook of Hisology," 11th ed., Saunders, Philadelphia (1986).

43. M. R. Kalt, Ultrastructural observations on the germ line of Xenopus laevis, Z. Zellforsch. 138:41 (1973).

44. L. J. Laughran, J. H. Larsen and P. C. Schroeder, Ultrastructure of developing ovarian follicles and ovulation in the lizard Anolis carolensos (Reptilia), Zoomorphology (Berl.) 98:191 (1981).

45. P. B. Moens and V. L. W. Go, Intercellular bridges and division patterns of rat spermatogonia, Z. Zellforsch. 127:201 (1972).

46. D. L. Odor and R. J. Blandau, Ultrastructural studies on fetal and early postnatal mouse ovaries. II. Cytodifferentiation, Am. J. Anat. 125:177 (1969).

47. J. R. Ruby, R. F. Dyer and R. G. Skalko, The occurence of intercellular bridges during oogenesis in the mouse, J. Morphol. 127:307 (1969).

48. J. A. H. V. Van Vorstenbosch, E. Spek, B. Colenbrander and C. J. G. Wensing, The ultrastructure of normal fetal and neonatal pig testis germ cells and the influence of fetal decapitation on the germ cell development, Development 99:553 (1987).

49. J. E. Weber and L. D. Russel, A study of intercellular bridges during spermatogenesis in the rat, Am. J. Anat. 180:1 (1987).

50. W. E. Foor, Cytoplasmic bridges in the ovary of Ascaris lumbricoides, Bulletin of the Tulane University Medical Faculty 27:23 (1968).

51. W. H. Telfer, Development and physiology of the oocyte-nurse cell syncytium. Adv. Insect Physiol. 11:223 (1975).

52. A. Fiil, Follicle cell bridges in the mosquito ovary: syncytia forma-tion and bridge morphology, J. Cell Sci. 31:137 (1978).

53. J. Billen, Ultrastructure of the worker ovarioles in Formica ants (Hymenoptera: Formicidae), Int. J. Insect Morphol. & Embryol. 14: 21 (1985).

54. I. Mandelbaum, Intercellular bridges and the fusome in the germ cells of the Cecropia moth, J. Morphol. 166:37 (1980).

55. S. M. Meola, H. H. Mollenhauer and J. M. Thompson, Cytoplasmic bridges within the follicular epithelium of the ovarioles of two Diptera, Aedes aegypti and Stomoxys calcitrans, J. Morphol. 153:81 (1977).

56. N. Richard-Mercier, Evolution des cellules germinales et mesodermiques des testicules larvaires du Doryphore, Leptinotarsa decemlineata Say (Coleoptera: Chrysomelidae), Int. J. Insect Morphol. Embryol. 8:335 (1979).

57. F. Sabelli Scanabissi and M. Trentini, Ultrastructural observations on oogenesis in Triops cancriformis (Crustacea, Notostraca), Cell Tiss. Res. 201:361 (1979).

58. H. -D. Pfannestiel and C. H. Grünig, Yolk formation in an annelid (Ophryotrocha puerilis, Polychaeta), Tiss. & Cell 14:669 (1982).

59. R. G. Kessel, The structure and function of annulate lamellae: porous cytoplasmic and intranuclear membranes, Int. Rev. Cytol. 82:181 (1982).

60. C. Campanella, P. Andreucetti and L. Bellini, Annulate lamellae in oogenesis of Discoglossus pictus (Anura), Bull. Zool. 50:79 (1983).

61. M. R. Dohmen, Nuage material the origin of dense-core vesicles in oocytes of Nassarius reticulatus (Mollusca Gastropoda), Int. J. Invert. Reprod. 8:117 (1985).

62. H. Imoh, Behavior of annulate lamellae during the maturation of oocytes in the newt, Cynops pyrrhogaster, J. Embryol. Exp. Morph. 70:153 (1982).

63. R. G. Kessel, Fibrogranular bodies, annulate lamellae, and polyribosomes in the dragonfly oocyte, J. Morphol. 176:171 (1983).

64. J. P. Stafstrom and L. A. Staehelin, Are annulate lamellae in the Drosophila embryo the result of overproduction of nuclear pore components?, J. Cell Biol. 98:699 (1984).

65. J. Arnaud, M. Brunet and J. Mazza, Etude de l'ovogénèse chez Centropagus typicus, Reprod. Nutrition Développement 22:537 (1982).

66. H. W. Beams and R. G. Kessel, Ultrastructure and vitellogenesis in the oocyte of the crustacean, Oniscus asellus, J. Submicrosc. Cytol. 12:17 (1980).

67. S. Bilinski, Oogenesis in Campodea sp. (Diplura). The ultrastructure of the egg chamber during vitellogenesis, Cell Tissue Res. 202:133 (1979).

68. S. Bilinski, Ultrastructural study of yolk formation in Porcellio scaber Latr. (Isopoda), Cytobios 26:123 (1979).

69. P. I. Blades-Eckelbarger and M. J. Youngbluth, The ultrastructure of oogenesis and yolk formation in Labidocera aestiva (Copepoda: Calanoida), J. Morphol. 179:33 (1984).

70. H. Charniaux-Cotton, Vitellogenesis and its control in Malacostracan Crustacea, Amer. Zool. 25:197 (1985).

71. J. N. Dumont and E. Anderson, Vitellogenesis in the horseshoe crab Limulus polyphemus, J. Microscopie 6:791 (1967).

72. L. Eurenius, An electron microscope study on the developing oocytes of the crab Cancer pagurus L. with special reference to yolk formation, Z. Morph. Tiere 75:243 (1973).

73. G. W. Hinsch and M. V. Cone, Ultrastructural observations of vitello-genesis in the spider crab, Libinia emarginata L, J. Cell Biol. 40:336 (1969).

74. R. G. Kessel, Mechanisms of yolk protein synthesis and deposition in crustacean oocytes, Zeitschr. Zellforsch. 89:17 (1968a).

75. B. S. Komm and G. Hinsch, Oogenesis in the terrestrial hermit crab, Coenobita clypeatus (Decapoda, Anomura): II. Vitellogenesis, J. Morphol. 192:269 (1987).

76. E. Papathanassiou and P. E. King, Ultrastructural studies on gameto-genesis of the prawn Palaemon serratus (Pennant). I. Oogenesis, Acta Zoologica (Stockh.) 65:17 (1984).

77. M. L. Schade and R.R. Shivers, Structural modulation of the surface and cytoplasm of oocytes during vitellogenesis in the lobster Homarus americanus. An electron microscope-protein tracer study, J. Morphol. 163:13 (1980).

78. E. M. Wolin, H. Laufer and D. F. Albertini, Uptake of yolk protein, lipovitellin, by developing crustacean oocytes, Develop. Biol. 35:160 (1973).

79. C. Zerbib, Contribution a l'etude ultrastructurale de l'ovocyte chez le crustace amphipode Orchestia gammarella (Pallas), C. R. Acad. Sci. (Paris) 277:1209 (1973).

80. C. Zerbib, Etude ultrastructuale de l'ovocyte en vitellogénèse chez les Ecrevisses Astacus astacus et A. leptodactylus, Int. J. Invert. Reprod. 1:289 (1979).

81. C. Zerbib, Ultrastructural observation of oogenesis in the crustacea amphipoda Orchestia gammarellus (Pallus), Tiss. & Cell 12:47 (1980).

82. A. Dhainaut and M. De Leersnijder, Etude cytochimique et ultra-structurale de l'evolution ovocytaire du crabe Eriocheir sinensis. I. Ovogénèse naturelle, Arch. Biol. 87:261 (1976).

83. C. Zerbib, Nature chimique des enclaves vitellines de l'ovocyte du crustace, l'amphipode Orchestia gammarellus (Pallas), Ann. Histochim. 21:279 (1976).

84. K. Bier, W. Kunz and D. Ribbert, Struktur und Funktion der Oocyten-chromosomen und Nucleolen sowie der Extra DNS während der Oogenese panoistischer und meroistischer Insekten, Chromosoma (Berl) 23:214 (1967).

85. A. Fiil, Oogenesis in the malaria mosquito Anopheles gambiae, Cell Tissue Res. 167:23 (1976).

86. M. N. Gruzova, Z. P. Zaichikova and I. I. Sokolov, Functional organization of the nucleus in the oogenesis of Chrysopa perla L. (Insecta, Neuroptera), Chromosoma (Berl.) 37:353 (1972).

87. G. Rue and M. Gontcharoff, Etude des modifications ultrastructurales du nucléole au cours de l'ovogénèse de Lineus ruber (Hétéronémertes), C. R. Acad. Sci. Paris 273:752 (1971).

88. U. Mothes-Wagner and K. -A. Seitz, Ultrahistology of oogenesis and vitellogenesis in the spider mite Tetranychus urticae, Tiss. & Cell 16:179 (1984).

89. P. Andreucetti and C. Campanella, Regional differences in the pattern of vitellogenesis in the painted frog Discoglossus pictus, Tissue & Cell 14:681 (1982).

90. M. L. Bonnenfant-Jais and P. Mentre, Study of oogenesis in the newt Pleurodeles watlii M. Ultrastructural study of different stages of oocyte development, J. Submicrosc. Cytol. 15:453 (1983).

91. H. Emanuelsson, Electronmicroscopical observations on yolk and yolk formation in Ophiotrocha labronica LaGreca and Bacci, Z. Zellforsch. 95:19 (1969).

92. A. Medina, J. C. Garcia, F. J. Moreno and J. L. Lopez-Campos, Comparative studies on the histology of the ovotestis in Hypselodoris tricolor and Godiva banyulensis (Gastropoda Opisthobranchia) with special reference to yolk formation, J. Morphol. 188:105 (1986).

93. R. Riehl, Licht- und elektronenmikroskopische Untersuchungen an den Oocyten der Süswasser-Teleosteer Noemacheilus barbatulus (L.) and Gobio gobio (L.) (Pisces, teleostei), Zool. Anz. 201:199 (1978).

94. D. A. Wall and S. Platel, Multivesicular bodies play a key role in vitellogenin endocytosis by Xenopus oocytes, Develop. Biol. 119: 275 (1987).

95. R. T. Ward, The origin of protein and fatty yolk in Rana pipiens. III. Intramitochondrial and primary vesicular yolk formation in frog oocytes, Tiss. & Cell 10:515 (1978a).

96. R. T. Ward, The origin of protein and fatty yolk in Rana pipiens. IV. Secondary vesicular yolk formation in frog oocytes. Tiss. & Cell 10:525 (1978b).

97. N. Garreau de Loubresse, Etude chronologique de la mise en place des enveloppes de l'oeuf d'un crustacé phyllopode: Tanymastyx lacunae, J. Micrscopie 20:21 (1974).

98. G. De Maeyer-Criel, N. Fautrez-Firlefyn and J. Fautrez, Formation de la mebrane de fécondation dans l'oeuf d'Artemia salina, W. Roux' Archiv. 183:223 (1977).

99. E. Van Beek, M. Van Brussel, G. Criel and A. De Loof, A possible extra-ovarian site for synthesis of lipovitellin during vitellogenesis in Artemia sp. (Crustacea; Anostraca). Int. J. Invert. Reprod. Develop. 12:227 (1987).

100. R. G. Kessel, The permeability of the crayfish oocyte-follicle complex as studied with peroxidase as a tracer, J. Cell Biol. 39:169A (1968b).

101. A. S. Raikhel and A. O. Lea, Internationalized proteins directed into accumulative compartments of mosquito oocytes by the specific ligand vitellogenin, Tiss. & Cell 18:559 (1986).

102. F. Roels, Localisation d'activités peroxidasiques dans l'oeuf d'Artemia salina à l'aide de 3,3' -diaminobenzidine et de pyrogallol, Arch. Biol. 81:229 (1970).

103. F. Roels and E. Wisse, Distinction cytochimique entre catalases et peroxidases, C. R. Acad. Sci. 276:391 (1973).

104. C. R. Hopkins, The histochemistry and fine structure of the accessory nuclei in the oocyte of Bombyx mori, Quart. J. Microscop. Sci. 105:475 (1964).

105. P. E. King and M. R. Fordy, The formation of "accessory nuclei" in the developing oocytes of the parasitoid hymenopterans Ophion luteus (L.) and Apanteles glomeratus (L.), Z. Zellforsch. 190:158 (1970).

106. J. D. Cassidy and R. C. King, Ovarian development in Habrobracon juglandis (Ashmead) (Hymenoptera: Braconidae) I. The origin and differentiation of the oocyte-nurse cell complex, Biol. Bull. 143: 483 (1972).

107. G. F. Meyer, S. Sokoloff, B. E. Wolf and D. Brand, Accessory nuclei (nuclear membrane balloons) in the oocyte of the dipteran Phryne, Chromosoma (Berl.) 75 (1979).

108. F. Blochmann, Über die Metamorphose der Kerne in den Ovarialeier und über den Beginn der Blastodermbildung bei den Ameisen, Verh. Naturh. Med. Verein Heidelberg 3:243 (1884).

109. K. Bier, Endomitose und Polytänie in der Nährzellkernen von Calliphora erythrocephala Meigen, Chromosoma (Berl.) 8:493 (1957).

110. J. Cardoen, L. Schoofs, D. Broekaert, H. Van Mellaert, B. Vanachtert and A. DeLoof, Polyploidization and localization of poly(a)$^+$ RNA in the different cell types of the vitellogenic meroistic ovary of the fleshfly, Sarcophaga bullata, Histochemistry 85:305 (1986).

111. W. C. Choi and W. Nagl, Patterns of DNA and RNA synthesis during the development of ovarian nurse cells in Gerris najas (Heteroptera), Develop. Biol. 61:262 (1977).

112. M. Guelin and M. Durand, Evolution des cellules nourricieres au cours de l'ovogenese chez Ephestia kuhniella Z. (Insecte, Lepidoprere), Annales des Sciences naturelles, Zoologie, 13ieme serie 2:167 (1980).

113. R. C. King, E. M. Rasch, S. F. Riley, P. M. O'Grady and P. D. Storto, Cytophotometric evidence for the transformation of oocytes into nurse cells in Drosophila melanogaster, Histochemistry 82:131 (1985).

114. M. Ksiazkiewics, Ultrastructure of the trophic chamber and nutritive cord of Aspidiotus hederae (Homoptera, coccoidea), Cell Tissue Res. 213:149 (1980).

115. D. A. Lutz and E. Huebner, Development of nurse cell-oocyte interactions in the insect telotrophic ovary (Rhodnius prolixus), Tissue & Cell 13:321 (1981).

116. E. Nour-Eddine and A. M. Laverdure, Etude du mechanisme de l'endomitose dans les noyaux trophocytes ovariens de Tenebrio molitor, C. R. Acad. Sci. Ser III 294:267 (1982).

117. B. Griffond and L. Gomot, Ultrastructural study of the follicle cells in the fresh water gastropod Viviparus viviparus L, Cell Tissue Res. 202:25 (1979).

118. J. -M. Arcier and M. Brehelin, Etude histologique et ultrastructurale du tissue folliculaire au cours des cycles de développement ovarien chez Palaemon adspersus (Rathke, 1837), Arch. Biol. 93:9 (1982).

119. P. Jugan and C. Zerbib, Follicle cell tubular system in the prawn Macrobrachium rosenbergii: a route for exchanges between haemolymph and vitellogenic oocytes?, Biol. Cell 51:395 (1984).

120. C. Zerbib and P. Jugan, Mise en évidence d'un mode nouveau de franchissement de l'épithelium folliculaire secondaire chez le crustaceé décapode Macrobrachium rosenbergii, Int. J. Invert. Reprod. Develop. 7:227 (1984).

121. A. H. Warner and J. C. Bagshaw, Absence of detectable 5-methylcytosine in DNA of embryos of the brine shrimp, Artemia. Develop. Biol. 102:264 (1984).

ENZYME MARKERS IN DEVELOPMENT: CHOLINESTERASE (ChE), ACID HYDROLASES,
ALKALINE PHOSPHATASE (ALP) AND AMINOPEPTIDASE (AP) IN EMBRYOS AND LARVAE
OF ARTEMIA

Margherita Raineri

Institute of Comparative Anatomy
University of Genova
16132 Genova, Italy

INTRODUCTION

Artemia is a convenient invertebrate model for developmental investigations owing to its easy breeding in laboratory conditions. These studies
are carried on by different approaches, including the study of enzyme activities, which give promising results from many points of view. At present,
most information on the developmental enzymology of Artemia consists of biochemical data [1-6], although some morphohistochemical findings concerning
yolk degradation are available [7].

In our laboratory, the embryonic and larval development of Artemia has
been the subject of parallel morphological and histoenzymological investigations carried out in concert with electrophoretic and quantitative analyses
and in vivo observations. The aim of the biochemical studies is to support
and better understand the microscopical data, and to relate them to the behaviour of the larvae, but not to characterize the enzymes in molecular
detail. We studied several enzymes which can be considered as developmental
markers of some organs and/or cell functions (Table 1) and the results are
summarized in this chapter.

METHODS

Unless specified otherwise, all results shown in Figures 1-7 and 9-45
illustrate histochemical stainings obtained with acetylthiocholine or
acetyl-β-methylthiocholine as substrates hydrolyzed preferentially by AChE
according to Raineri and Falugi [8].

RESULTS AND CONCLUSIONS

Cholinesterase

Research on ChE activity in the development of Artemia and other
crustaceans such as Balanus was started mainly to get information on
neurogenesis, since the enzyme is a good marker of nerve cells in addition to those which release acetylcholine (ACh) as a neurotransmitter.
Myogenic cells are generally characterized by ChE activity, providing an
opportunity to investigate the differentiation of muscles. Very soon,
however, we became interested in the wider role of ACh and acetylcholinesterase (AChE) in excitable cell membranes [9], concerned, for instance,

with epithelial conduction [10], and the possibility that a cholinergic-like system different from that of synaptic transmission takes part in the regulation of development [11]. Several findings showed that morphogenetic cell activities and interactions have "embryonic ChE" [12] as a marker. Usually the activity detected has properties of both AChE and pseudocholinesterase (BuChE) [8] and is inactivated by fixatives more easily than the nervous or muscular ChE [13]. Histochemical and biochemical investigations were extended to the early embryonic stages of Artemia, and eventually to the egg and sperm cells, as more information pointed to a role for a cholinergic-like system at fertilization.

Our findings can be described as if they concern two different views of Artemia development. On one hand, ChE activity characterizes neuroblasts and myoblasts remaining, in most cases prevalent, in differentiated nervous and muscular cells. On the other hand, "embryonic ChE" is temporarily expressed by different types of cells as they divide, change shape, migrate, interact with each other and differentiate. It is not possible, however, to completely discriminate between the two enzyme activities by the techniques employed here, that is, on the basis of substrate and inhibitor specificities and electrophoretic mobilities [8]. Developing nervous and muscular cells of Artemia possess an enzyme with most of the features of AChE, but "embryonic ChE" has AChE and BuChE properties, and hydrolyzes propionylthiocholine (PrThCh) particularly actively. Moreover, in the quantitative enzyme determinations, the earliest stage employed was the encysted gastrula, which already shows histochemically detectable stronger ChE activity, both in the mesodermal cells (future myoblasts) and in the dorsal ectoderm where neuroblasts develop. Disk electrophoresis indicates the same enzymes with most of the properties of AChE in the egg, embryos, nauplii and early metanauplii. Another more anodal band was easily detected with butyrylthiocholine (BuThCh) only in concentrated supernatants of homogenized adult heads that were employed for comparison (unpublished results). This may indicate that some nervous structures of Artemia are rich in BuChE activity, as found in other crustaceans [14,15]. The failure to demonstrate a BuChE-like activity in developing embryos and early larval stages by electrophoresis may result from the small amounts of sample employed and the low ChE activity of the material.

Neurogenesis and myogenesis. ChE appeared to be a good marker of morpho-functional nervous and muscular development in Artemia. Early in development the histochemical stainings reveal bilaterally symmetrical neuroblasts which divide and grow axons to establish the basic pattern of the nervous system (segmental ganglia joined by longitudinal connectives and transversal commissures) and the innervation of the appendages (Fig. 1-3). On the basis of their features and their similarities with those in the grasshopper [16] these cells are called pioneer neurons. Very soon in development their fibres are accompanied by small ChE-active, glial-like cells. The existence of pioneer neurons in Artemia was confirmed by light and electron microscopic investigations [17]. The pioneer neurons of the naupliar ganglia are detected during pre-emergent development. From the rudiment of each antennal ganglion two bigger ovoidal neurons, more intensely stained for ChE activity, send their axons into the buds of the antennae (Fig. 2). These appendages are the most developed of the nauplius and they are equipped with strong muscles for swimming. On the other hand, from the tip of the antennulae and antennae bipolar sensory-like neurons (Fig. 16) grow the axons centripetally to the corresponding ganglia. Since these cells are not detected in the metanauplii it seems, as in grasshopper [18,19], one of their functions is to guide innervation of the growing limbs.

The map of the pioneer neurons of Artemia has not been drawn in detail. Histochemically localized ChE, however, has proven to be useful for such

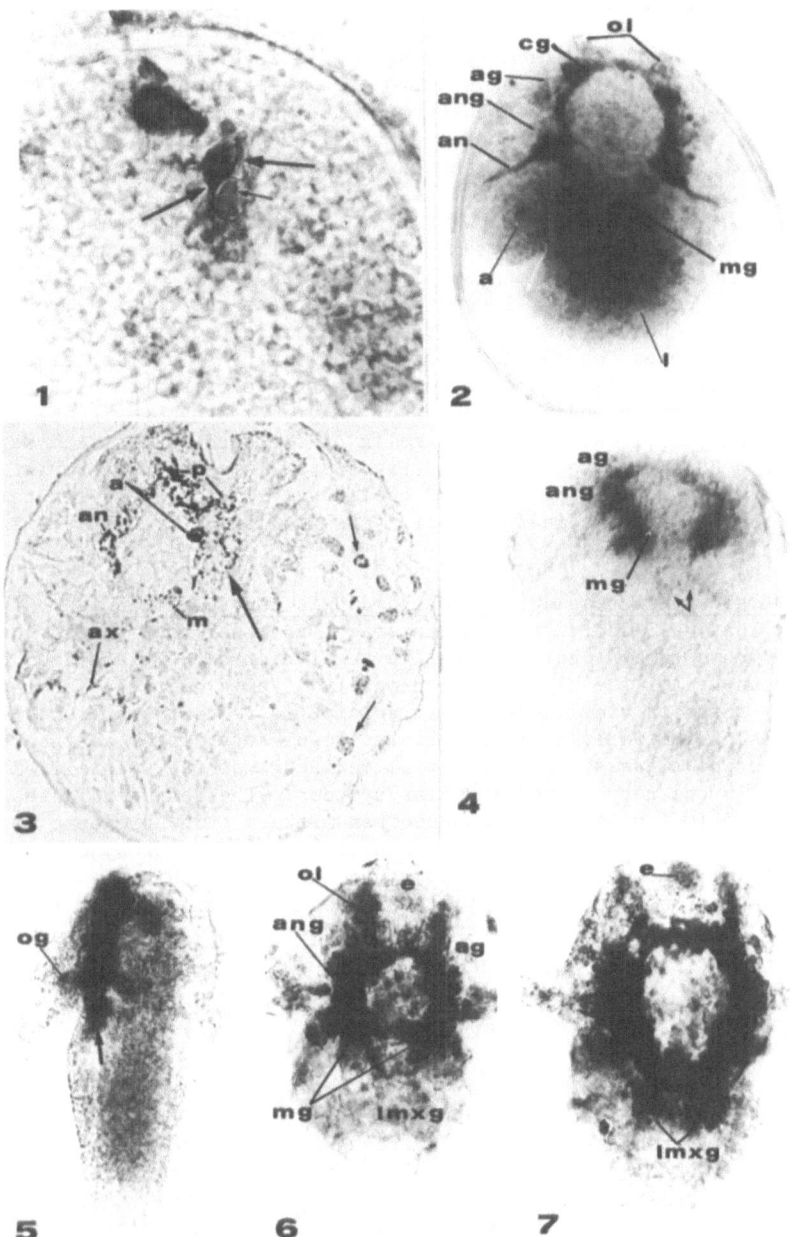

Fig. 1. In toto view of a cerebral ganglion primordium. Two closely
         associated ChE-active neuroblasts grow axons with opposite
         polarities (large arrows). Small arrow, enzyme staining
         associated with nuclear envelope, X 600.
Fig. 2. In toto frontal view of an encysted prenauplius. a, antennal
         bud; ag, antennular ganglion; an, antennal nerve; ang, an-
         tennal ganglion; cg, cerebroid ganglion; l, lip bud; mg,
         primordia of the mandibular ganglia; ol, optic lobes, X 300.
Fig. 3. Prenauplius. ChE activity is strong in the rudiment of the
         nervous system (large arrow) and evident in the myoblasts of

(Continued)

133

research, and possibly, to follow the development of the simplest sensory-motory connections which determine the behaviour of the hatched nauplius. In the prenaupliar stages, for instance, the synchronous early appearance of neuroblasts connected with the rudiment of the naupliar eye (optic lobes) and others which innervate the developing myoblasts of the limbs (particularly the antennae) (Fig. 2,3) have a probable relationship to the swimming activity guided by the positive phototropism of the nauplius. Development of nerve cells and fibres is accompanied by a fast quantitative ChE increase (Fig. 8) shown histochemically in the larval (Figs. 4-7,9-10,12-14,17) and the more differentiated, adult-like nervous structures (Figs.11,18-19). In each developing post-naupliar segment, ChE activity marks the pioneer neurons of the two bilaterally symmetrical ventral ganglia (Fig.15) and the myoblasts of the appendages which start to differentiate at the same time. As a consequence, elongating axons reach the rows of myoblasts before they fuse. Later, synaptic nerve endings wrapped by ChE-active glial-like cells appear on the developing muscle fibres, and ChE staining of nerve and muscle cells becomes stronger as they differentiate (Figs. 21,22). In contrast, a moderate amount of enzyme also showing BuChE properties is localized in all the other cells of the early bud of the segment and disappears during differentiation (Fig. 20). This is characteristic of embryonic ChE which, as a marker of cholinergic-like systems, is temporarily expressed in non-cholinergic cell types and is involved in their growth and differentiation.

Cholinergic regulation during embryogenesis and fertilization. The possibility that cell activities and interactions other than the nervous and muscular ones have a cholinergic-like regulation can explain the ChE activity shown by all the blastomeres of early Artemia embryos. From an evolutionary point of view there was an attempt to identify molecules which would be present initially as intracellular regulators, later as intercellular chemical signals, and finally as mediators of the specialized cell excitability in the nervous and muscular systems. The best candidates for this wide biological role were the neurotransmitters (ACh, biogenic amines) and, to some extent, the cyclic nucleotides adenosine monophosphate (cAMP) and guanosine monophosphate (cGMP) [20]. In support of this idea the amounts of neurotransmitters and cyclic nucleotides inside the cell appeared to be related to each other and they varied according to the phases of the

---

Fig. 4,5,6,7.

(Continued)

the appendages (small arrows), but not significant in the other tissues. The first neurons of the protocerebral (p), antennular (a), antennal (an) and mandibular (m) ganglia are distinguished. ax, ChE-active axons grow in the ventral area of the embryo. 3 μm section, X 350.

Fig. 4. Emerging nauplius enclosed in the hatching membrane. ag, antennular ganglion; ang, antennal ganglion; mg, mandibular ganglion. Arrows, early primordia of the I maxillary ganglia, X 100.

Fig. 5. Lateral view of a first instar nauplius. Besides the major nerve ganglia, the small oesophageal ganglion (og) is distinguished. Arrow, I maxillary ganglia, X 80.

Fig. 6. Dorsal view of a first instar nauplius. ag, antennular ganglia; ang, antennal ganglia; e, middle eye; mg, mandibular ganglia; Imxg, I maxillary ganglia; ol, optic lobes, X 100.

Fig. 7. Dorsal view of a 2nd instar nauplius. The ganglia of the I maxillae (Imxg) are more differentiated and the naupliar eye (e) more intensely pigmented, X 100.

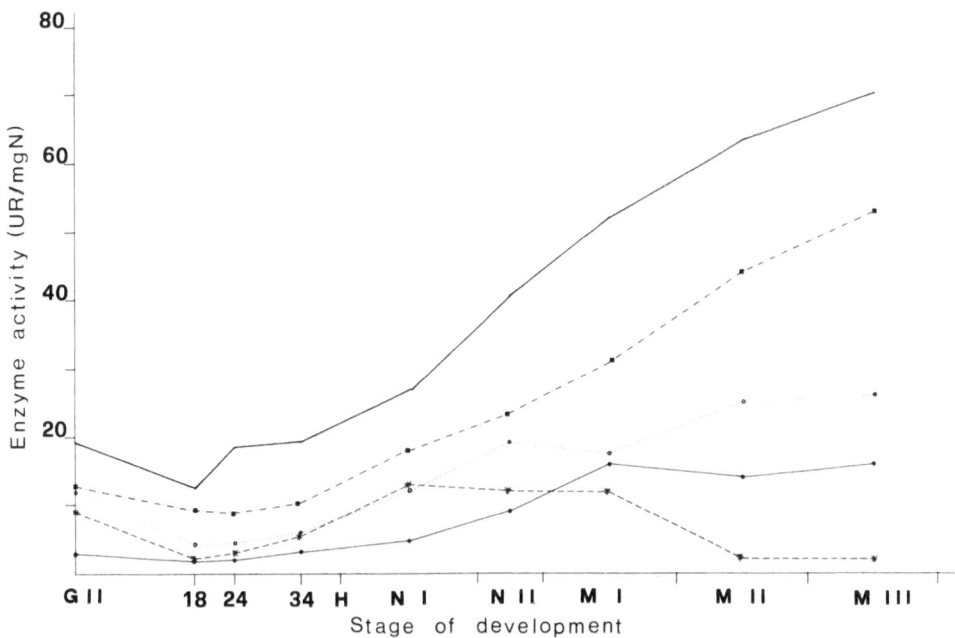

Fig. 8.   ChE activity (——) during <u>Artemia</u> development
          according to Raineri and Falugi [8].   The effects
          of ChE inhibitors (10^{-5}M) added to the homogenates
          are shown.   Eserine (-●-) and diisopropylfluoro-
          phosphate (DFP) (--✳--) are inhibitors of AChE
          and BuChE; 1,5-bis-(allydimethylammoniumphenyl)-
          pentan-3-one dibromide (BW284c51) (---○---) is
          a specific inhibitor of AChE; tetraisopropylpyro-
          phosphoroamide (iso-OMPA) (-·-■-·-) is a specific
          inhibitor of BuChE; GII, encysted gastrula; NI
          and NII are first and second instar nauplii; MI
          to MIII are metanaupliar stages.

cell cycle [21].   A possible explanation was that the changes in cyclic
nucleotides were a direct or indirect response to the binding of neurotransmitters to a receptor.   A fairly large amount of information is now
available on neurotransmitters present in embryos, plants, fungi, protozoa
and bacteria [22].   The term "prenervous neurotransmitters" [23] indicates
the more primitive and general functions played by these molecules, which
may have cytoskeletal elements and various enzymes as their targets [24].
For many years, however, the exact interrelationships remained unknown.
Basic questions still open are the identity of the neurotransmitter
receptors and the intermediate steps coupling their activation with the
biological effects observed, such as variations in ion (Ca^{2+}, Na^{+}, K^{+}, H^{+},
Cl^{-}) and cyclic nucleotide concentrations.

     At present, the physiological role of receptor activation can be in-
vestigated at the molecular level by DNA cloning and sequencing.   An "embryonic" ACh receptor that belongs to the family of muscarinic receptors
has been partly characterized [25,26].   Several subtypes of muscarinic
receptors have been identified at the gene level and all of them seem to
be derived from the gene for the photopigment rhodopsin, which probably also

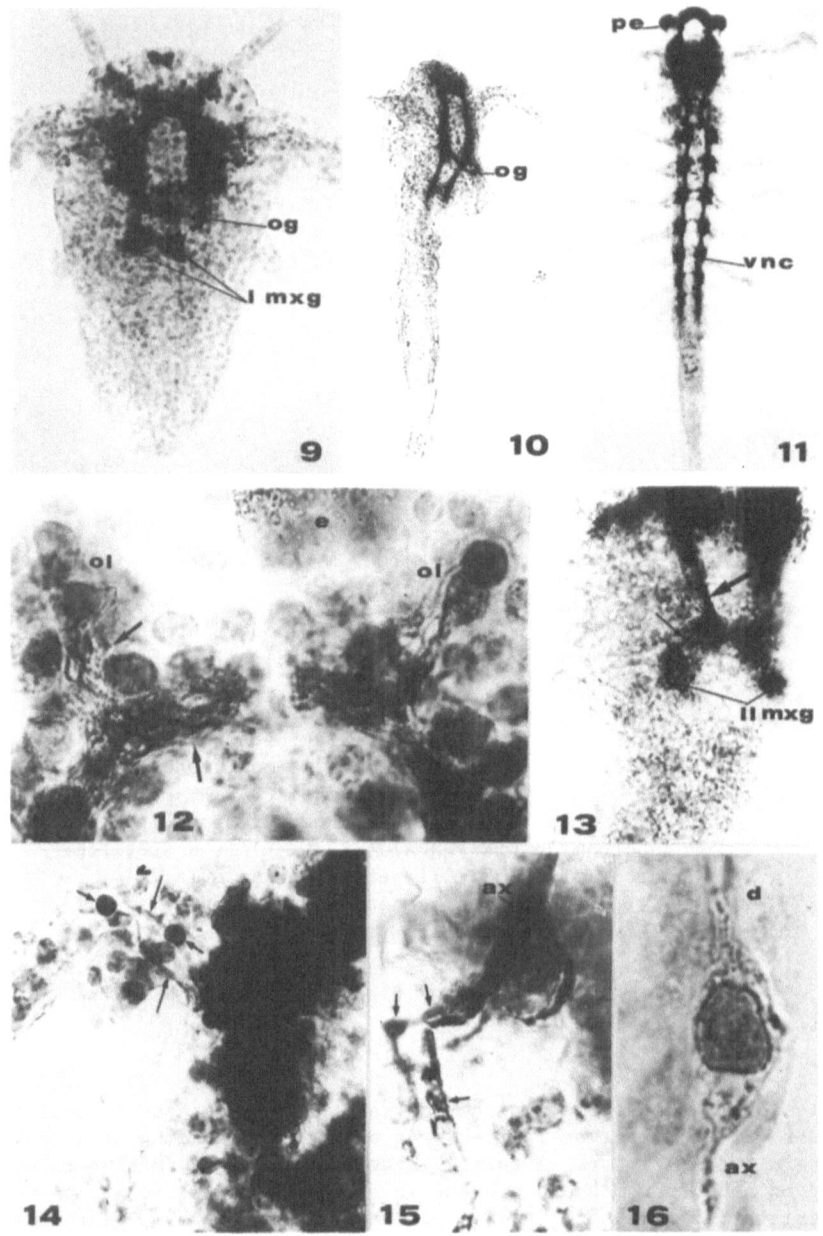

Fig. 9. Ventral view of a 2nd instar nauplius. Imxg, I maxillary
        ganglia; og, oesophageal ganglion, X 80.
Fig. 10. Ventral view of a metanauplius II. og, oesophageal gang-
        lion, X 40.
Fig. 11. Juvenile stage. pe, paired eyes; vnc, ventral nerve cord
        with longitudinal connectives and double transversal com-
        missures between the segmental ganglia, X 40.
Fig. 12. First instar nauplius, showing neuropilar medulla of the
        protocerebrum (arrows). The nerve cell bodies are located
        at the periphery. e, middle eye; ol, optic lobes, X 900.

(Continued)

Table 1. Summary of Enzymes Studied

Enzymes	Cell Organelles	Organs, Cell Functions
Cholinesterase (ChE)	cell membranes	nervous and muscular cells, ion fluxes, cell excitability
Acid phosphatase (Ac-P) β-glucuronidase (β-Glu) β-galactosidase (β-Gal)	lysosomes	alimentary and excretory organs, yolk degradation, cell digestion
Cytochrome oxidase (Cy-O) Succinico dehydrogenase (Su-D)	mitochondria	muscles, epithelia specialized for active transport
Alkaline phosphatase (ALP) Aminopeptidase (AP)	microvilli	alimentary and excretory organs, contact digestion, absorption

Fig. 13,14,15,16.
(Continued)

Fig. 13. Metanauplius II, ventral side. The developing ganglia of the II maxillae (IImxg) are joined to those of the I maxillae by a connective (small arrow) less conspicuous than that between I maxillary and mandibular ganglia (large arrow), but they are not yet connected by a transversal commissure, X 300.

Fig. 14. In a metanauplius I the nerves (large arrows) to the antennae are accompanied by ChE-active cells (small arrows, X 350.

Fig. 15. Early development of a II maxillary ganglion. Two big pioneer-like neurons (one of them is focused) have established the longitudinal connective by their axons (ax) accompanied by ChE-active glial-like cells (arrows), X 1,100.

Fig. 16. ChE-active sensory-like neuron at the tip of an antenna in a first instar nauplius, in toto. Its centripetally directed axon (ax) may guide the innervation inside the limb. d, dendrite, X 2,000.

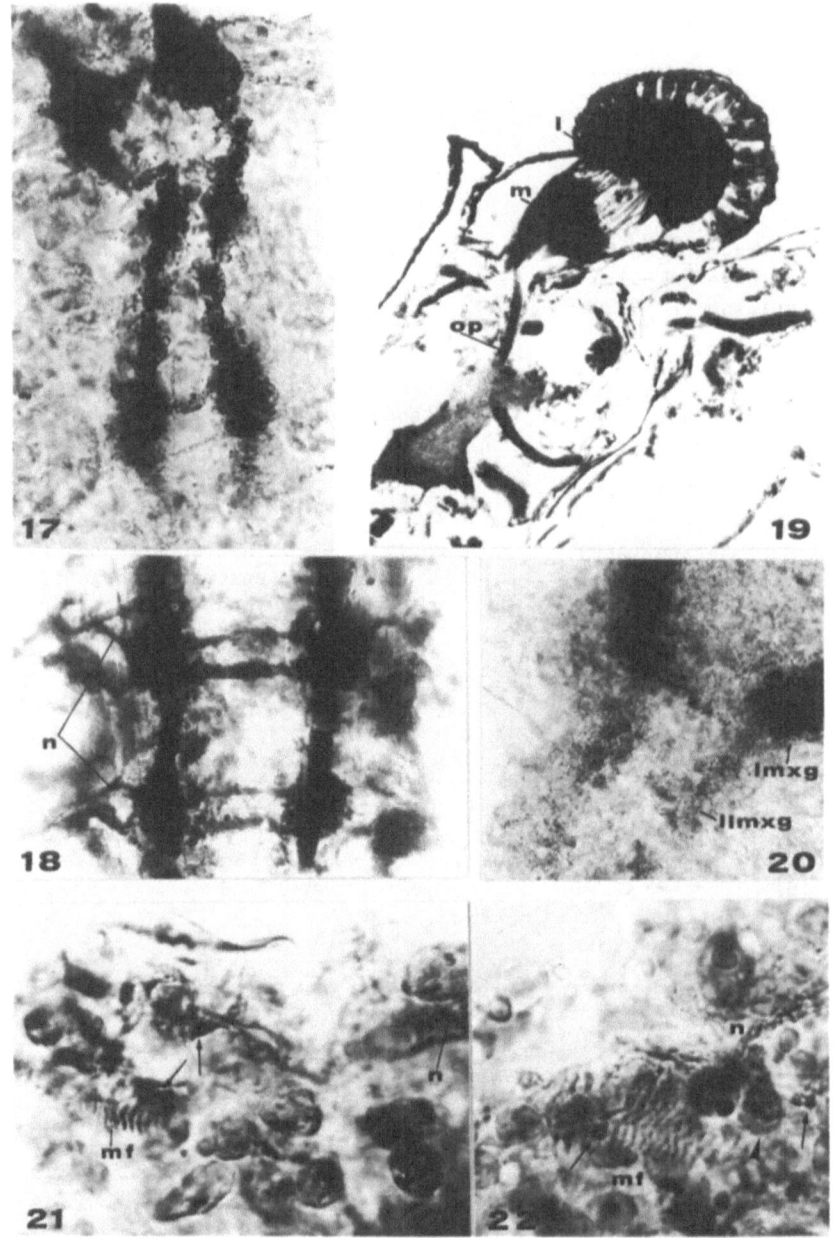

Fig. 17, Development of the ventral nerve cord in a metanauplius
and 18 III (Fig. 17) and a juvenile stage (Fig. 18). Fully dif-
        ferentiated segmental ganglia are transversely linked by
        a double commissure and send two major nerves (n) to each
        corresponding appendage, X 300.
Fig. 19. Paired eye in a juvenile stage. The neuropilar lamina
        (l), the medulla (m) and the optic peduncle (op) are
        strongly ChE-active. 3 μm section, X 270.
Fig. 20. The ventral nerve cord of a metanauplius stained in toto
        by butyrylthiocholine as substrate. The BuChE-like acti-

gave rise to the β-adrenergic receptor. All of these receptor proteins are structurally related to each other. When activated, they interact with a guanine nucleotide-binding protein (G-protein), which in turn regulates adenyl cyclase and phospholipase C activities, and possibly, in some cases, K^+ transport proteins [27-30].

Most embryologists interested in the cholinergic regulation of mor-phogenesis focus attention on the steps coupling the activation of the "embryonic" ACh receptor to the rise of free intracellular Ca^{2+} [26,30]. Calcium modulates enzyme activities and cytoskeleton dynamics, thus playing a key role in animal and plant development [32]. Ligand binding to a sub-type of muscarinic ACh receptor activates phospholipase C, which in turn splits plasma membrane polyphosphoinositide into diacylglycerol and inositol-1,4,5-trisphosphate ($InsP_3$). These molecules act synergistically as second messengers: the former activates protein kinase C, the latter triggers Ca^{2+} release from intracellular storage vesicles [33]. Probably a cholinergic-like system operates in many cell processes by this mechanism. The term "cholinergic-like system" is justified since in some cases, besides ChE, the ACh-synthesizing enzyme, choline acetyltransferase (CAT), and a muscarinic ACh receptor have been demonstrated in non-nervous and non-muscular cells such as oocytes, eggs, early blastomeres, embryonic tissues, and some normal adult and tumor cells [22,34].

The possible intracellular localization of the embryonic muscarinic receptor is the membrane of a peculiar type of smooth endoplasmic reticulum which is highly developed and ChE-active [35,36]. A comparable system is localized in the cortex and the mitotic spindle of different cells [37,39]. From an evolutionary perspective this could be the ancestor of the sarco-plasmic reticulum of muscle fibres and have the same functions, that is, to release and sequester Ca^{2+} [36].

In the egg and early blastomeres of Artemia most ChE activity is visualized as cytoplasmic granules which, from preliminary electron micro-scopic results, appear to be membrane vesicles [24]. By regulating Ca^{2+} levels, the vesicular system could modulate cytoskeleton functions such as those which determine mitosis, cell shaping and gastrulation. On the other hand, the activation of protein kinase C starts different metabolic path-ways. This, together with Ca^{2+}, could play a role in cell differentiation and cell-cell interactions during development.

Fig. 20,21,22.

(Continued)

                vity is much weaker than that of AChE (compare with Fig. 18). Imxg, IImxg, ganglia of the I and II maxillae, X 300.

Fig. 21, Development of the innervation of an antennal muscle in and 22 toto, in an emerging (Fig. 21) and first instar nauplius (Fig. 22). The nerve synaptic endings associated with ChE-active small cells (arrows) increase in number with differentiation. A histochemically stained nervous-like cell (sensory neuron?) (arrow head) is distinguished at the branching point of a nerve (n) on a muscle fibre (mf). The cross-striated enzyme reaction of the muscle may depend on ChE activity of the sarcoplasmic reticulum assoc-iated with the T-system, X 800.

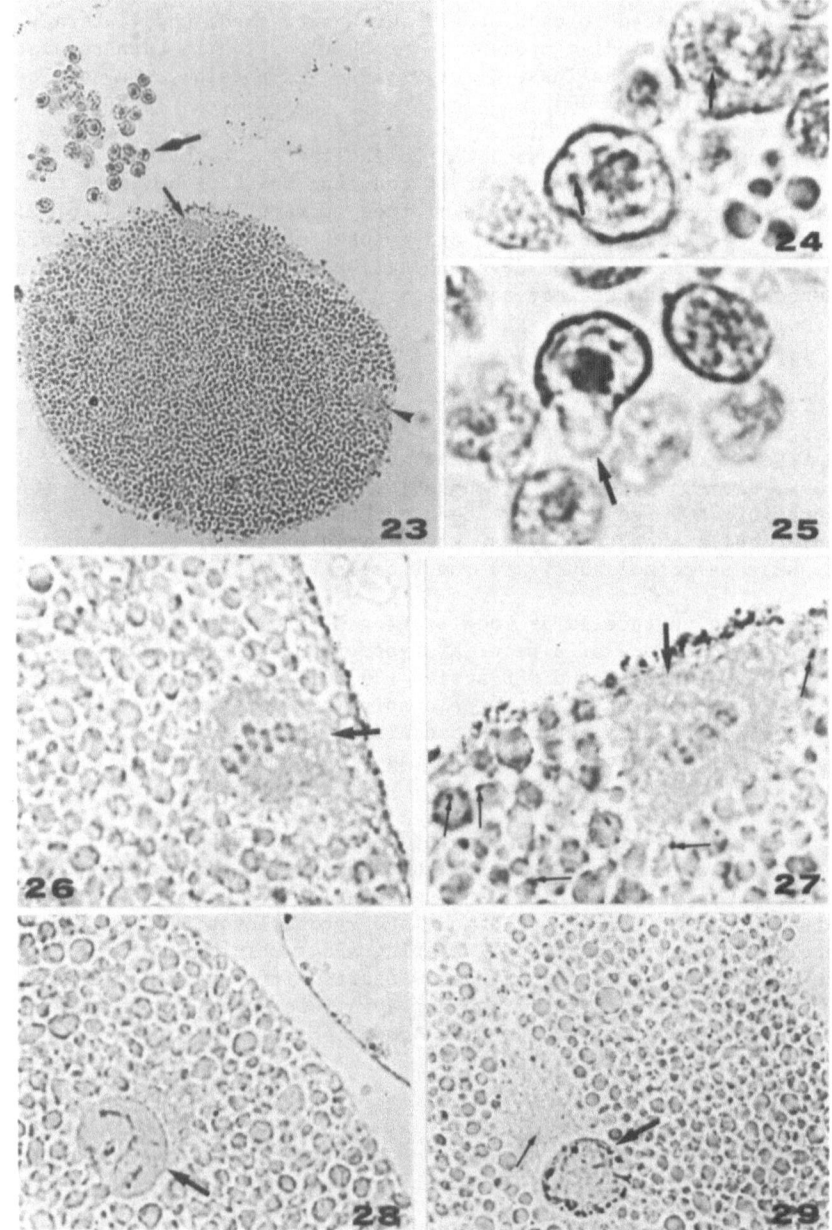

Fig. 23. Egg just after sperm entry. ChE activity is not significant in the egg but is evident in the sperm cells (large
arrow). Small arrow, sperm entry point; arrow head,
meiotic spindle. 3 μm section, X 300.

Fig. 24, In the sperm cell, ChE activity is localized at the plasma
25 membrane and the nuclear envelope where it is joined to
enzyme-active cytoplasmic structures which may correspond
to endoplasmic reticulum (small arrows). The everted
bulge shown by some sperm cells (large arrow) may correspond to the reacted acrosome. 3 μm sections, X 3,000.

A related question is the possible cholinergic regulation of fertilization. Vertebrate and invertebrate eggs have a muscarinic ACh receptor and are activated by InsP$_3$ microinjection which triggers a cytoplasmic rise in free Ca^{2+}. All sperm cells so far investigated seem to have a nicotinic ACh receptor, and in some cases the other components of the cholinergic system have been identified [40-42]. The typical, tail-bearing sperm cell is suggested to be "a completely self-contained, functional cholinergic unit" [43], possibly regulating the tail beating. Such a function, however, is not possible in the amoeboid, motionless sperm cell of Artemia which, along with the male pronucleus, stains much stronger than the mature egg (Figs. 23-29). The staining is mostly localized at the plasma membrane and the nuclear envelope (Figs. 24,25), and the enzyme has substrate and inhibitor specificities of AChE. The sperm cell of Artemia, then, seems to be a true excitable cell. The modifications that such a cell undergoes during the early phases of fertilization are unknown. However, my results indicate that sperm cells close to the egg may show an everted or "exploded" bulge (Fig. 25) which in other crustaceans corresponds to the reacted acrosome. As a hypothesis, the cholinergic system may trigger the fast intracellular movements of the acrosomal reaction, and/or take part in sperm-egg interactions. Another crucial point is what the highly excitable sperm plasma membrane does when inserted into that of the egg. If it contains nicotinic ACh receptors similar to those of the muscle cells, they could directly gate ion channels [44]. There is some evidence that the earliest electrical changes of the fertilized egg membrane depend on ion channels located in the plasma membrane of the sperm [45], and the "fertilization wave" that modifies the cortex and the cell membrane of the egg starts from the point of the sperm entry [46,47]. As suggested by some observations, one of the fast post-fertilization changes could be the activation of the cholinergic system of the egg, accompanied by a conspicuous increase in the turnover of polyphosphoinositides [48].

The fertilized egg of Artemia shows increased histochemically detected ChE activity at the plasma membrane and in the cytoplasm (Figs. 27,32). At the same time intracellular movements take place: the 90° rotation of the meiotic spindle (Figs. 26,27), the completion of meiosis, the migration of the pronuclei to join each other and then towards the centre of the egg (Figs. 28-30,32-34), and eventually mitosis (Fig. 35). The migration of the pronuclei in Artemia is similar to the process described in other species but best known in the sea urchin [49]. In particular, an aster develops, which, judging from its location, is organized around a centriole contributed by the sperm (Figs. 29,31). The female pronucleus (Fig. 28) may "glide" along the microtubules of the sperm aster to reach the male pronucleus. The two pronuclei, as they shift to the centre of the egg

---

Fig. 23,24,25,26,27,28,29.

(Continued)

Fig. 26, Rotation of the meiotic spindle of the egg (large arrows)
27 which soon after sperm entry is parallel (Fig. 26) and later perpendicular (Fig. 27) to the surface. At the same time ChE-active granules (small arrows), mostly associated with yolk, are detected in the cytoplasm. 3 μm sections, X 2,000.
Fig. 28, Migration of the female (Fig. 28) and male (Fig. 29) pro-
29 nuclei (large arrows). The male pronucleus shows stronger ChE activity and is accompanied by a developing aster (small arrow). 3 μm sections, X 1,300.

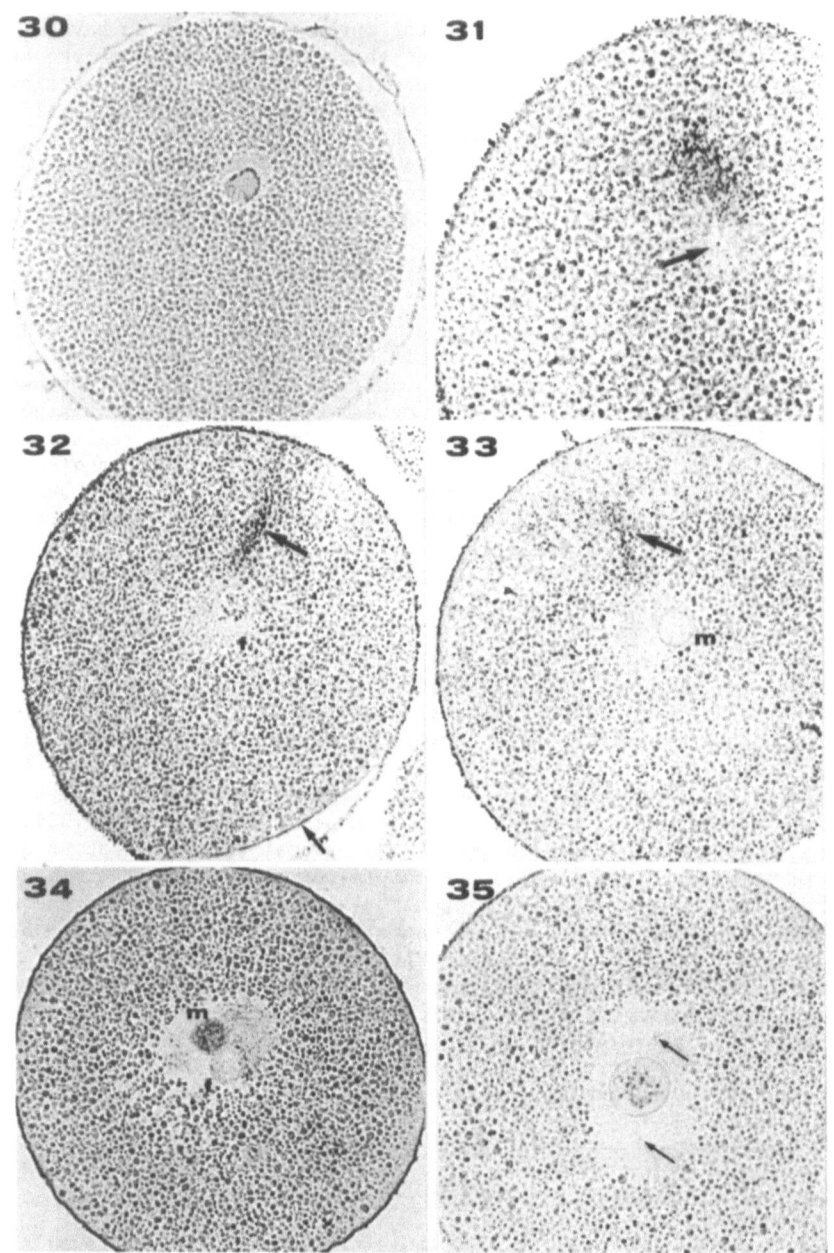

Fig. 30- Centration of the pronuclei. Fig. 30; early phase of the cen-
35.   tripetal migration; the ChE-active male pronucleus is evident,
      X 350. Fig. 31; development of the sperm aster (arrow) associ-
      ated with many ChE-active granules, X 500. Figs. 32-34; the
      migrating pronuclei (f:female; m:male) are followed by a
      "streak" of filaments (microtubules?) and ChE-active gran-
      ules (large arrows). ChE staining is also associated with
      the egg surface (small arrow). At centration (Fig. 34) the
      "streak" is no longer evident, X 370. Fig. 35; early prophase.
      Arrows, centrioles, X 450. 3 μm sections.

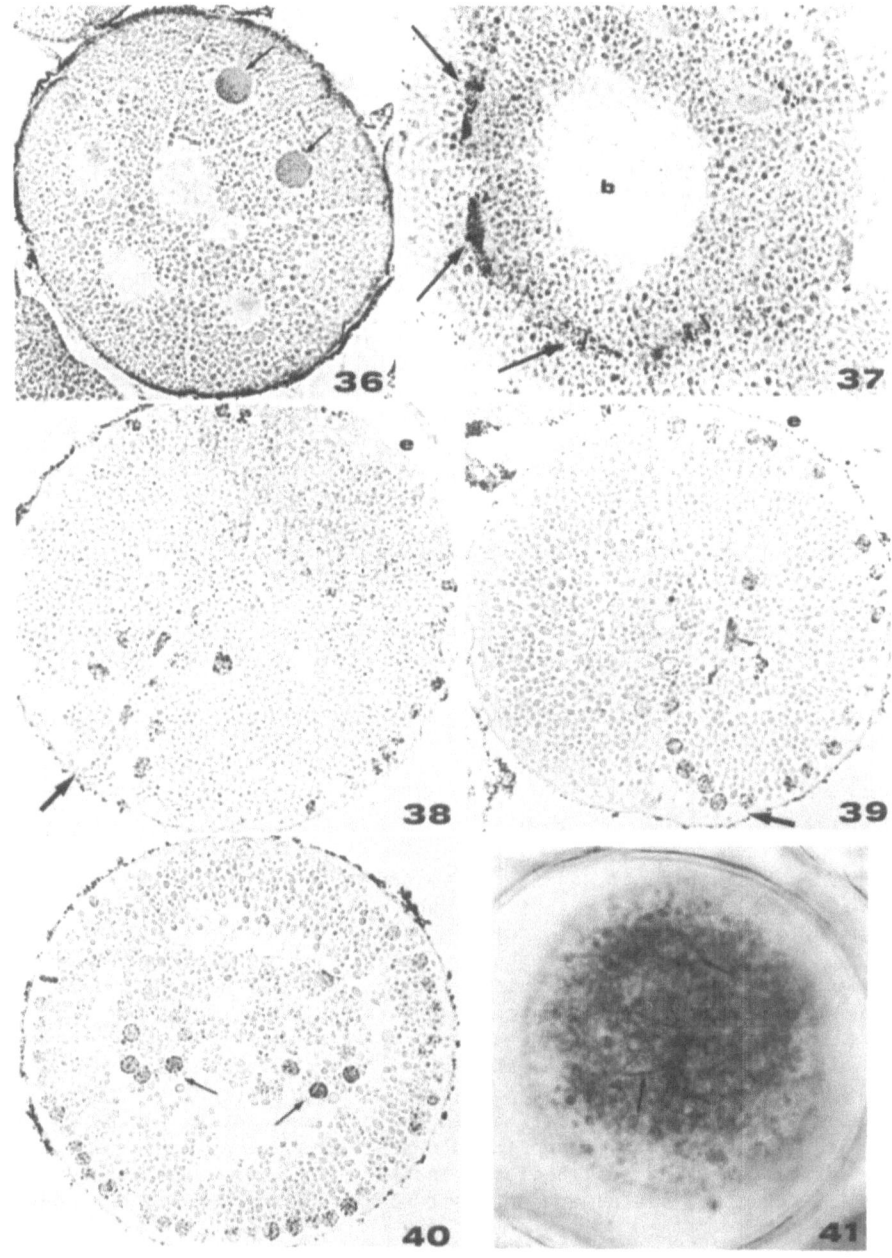

Figs. 36, During segmentation, ChE is associated with the nuclear
     37. envelope (small arrows) at interphase (Fig. 36, X 450)
          and the spindles (large arrows) at mitosis (Fig. 37,
          metaphase, X 350). b, blastocoel; 3 μm sections.
Figs. 38, Early (Fig. 38) and late (Fig. 39) gastrulation I, that
     39. is, the migration of the mesodermal cells. The presump-
          tive endoderm (e) is still at the embryo surface and will
          move inside at gastrulation II. The migrating mesodermal
          cells (arrows) show ChE activity. 3 μm sections, X 350.
Fig. 40. Transverse 3 μm section of a gastrula I at the level of

(Fig. 30), are followed by a trail rich in ChE-active granules and filaments which look like microtubules (Figs. 31-33). The streak disappears
by centration (Fig. 34), two centrioles are distinguished (Fig. 35), the
nuclear envelopes disassemble, and the egg enters division. Such cytoskeleton reorganization may be Ca^{2+} -dependent and modulated by the cholinergic system.

During segmentation, ChE staining varies in each blastomere according
to the phase of mitosis and the stain is mostly associated with the nuclear
envelope or the spindle (Figs. 36,37). Cell division, then, seems to involve cholinergic regulation. The ChE-active vesicles also associate with
yolk platelets (Figs. 27,42,44,45) as found in vertebrates by light microscopical and biochemical methods [50], and ACh may take part in the activation of lysosomal enzymes [51]. This is in agreement with the fact that
Artemia vitellolysis depends on lysosomal digestion [7], a process activated when cells divide and differentiate in embryos and early larvae.

At gastrulation, the ChE reaction is strongest in the immigrating
mesodermal (Figs. 38,39), and later in the endodermal cells, possibly in
relation to their movements. Afterwards, the hisochemical staining becomes
weaker and weaker in the endoderm, but increases in the presumptive myoblasts of the naupliar segments (Fig. 40) and contemporarily in the dorsal
surface of the embryo (presumptive neuroectoderm) (Figs. 41,43). Studies
of vertebrate neurogenesis and myogenesis have shown changes in the
cholinergic system during differentiation [52]. Up to now, however, there
is no detailed biochemical information about the transition from the "embryonic" to the adult nervous or muscular cholinergic system. The same
is true for the somewhat related question of how differentiated, non-
cholinergic cells can express embryonic-like ChE when they undergo tumor
transformation [53].

Acid Phosphatase

Detailed histochemical, quantitative and electrophoretic investigations
have been carried out on Ac-P activity, while β -Glu and β -Gal have been
studied mostly by histochemical staining with only a few electrophoretic
analyses and quantitative measurements. These studies concerned the larval
development of Artemia, starting with the nauplius protruding from the
cyst and still enclosed in the hatching membrane (emergence stages I and
II) [54], up to juvenile forms and adults. On the whole, the three acid
hydrolase activities investigated show significant developmental similarities. In the emerging and first instar nauplius the strongest enzyme
histochemical stainings are localized in the differentiating endodermal
midgut (Fig. 46), though some appreciable reactions are evident in other
larval territories. The yolk bodies, visualized by the periodic acid-
shiff (PAS) reaction, are numerous in all the naupliar tissues, but mostly

---

Fig. 36,37,38,39,40,41
(Continued)

       the presumptive antennal segment. Stronger ChE staining
       is localized in the myogenic cells (arrows), X 350.

Fig. 41.   Encysted gastrula in toto showing the ChE-active
       dorsal area where neuroblasts will differentiate. AChE-
       like histochemical staining is most evident at the plasma
       membranes (arrows), X 350.

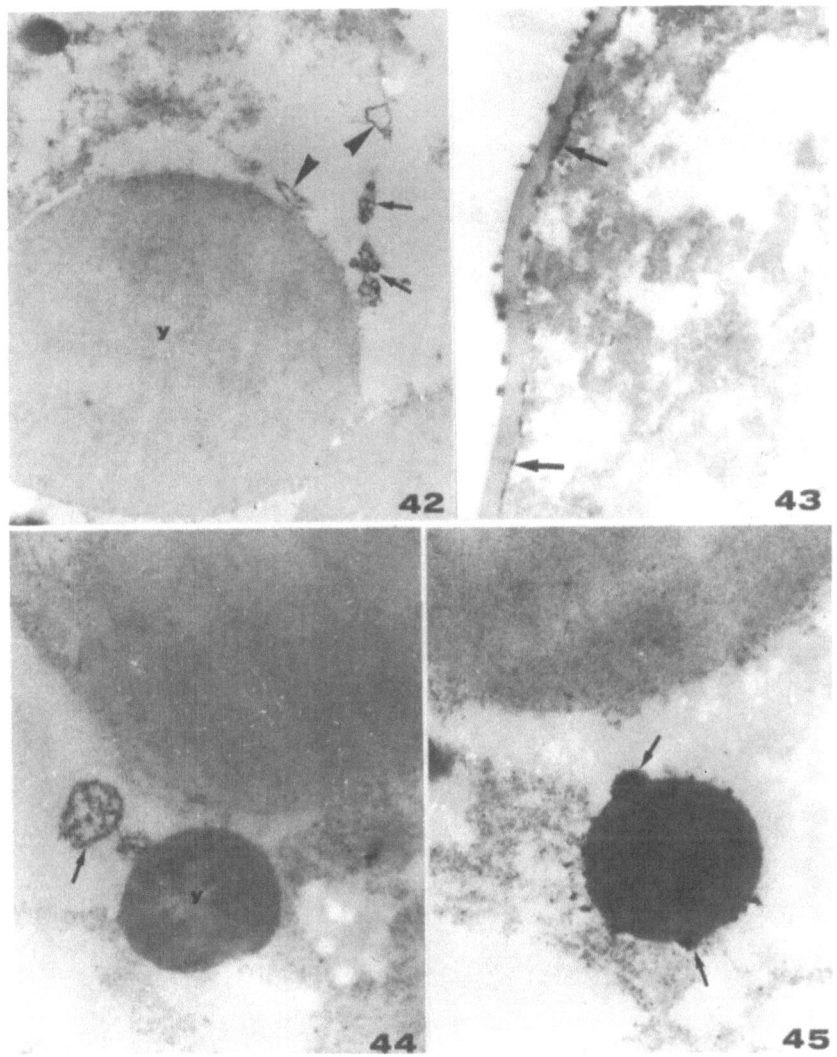

Figs. 42-45. Electron microscopic localization of ChE activity.
Fig. 43: the enzyme reaction is associated with
the surface cell membrane (large arrows) in the
dorsal area of an encysted gastrula, X 45,000.
Fig. 42, X 32,000 and Figs. 44,45, X 70,000, fertilized egg during pronuclear migration. Some ChE-
positive cytoplasmic vesicles (small arrows) are
morphologically similar to the primary lysosomes
observed by Perona et al. [7], and their ChE
activity might be involved in regulating their
functions of yolk (y) degradation. In this hypothesis the pattern of Fig. 44 should correspond
to an earlier stage than that of Fig. 45. Other
histochemically stained membrane vesicles (arrow
heads) might be involved in cytoplasmic free
Ca^{2+} regulation.

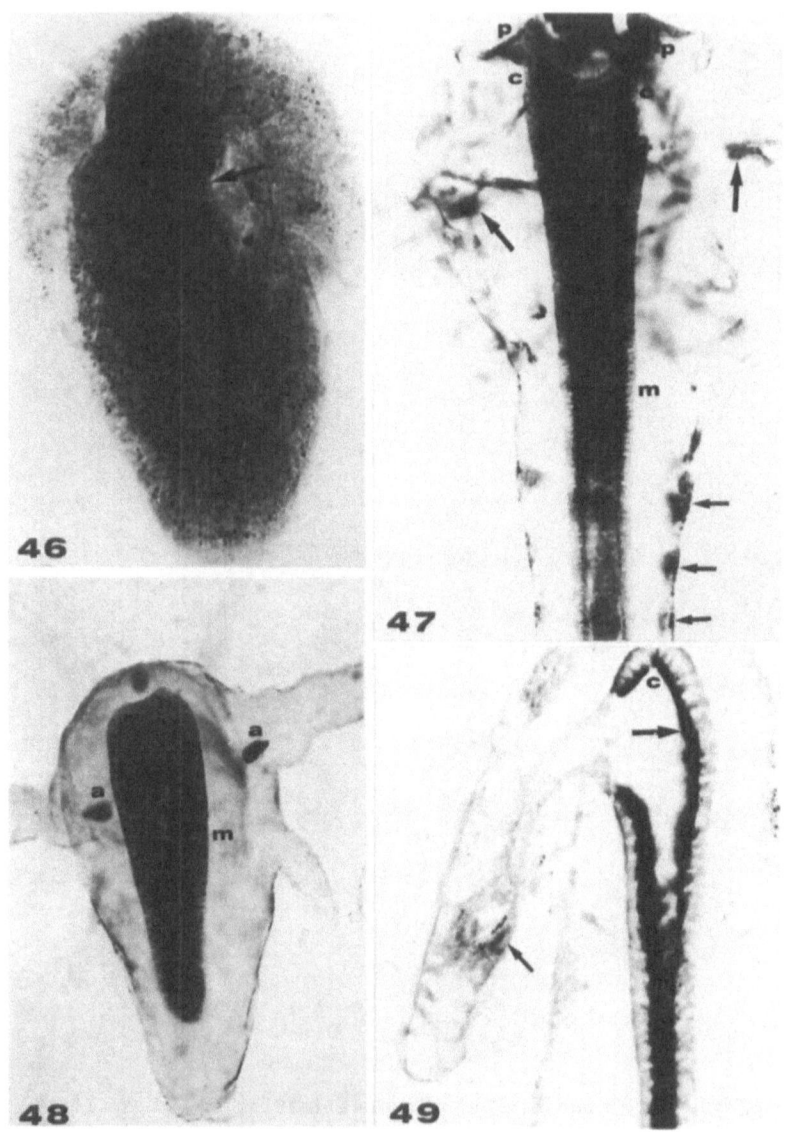

Fig. 46.　Emerging nauplius stained <u>in</u> <u>toto</u> for Ac-P activity.　The
　　　　　enzyme reaction is detected in all the tissues but is
　　　　　stronger in the midgut (arrow), X 300.

Fig. 47.　Metanauplius II stained <u>in</u> <u>toto</u> for AP activity.　The
　　　　　enzyme reaction is most evident in the gastric caeca (c)
　　　　　and the anterior midgut (m) and less in the protocerebral
　　　　　ganglia (p), antennal glands (large arrows) and rudiments
　　　　　of the postnaupliar segments (small arrows), X 250.

Fig. 48.　Just hatched nauplius stained <u>in</u> <u>toto</u> for ALP activity.
　　　　　a, antennal glands; m, midgut, X 200.

Fig. 49.　Metanauplius II, 5 μm longitudinal section stained for AP
　　　　　activity, to be compared with Fig. 47.　The enzyme reaction
　　　　　is very evident in the brush-border (large arrow) of the
　　　　　gastric caeca (c) and middle midgut, and is localized in
　　　　　some nerve cells inside the lip (small arrow), X 300.

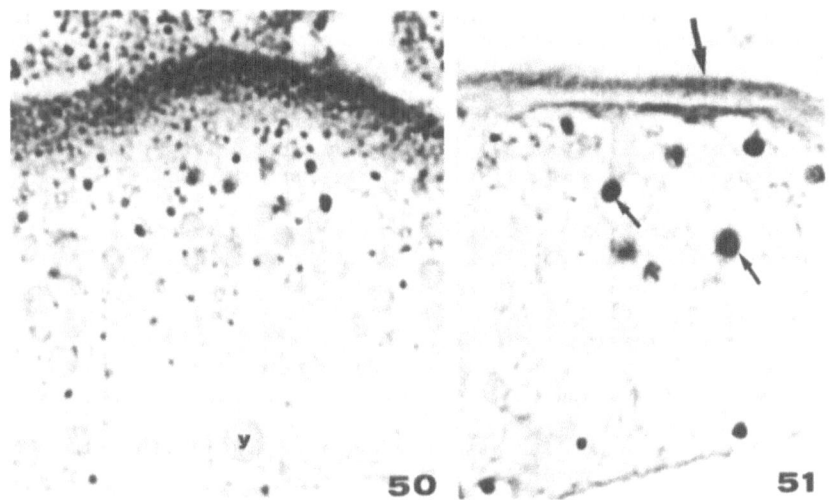

Fig. 50, 51. Developmental pattern of ALP associated
with the brush-border differentiation of
the midgut epithelium in an early emerging (Fig. 50) and just hatched nauplius
(Fig. 51). The distribution of AP is
similar. The histochemically stained
granules or vesicles at the apex of the
cells disappear progressively while the
enzyme-active brush-border develops
(large arrow). The yolk bodies (y) decrease in number and their staining
(small arrows) probably indicates breakdown by Ac-P. 3 μm sections X 2,000.

in the midgut cells, and they are associated with the greatest amount of
histochemical product. As a hypothesis, we suggest that Artemia yolk
contains lytic enzymes which are activated for its breakdown in a way coordinated with other metabolic developmental events [55]. At present, such
a possibility is strongly supported by electron microscopic results obtained
by other authors. Acid hydrolases, cathepsin B for instance, become associated with Artemia yolk owing to the fusion of primary lysosomes with the
vitelline granule. A phagosome is formed, and most probably the degradation products are released into the cytosol [7]. The regulatory mechanisms
of lytic enzyme activation have been investigated at the molecular level,
and the acid proteinase which takes part in the process is extensively
characterized [6]. The functional role of a true lysosomal system is supported by our electrophoretic and quantitative results. In emerging
nauplii, the appearance of faster migrating electrophoretic bands of Ac-P
and β-Glu concomitantly with a conspicuous quantitative increase of the
same enzymes can be interpreted as being due to lysosomal activation [56,
57].

In the hatched nauplius the yolk content of the midgut diminishes
progressively starting from the cephalad region (primordia of the gastric
caeca) and the histochemical distribution patterns of Ac-P, β-Glu and
β-Gal undergo parallel changes. As a consequence, at the end of the
naupliar stage the enzyme stainings are stronger in the posterior half of

the midgut and the hindgut region shows acid hydrolase activities. This could be related to cell rearrangements leading to the establishment of a continuous lumen through the midgut and the ectodermal hindgut with the opening of a functional anus. In fact, morphogenetic processes based upon tissue lysis and reshaping generally involve lysosomal activation. The same enzyme histochemical stainings indicate activated lysosomes play a role in the regression of the salt gland in metanauplii [58].

Ac-P, β-Glu and β-Gal activities are temporarily localized in the post-mandibular area of the nauplius and, during the following larval stages, in the primordia of the post-naupliar segments, which might be ascribed to differentiation processes. On the other hand, the same enzymes are detected in gangliar neurons and the gonads up to the adult stage. Fully differentiated gonads particularly show intense β-Glu staining but the role of these enzymes and their developmental changes are unknown.

As the yolk content and its associated enzyme stainings decrease in the naupliar tissues, acid hydrolases, which remain localized in some cells, can be more and more appreciated. Particularly, these enzymes are evident in the excretory duct of the antennal glands and in the midgut cells, where lysosomes have been detected ultrastructurally [59]. Two faster migrating electrophoretic bands of Ac-P, which are not present in nauplii, are detected in the metanaupliar stages. On the contrary, the more conspicuous anodal β-Glu band appearing at emergence is most evident in the first instar nauplius, but less so in early metanauplii, until it disappears in late metanauplii. These investigations should be repeated on carefully synchronized material. If acid phosphatases take part in intracellular digestion and excretion in the midgut and the antennal/maxillary glands respectively, then their activities may vary with differences in development and function of these organs. Lysosomal phosphatases may be involved in tissue lysis at molt [55] but the β-Glu isozyme with the highest activity appears to be more closely related to yolk breakdown. The histochemical data on β-Gal are very similar to those on Ac-P, suggesting comparable functions of the enzymes. Particularly, the staining is strong in the midgut of nauplii and early metanauplii, but reduced in late metanauplii and juvenile forms. This may reflect differences in the digestion mechanisms.

Cytochrome Oxidase and Succinic Dehydrogenase

Some unpublished histochemical results on Su-D and Cy-O have been obtained for nauplii and metanauplii and can be compared with those concerning the other enzymes discussed herein. Strong histochemical staining for the two mitochondrial markers is localized in the salt gland of the first instar nauplius, but decreases in metanauplii when acid hydrolases become significant in the same organ. This indicates that the ATP-dependent active transport related to the osmoregulatory role of the gland [60] ceases with the onset of cell regression and autolytic digestion. In contrast, Su-D and Cy-O reactions become more evident in the excretory ducts of the antennal (Fig. 56), and later the maxillary glands, as well as in the exopodite of juvenile stage thoracopods. This reflects the osmoregulatory functions performed by these structures in different developmental stages [58]. There is also increased staining for the mitochondrial enzymes in the larval striated muscles (Fig. 56) (most evident in the antennae), as their AChE reaction and neuro-muscular synapses increase [8]. This reflects the morpho-functional development of the nervous and muscular cells which, as previously mentioned, goes on in a synchronous coordinated way.

148

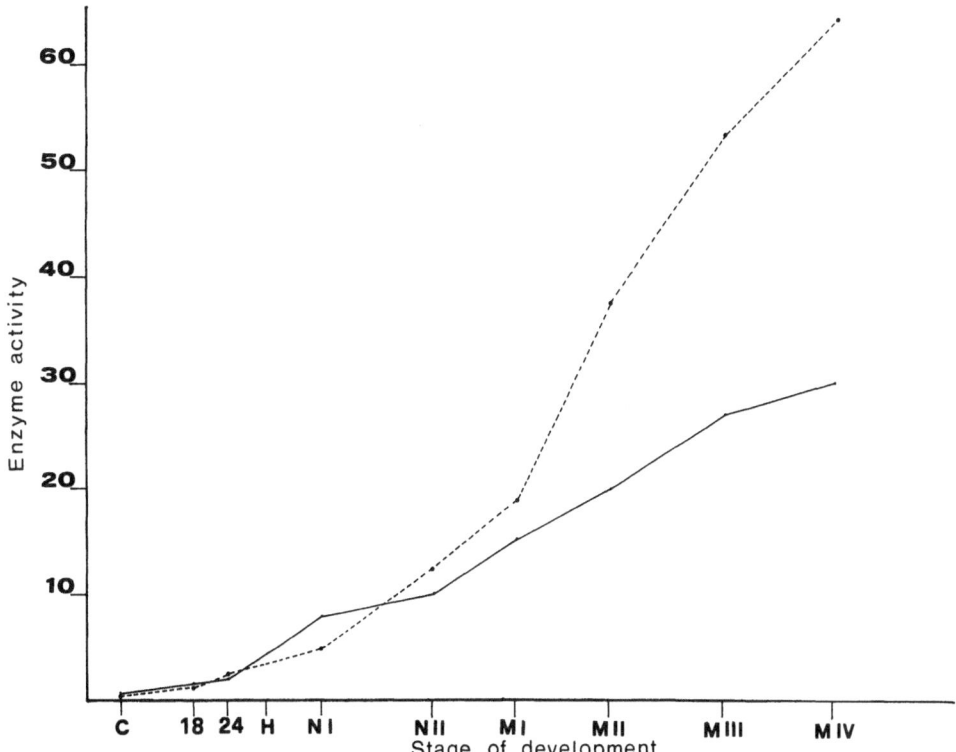

Fig. 52. ALP (-- ● --) and AP (——) activities during Artemia
development according to Raineri [55] and Falugi
et al. [62]. The enzyme units of AP were reduced
10-fold to allow a diagramatic comparison with ALP.

Alkaline Phosphatase and Aminopeptidase

Histochemical, quantitative and electrophoretic investigations of AP
and ALP were carried out mostly to follow the differentiation of brush-
border bearing epithelia, where these enzymes are generally codistributed.
However, since AP and ALP are widespread in different cells and tissues
with a variety of functional roles, the studies were extended to encysted
gastrulae and developing pre-nauplii. The use of different substrates
and inhibitors allowed characterization of the enzymes revealing significant
developmental similarities between AP and ALP [55,61,62].

The very low activities measured in the dehydrated cyst (5 and 0.44
units/mg protein nitrogen of AP and ALP respectively) [55,62] correspond
to a single cathodic band upon electrophoresis and to a very scarce histochemical precipitate distributed as cytoplasmic granules in the various
cell lines without significant differences from one to another. At emergence I and II, while AP and ALP activities increase up to 10 to 20 and
1.28 to 2.55 units respectively, without any change in their electrophoretic
patterns, the histochemical reactions become more and more visible in the
rudiments of the midgut and the antennal glands. For other tissues some
AP reaction is associated with the yolk granules in a pattern comparable
to the one previously described for Ac-P. This may be related to proteolytic yolk digestion, since some proteases can be detected by the histochemical method for AP employed here [63].

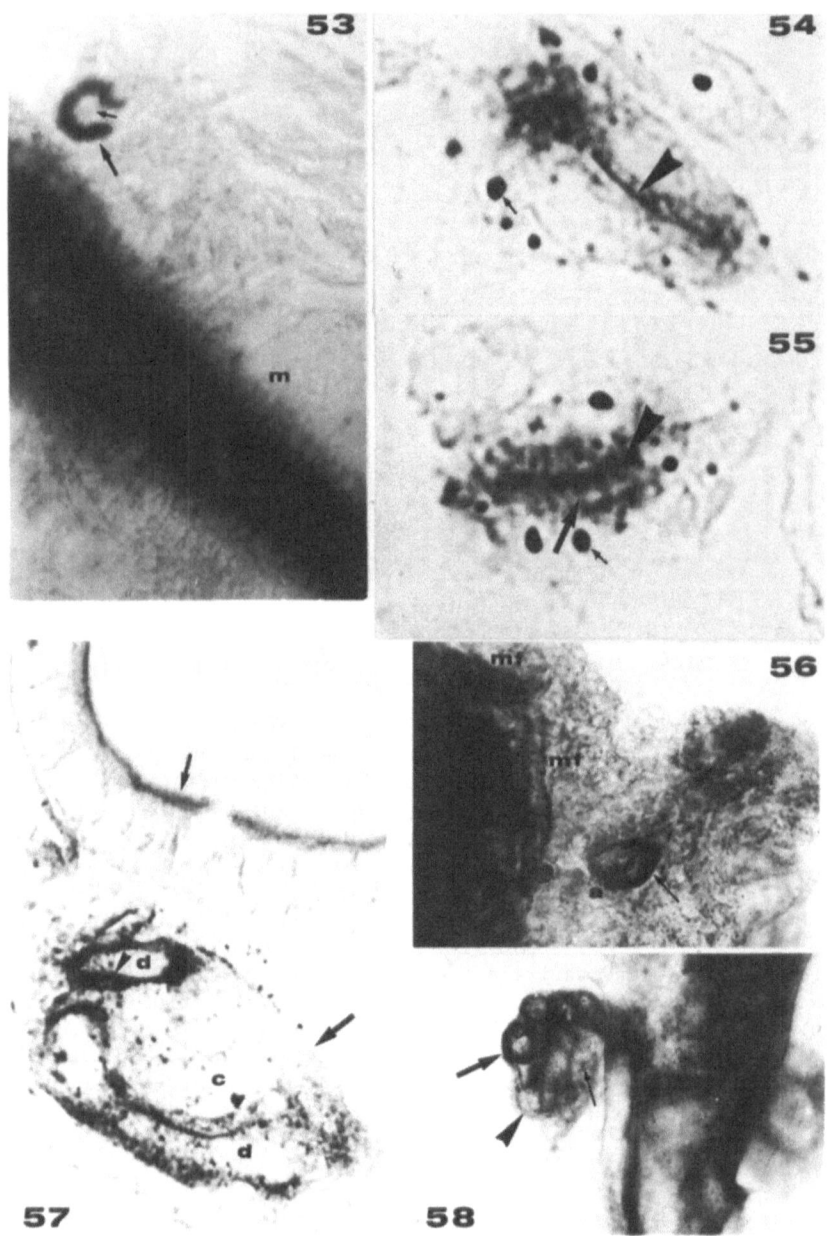

Fig. 53.   Metanauplius II stained in toto for ALP activity.   In
           an antennal gland the coelomic sac (small arrow) and
           the strongly ALP-active excretory duct (large arrow)
           are distinguished.   Very evident ALP staining is lo-
           calized in the midgut (m), X 350.
Figs. 54, Emerging nauplius, consecutive 3 μm sections of an
      55. antennal gland stained for ALP activity.   In the dif-
           ferentiating excretory duct the enzyme is localized
           at the lumen (arrow heads) and as a striation in the
           cell basal region (large arrow, tubular membrane in-

In the first instar nauplius AP and ALP undergo a further quantitative increase of 45 and 4.8 units respectively and there is the appearance of a more anodal and conspicuous band upon electrophoresis of AP and ALP. At the same time, contrary to the foregut and hindgut which remain unstained, increasing AP and ALP reactions become localized in the midgut epithelium (Fig. 48). In the newly hatched nauplius, the enzymes are visualized mostly in the apical region of the cells as numerous granules or vesicles which later decrease in number, while an enzyme-active brush-border develops (Fig. 50,51). It seems probable, therefore, that cytoplasmic AP and ALP are the precursors of the microvillous enzymes. The use of L-phenylalanine (LPA), an inhibitor of intestinal microvillous ALP, indicates most apical cytoplasmic and microvillous ALP histochemical reactions depend on a midgut-specific, intestinal-like brush-border enzyme. The enzyme probably corresponds to the more anodal electrophoretic band, since it is sensitive to the same inhibitor and detected in the same developmental stage.

Some cytoplasmic ALP histochemical staining appears to depend on other types of phosphatases, since it is insensitive to LPA, but abolished by p-bromotetramisole oxalate (BTO), a general inhibitor of ALP. In the basal area of cells, the BTO-sensitive histochemical reaction, seen as a vertical striation, suggests an association of ALP with plasma membrane infolds. Such a cell specialization could be related to active transport performed by the midgut epithelium for osmoregulation [58,64]. A comparable pattern is seen in the excretory duct of the antennal glands (Figs. 48,53), where ALP staining is localized in both the brush-border and, as a vertical striation, in the basal region of the cells (Figs. 54,55).

Microvillous AP and ALP can cooperate in contact breakdown of molecules, absorption and/or secretion. In _Artemia_ larvae they may play a digestive role in the midgut and an excretory one in the antennal and the maxillary glands. On the other hand, the functions of cytoplasmic AP and ALP visualized in some larval tissues and organs are obscure. This is true, for instance, for the enzyme reactions localized in the developing post-naupliar segments and gonads, and for AP in several gangliar neurons. As a common feature all these structures have high metabolic levels which

---

Fig. 53,54,55,56,57,58.

(Continued)

vaginations?). Small arrows, the enzyme reaction of yolk bodies is more evident at acidic pH and probably depends on Ac-P activity. See also Figs. 50, 51, X 2,000.

Fig. 56. Metanauplius I stained _in toto_ for Cy-O activity. The mitochondrial marker is evident in striated muscle fibres (mf) and the excretory duct (arrow) of the antennal gland (a), X 350.

Fig. 57. Juvenile stage, 3 μm section stained for ALP activity. Small arrow, brush-border of the middle midgut. Large arrow, maxillary gland with its coelomic sac (c) wrapped by the coiled excretory duct (d). This shows an ALP activity that is strongest in the brush-border (arrow head) of its proximal portion, X 400.

Fig. 58. Metanauplius IV, _in toto_ view of a maxillary gland. ALP staining is visualized in the proximal tubule (large arrow) coiled around the coelomic sac (small arrow) and is less intense in the distal tubule (arrow head), X 120.

would be based upon high activities of different enzymes [62].

In the second instar nauplius the differentiation of a functional digestive tract corresponds to stronger AP and ALP histochemical stainings of the midgut epithelium, particularly in the brush-border. The antennal glands show increased enzyme reactions and such a parallel development may reflect the cooperation of the gut and the glands in osmoregulation [58]. This is accompanied by a fast quantitative increase in AP and ALP of 80-100 and 12.5 units respectively (Fig. 53). Similarly in metanaupliar stages, quantitative increases of AP and ALP, accompanied by the appearance of new electrophoretic bands, may be related to further differentiation of the midgut and the antennal glands as well as to the development of the maxillary glands. This is supported by the histochemical findings, which indicate that portions of the midgut and the excretory-osmoregulatory glands become specialized to perform different functions, in agreement with ultra-structural observations of other authors [57,65,66].

Even though the midgut epithelium consists of a single basic cell type, the gastric caeca are specialized for lipid storage and possibly enzyme secretion for extracellular digestion, while glycogen storage and osmo-regulation are performed mainly by the midgut cells [59]. The histochemical distribution pattern of AP and ALP, the former more evident in the gastric caeca, the latter in the middle midgut, may reflect local functional specialization in contact digestion and absorption by the brush-border. The microvilli are longer in the gastric caeca than in the middle midgut, and shorter in the posterior midgut [59]. As for the antennal and maxillary glands, the coelomic sac is the least AP- and ALP-positive portion of these organs, while, as described previously, it shows some lysosomal hydrolase activities. Therefore the production of primary urine seems not to involve significant active transport, while some intracellular digestion may take place. Most probably the major modifications of the urine occur in the proximal tubule of the excretory duct, which has much more evident AP and ALP activities, mostly localized in the brush-border. This pattern may be related to contact breakdown and absorption/secretion of molecules, as suggested by the detection of cytoplasmic acid hydrolases. These glands, then, seem to perform excretory and osmoregulatory roles.

All of the observations I report indicate that during Artemia develop-ment the appearance of most new molecular forms of AP and ALP is related to the function, rather than the early differentiation, of the organs where they are localized. Such a production of stage-specific enzymes may be of interest for studies on selective gene expression.

REFERENCES

1. J. S. Clegg and F. P. Conte, A review of the cellular and developmental biology of Artemia, in: "The Brine Shrimp Artemia", Vol. 2, G. Persoone, P. Sorgeloos, O. Roels and E. Jaspers, eds., Universa Press, Wetteren (1980).
2. M. G. Cacace, M. Bergami and A. Sada, Developmental regulation of Artemia glycogen phosphorylase, in: "Artemia Research and its Ap-plication", Vol. 2, W. Decleir, L. Moens, H. Slegers, E. Jaspers and P. Sorgeloos eds., Universa Press, Wetteren (1987).
3. B. Funke and K-D. Spindler, Developmental changes of chitinolytic enzymes and ecdysteroid levels during the early development of the brine shrimp Artemia, in: "Artemia Research and its Application", Vol. 2, W. Decleir, L. Moens, H. Slegers, E. Jaspers and P. Sorgeloos, eds., Universa Press, Wetteren (1987).
4. P. Llorente and A. Ruíz-Cárdaba, Carbamyl phosphate synthetase activi-ties during Artemia development, in: "Artemia Research and its Ap-

plication", Vol. 2, W. Decleir, L. Moens, H. Slegers, E. Jaspers and P. Sorgeloos, eds., Universa Press, Wetteren (1987).

5. A. H. Warner, The role of proteases and their control in Artemia development, in: "Artemia Research and its Application", Vol. 2, W. Decleir, L. Moens, H. Slegers, E. Jaspers and P. Sorgeloos, eds., Universa Press, Wetteren (1987).

6. B. Ezquieta and C. G. Vallejo, Artemia tripsin-like proteinase: a developmentally regulated proteinase, in: "Artemia Research and its Application", Vol. 2, W. Decleir, L. Moens, H. Slegers, E. Jaspers and P. Sorgeloos, eds., Universa Press, Wetteren (1987).

7. R. B. Perona, B. Ezquieta and C. G. Vallejo, The degradation of yolk in Artemia, in: "Artemia Research and its Application", Vol. 2, W. Decleir, L. Moens, H. Slegers, E. Jaspers and P. Sorgeloos, eds., Universa Press, Wetteren (1987).

8. M. Raineri and C. Falugi, Acetylcholinesterase activity in embryonic and larval development of Artemia salina Leach (Crustacea Phyllopoda), J. Exp. Zool. 227:229 (1983).

9. E. Neumann and D. Nachmansohn, Nerve excitability. Towards an integrating concept, in: "Biomembranes", Vol. 7, L. A. Manson, ed., Plenum Press, New York (1975).

10. A. Minganti and C. Falugi, An epithelial localization of acetylcholinesterase in the ascidian Ciona intestinalis embryos and larvae, Acta Embryol. Exper. n.s., 1:143 (1980).

11. A. Minganti, C. Falugi, M. Raineri and M. Pestarino, Acetylcholinesterase in the embryonic development: an invitation to a hypothesis, Acta Embryol. Exper. n.s., 2:30 (1981).

12. U. Drews, Cholinesterase in embryonic development, "Progress in Histochemistry and Cytochemistry", Vol. 7, Gustav Fisher Verlag, Stuttgart (1975).

13. C. Falugi and M. Raineri, Acetylcholinesterase (AChE) and pseudocholinesterase (BuChE) activity distribution pattern in early developing chick limbs, J. Embryol. Exp. Morphol. 86:89 (1985).

14. E. Maynard, Esterases in crustacean nervous system. I. Electrophoretic studies in lobsters, J. Exp. Zool. 157:251 (1964).

15. E. Maynard, Microscopical localization of cholinesterases in the nervous system of the lobsters, Panulicus argus and Homarus americanus, Tiss. Cell 3:215 (1971).

16. C. M. Bate and E. B. Grunewald, Embryogenesis of an insect nervous system. II. A second class of neuron precursor cells and the origin of the intersegmental connectives, J. Embryol. Exp. Morphol. 61:317 (1981).

17. C. E. Blanchard, Pioneer neurons and the early development of the nervous system in Artemia, in: "Artemia Research and its Application", Vol. 1, P. Sorgeloos, D. A. Bengtson, W. Decleir and E. Jaspers, eds., Universa Press, Wetteren (1987).

18. H. Keshishian, The origin and morphogenesis of pioneer neurons in the grasshopper metathoracic leg, Dev. Biol. 80:388 (1980).

19. P. M. Whitington and E. Seifert, Axon growth from limb motoneurons in the locust embryo: the effect of target limb removal on the path taken out of the central nervous system, Dev. Biol. 93:206 (1982).

20. D. McMahon, Chemical messengers in development: a hypothesis, Science 185:1012 (1974).

21. G. A. Buznikov, I. V. Chudakova, L. V. Berdysheva and N. M. Vyazmina, The role of neurohumors in early embryogenesis. II. Acetylcholine and catecholamine content in developing embryos of sea urchin, J. Embryol. Exp. Morphol. 20:119 (1968).

22. G. A. Buznikov, Biogenic monoamines and acetylcholine in Protozoa and metazoan embryos, in: "Neurotransmitters: Comparative Aspects", J. Salanki and T. M. Turpaev, eds., Akadémiai Kiadó, Budapest (1980).

23. G. A. Buznikov and Y. B. Shmukler, Possible role of "prenervous" neurotransmitters in cellular interactions of early morphogenesis:

a hypothesis, <u>Neurochem. Res.</u> 6:55 (1981).

24. M. Raineri and P. Modenesi, The cholinergic system: a hypothesis of its general role in living cells, in: "Cellular and Molecular Control of Direct Cell Interactions in Developing Systems", (Abstracts), ASI-NATO, Banyuls-sur-Mer (1984).

25. H. Schmidt, Muscarinic acetylcholine receptor in chick limb bud during morphogenesis, <u>Histochemistry</u> 71:89 (1981).

26. H. Schmidt, G. Oettling, T. Kaufenstein, G. Hartung and U. Drews, Intracellular calcium mobilization on stimulation of the muscarinic cholinergic receptor in chick limb bud cells, <u>Roux's Arch. Dev. Biol.</u> 194:44 (1984).

27. T. Kubo, K. Fukuda, A. Mikami, A. Maeda, H. Takahashi, M. Mishina, K. Haga, A. Ichiyama, K. Kangawa, M. Kojima, H. Matsuo, T. Hirose and S. Numa, Cloning, sequencing and expression of complementary DNA encoding the muscarinic acetylcholine receptor, <u>Nature</u> 323:411 (1986).

28. E. G. Peralta, J. W. Winslow, G. L. Peterson, D. N. Smith, A. Ashkenazi, J. Ramachandran, M. I. Schimerlik and D. J. Capon, Primary structure and biochemical properties of an M_2 muscarinic receptor, <u>Science</u> 236:600 (1987).

29. K. Fukuda, T. Kubo, I. Akiba, A. Maeda, M. Mishina and S. Numa, Molecular distinction between muscarinic acetylcholine receptor subtypes, <u>Nature</u> 327:623 (1987).

30. A. Ashkenazi, J. W. Winslow, E. G. Peralta, G. L. Peterson, M. I. Schimerlik, D. J. Capon and J. Ramachandran, An M_2 muscarinic receptor subtype coupled to both adenyl cyclase and phosphoinisitide turnover, <u>Science</u> 238:672 (1987).

31. G. Oettling, H. Schmidt and U. Drews, The muscarinic receptor of chick embryo cells: correlation between ligand binding and calcium mobilization, <u>J. Cell Biol.</u> 100:1073 (1985).

32. A.J. Trewavas, R. Sexton and P. Kelly, Polarity, calcium and abscission: molecular bases for developmental plasticity in plants, <u>J. Embryol. Exp. Morphol.</u> 83 (Suppl.):179 (1984).

33. M. J. Berridge and R. F. Irvine, Inositol trisphosphate, a novel second messenger in cellular signal transduction, <u>Nature</u> 312:315 (1984).

34. G. A. Buznikov, Acetylcholine and biogenic monoamines as intracellular regulators of early embryogenesis, <u>Sov. Sci. Rev. F. Physiol. Gen. Biol.</u> 1:137 (1987).

35. P. Vanittanakom and U. Drews, Ultrastructural localization of cholinesterase during chondrogenesis and myogenesis in the chick limb bud, <u>Anat. Embryol.</u> 172:183 (1985).

36. K.-U. Thiedemann, P. Vanittanakom, F.-M. Schweers and U. Drews, Embryonic cholinesterase activity during morphogenesis of the mouse genital tract. Light- and electron-microscopic observations, <u>Cell Tiss. Res.</u> 244:153 (1986).

37. D. M. Gardiner and R. D. Grey, Membrane junctions in <u>Xenopus</u> eggs: their distribution suggests a role in calcium regulation, <u>J. Cell Biol.</u> 96:1159 (1983).

38. A. Cartaud, J. Boyer and R. Ozon, Calcium sequestring activities of reticulum vesicles from <u>Xenopus laevis</u> oocytes, <u>Exp. Cell Res.</u> 155:565 (1984).

39. A. Forer and P. J. Sillers, The role of the phosphatidylinositol cycle in mitosis in sea urchin zygotes. Lithium inhibition is overcome by myo-inositol but not by other cyclitols or sugars, <u>Exp. Cell Res.</u> 170:42 (1987).

40. M. Whitaker, Inositol 1,4,5-trisphosphate microinjection activities sea urchin eggs, <u>Nature</u> 312:636 (1984).

41. W. B. Busa, J. E. Ferguson, S. K. Joseph, J. R. Williamson and R. Nuccitelli, Activation of frog (<u>Xenopus laevis</u>) eggs by inositol trisphosphate. I. Characterization of Ca^{2+} release from intracellular stores, <u>J. Cell Biol.</u> 101:677 (1985).

42. P. R. Turner, L. A. Jaffe and A. Fein, Regulation of cortical vesicle exocytosis in sea urchin eggs by inositol 1,4,5-trisphosphate and GTP-binding protein, J. Cell Biol. 102:70 (1986).
43. L. Cariello, G. Romano and L. Nelson, Acetylcholinesterase in sea urchin spermatozoa, Gamete Res. 14:323 (1986).
44. B. Sakmann, C. Methfessel, M. Mishina, T. Takahashi, T. Takai, M. Kurasaki, K. Fukuda and S. Numa, Role of acetylcholine receptor subunits in gating of the channel, Nature 318:538 (1985).
45. B. Dale and L. Santella, Sperm-oocyte interaction in the sea-urchin, J. Cell Sci. 74:153 (1985).
46. B. Picheral and M. Charbonneau, Anuran fertilization: a morphological reinvestigation of some early events, J. Ultrastr. Res. 81:306 (1982).
47. A. Eisen, D. P. Kiehart, S. J. Wieland and G. T. Reynolds, Temporal sequence and spatial distribution of early events of fertilization in single sea urchin eggs, J. Cell Biol. 99:1647 (1984).
48. P. R. Turner, M. P. Sheetz and L. A. Jaffe, Fertilization increases the polyphosphoinositide content of sea urchin eggs, Nature 310:414 (1984).
49. G. Schatten, Motility during fertilization, Int. Rev. Cytol. 79:60 (1982).
50. R. A. Fluck, Localization of acetylcholinesterase activity in young embryos of the medaka Oryzias latipes, a teleost, Comp. Biochem. Physiol. 72C:59 (1982).
51. T. Laasberg, A. Pedak and T. Neuman, The muscarinic receptor-mediated action of acetylcholine in the gastrulating chick embryo, Comp. Biochem. Physiol. 86C:313 (1987).
52. A. Miki and H. Mizoguti, Acetylcholinesterase activity in the myotome of the early chick embryo, Cell Tiss. Res. 227:23 (1982).
53. C. Falugi, E. Balza and L. Zardi, Localization of acetylcholinesterase in normal human fibroblasts and a human fibrosarcoma cell line, Bas. Appl. Histochem. 27:205 (1983).
54. Y. H. Nakanishi, T. Iwasaki, T. Okigaki and H. Kato, Cytological studies of Artemia salina. I. Embryonic development without cell multiplication after the blastula stage in encysted dry eggs, Annot. Zool. Jpn. 35:223 (1962).
55. M. Raineri, Histochemical and biochemical study of alkaline phosphatase (ALP) activity in developing embryos and larvae of Artemia, in: "Artemia Research and its Application", Vol. 2, W. Decleir, L. Moens, H. Slegers, E. Jaspers and P. Sorgeloos, eds., Universa Press, Wetteren (1987).
56. R. T. Swank and K. Paigen, Biochemical and genetic evidence for a macromolecular β-glucuronidase complex in microsomal membranes, J. Mol. Biol. 77:371 (1973).
57. C. Falugi and M. Raineri, Fosfatasi nello sviluppo dei Crostacei, Boll. Zool. 45:210 (1978).
58. F. P. Conte, Structure and function of the crustacean larval salt gland, Int. Rev. Cytol. 91:45 (1984).
59. A. Schrehardt, Ultrastructural investigations of the filter-feeding apparatus and the alimentary canal of Artemia, in: "Artemia Research and its Application", Vol. 1, P. Sorgeloos, D. A. Bengtson, W. Decleir and E. Jaspers, eds., Universa Press, Wetteren (1987).
60. R. J. Lowy and F. P. Conte, Isolation and functional characterization of crustacean larval salt gland, Am. J. Physiol. 248:R702 (1985).
61. C. Falugi, Histochemical investigations on aminopeptidases in Artemia salina (Phyllopoda) embryos and larvae, Acta Embryol. Exper. 2:171 (1978).
62. C. Falugi, M. Raineri and E. Vanara, Aminopeptidasi nello sviluppo di Artemia salina Leach, in: "Ricerca Scientifica ed Educazione Perman-ente," Suppl. n.6, Atti del XLVII Convegno dell'Unione Zoologica Italiana, ed. Universita degli Studi di Milano (1979).

63. M. M. Nachlas, B. Monis, D. Rosenblatt and A. M. Seligman, Improvement in the histochemical localization of leucine aminopeptidase with a new substrate, L-leucyl-4-methoxy-β-naphthylamide, J. Biophys. Biochem. Cytol. 7:261 (1960).
64. S. R. Hootman and F. P. Conte, Fine structure and function of the alimentary epithelium in Artemia salina nauplii, Cell Tiss. Res. 155:423 (1974).
65. G. E. Tyson, The fine structure of the maxillary gland of the brine shrimp, Artemia salina: the end sac, Z. Zellforsch. 86:129 (1968).
66. G. E. Tyson, The fine structure of the maxillary gland of the brine shrimp, Artemia salina: the efferent duct, Z. Zellforsch. 93:151 (1969).

MULTIPLE PROTEASE INHIBITORS IN ARTEMIA CYSTS

Alden H. Warner

Department of Biological Sciences
University of Windsor
Windsor, Ontario, Canada N9B 3P4

INTRODUCTION

Proteolysis in developing systems requires that intracellular proteases operate within certain physiological limits. Several factors appear to regulate intracellular protease activity including compartmentalization, membrane association, cofactor levels intracellular pH (pH_i) and the presence of endogenous protease inhibitors [1]. It now seems clear that proteolytic activity is required for yolk utilization, at least in embryos of amphibia [2] and crustacea [3,4], and perhaps in the unmasking of ribosomes [5,6] and mRNP particles [7] following fertilization or egg activation. However, in all cases it is imperative that proteolytic activity be controlled closely, regardless of its physiological purpose, so that inappropriate proteolysis doesn't occur with resulting cytolysis and loss of embryo viability.

Encysted embryos of Artemia possess great plasticity and the ability to resist harsh environmental conditions such as high salinity and low oxygen pressures [8,9]. Consequently, the embryos are able to withstand several months of anoxia while fully hydrated, without irreversible cellular damage to the embryo, during which time the pH_i becomes quite acidic [10-12]. The mechanism which operates in Artemia to protect the embryo from anoxia-induced cytolysis is not known, but it must, arguably, include intracellular protease regulation.

In this chapter I describe the occurrence, partial purification and some general properties of the multiple protease inhibitor proteins (isoinhibitors) from dormant cysts of Artemia. More detailed analyses will be published elsewhere. Current data suggest that these proteins may be important regulatory molecules in protease control during early development of Artemia.

MATERIALS AND METHODS

The Artemia cysts used in all experiments were from the Utah salterns (lot number 12715, Sanders Brine Shrimp Company). DEAE-Sephadex, Sephadex G-75 and Mono S were from Pharmacia; the microconcentrators (Centricon-3) were from Amicon. The C-18 reverse phase column was 4.6 x 250 mm of Chemopack and prepared by Dr. F. L. Huang, National University of Tiawan.

Protamine sulfate was from Calbiochem and 2,4,6-tri-nitrobenzene sulfonic acid (TNBS) was from ICN Biomedicals. All other chemicals were of reagent grade or better.

Assay for Protease Inhibitor Activity

All reaction vessels used to measure protease activity contained the following components in 200 μl final volume unless stated otherwise: 0.1 M sodium acetate, pH 5.0, 1 mM EDTA, 1 mM dithiothreitol (DTT), 7.5% glycerol, 4 mg/ml protamine sulfate, 10-14 enzyme units of Artemia cyst thiol (acid) protease and fractions to be tested for inhibitor activity. The thiol protease was a partially purified enzyme preparation from Artemia cysts which had been through a DEAE-Sephadex column [4]. All reaction vessels were incubated at 30°C and at the desired times 50 μl were removed for the determination of enzyme activity using the TNBS method of Nagainis and Warner [13]. From the measurement of protease activity, the inhibitor activity was calculated and expressed either as percent inhibition (%I) or inhibitor units (IU) compared to controls containing only enzyme. One IU is defined as the amount of protein which prevents (completely) the ex-pression of one enzyme unit (EU). One EU is equivalent to the amount of enzyme that liberates 1 nmol/min of amino groups using l-arginine-HCl as the standard.

Preparation of Cysts for Inhibitor Isolation

Prior to use dormant cysts were hydrated overnight in 0.25 M NaCl in an ice-bath. Floating cysts were removed by suction and the fully hydrated cysts were collected on a sintered-glass filter, washed well with ice cold distilled water and finally with homogenizing buffer, all with suction. The yield from 100 grams dry cysts as starting material was generally between 270-290 grams wet weight.

RESULTS

Isolation of Acid Protease Inhibitors from Artemia Cysts

Extracts from hydrated dormant cysts of Artemia show considerable protease activity when assayed under acidic conditions using a variety of substrates including BSA and protamine sulfate [4]. The enzyme responsible for this activity has been characterized and shown to be an acid (thiol) protease with several properties similar to mammalian cathepsin B. When the embryo extract is passed through a DEAE-Sephadex or DEAE-cellulose column, the total acid protease activity is increased by about 125% [4]. This increase in activity is due to removal of an inhibitor(s) from the protease preparation. The results in Figure 1 show the pattern of inhibitor and enzyme separation using one of the above columns. At pH 6.8 the protease inhibitor fails to bind to DEAE-Sephadex, while the acid protease (AP) is tightly bound and requires moderate concentrations of NaCl, either stepwise or by gradient, to elute from the column. Using a gradient the protease elutes from the column around 0.32 to 0.38 M NaCl.

To further purify the protease inhibitor(s) the unbound fraction from the DEAE-Sephadex column was concentrated by ammonium sulfate (75%) and passed through a Sephadex G-75 column equilibrated with a phosphate buffer at pH 6.1 containing 25 mM KCl and 10% glycerol (Fig. 2). This treatment increases the purity of the inhibitor(s) by over 14-fold and indicates that the Artemia cyst protease inhibitor(s) has (have) a relatively low molecular weight compared to most of the other proteins in the preparation but similar to other intracellular protease inhibitors [14,15].

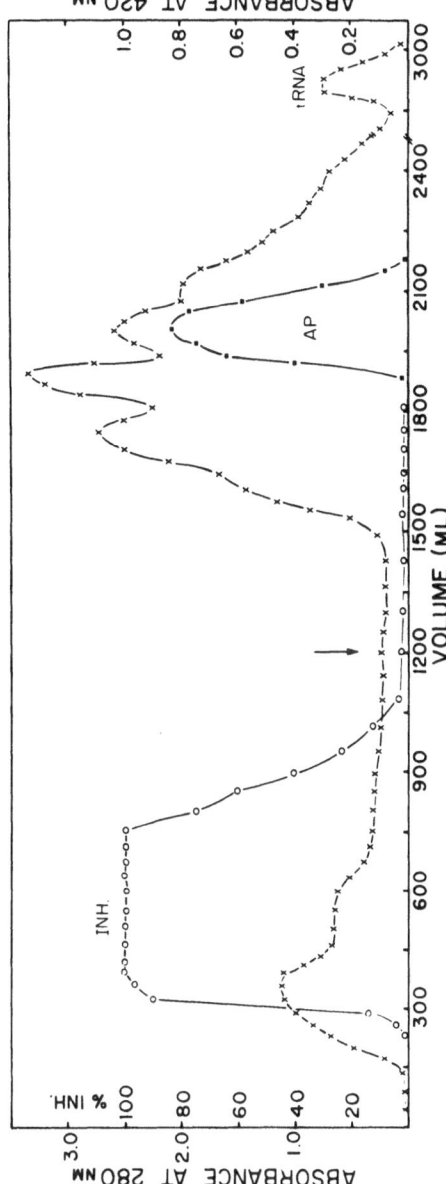

Fig. 1. Separation of protease inhibitor from the acid (thiol) protease on DEAE-Sephadex. A 25-75% ammonium sulfate fraction from the post-ribosomal supernatate from 275 grams wet weight cysts was desalted on a G-25 Sephadex column and applied to a DEAE-Sephadex column (3.5 x 45 cm) equilibrated with the starting buffer (15 mM K-phosphate plus 25 mM KCl, pH 6.8). After the column had been washed with about 1 liter of the starting buffer, the bound protein was eluted (starting at the arrow) with a 2 liter linear gradient of KCl (to 0.75M) in the starting buffer. Protein was monitored at 280 nm (-X-) while acid protease activity was monitored at 420 nm (-●-). The open circles represent inhibitor activity (as % Inh.) in the standard TNBS assay (see Materials and Methods).

In a previous report we indicated that <u>Artemia</u> cysts contain two thiol protease inhibitors [16]. More recent results using a different purification protocol suggest that <u>Artemia</u> cysts contain more than two low molecular weight protease inhibitors. This conclusion is based on results obtained using fast protein liquid chromatography (FPLC). When a Mono S column (cationic, Pharmacia) was used to purify further the inhibitors in the Sephadex G-75 fraction, the results shown in Figure 3 were obtained. While this column did not completely resolve all of the inhibitor-containing fractions, the data suggest that at least four inhibitors are present in partially purified extracts of <u>Artemia</u> cysts. Further chromatography of the Mono S fractions containing inhibitor activity (see small arrows in

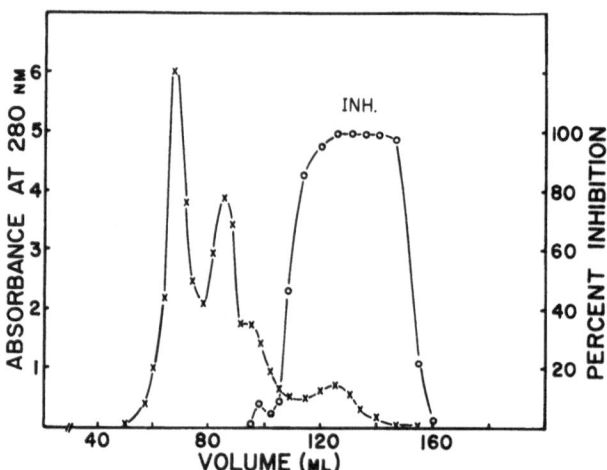

Fig. 2. Purification of the protease inhibitor(s) on Sephadex G-75. The concentrated protease inhibitor fraction from the previous DEAE-Sephadex step was applied to a column of Sephadex G-75 (2.2 x 54 cm) equilibrated with the same buffer used to equilibrate the DEAE-Sephadex column and the column was developed with this buffer. Fractions were assayed for protein at 280 nm (-X-) and for thiol protease inhibitor activity (-O-) as described in Materials and Methods.

Fig. 3) showed four distinct fractions although some overlap in inhibitor activity still existed (data not shown). Since further analyses of each inhibitor fraction on SDS polyacrylamide gels showed multiple bands in each fraction, additional work will be required to completely purify the proteins. Nevertheless, we can conclude that <u>Artemia</u> cysts contain multiple low molecular weight thiol protease inhibitors.

Biophysical Properties of the Thiol Protease Inhibitors in Artemia Cysts

The six fractions from the Mono S column which contained protease

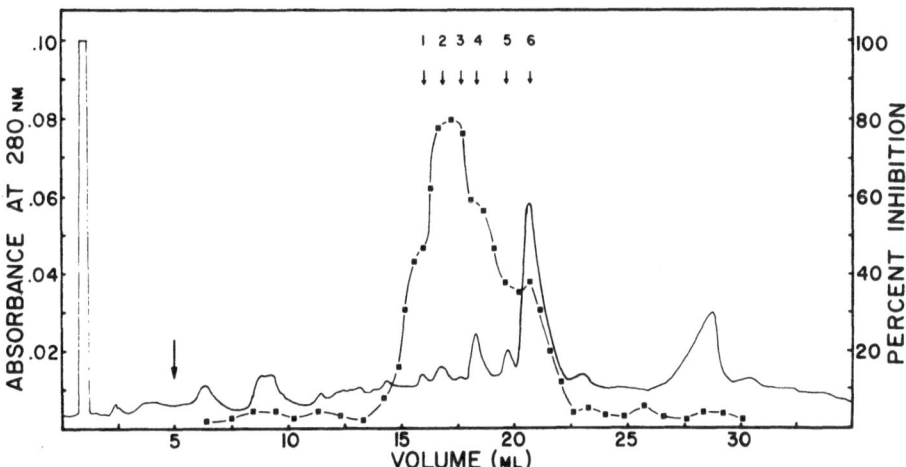

Fig. 3. Purification of Artemia cyst thiol protease inhibitors by fast protein liquid chromatography.
An aliquot (0.5 mg) from the concentrated Sephadex
G-75 inhibitor fraction was applied to a FPLC
column (Mono S, 0.5 x 5 cm) equilibrated with a
buffer containing 20 mM sodium acetate, 10% glycerol and 1 mM EDTA, pH 5.0. Following a 6 ml wash
of the column with the starting buffer, the protein
was eluted with a linear gradient of NaCl (to 0.3
molar) in the starting buffer. The solid line
represents the actual tracing from the recorder of
material eluting at 280 nm while the broken line
(- ■ -) represents inhibitor activity in aliquots
of each column fraction. The large arrow indicates
the start of the NaCl gradient while the small
arrows indicate the column fractions saved for
further analysis.

inhibitor activity were analyzed on a 7-18% polyacrylamide gel in the
presence of 0.1% sodium dodecylsulfate (SDS) and stained first with
Coomassie Blue and then with silver. The results of this analysis showed
several bands in each lane with the major bands migrating at positions of
5, 7 and 14 kilodaltons (data not shown). Unfortunately we are not yet
able to state which band(s) on the gel has inhibitor activity. However,
using gel filtration chromatography on Sephadex G-50 we have been able to
show that the native protease inhibitor protein(s) has (have) a molecular
mass of 8.7 - 9.3 kilodaltons [16]. At the present time we are attempting
to obtain more precise molecular mass determinations of these inhibitors
using SDS-urea gels to better resolve the low molecular mass proteins.

    In two attempts to purify the protease inhibitor(s) using a preparative
isoelectric focusing technique, we found that inhibitor activity focused
in multiple peaks ranging from pH 5.2 to 5.9 (data not shown). The data
suggest that Artemia thiol protease inhibitors are acidic proteins, but
given that they fail to bind to anionic exchangers at neutral pH (see Fig.
1), and bind to cationic exchangers at pH 5, the proteins may, in fact, be
basic molecules. Clearly, further work is required to more clearly establish this chemical property of the inhibitors.

Many low molecule weight intracellular thiol protease inhibitors are relatively stable to heat and extremes of pH [14,17]. When we investigated the sensitivity of the major low molecular weight inhibitor of <u>Artemia</u> cysts to varying pH at two temperatures the results shown in Figure 4 were obtained. At 0°C (ice) the inhibitor is relatively stable to extremes of pH, but at 40°C the inhibitor is unstable at acidic pH and virtually inactive after 30 minutes incubation at pH 3. In other experiments using the (total) inhibitor fraction purified through Sephadex G-75, we found that heating at 75°C and pH 6.0 destroyed all inhibitor activity after 45 minutes incubation (data not shown). Therefore, the <u>Artemia</u> cyst protease inhibitors are relatively stable at neutral and alkaline conditions, but inactivated rapidly at 40°C if the pH is acidic. Whether the inhibitor is unstable in the presence of the cyst thiol protease or other proteins under the above conditions is not known.

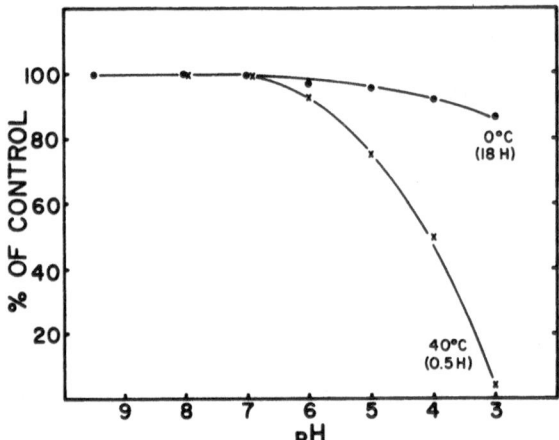

Fig. 4. The inactivation of the major thiol protease inhibitor from <u>Artemia</u> cysts at acidic pH. Pooled inhibitor proteins from the Mono S column (fractions 2 and 3 in Fig. 3) were incubated in either 25 mM sodium phosphate buffer (pH 6,7,8), 25 mM sodium acetate (pH 3,4,5) or in 25 mM NaOH/sodium phosphate (pH 9.5) containing 10% glycerol and 1 mM EDTA at the temperature and times indicated, then assayed for inhibitor activity in a standard reaction mixture.

Biochemical Properties of the Thiol Protease Inhibitors in Artemia Cysts

In an early series of experiments we investigated the kinetics of inhibition of the cyst thiol protease using both dimethyl BSA and protamine sulfate as enzyme substrate at various pH's. The results shown in Figure 5, using inhibitor purified through the Sephadex G-75 step (see Fig. 2) are representative of the type of kinetics observed with protease-free inhibitor preparations. These data show that the extent of inhibition is linear up to about 50-55% inhibition (at pH 5) of the theoretical maximum level of inhibition.

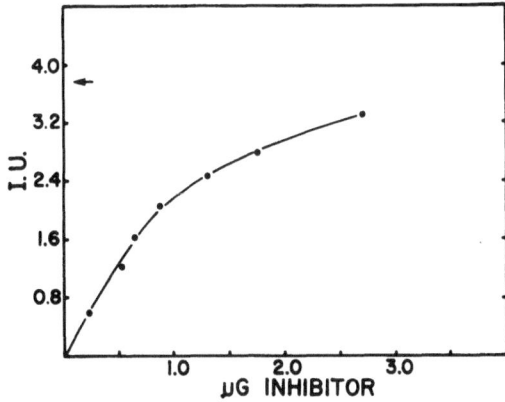

Fig. 5.   Kinetics of Artemia cyst thiol protease inhibition.
Aliquots (50 µl) were removed from a 400 µl
reaction vessel containing various amounts of
total thiol protease inhibitor from the Mono S
column, and the amount of inhibitor units (IU)
were determined. The arrow indicates the amount
of partially purified cyst thiol protease (3.8
EU) which if completely inhibited, would yield
maximum inhibition (100%) or 3.8 IU.

    We also found that the extent of inhibition of the cyst thiol protease
is pH dependent (see Fig. 6). Thus while the acidity optimum of the cyst
acid (thiol) protease is pH 3.6-3.8 [18], the inhibitor shows increased
potency (as measured by % inhibition) at pH 5 and 6 compared to pH 3.8.
However at pH 4, the potency of the inhibitor as defined by inhibitor units
(where 1 IU is that amount of protein which inhibits 1 EU), is about 3-
fold greater than at pH 6. Since the protease inhibitor is less stable
at pH 4 than at pH 5 or 6 (see Fig. 4), we now routinely conduct inhibitor
assays at pH 5 rather than at the pH optimum of the enzyme (pH 3.6-3.8) to
avoid, as much as possible, heat/pH inactivation during the assay.

    In addition to the above properties, the cyst thiol protease inhibi-
tors are sensitive to trypsin inactivation and they are unable to bind
concanavalin A. Thus the Artemia cyst thiol protease inhibitors appear to
be carbohydrate-free proteins and in some respects similar to the low
molecular weight intracellular protease inhibitors in mammalian cells.
However, the Artemia inhibitors appear to be less heat/pH stable than most
mammalian intracellular thiol proteases.

    Low molecular weight intracellular thiol protease inhibitors in verte-
brates are either competitive or non-competitive, depending on their source,
with K_i values in the range of 5×10^{-12} M to 1.2×10^{-10} M [17,19,20]. In
Artemia cysts the most abundant of the thiol protease inhibitors (fractions
2 plus 3, Fig. 3) is non-competitive with a K_i of 1.7×10^{-11} M (see Fig.
7). From purification data, we have estimated the concentration of the
inhibitors to be about 1-2 µM in dormant cyst and less than 0.1 µM in 36-hr
embryos. These calculations are based on an assumed average molecular mass

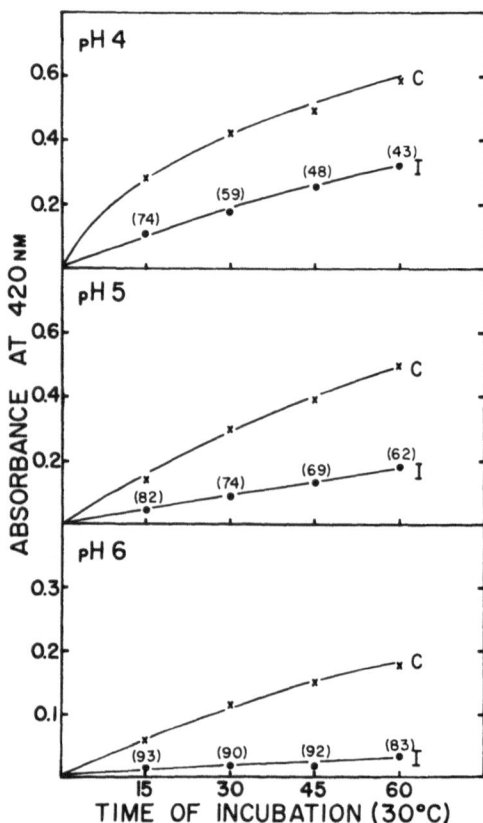

Fig. 6. Effect of pH on Artemia cyst thiol protease-protease
inhibitor interaction. Artemia cyst thiol protease
activity was measured in the absence (C) and presence
(I) of Mono S-purified thiol protease inhibitor (total)
in acetate buffer at pH 4, 5, and 6 as indicated. All
control vessels (C) contained 65 µg of partially purified cyst thiol protease while the treated vessels (I)
contained 0.56 µg inhibitor protein in addition to the
protease in 300 µl total volume. The numbers in par-
enthesis are % inhibition.

of 9000 daltons for the inhibitors (from gel filtration data, see ref. 16), and an estimate, based on SDS-polyacrylamide gel electrophoresis data, that the inhibitors in the Mono S fractions are still only 50% pure. Given these assumptions it appears that the major Artemia cysts thiol protease inhibitor exhibits pseudoirreversible behavior in vivo, at least during early development and prior to hatching.

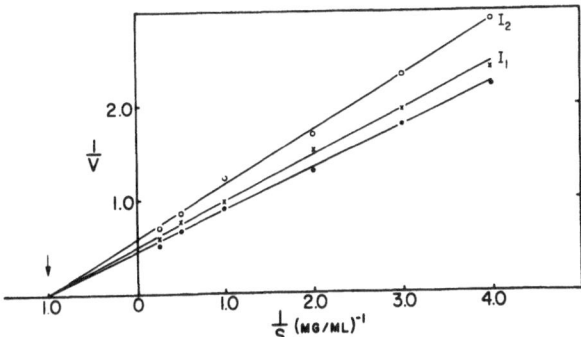

Fig. 7.   Kinetic analysis of the reaction between the Artemia cyst thiol protease and thiol protease inhibitor. Each reaction vessel contained 4 µg of purified Artemia thiol protease [4], protamine sulfate as substrate and either buffer (- ● -), 0.18 µg inhibitor (- X -) or 0.36 µg inhibitor (- O -) in an otherwise standard 200 µl reaction vessel at pH 5.0. The inhibitor was from the Mono S column (fractions 2 and 3 pooled, see Fig. 3). The temperature of the reaction was 30°C. The equation for each of the slopes was generated by regression analysis of all data points in each set. An average mass of 7500 daltons was used for protamine sulfate to calculate the kinetic constants. The K_m for the uninhibited reaction was calculated to be 1.45×10^{-4} M.

Protease inhibitors are considered to behave in an irreversible manner if the $[I^0]/K_i$ is > 1000 where $[I^0]$ is equal to the in vivo concentration of the inhibitor [21]. This value predicts that 0.1% or less of the total protease is "free" in vivo to act on substrates [21]. Given the above relationship, we have determined that the $[I^0]/K_i$ is greater than 50,000 for the major Artemia protease inhibitor suggesting that virtually all of the thiol protease is complexed and inactive in Artemia cysts.

Previously we determined that the ratio of thiol protease units to inhibitor units is about 2 in dormant cysts on the basis of specific activity data [16,22]. Given that about 50% of the thiol protease activity may be lysosomal [23] while all of the inhibitor is cytosolic [18,22], the protease and inhibitor appear to be present in stoichiometric amounts in the cytosol of dormant cysts. These data, together with the high K_i for the inhibitor, suggest that the thiol protease is highly regulated in encysted embryos of Artemia.

Attempts to identify intracellular factors in Artemia embryos that might influence the thiol protease-inhibitor equilibrium have revealed that physiological concentrations of NaCl, KCl and $CaCl_2$ have no effect on the Vmax of partially inhibited thiol protease reactions, while glycerol enhances the inhibitor effect. The glycerol effect is shown in Table 1.

SUMMARY AND CONCLUSIONS

The purification of the Artemia embryo thiol protease inhibitors has
proven to be a major task because of the numerous steps which are required
to purify the various isoforms of the primary inhibitor. While the individual isoinhibitors have yet to be purified to homogeneity and studied,
the results of our research thus far suggest that the various forms of the
inhibitor have similar biophysical and biochemical properties. The extent
of our current knowledge of the Artemia cyst thiol protease inhibitors
are summarized in Table 2.

In Artemia embryos the protease inhibitors appear to be important
proteins in preventing inappropriate proteolysis during early development
and especially during periods of environmental stress such as changing
salinity and oxygen levels. Whether the loss of embryo resistance to
periods of oxygen deprevation following hatching from the cysts is related
to an increase in enzyme/inhibitor ratio that accompanies development of
swimming nauplii, the resumption of cell division following hatching, or
the remodeling of embryonic tissues in the nauplius remains to be determined.

Table 1. Effect of Glycerol on the Potency of the Thiol Protease Inhibitor

Glycerol conc.	I.U.	% change
0% (control)	0.85	-
3.1%	1.35	+59
6.3%	1.75	+106
12.5%	1.78	+109

Each reaction vessel contained 1 μg of total Artemia
thiol protease inhibitor (after Sephadex G-75 step),
8 units of DEAE-purified thiol protease and glycerol
as indicated in a standard 200 μl reaction at pH 5.0
and 30°C.

Table 2. Characteristics of <u>Artemia</u> Cyst Thiol Protease Inhibitors

---

Biophysical properties:

Molecular mass:	8700 to 9000 da. (gel filtration)
	5000 to 14000 da. (SDS-PAGE)
Isoelectric point:	pH 5.2 to 5.9 (85%)
Temperature stability:	pH and time dependent; 100% activity at 60°C (pH 6, 30 min.); 5% activity at 75°C (pH 6, 45 min.); stable at -10°C (pH 6) for at least three years.
pH stability:	stable at pH 6 to 9.5 (4°C); unstable below pH 5
Composition:	trypsin sensitive protein; sensitive to <u>Artemia</u> thiol (acid) protease at pH 4; lacks carbohydrate

Biochemical properties:

Type of inhibition:	non-competitive; $K_i = 1.7 \times 10^{-11}$ M; $[I^o]/K_i > 50{,}000$ suggests pseudoirreversible behavior
Concentration <u>in vivo</u>:	1-2 μM in fully hydrated cysts; undetectable in 36-hr larvae
Modifying agents:	10-12% glycerol enhances the activity by about 100%
	N-ethylmalemide (5mM) increases the inhibitor activity slightly
Specificity:	<u>Artemia</u> acid (thiol) protease, $ID_{50} = 1.2 - 1.8$ μg/ml
	mammalian cathepsin B (heart), $ID_{50} = 2.5$ μg/ml

ACKNOWLEDGEMENTS

I wish to thank the Natural Sciences and Engineering Research Council

167

of Canada (grant A2909) for the financial support needed to carry out these studies. Thanks are also due to Viji Shridhar for her expert technical assistance during the early part of this work, to Dr. Paul Taylor for making his computer programs available for the enzyme analyses and to Dr. M. J. Dufresne for his critical reading of the Manuscript.

REFERENCES

1. H. Holtzer and P. C. Heinrich, Control of proteolysis, Ann. Rev. Biochem. 49:63 (1980).
2. D. Slaughter and E. Triplett, Amphibian embryos protease inhibitor. II. Biological properties of the inhibitor and its associated protease, Cell Differen. 4:23 (1975).
3. B. Ezquieta and C. G. Vallejo, The trypsin-like proteinase of Artemia. Yolk localization and developmental activation, Comp. Biochem. Physiol. 82B:731 (1985).
4. A. H. Warner and V. Shridhar, Purification and characterization of a cytosol protease from dormant cysts of the brine shrimp Artemia, J. Biol. Chem. 260:7008 (1985).
5. A. Monroy, R. Maggio and A. M. Rinaldi, Experimentally induced activation of the ribosomes of the unfertilized sea urchin egg, Proc. Nat. Acad. Sci. USA 54:107 (1965).
6. F. L. Huang and A. H. Warner, Control of protein synthesis in brine shrimp embryos by repression of ribosomal activity, Arch. Biochem. Biophys. 163:716 (1974).
7. A. S. Spirin, On "masked" forms of messenger RNA in early embryogenesis and in differentiating systems, Curr. Top. Develop. Biol. 1:1 (1966).
8. J. Dutrieu and D. Christia-Blanchine, Resistance des oeufs durables hydrates d'Artemia salina a l'anoxia, C. R. Acad. Sci. Ser. D. 263: 998 (1966).
9. F. P. Conte, J. Lowry, J. Carpenter, A. Edwards, R. Smith and R. D. Ewing, Aerobic and anaerobic metabolism of Artemia nauplii as a function of salinity, in "The Brine Shrimp Artemia", G. Persoone, P. Sorgeloos, O. Roels and E. Jaspers, eds., Universa Press, Wettren (1980).
10. D. M. Stocco, P. C. Beers and A. H. Warner, Effect of anoxia on nucleotide metabolism in encysted embryos of the brine shrimp, Develop. Biol. 27:479 (1972).
11. W. B. Busa and J. H. Crowe, Intracellular pH regulates transitions between dormancy and development of brine shrimp (Artemia salina) embryos, Science 221:366 (1983).
12. W. B. Busa, J. H. Crowe and G. B. Matson, Intracellular pH and the metabolic status of dormant and developing Artemia embryos, Arch. Biochem. Biophys. 216:711 (1982).
13. P. A. Nagainis and A. H. Warner, Evidence for the presence of an acid protease and protease inhibitors in dormant embryos of Artemia salina, Develop. Biol. 68:259 (1979).
14. J. F. Lenney, Inhibitors associated with the proteinases of mammalian cells and tissues, Curr. Top. Cell Reg. 17:25 (1980).
15. N. Katunuma and E. Kominami, Molecular basis of intracellular regulation of thiol protease inhibitors, Curr. Top. Cell Reg. 27:345 (1985).
16. A. H. Warner, The role of proteases and their control in Artemia development, in "Artemia Research and its Applications", Vol. 2, W. Decleir, L. Moens, H. Slegers, E. Jaspers and P. Sorgeloos, eds., Universa Press, Wetteren (1987).
17. E. Kominami, N. Wakamatsu and N. Katunuma, Purification and characterization of thiol proteinase inhibitor from rat liver, J. Biol. Chem. 257:14648 (1982)
18. A. H. Warner and V. Shridhar, Characterization of an acid protease from

encysted embryos of <u>Artemia</u>, in "The Brine Shrimp," Vol. 2, G.
Persoone, P. Sorgeloos, O. Roels and E. Jaspers, eds., Universa
Press, Wettren (1980).

19. A. J. Barrett, The cystatins: small protein inhibitors of cysteine
proteinases, in "Intracellular Protein Catabolism," E. A. Khairallah,
J. S. Bond and J. W. C. Bird, eds., Allan R. Liss, Inc., New York
(1985).

20. L. Waxman and E. G. Krebs, Identification of two protease inhibitors
from bovine cardiac muscle, <u>J. Biol. Chem.</u> 253:5888 (1978).

21. J. G. Bieth, Possible biological functions of protein proteinase inhibitors, in "Cysteine Proteinases and their Inhibitors", V. Turk,
ed., Walter de Gruyter and Co., Berlin (1986).

22. A. H. Warner, Protease and protease inhibitors in <u>Artemia</u> and their
role in the developmental process, in "Biochemistry and Cell
Biology", T. H. MacRae, J. C. Bagshaw and A. H. Warner, eds., CRC
Press, Boca Raton (1988).

23. R. Perona and C. G. Vallejo, The lysosomal proteinase of <u>Artemia</u>:
purification and characterization, <u>Eur. J. Biochem.</u> 124:357 (1982).

EFFECT OF STARVATION ON AMYLASE AND TRYPSIN ACTIVITIES AT TWO DEVELOPMENTAL

STAGES OF ARTEMIA (SAN FRANCISCO)

J. F. Samain, J. Moal, J. Y. Daniel and J. R. Le Coz

IFREMER DRV/PA
BP 70
29263 Plouzane, France

ABSTRACT

An apparent repressive control of amylase and trypsin respectively by starch and proteins of the diet were precedently demonstrated on artificial axenic media [1]. Such a mechanism has been improved also on Artemia fed Platymonas sueccica in more usual rearing conditions. Artemia face a starvation period of 4-5 days during growth or at adult stage by increasing enzymatic activities when food was lacking. This compensatory mechanism was immediate for amylase but was delayed for trypsin depending on the duration of starvation and the decrease of internal protein content. This response time difference is hypothesized to be related to differences in circulating carbohydrates, proteins or amino acid pools in the hemolymph.

These results corroborate observations using artificial axenic media. They suggest that digestive enzyme synthesis is mediated through circulating carbohydrates, proteins or amino acids by a neuroendocrine system. They improve the compensatory model presently proposed and its use as a tool in nutritional requirement studies.

REFERENCE

1. J. F. Samain, A. Hernandorena, J. Moal, J. Y. Daniel and J. R. Le Coz, Amylase and trypsin activities during Artemia development on artifical axenic media: effect of starvation on specific deletions, J. Exp. Mar. Biol. Ecol. 86:255 (1985).

ARTEMIA TREHALASE: REGULATION BY FACTORS THAT ALSO CONTROL RESUMPTION OF

DEVELOPMENT

Carmen G. Vallejo

Instituto de Investigaciones Biomedicas
CSIC. Facultad de Medicina, UAM
Arzobispo Morcillo, 4. 28029-Madrid, Spain

INTRODUCTION

The ability of the dry dormant embryo of Artemia to survive for dec-
ades has been related to the presence in the cyst of a high concentration
of trehalose[1]. This sugar apparently allows the organism to escape the
irreversible damage that complete dehydration produces in membranes[2].
On the other hand, after resumption of development trehalose is used as the
bulk source of energy and the bulk substrate of respiration[3].

An interesting feature of Artemia development is that its resumption
can be reversible in order to cope with the changing environment. The
transition is accompanied by a reversible shift in pH_i from 6.3 to at least
7.9[1]. The transition depends on environmental factors such as hydration,
oxygenation, temperature and salinity which work in a reversible manner[3,4].
We have reported previously that cytochrome oxidase is probably the molecular sensor of the environmental oxygen. I propose here that trehalase
may be a molecular sensor of the environmental temperature.

Since trehalose is used as the source of energy and substrate of respiration, trehalase must (apparently) be activated immediately after resumption of development. The activation could result from enzyme already present in the dormant embryo[5]. In this chapter I present evidence indicating that cyst trehalase is in fact susceptible to activation in vitro.

The changes that trehalase and trehalose experience during Artemia
development have been studied as well as the subcellular localization of
both the enzyme and the substrate. The in vivo data support the interpretation of the data observed in vitro.

MATERIALS AND METHODS

Artemia dormant embryos (cysts) from San Francisco Bay Brand, Newark,
Calif. were used in all experiments.

Culture, Handling, Homogenization, Subcellular Fractionation and Solubilization of Trehalase

Culture, counting and handling of embryos and nauplii were as described previously[6]. Dormant embryos were homogenized in 3 vol/g and nauplii in 1 vol/g of Ficoll medium[7] as described before[8]. This medium has proven to preserve the subcellular structures of _Artemia_ dormant embryos[7,9]. The subcellular fractionation of homogenates was accomplished as described by Vallejo et al.[6].

The particulate fractions were prepared to solubilize trehalase. When no subcellular fractionation was required, trehalase was extracted from the whole particulate fraction obtained by centrifugation of the homogenate at 150,000xg for 75 min. Trehalase was solubilized by extraction in 3-10 vols the original homogenate of a medium containing 1% Triton X-100, as described before[10]. The preparation of trehalase thus obtained was used for activation and inactivation studies. To obtain the least active form of the enzyme, the preparation was dialysed in the presence of polyethyleneglycol (PEG) versus 50mM sodium phosphate pH 6.3 for 4 h. To activate the enzyme at alkaline pH, the preparation was dialysed at pH 8.6 for 4 h in a phosphate buffer. To activate the enzyme by heat, the preparation was first dialysed in 50mM sodium phosphate, pH 7.0 for 2 h.

Gradients

A continuous sucrose gradient was obtained using 35% and 60% sucrose solutions (17+17 ml) containing 100 μg/ml soybean trypsin inhibitor. After layering of the samples, the gradients were centrifuged in a SW27 rotor at 25,000 rpm for about 15 h.

Assay of Trehalase Activity

The reaction of trehalase was followed continuously at 340 nm and 37°C. The routine assay contained in 1 ml, 50mM Imidazol buffer pH 7.0, 40 mM trehalose and the auxiliary system contained 0.5 mM ATP, 0.67 mM $MgCl_2$, 0.5 mM NADP and 2.6 units hexokinase (Sigma, Type C-300) and 4.4 units glucose 6 phosphate dehydrogenase (Boehringer, Grade I). When trehalase was assayed at pH 6.3 or pH 8.6, 50 mM sodium phosphate buffer was used. Occasionally trehalase was determined by measuring the production of glucose after incubation of the enzyme with the substrate and followed by destruction of trehalase at 100°C for 10 min. This assay was used to study the effect of ATP on the trehalase reaction. Glucose was determined with the routine assay (as described above), except for the omission of trehalose.

Extraction and Determination of Trehalose

To extract trehalose, embryos or nauplii were homogenized with 3 volumes (v/w) of cold 0.5 M trichloroacetic and the resulting homogenate was agitation at room temperature for 40 min and centrifuged for 10 min at 10,000 rpm. The pellet thus obtained was reextracted in the same way. The two supernatants were combined for analysis.

The subcellular localization of trehalose was studied by extraction of the subcellular fractions obtained in Ficoll medium as described above. Trehalose was determined with the anthrone method[11]. No glucose (as determined by the hexokinase glucose-6-phosphate dehydrogenase system) was detected in extracts of embryos and low concentrations were found in nauplii. Fractionation in a buffered medium (Ficoll medium, see above) to obtain subcellular fractions allowed a certain amount of trehalose hydrolysis which was taken into consideration in the calculations.

Determination of the Molecular Weight of Artemia Trehalase

A Sephacryl S-200 (Pharmacia) column (65 x 1.3 cm) was equilibrated with 50 mM sodium phosphate (pH 6.3, 7.0, 8.6) in the presence or absence of 1.0 M KCl. The volume of the sample was 1.2 ml. The calibration was accomplished with blue dextran 2000 and the marker proteins catalase (232 KDa), alkaline phosphatase (100 KDa), bovine serum albumin (67 KDa), ovalbumin (43 KDa), and ribonuclease (13.7 KDa). All of these markers were from Pharmacia except the alkaline phosphatase which was from Boehringer.

RESULTS

The Trehalase of Artemia Dormant Embryos can be Activated in Vitro

We have observed that the cyst trehalase can be activated at neutral pH under temperatures which also control activation of the cryptobiotic embryo[4]. We found that at pH 7.0 the enzyme was rapidly and extensively activated at temperatures ranging from 30° to 45°C (Fig. 1). In fact the maximum activation was observed at about 37°C. Previously we reported that 35°C is the optimum temperature for rapid development and that the hatching time at 20°C is twice the optimum[4]. In Fig. 1, it can be seen that rate of trehalase activation is temperature dependent and about one half at 21°C when compared with 37°C.

We have also observed that activation of trehalase is highly dependent on the concentration of the enzyme and that at high concentrations, the enzyme is not activated (Fig. 2). Polyethyleneglycol (PEG) has been used to mimic high concentrations of enzyme in diluted solutions. PEG apparently operates through a "water trapping" mechanism[12,13]. As can be seen in Fig. 2, the effect of PEG on activation was reversible upon dilution of PEG down to 6%. Under these conditions the enzyme was activated to the level observed in the absence of PEG. However, the velocity of the activation was somewhat lower probably due to the remaining 6% PEG. In Fig. 3, the kinetics of trehalase activation at 37°C in the presence of increasing concentrations of PEG is shown. The activation was blocked at concentrations of PEG above 10% PEG and at 25% PEG, the original activity was reduced to about 50%. This observation indicated that the trehalase in this particular experiment was already partially activated. Since the preparation had not been heated previously, this and other reults (not shown) suggest that the trehalase can be activated partially be mere dilution. The effect of other cryoprotectants on the activation was tested. Only dextran (8%) was found to inhibit partially the activation, to about the extent of 10% PEG (data not shown). Due to its low solubility, higher concentrations were not tried. The other cryoprotectants tested were glycerol and dimethyl sulfoxide, at up to 40% and 20% respectively, but these agents did not prevent activation; instead, they enhanced it. Although we do not know the reason for the marked differences observed with the different cryoprotectants, they are probably due to the different properties of the cryoprotectants. Glycerol and dimethyl sulfoxide are low molecular weight compounds that mix with water, while PEG and dextran are polymers that trap water, thus increasing the local concentration of protein.

Activation of Artemia trehalase can be also achieved in the cold at alkaline pH as recently reported[14]. Results from my laboratory indicate that the enzyme can be activated to a similar degree by one or the other procedure (Figs. 2 and 4) but that the two maximum activations are not

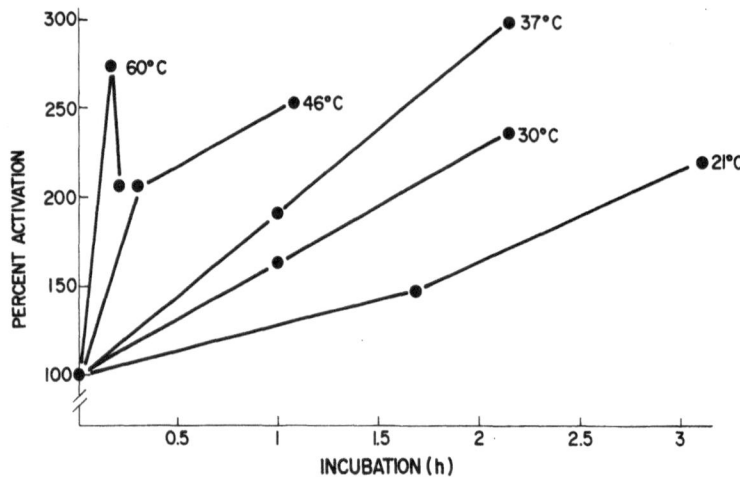

Fig. 1. Activation of <u>Artemia</u> trehalase by heating at different
temperatures. Trehalase was solubilized from the
whole particulate fraction of dormant embryos. Samples
diluted 55-fold (vol./wet weight of cysts) were heated
at the temperatures indicated and aliquots were taken
at the times indicated for the assay of activity. The
activity has been referred to the unincubated sample
(100%).

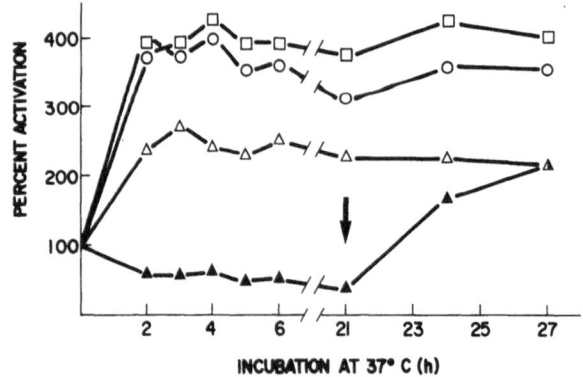

Fig. 2. The activation by heat of <u>Artemia</u> trehalase is.
enzyme concentration-dependent. Samples of
trehalase obtained as in Fig. 1 and diluted
16.5-fold (Δ-Δ), 33-fold (O-O) and 66-fold
(□-□) were incubated at 37°C for about 27
h and aliquots taken at the times indicated
for assay of activity. A sample diluted 16.5
fold was also incubated in the presence of 25%
polyethyleneglycol (PEG) (▲-▲) and after 21
h of incubation (arrow) diluted 4-fold to allow
trehalase to activate.

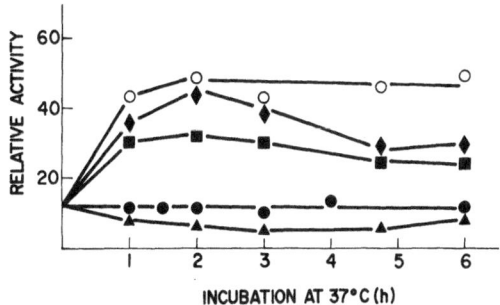

Fig. 3. The kinetics of activation of <u>Artemia</u> trehalase in the
presence of varying concentrations of PEG. Samples of
trehalase, obtained as in Fig. 1, were diluted 66-fold
(v/wt w) and incubated at 37°C in the absence of PEG
(O-O) or in the presence of 6% PEG (♦-♦), 10% PEG
(■-■), 15% PEG (●-●) and 25% PEG (▲-▲) for up to
6 hours. Aliquots were taken for trehalase activity
measurements at the times indicated.

additive (data not shown). The possible dependence of the activation at
alkaline pH on enzyme concentration was also investigated. The data indicated that the activation at pH 8.6 is concentration-dependent as it is
at 37°C (Figs. 2, 3). The data in Fig. 4 also show that in the presence
of PEG, activation at pH 8.6 is not achieved. Therefore the activation
observed either by heating the enzyme at 37°C and pH 7.0, or by dialysing
at pH 8.6, involve a concentration-dependent mechanism(s).

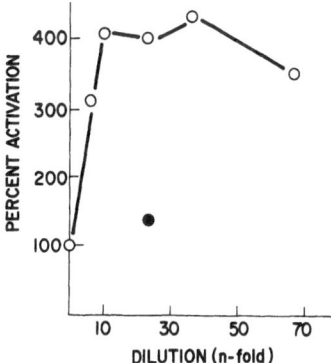

Fig. 4. The activation of <u>Artemia</u> trehalase at
alkaline pH is also enzyme concentration-dependent. Samples of trehalase
obtained as in Fig. 1 were diluted as
indicated, dialysed versus 50 mM sodium
phosphate pH 8.6 for 4 h and assayed (O).
A sample diluted 16.5-fold in the presence
of 25% PEG (●) was also dialyzed and
analyzed in the same way.

ATP has been reported to inhibit yeast trehalase [15]. The effect of ATP on Artemia trehalase was investigated and it was found that ATP activates further the active form of the enzyme at pH 8.6. The effect of varying the ATP concentration on trehalase activity at pH 8.6 is shown in Fig. 5. Clearly the extent of trehalase activation increased between 0 and 5 mM ATP; higher concentrations of ATP were not tested. The enzyme activated by heating at pH 7.0 was not affected by ATP; neither was the form activated at pH 6.3. Therefore it appears that ATP is an effector of the active form of Artemia trehalase at alkaline pH, but it is not able to activate the enzyme at neutral or acid pH. These kinetic studies of trehalase were made in the presence of 0.5 mM ATP, the concentration found in the dormant embryo [14,16].

We have studied the kinetic behaviour of the enzyme obtained by dialysis at pH 6.3 in the presence of PEG and the two activated forms of the enzyme obtained either at pH 7.0 and 37°C or at pH 8.6 in the cold. The saturation curves obtained with the three preparations of enzyme are shown in Fig. 6. The three forms of the enzyme presented a different kinetic behaviour. The form at pH 6.3 presented hyperbolic kinetics (Fig. 6a) and was inhibited by high concentrations of substrate which are physiological [14,17]. The form activated at pH 7.0 and 37°C presented also hyperbolic kinetics (Fig. 6b), but it was no longer inhibited by high substrate concentrations. The form activated at alkaline pH (Fig. 6c) showed kinetics that suggested negative cooperativity towards the substrate. The $\underline{n}$ value calculated from a Hill plot was however 0.9, compared to 1.1 obtained for the form at pH 7.0. We have observed that the kinetics of the enzyme at pH 8.6 can be transformed into that of the enzyme activated at pH 7.0 and 37°C if the preparation is previously frozen and thawed (data not shown). These results suggest that the active form at alkaline pH can be transformed into that obtained by heating at pH 7.0. The form activated at alkaline pH can be also reversibly transformed into that observed at acid pH (not shown). The Vmax and $S_{0.5}$ of the three forms of Artemia trehalase are given in Table 1.

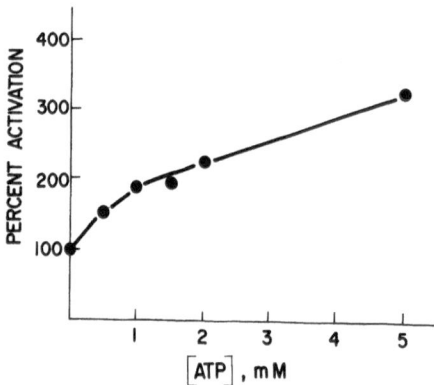

Fig. 5. ATP activates the active form of Artemia trehalase at alkaline pH. A sample of trehalase, obtained as in Fig. 1, was diluted 20-fold (v/wet-wt), dialysed at pH 8.6 and assayed at pH 8.6 with 40 mM trehalose in the presence of increasing concentrations of ATP. The reaction was stopped by boiling the reaction samples for 10 min. The glucose produced was determined with HK-G6PDH as indicated in Materials and Methods. Similar results were obtained with 320 mM trehalose.

Table 1. Kinetic Constants of the Different Forms of <u>Artemia</u> Trehalase

V_{max}			$S_{0.5}$		
pH 6.3	pH 7.0	pH 8.6	pH 6.3	pH 7.0	pH 8.6
20	1000	700	5 mM	28 mM	24 mM

The experimental conditions were as in Fig. 6.

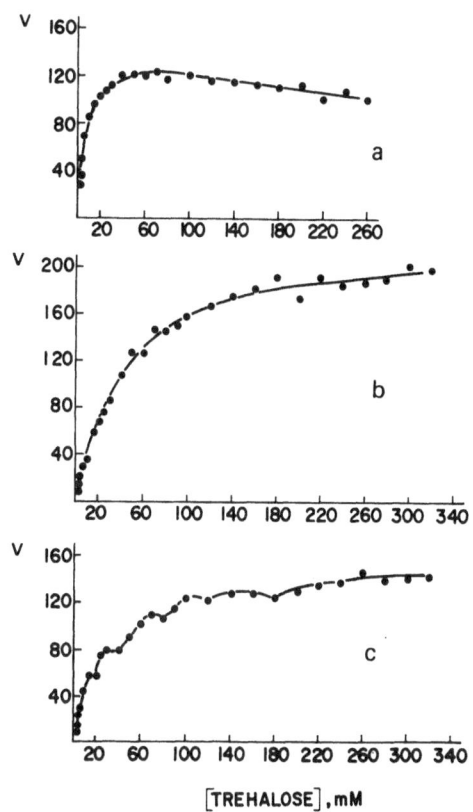

[TREHALOSE],mM

Fig. 6. The kinetic behaviour of the different forms of <u>Artemia</u>
         trehalase. Trehalase extracted from the whole particu-
         late fraction (Materials and Methods) at 3.3-fold dilu-
         tion (volume per wet weight) was dialysed at pH 6.3 for
         4 h in the presence of 25% PEG. The kinetic study was
         carried out at pH 6.3 in 50 mM Pipes buffer (a) in the
         presence of 1.5% PEG; no activation was detected during
         the time of the assay. In the middle panel (b) the
         saturation curve was obtained at pH 7.0 with an enzyme
         sample diluted 33-fold, dialysed at pH 7.0 for 4 h and
         heated at 37°C for 3 h. The kinetics were obtained at
         pH 7.0 with 50 mM Imidazole buffer. In the bottom panel
         (c) the active form at alkaline pH was obtained and
         dialyzed at pH 8.6 for 4 h. A sample of the trehalase
         diluted 33-fold and the kinetic assay was carried out at
         pH 8.6 with 50 mM Tricine buffer.

Molecular Weight of Artemia Trehalase

The size determined for trehalase was about 83 kd at pH 6.3 and about 77 kd at pH 8.6 (Fig. 7a and b, respectively). The form activated by heat at pH 7.0 presented two peaks of activity of 61 kd and 23 kd (Fig. 7c). This result suggests a dissociation of the form of 83 kd (pH 6.3) as the basis of the activation. Frequently only the 61 kd form can be found and the recovery is low (about 60%), suggesting that the low molecular weight form is unstable.

Developmental Changes in the Activity of Artemia Trehalase

The activity of trehalase was determined in embryos and nauplii of different times of development and the data obtained are represented in Fig. 8. As can be seen, the activity starts increasing immediately upon hydration (zero time is the dry dormant embryo) to reach a plateau at around hatching of the swimming larva. After about 40 h of incubation, the activity increased again to reach a maximum at about 50 h. The maximum value of this peak was somewhat variable (probably depending on the batch of cysts), but the general trend of the profile was reproducible. The first, slow increase in activity is probably the result of the activation of the enzyme stored in the dormant embryo as shown before. However, the nature of the late peak of activity was investigated further. As seen in Fig. 9, this late activation was prevented in the presence of the drug cycloheximide which inhibits protein synthesis, suggesting that this activation was due to synthesis de novo of the enzyme.

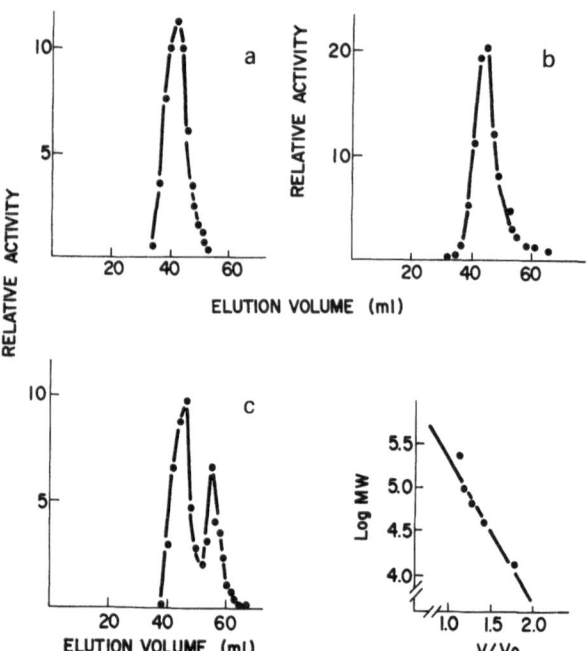

Fig. 7. The molecular weight of the different forms of Artemia trehalase. a) the elution profile of the form at pH 6.3, b) the active form at pH 8.6, and c) the activated form at pH 7.0 and 37°C. The calibration of the column, generated with protein markers as indicated in Materials and Methods, is also shown. The molecular mass calculated for a) was 83 kd, for b) 77 kd and for c) 61 kd and 23 kd.

Table 2. Subcellular Localization of Artemia Trehalase

	Dormant Embryos	Late Larvae
2,000 x g, 10 min fraction	83%	75%
7,500 x g, 20 min fraction	14%	5%
27,000 x g, 30 min fraction	2%	4%
150,000 x g, 75 min fraction	< 1%	7%
150,000 x g, supernatant	1%	9%

Embryos (0-h) and larvae (about 55-h) were homogenized as indicated in
Materials and Methods and the homogenates subjectd to the subcellular
fractionation as indicated. The value of the corresponding homogenate
is represented in Fig. 8 (0.1 unit/10[6] animals and 6.2 units/10[6] animals
for 0-h and 55-h, respectively). The data given are the average of two
experiments and stand for the percent distribution among the fractions.
The particulate fractions were extracted in the presence of Triton X-100
as indicated also in Materials and Methods. The recoveries of the fractionation were high (about 200%) thus reflecting the latency of the
particulate fractions.

Particulate Localization of Artemia Trehalase

The distribution of trehalase activity among the different subcellular
fractions of homogenates of cysts and late nauplii is shown in the Table 2.
As can be seen, the majority of the activity is found in the heavy 2000 x
g fraction. We have shown previously that this fraction contains yolk
granules and nuclei [7,18]. The data suggested that trehalase might be
associated with yolk granules. In order to test directly if trehalase
is associated with these organelles, we used a method reported previously
to obtain a pure preparation of Artemia yolk granules [7]. The 500 x g
fraction of dormant embryos, which is composed mainly of yolk granules
was sedimented through a gradient. Yolk granules have a density 1.3 and
sediment to the bottom in sucrose isopycnic gradients (0.9 - 2.0 M) [7].
Most of the enzymatic activity was found in the bottom of the gradient
while a small fraction of the activity appeared at the top of the gradient,
apparently solubilized during the run (see Fig. 10a). Trehalase from late
nauplii behaved similarly (see Fig. 10b).

Subcellular Localization and Changes in the Concentration of Trehalose during Artemia Development

The subcellular distribution of trehalose in embryos and nauplii is
shown in Table 3. The majority of trehalose was found in the heavy 2000
x g fraction. Our results support previous data on the presence of
trehalose in purified yolk granules of the dormant embryo [19]. In addition they indicate that yolk granules are the main site of localization
of trehalose in both embryos and nauplii, despite the fact that most yolk
granules are degraded during the larval stage of development [20,21].

The changes in concentration of trehalose observed during development
are shown in the Fig. 11. The trehalose concentration found in dry embryos
was about 300 umoles/g and similar to previously published data [14,17,22].
About 70% of the cysts trehalose was hydrolyzed (in our conditions)

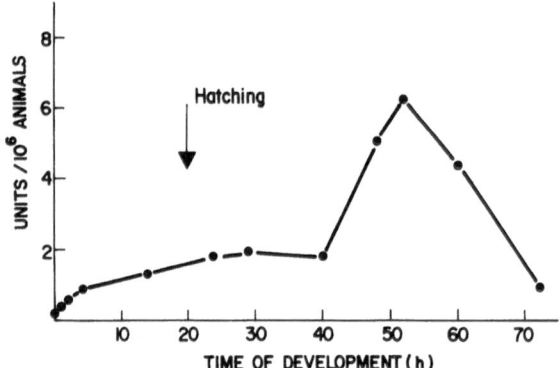

Fig. 8. Development changes in the activity of
         Artemia trehalase. Dormant cysts were
         incubated at 30°C for the time periods
         indicated. Larvae were synchronized at
         hatching and further incubated for the
         time required. Trehalase was extracted
         from the whole particulate fraction of the
         homogenate and determined as indicated in
         Materials and Methods.

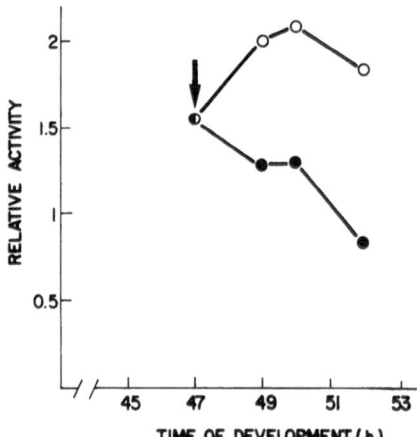

Fig. 9. The increase in trehalase activity ob-
         served in late larvae can be prevented
         by cycloheximide. At the time indicat-
         ed by the arrow, cycloheximide (50
         μg/ml) (●-●) was added to the cultures
         and the incubation was continued for
         the times indicated. (O-O), control
         cultures. Other details are as in Fig. 2.

182

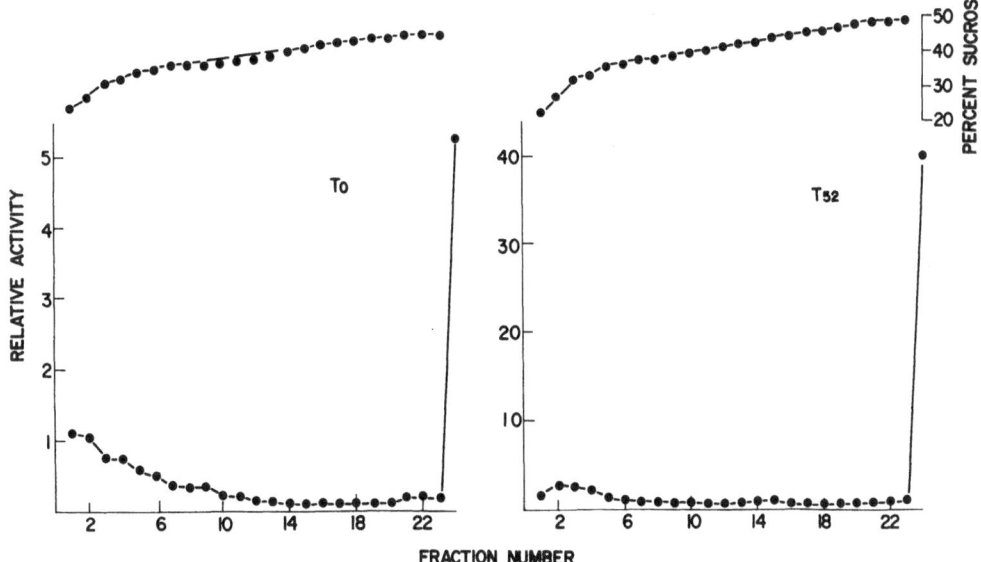

Fig. 10. Apparent localization of _Artemia_ trehalase
in yolk granules. Dormant embryos (0-h)
and late larvae (52-h) were homogenized and
the 500 x g subcellular fraction was layered
on the top of a continuous sucrose gradient
(see Materials and Methods). In this
gradient yolk granules sediment to the bottom
(Vallejo et al., 1981) while mitochondria
and lysosomes band at around fraction 12
(not shown), the latter appearing in a somewhat lighter position.

during early embryonic development while the rest occurred in the late
nauplii. By 70 h incubation, both trehalase activity and trehalose concentration were low (see Fig. 8 and 11).

The Larval Trehalase has the same Characteristics as the Embryo Enzyme

The appearance of a relatively high peak of trehalase activity in
late larvae (see Fig. 8) posed the question of whether the _de novo_ synthesized trehalase (see Fig. 9) had the same characteristics as the embryo
enzyme. The following results support the view that both enzymes are the
same.

a). The larval trehalase is localized in the same subcellular fraction.
Trehalase is found associated with the remaining yolk granules in the
larval stage (Table 2 and Fig. 10), despite the fact that the majority of
yolk granules are degraded in this stage [20,21]. We have described a
different behaviour for another enzyme associated with yolk granules, a
trypsin-like proteinase, that was found to become increasingly more soluble
in parallel with yolk degradation [8].

b). The larval trehalase can be also activated in vitro by alkaline
pH or by heating. As already indicated for the embryo enzyme, the enzyme

Table 3. Subcellular Localization of Trehalose in *Artemia*

	Dormant Embryos	Late Larvae
2,000 x g, 10 min fraction	82	54
7,500 x g, 20 min fraction	12	7
150,000 x g, 75 min fraction	1	3
150,000 x g, supernatant	5	36

Dormant embryos (0-h) and larvae (41-h total incubation) were homogenized and the homogenates subjected to the sub-cellular fractionation indicated (Materials and Methods). The particulate fractions were extracted with 0.5 M tri-chloroacetic acid, also as indicated in Materials and Methods. To the 150,000 x g supernatant, acid was added up to 0.5 M and the sample centrifuged before assay. Trehalose was determined by the anthrone method (Herbert et al., 1971). The data stand for the percent distribution among the fractions. The corresponding value of the whole homogenate, as in Fig. 4. The data given are the average of two experiments.

from larvae can be inhibited at pH 6.3 (not shown) and activated by dilution at pH 8.6 (Fig. 12a) as well as by heating (37°C) at pH 7.0 (Fig. 12b). In both cases the activation was enzyme concentration-dependent as reported above for the embryo enzyme (Figs. 3 and 4). In the presence of poly-ethylenglycol (PEG), activation was not observed.

The study of the three forms of the larval trehalase obtained at pH 6.3, at pH 8.6 and by heating at pH 7.0 showed the characteristics reported above for the embryo enzyme. The kinetic constants for the three forms of the larval enzyme were also very similar to those of the embryo enzyme (Table 1). Furthermore, ATP activated the pH 8.6 form but had no effect at pH 6.3 or 7.0 (see Fig. 13). The kinetic study was performed in the presence of 0.5 mM ATP.

The determination of the molecular weight of the three forms of the larval trehalase indicated differences in mass as found for the embryo trehalase (data not shown).

DISCUSSION

The trehalase of *Artemia* appears to be well suited to respond immediately to environmental signals that also control development. The enzyme isolated from the cyst can be activated in vitro at neutral pH by heating in the viable temperature range of *Artemia* development [4]. It can be also activated in the cold at alkaline pH in the range of intracellular pH observed during the activation of the dormant embryo [1]. The activation at alkaline pH was also observed by Hand and Carpenter [14]. It is shown here that both activation procedures are enzyme concentration dependent, being suppressed at high concentrations.

The three factors of pH, temperature and concentration seem to have physiological meaning. The pH of the dormant embryo becomes alkaline upon activation [1] and the resumption of development depends on temperature [4]

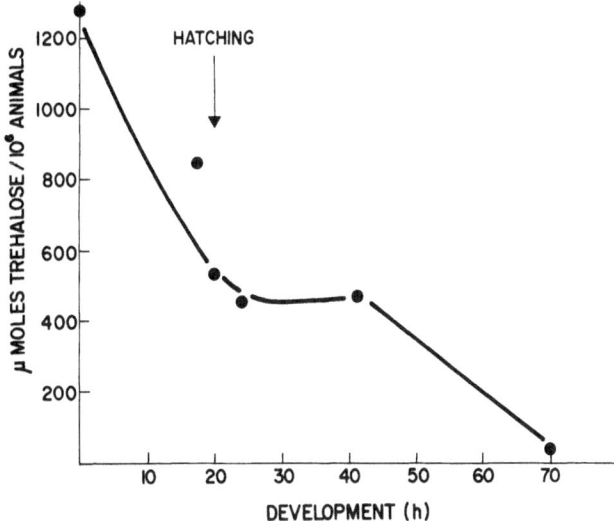

Fig. 11. Developmental changes in the concentration of trehalose
in <u>Artemia</u>. Trehalose was extracted from embryos and
larvae at various times of development (as indicated
in Materials and Methods). Trehalose was determined
by the anthrone method (Herbert et al., 1971). The
data given are for the average of 2-6 experiments.
The deviation from the curve of the value observed at
17-h development probably reflects the asynchrony of
the culture. The cultures were synchronized at hatching.

in the range in which trehalase is activated. The enzyme is also concentrated in yolk granules, organelles that contain at least 80% of the total
protein of the cyst [7]. It is presumed that metabolism of the yolk
granules which starts upon activation of development [21] also leads to the
disaggregation or dilution of trehalase and hence its activation.

The kinetics of the non-activated enzyme at pH 6.3 and of the enzyme
activated by heating or at alkaline pH were found to be different.
Trehalase at pH 6.3 was inhibited at high but physiological concentrations
of trehalose while the two activated forms were not. The maximum velocity
of these two forms increased about 50-fold although the affinity decreased
around 6-fold. This, however, should not affect the hydrolysis of trehalose
since the substrate is in high concentration (about 320 mM) in cysts
[14,17]. ATP on the other hand activated the enzyme at alkaline pH but it
had no effect at neutral or acid pH.

The kinetics data presented here are at variance with those described
by Hand and Carpenter [14] except for the inhibition of trehalase at acid
pH by high concentrations of trehalose. Also these authors stated that
the activation at alkaline pH was accompanied by a dissociation of a dimer
to a monomer. However, we have not been able to detect any dimer even
using their conditions. Therefore, disaggregation of trehalase does not
appear to be a requisite for its activation. These discrepancies are

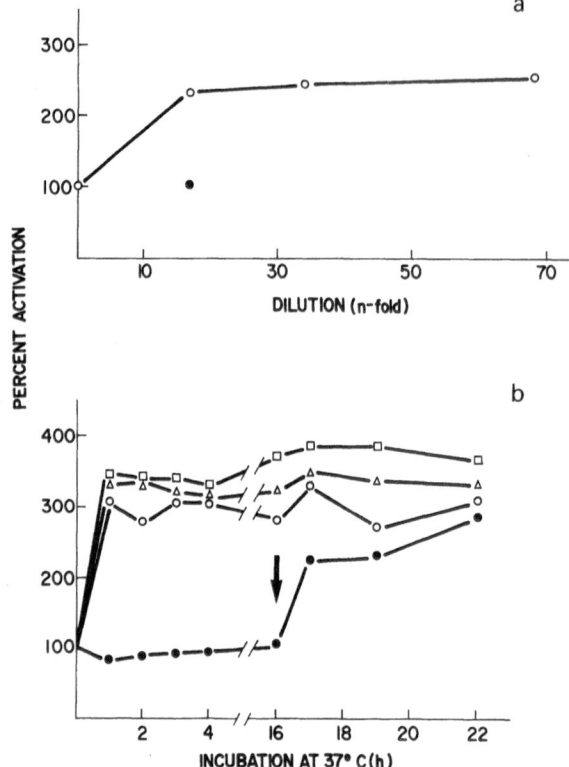

Fig. 12. The activation of the larval trehalase is enzyme concentration dependent. The larval (52-h) trehalase can be activated by dilution at pH 8.6 (a) and by heating at pH 7.0 (b). In both cases the activation can be prevented by the presence of 25% PEG (●). In a), the activation observed versus the dilution of the extract (volume per wet weight) is presented. In b) samples diluted 17-fold (O-O), 34-fold (Δ-Δ), 68-fold (□-□) and 17-fold in the presence of 25% PEG (●-●) were incubated at 37°C for about 22 h, and aliquots taken for trehalase assay at the times indicated. The sample in the presence of PEG was diluted 4-fold (arrow) at about 16 h incubation and incubated further to allow activation.

difficult to explain except for the fact that the cysts employed seem to be of a different geographic origin. Their cysts were from Utah, i.e., of American origin while ours, although purchased from California, appear to be of Euroasiatic origin as suggested by the genomic organization of rDNA (Medina et al., personal communication).

The developmental changes and subcellular localization of trehalase and trehalose have been also studied. The developmental profile of trehalase activity and the variations in the concentration of trehalose suggest that the enzyme is in fact activated at the time of activation of

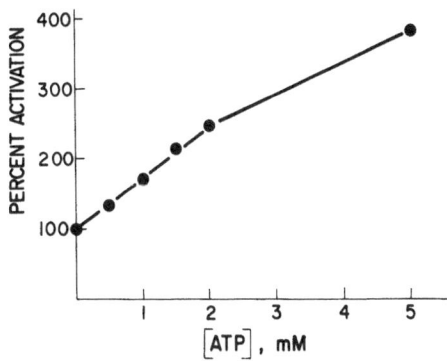

Fig. 13. Activation of the active form of larval trehalase
by ATP at alkaline pH. A sample of trehalase from
late larvae (52 h incubation) was diluted 35-fold
(v/wet weight), dialyzed at pH 8.6 versus 50 mM
sodium phosphate for 4 h and assayed in the same
buffer in the presence of varing concentrations
of ATP. The reaction was stopped by boiling the
samples for 10 min. The glucose produced was
determined with HK-G6PDH as indicated in Materials
and Methods.

development. The activation was observed at or before 0.5 h and before
protein synthesis is initiated following dormancy [3]. The activity continued to increase until 30 h of incubation, and at 40 h incubation the
activity increased again reaching a maximum at about 50 h. This peak of
activity was dependent on protein synthesis. Thus, synthesis de novo may
also account for the activity of trehalase. However, this may be so only
in emergency situations since the nauplii are able to feed after 30-35 h
of incubation [3]. These situations can be encountered if the cyst
undergoes several transitions between dormancy and development with a
corresponding utilization of trehalose. If the trehalose content is low,
the animal may need to synthesize additional trehalase to compensate for
the relatively low affinity of the activated forms of the enzyme. De novo
synthesized trehalase has the same characteristics as the embryo enzyme.

The study of trehalose concentration during development indicated
that about 70% is hydrolyzed during early embryonic development and the
rest in late nauplii at the time when trehalase acitivity peaks. The data
suggest that shortly after hatching of nauplii, trehalose is not consumed
for nearly 20 h. During this period of time degradation of yolk is
intense [20,21] and it may be that the source of energy is shifted from
trehalose to yolk proteins. The remaining trehalose is consumed by late
nauplii when most yolk granules are already metabolized.

The fact that trehalase and trehalose are localized in yolk granules
may link the activation of trehalase with the degradation of these organelles. As discussed above, the metabolism of yolk granules would favor
trehalase activation. We have suggested that ATP may be involved in the
positive regulation of yolk degradation [20], and data presented here
show that it can activate trehalase, at least in vitro. The fact that the
concentration of ATP increases from the resumption of development [16]

may support its involvement in both processes. It is possible therefore that the activation of trehalase upon resumption of development depends not only on temperature and the initial alkalinization of the internal pH, but also on the metabolism of yolk granules.

ACKNOWLEDGEMENTS

This work was supported by a grant from Comisión Asesora de Investigación Científica y Técnica. I thank Carmen Moratilla for technical assistance during this work.

REFERENCES

1. W. B. Busa, J. H. Crowe and G. B. Matson, Intracellular pH and the metabolic status of dormant end developing Artemia embryos, Arch. Biochem. Biophys. 216:711 (1982).
2. J. H. Crowe, L. M. Crowe, J. F. Carpenter and C. Aurell Wistrom, Stabilization of dry phospholipid bilayers and proteins by sugars, Biochem. J. 242:1 (1987).
3. J. S. Clegg and F. P. Conte, A review of the cellular and developmental biology of Artemia, in "The Brine Shrimp Artemia", Vol 2, G. Persoone, P. Sorgeloos, O. Roels and E. Jaspers eds., Universa Press, Wetteren (1980).
4. C. G. Vallejo, F. De Luchi and R. Marco, The role of cytochrome oxidase in the resumption of the development of Artemia dormant cysts, in "The Brine Shrimp Artemia", Vol. 2, G. Persoone, P. Sorgeloos, O. Roels and E. Jaspers, eds., Universa Press, Wetteren (1980).
5. P. Ballario, M. Bergami, M. G. Cacace, F. Scala and L. Silvestri, $\alpha\alpha$-Trehalase from the brine shrimp Artemia salina. Purification and properties, Comp. Biochem. Physiol. 61B:265 (1978).
6. C. G. Vallejo, M. A. G. Sillero and R. Marco, Mitochondrial maturation during Artemia salina embryogenesis. General description of the process, Cell. Mol. Biol. 25:113 (1979).
7. C. G. Vallejo, R. Perona, R. Garesse and R. Marco, The stability of the yolk granules of Artemia. An improved method for their isolation and study, Cell Diff. 10:343 (1981).
8. B. Ezquieta and C. G. Vallejo, The trypsin-like proteinase of Artemia: yolk localization and developmental activation, Comp. Biochem. Physiol. 82B:731 (1985).
9. E. Roggen and H. Slegers, Isolation and characterization of cytoplasmic poly(A)polymerase from cryptobiotic gastrulae of Artemia salina, Eur. J. Biochem. 147:225 (1985).
10. R. Perona and C. G. Vallejo, The lysosomal proteinase of Artemia. Purification and characterization, Eur. J. Biochem. 124:357 (1982).
11. D. Herbert, P. J. Phipss and R. E. Strange, Chemical analysis of microbial cells, in "Methods in Microbiology", Vol. 5B, Academic Press, London (1971).
12. G. D. Reinhart, Influence of polyethylene glycols on the kinetics of rat liver phosphofructokinase, J. Biol. Chem. 255: 10576 (1980).
13. L. Bosca, J. J. Aragon and A. Sols, Modulation of muscle phosphofructokinase at physiological concentration of enzyme, J. Biol. Chem. 260:2100 (1985).
14. S. C. Hand and J. F. Carpenter, pH-induced metabolic transitions in Artemia embryos mediated by a novel hysteretic trehalase, Science 232:1535 (1986).
15. J. M. Thevelein, J. A. den Hollander and R. G. Shulman, Trehalase and the control of dormancy and induction of germination in fungal spores, Trends Biochem. Sci. 11:495 (1984).

16. A. H. Warner and F. J. Finamore, Nucleotide metabolism during the brine shrimp embryogenesis, J. Biol. Chem. 242:1933 (1967).

17. J. S. Clegg, The origin of trehalose and its significance during the formation of encysted dormant embryos of Artemia salina, Comp. Biochem. Physiol. 14:135 (1965).

18. R. Marco, R. Garesse and C. G. Vallejo, Storage of mitochondria in the yolk platelets of Artemia dormant gastrulae, Cell. Mol. Biol. 27:515 (1981).

19. A. H. Warner, J. G. Puodziukas and F. J. Finamore, Yolk platelets in brine shrimp embryos. Site of biosynthesis and storage of the diguanosine nucleotides, Exp. Cell. Res. 70:365 (1972).

20. R. Perona, B. Ezquieta and C. G. Vallejo, The degradation of yolk in Artemia, in "Artemia Research and Applications", Vol. 2, W. Decleir, L. Moens, H. Slegers, E. Jaspers and P. Sorgeloos, eds., Universa Press, Wetteren (1987).

21. R. Perona, J. C. Bes and C. G. Vallejo, The degradation of yolk in the brine shrimp Artemia. Biochemical and morphological studies on the involvement of the lysosomal system, Biol. Cell. 63:361 (1988).

22. A. P. Boulton and A. K. Huggins, Biochemical changes occurring during morphogenesis of the brine shrimp Artemia salina and the effect of alterations in salinity, Comp. Biochem. Physiol. 57A:17 (1977).

CHITIN DEGRADING ENZYMES: CHARACTERISTICS AND FUNCTIONS DURING <u>ARTEMIA</u>

DEVELOPMENT

Brigitte Funke[1], Godelieve Criel[2] and Klaus-Dieter Spindler[1]

1. Institut fur Zoologie, Lehrstuhl fur Hormon- und Entwick-
   lungsphysiologie, Universitaet Duesseldorf, Universit-
   aetsstr. 1, D-4000 Duesseldorf, F. R. G.
2. Laboratorium voor Anatomie, Rijksuniversiteit Gent,
   Ledeganckstraat 35, B-9000 Gent, Belgium

INTRODUCTION

In a preceeding paper we were able to demonstrate chitin and the chitin degrading enzymes, chitinase and Nacetyl-β-D-glucosaminidase, in cysts of <u>Artemia</u>[1]. The titers of the enzymes change during early development increasing to high levels at the beginning of emergence. Both chitinase and N-acetyl-β-D-glucosaminidase can be detected in the growth medium with a five-fold higher specific activity at the time when most of the animals reach the E_2 or fully emerged stage. The regulation of the titers of chitinolytic enzymes during the early development of <u>Artemia</u> is presumably under the control of moulting hormones and thus resembles the situation in arthropods where a close correlation exists between chitinolytic activity and moulting hormones[2-5]. The results were taken as an indication for participation of chitin degrading enzymes, in addition to osmotic mechanisms, in the emergence process[1]. A direct proof of this hypothesis is still lacking. We therefore tried to answer the following questions:

1) Do chitinolytic enzymes accelerate the emergence process?
2) Are chitinolytic enzymes present during the whole life of <u>Artemia</u> and if so, is the pattern of enzymes identical from dormant cysts up to adulthood?
3) Does the enzymatic activity change during a moulting cycle and if so, is the shape of the titer curve comparable to those of other arthropods?

MATERIALS AND METHODS

Methods

Cysts from Sera Aquaristic were used for all experiments except the determination of enzyme titer during the moulting cycle of adults. For this purpose animals from a permanent culture in the <u>Artemia</u> Reference Center in Ghent where used. The moulting stage of the adults was determined according to Criel and Walgraeve (submitted). Two g dry weight of cysts in 1 litre of medium (400 mM NaCl, 10 mM $MgCl_2$, 10 mM $CaCl_2$) were incubated at 25°C with continuous aeration. After 2 days, nauplii were collected making use of the positive phototactic behaviour, thoroughly washed with distilled water and shock-frozen. For continuous rearing Artemia was fed with Mikrozell according to demand.

Influence of Chitinolytic Enzymes on the Hatching Rate

For these experiments between 50 and 80 intact dry cysts were selected under the microscope and incubated in a total of 1 ml of medium (control) or medium with various enzymes (Table 3) at room temperature.

Preparation of the Cytosol for the Assays

Deep-frozen nauplii and adults were lyophilized and powdered. The powder can be stored at -20°C for several months without loss of activity. The dry powder was extracted with acetone at -10°C, centrifuged at -10°C (5 min, 7,000g), the supernatant decanted and the pellet reextracted 3 to 4 times. The final pellet was dried at 4°C. The dry powder can be stored at -20°C for several months. For all chromatographic purposes, the powder was suspended in 10 mM Na/K phosphate buffer, pH 6.2, and homogenized on ice with an Ultra-Turrax (18 times each 10 sec). The homogenate was centrifuged at 4°C for one hour at 150,000g. The supernatant was filtered, the pellet reextracted as described and the supernatants combined. Dry cysts were treated as described for the nauplii and adults except that acetone treatment was omitted and homogenization with the Ultra-Turrax was at 60 times each 10 sec.

Anion-Exchange Chromatography

The cytosol was separated on DEAE-Sepharose CL-6B (Pharmacia) equilibrated with 10 mM Na/K-phosphate buffer, pH 6.2. The bed volume was 10 ml and the flow rate was 18 ml/h. The enzymes were eluted from the column by a NaCl gradient and fractions of about 0.7 ml were collected.

Enzyme Assays

N-acetyl-B-D-glucosaminidase activity was determined by mixing between 50 and 100 µl of the samples with 50 µl of 0.3 % p-nitrophenyl-N-acetyl-B-D-glucosamine in 200 mM sodium citrate-phosphate buffer, pH 5.5, and incubating at 40°C. After 30 min the reaction was stopped by addition of 2.5 ml of 10 mM NaOH and the absorbance measured at 410 nm.

Chitinase activity was measured by using tritiated chitin as the substrate. ^{3}H-chitin, prepared according to Molano et al.[6], was suspended at a concentration of 12.5 mg/ml. Forty µl of the suspension were mixed with 60 µl of sample and incubated between 1 and 5 h at 35°C under continuous stirring. The reaction was stopped by the addition of 300 µl of ice-cold 10% trichloroacetic acid and the solution was centrifuged at 4°C for 5 min at 12,000g. Two hundred µl of the supernatant were mixed with 2.5 ml of scintillation cocktail (Aqua Luma, LKB) and the amount of radioactivity determined in a liquid scintillation counter (Packard, Tricarb 460).

Protein determination

Protein was determined according to Bradford[7] using bovine serum albumin as a standard.

RESULTS

The chitinolytic enzymes, chitinase and N-acetyl-β-D-glucosaminidase, can be separated by anion-exchange chromatography. Both enzymes are present in all developmental stages of Artemia as well as in the growth medium after emergence of the nauplii (Figs. 1 a - e). Chitinase elutes at NaCl concentrations of 0.33 to 0.36 M with the main peak of N-acetyl-β-D-glucosaminidase eluting between 0.18 - 0.2 M NaCl for all stages investigated. In addition to the main peak of N-acetylglucosaminidase I, another one appears specifically during early development. The N-acetylglucosaminidase II is present in developing cysts and freshly emerged nauplii (Figs. 1b,c) and maybe present in dry cysts as indicated by the shoulder of activity eluting at salt concentrations higher than N-acetylglucosaminidase I (Fig. 1a). N-acetylglucosaminidase II elutes at a salt concentration of 0.27 M NaCl and the peak of enzymatic activity can be further separated, according to molecular mass, into N-acetyl-β-D-glucosaminidase II_1 and II_2 (results not shown). The three types of N-acetyl-β-D-glucosaminidase not only differ in their molecular masses and elution patterns from anion-exchange chromatography but also in physico-chemical aspects and kinetic behaviour. The differences are summarized in Table 1. A detailed description of the purification and the kinetic properties of chitinase and the three N-acetyl-β-D-glucosaminidases will be given elsewhere (Funke and Spindler, in preparation).

The physiological relevance of chitin degrading enzymes was tested in two different developmental stages and processes: 1) during emergence of nauplii and 2) in relation to the moulting cycle. The presence of chitin in the embryonic cuticle of the shell of Artemia cysts[1] and the appearance of chitinolytic enzymes, preferentially chitinase, in the medium after hatching of nauplii (Table 2; Fig. 1d) indicates that chitinolysis is a necessary part of the emergence process. This is further substantiated by studies on the influence of chitinolytic enzymes on the amount (Table 3). After 50 h total hatching occurred when the medium contained chitinase, independently of whether a fungal or the Artemia chitinase, both applied with the same activities, was used. N-acetyl-β-D-glucosaminidase does not increase hatching as compared to the control with maximal hatching reached at about 70 h.

The level of chitinase and N-acetyl-β-D-glucosaminidase during the moulting cycle of adult males is shown in Fig. 2. Total activity per animal of both enzymes increases about 3 to 4-fold during the moulting cycle. Chitinolytic activity per animal steeply increases and reaches a maximum at stage D_0. The activity then decreases and reaches about the same levels in late premoult as in early postmoult. The protein content per animal reaches a maximum of about 40 µg at late intermoult. If the values are expressed as specific activity, the increase in activity of chitinase and N-acetyl-β-D-glucosaminidase starts at the beginning of the premoult period and the peaks are somewhat broader.

DISCUSSION

In the present paper we were able to demonstrate chitinase and N-acetyl-β-D-glucosaminidase in Artemia from dry cysts up to the adults. The pattern of N-acetyl-β-D-glucosaminidase isoenzymes is characteristic for a certain developmental stage. In all stages, chitinase and the predominant N-acetyl-β-D-glucosaminidase I are present and are certainly involved in chitin degradation, independently of whether this occurs as part of the emergence and hatching process of nauplii or in correlation with moulting events. The similarities in emergence and moulting and the involvement of chitinolytic degradation in these processes have already

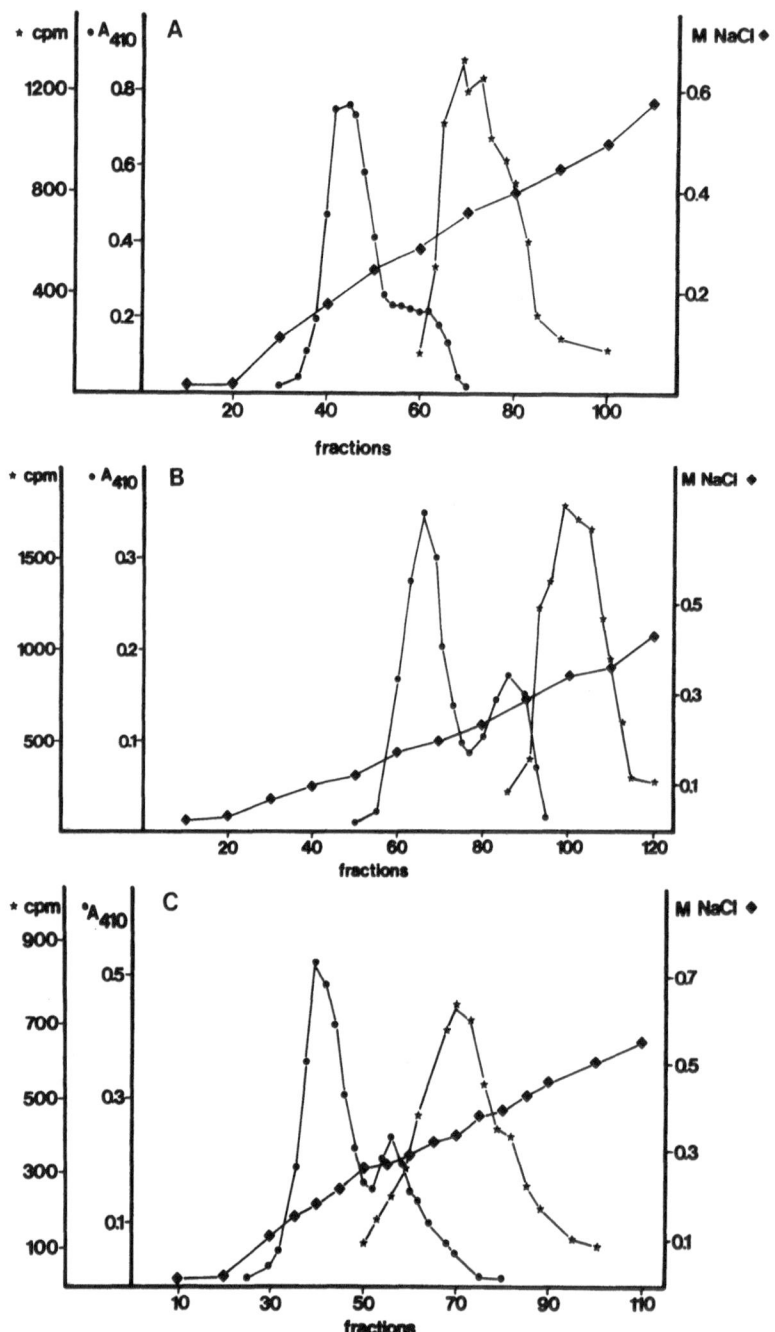

Fig. la,b,c. Separation of chitinase (-★-) and N-acetyl-β-D-glucosami-
nidase (-●-) from <u>Artemia</u>. (-◆-) salt gradient. The
separation conditions are as described in Materials and
Methods, with the exception of 1b, where a bigger column
(bed volume, 200ml; flow rate, 45ml/h; fraction size,
4.2ml) was used for a better separation. a) dry cysts;
b) cysts after 5 hours of incubation in medium; c) newly
hatched nauplii.

Table 1. Properties of Chitinolytic Enzymes from Newly Hatched <u>Artemia</u> Nauplii

	chitinase	N-acetyl-β-D-glucosaminidases		
		I	II_1	II_2
approximate molecular mass (Da, gelfiltration)	32,000	83,000	>100,000	56,000
S-value	5.9	8.6	11.9	7.9
isoelectric point	3.8	5.4	5.9	5.9
pH optimum	5.8	5.1	4.5	6.1
approximate temperature optimum, °C	55	60	65	55
temperature stability (100% activity)	50	50	55	20
binding to Con A-Sepharose	+	+	+	+
K_m-values				
^{3}H-chitin (mg/ml)	0.83	–	–	–
pNpGlcNAc (mM)	–	0.16	0.72	0.63
pNpGalNAc (mM)	–	0.21	0.31	0.40
GlcNAc$_2$ (mM)	–	0.23	0.65	0.62
inhibition, (mM)				
pNpGlcNAc	–	–	0.20	0.23
GlcNac$_2$	–	–	K_{ss},0.41	K_{ss},0.52
GlcNAc	–	c, 5.3	nc, 9	nc, 9
GalNAc	–	nc, 0.62 and 2.66	–	–
2-acetamido-β-D-deoxy-galactono-lactone	–	nc,0.013	nc, 0.95	nc, 0.70

+, positive effect; –, no substrate; c, competitive inhibitor; nc, = non-competitive inhibitor. If the two inhibitor constants K_i and K_{ii} are nearly equal, only one value is given, otherwise both. K_{ss} is the constant for the formation of the enzymatically inactive dead-end complex.

been discussed[1], but a direct proof was missing. The agreement in basic mechanisms for emergence and moulting probably also includes the regulation of chitinolytic activities by moulting hormones. There exists a correlation between the titers of chitinolytic enzymes and ecdysteroids during development up to stage I nauplius[1]. The same relationship must exist during the moulting cycles of adults. Changes in ecdysteroid concentrations during the moulting cycle of adult female <u>Artemia</u> have already been published[8]. If we compare the titer curves of chitinolytic enzymes with those of ecdysteroids it becomes evident that an increase in ecdysteroid concentration coincides with an increase in chitinolytic activity. Maximal enzymatic activities are reached between the time of apolysis and the middle of the premoult stage. Titer curves for chitinolytic enzymes during the moulting cycle comparable to the results found in <u>Artemia</u> have been described in only a few arthropods[2-4] (Buchholz, personal communication). A direct stimulation of chitin degradation by ecdysteroids has only been

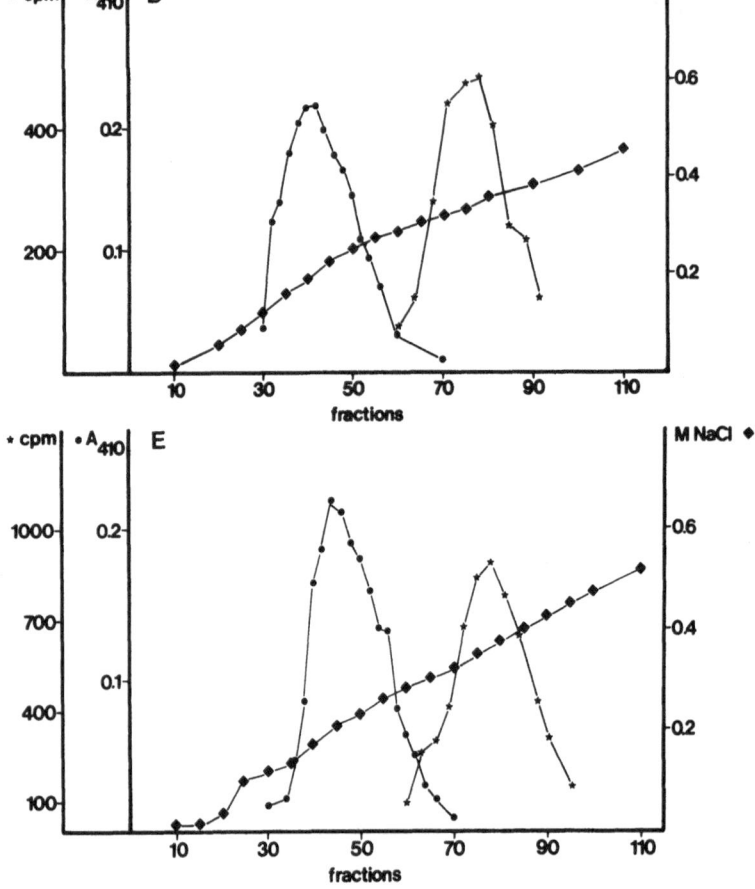

Fig. 1d,e Separation of chitinase (-★-) and N-acetyl-β-D-glucosaminidase (-●-) from <u>Artemia</u>. (-◆-) salt gradient. The separation conditions were as described in Materials and Methods. d) medium at the time of hatching; e) adults.

shown in <u>Balanus amphitrite</u>[9,10] and in the insects <u>Bombyx mori</u>[11], <u>Manduca sexta</u>[5] and the <u>Drosophila</u> K_c cell line (Spindler-Barth and Spindler, unpublished results).

The presence of chitinolytic enzymes in the growth medium at the time of emergence and hatching as well as the increase of the enzymatic activities prior to hatching indicated that chitin degradation is part of the emergence process[1]. Our present investigation supports this hypothesis. The concomitant presence of proteolytic and chitinolytic enzymes in the medium, beginning from emergence and reaching a maximum prior to hatching, does not mean that these enzymes play a special role in these processes but instead means partial degradation of the embryonic cuticle by the enzymes leads to their translocation into the extraembryonic space. As soon as the nauplii emerge the content of the extraembryonic space is released into the outer medium. The simultaneous presence of proteolytic and chitinolytic enzymes in the medium is easily explained by the fact that chitin does not exist in a pure form in nature but is always complexed with proteins[2]. In addition, native chitin containing structures can

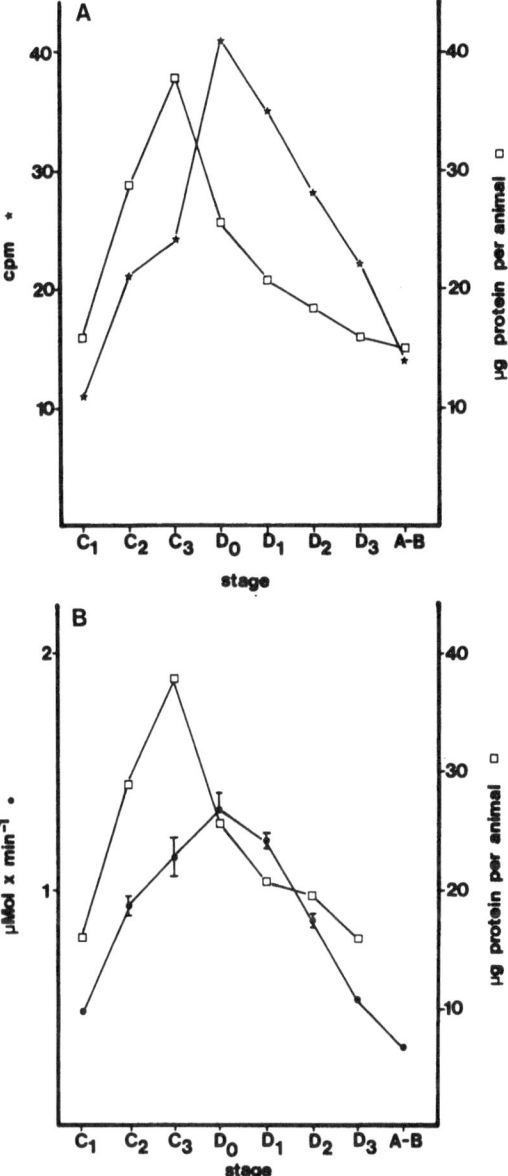

Fig. 2a,b. Titer of chitinolytic enzymes and the
amount of protein during the moulting
cycle of males. C_1 to C_3 (intermoult),
D_0 to D_3 (premoult) and A-B (postmoult)
are stages of the moulting cycle. (- □ -),
protein a) chitinase (-★-); b) N-acetyl-
β-D-glucosaminidase (- ● -).

only be degraded by chitinolytic enzymes if partial protein degradation
precedes[12]. The addition of chitinase, but not of N-acetyl-β-D-glucos-
aminidase which is not able to degrade chitin, to the outer medium only
shortens the later phases of emergence and hatching, but does not alter the

Table 2. Protein Content, Chitinolytic and Proteolytic Activity of Newly Hatched Nauplii and the Medium just after Hatching

		Nauplii	%	Medium	%
Protein		66.8	97	1.8	3
β-NAGase		1562	96	65	4
β-NAGase spec. act.		23.4		36.1	
Chitinase		5678	88	778	12
Chitinase spec. act.		85		432	
Protease, substrate	pH				
^{14}C-methyl-insulin	5.5	1662	98	33	2
"	7.0	1277	95	64	5
"	8.5	1079	92	86	8
azocasein	9.0	0.439	94	0.026	6

Values for protein are expressed in mg and those for the enzymes are corrected for one minute of incubation. N-acetyl-β-D-glucosaminidase (β-NAGase) is expressed as µM, chitinase and protease in cpm if tested with radiolabelled substrate, or as optical density at 366 nm if azocasein was used as substrate.

percent of hatched animals. Horne[13] has already demonstrated that egg hatchability is not affected by lipase, trypsin or chitinase, but he did not investigate hatching velocity.

As concerns the properties of the chitin degrading enzymes only a few comments will be given since the physico-chemical parameters and kinetic data will be described in detail (Funke and Spindler, in preparation). With the exception of N-acetyl-β-D-glucosaminidase II, all other chitinolytic enzymes have quite high apparent temperature optima and a pronounced temperature stability. Comparable high values have been described for Locusta migratoria[14], but even in the Antarctic krill, Euphausia superba, which always lives at temperatures below +2°C, rather high values were described[15]. The pH optima are in the same range as found in other arthropods, which is also true for the molecular mass of the chitinase, but not for all isoenzymes of the N-acetyl-β-D-glucosaminidases. Most described N-acetyl-β-D- glucosaminidases have molecular masses of about 120 - 140kDa, but there exist a few examples for molecular masses of about 60kDa[15,16]. The intermediate size class of about 80kDa, as described for N-acetyl-β-D-glucosaminidase I in Artemia, is unique, but must not be overinterpreted, since all N-acetyl-β-D-glucosaminidases from Artemia are glycoproteins, which may affect the determination of the molecular mass by gel filtration. In addition, nothing is known about the three-dimensional structure of the enzymes, not only from Artemia, but from other

Table 3.   Influence of Chitinases and N-acetyl-β-D-glucosaminidases on the Hatching Events in _Artemia_

Treatment	Hatching (%) after 50 h*
control	61 ± 11
chitinase from _Streptomyces_	96 ± 8
chitinase from _Artemia_, purified	100 ± 6
N-acetyl-β-D-glucosamindase from _Artemia_, purified	52 ± 10

*After 50 hours incubation at room temperature all hatchable animals appeared when treated with _Artemia_ chitinase. This value was set as 100% and the amount of hatching at this time was determined for the different treatments. (means ± S. D.; n = 3).

arthropod species. This again may lead to incorrect size determinations.

Our present investigations clearly demonstrate that chitin degrading enzymes are present during the whole life of _Artemia_ and they are involved in emergence and hatching of nauplii as well as in moulting of adults. If one compares the titer curves of moulting hormones and chitinolytic enzymes at both developmental stages of _Artemia_ with the situation in other arthropod species it is highly probable that the regulation of chitinolytic activity in _Artemia_ is due to moulting hormones but there is no direct proof for this hypothesis.

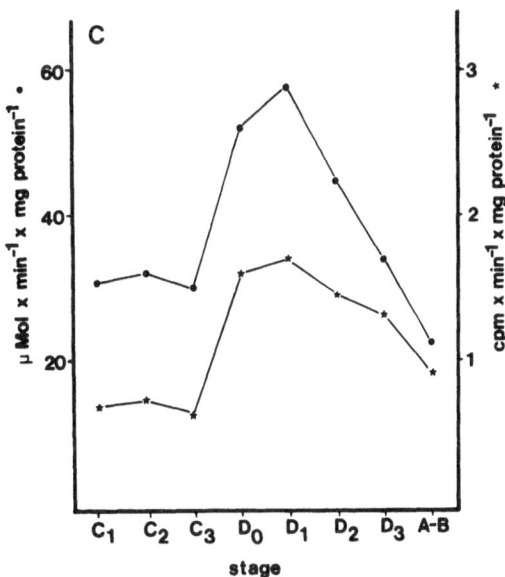

Fig. 2c. The specific activities of chitinase (-★-) and N-acetyl-β-D-glucosaminidase (-●-). Calculated from data in Fig. 2a,b.

REFERENCES

1. B. Funke and K. D. Spindler, Developmental changes of chitinolytic
    enzymes and ecdysteroid levels during the early development of the
    brine shrimp Artemia, in: "Artemia Research Applications," Vol 2,
    W. Decleir, L. Moens, H. Slegers, E. Jaspers and P. Sorgeloos, eds.,
    Universa Press, Wetteren (1987).
2. K.-D. Spindler, Chitin: Its synthesis and degradation in arthropods,
    in "The Larval Serum Proteins of Insects," K. Scheller, ed., Thieme
    Verlag, Stuttgart (1983).
3. U. Baier and H. Scheffel, Chitinaseaktivität während larvaler
    Häutungszyklen des Chilopoden Lithobius forficatus (L.), Zool. Jb.
    Physiol. 88:25 (1984).
4. M. Spindler-Barth, E. Shaaya and K.-D. Spindler, The level of chitin-
    olytic enzymes and ecdysteroids during larval-pupal development in
    Ephestia cautella and their modifications by a juvenile hormone
    analogue, Insect Biochem. 16:187 (1986).
5. T. Fukamizo and K. J. Kramer, Effect of 20-hydroxyecdysone on chitinase
    and β-N-acetylglucosaminidase during the larval-pupal transforma-
    tion of Manduca sexta (L.), Insect. Biochem. 17:547 (1987).
6. J. Molano, A. Duran and E. Cabib, A rapid and sensitive assay for
    chitinase using tritiated chitin, Anal. Biochem. 83:648 (1977).
7. M. N. Bradford, A rapid and sensitive method for the quantitation of
    microgramm quantities of protein utilizing the principle of protein-
    dye binding, Anal. Biochem. 72:248 (1976).
8. H. Walgraeve, K. Van Brussel and A. De Leenheer, Comparison of three
    separation systems for the radioimmunoassay of ecdysteroids in
    Artemia (Crustacea: Branchiopoda), Insect Biochem. 16:41 (1986).
9. J. A. Freeman and J. D. Costlow, Hormonal control of apolysis in
    barnacle mantle tissue epidermis in vitro, J. Exp. Zool. 210:333
    (1979).
10. J. A. Freeman, Hormonal control of chitinolytic activity in the integu-
    ment of Balanus amphitrite in vitro, Comp. Biochem. Physiol. 65A:13
    (1980).
11. S. Kimura, The control of chitinase activity by ecdysterone in larvae
    of Bombyx mori, J. Insect Physiol. 19:115 (1973).
12. M. L. Bade and A. Stinson, Activation of old cuticle as a substrate
    for chitinase in the molt of Manduca, Biochem. Biophys. Res. Comm.
    87:349 (1978).
13. F. R. Horne, The effect of digestive enzymes on the hatchability of
    Artemia salina eggs, Trans. Am. Microsc. Soc. 85:271 (1966).
14. R. Zielkowski and K.-D. Spindler, Chitinase and chitobiase from the
    integument of Locusta migratoria: characterization and titer during
    the fifth larval instar, Insect biochem. 8:67 (1978).
15. K.-D. Spindler and F. Buchholz, Partial characterization of chitin
    degrading enzymes in two euphasiids, Euphausia superba and
    Meganyctiphanes norvegica, Polar Biol. (in press) (1988).
16. C. Dziadik-Turner, M. S. Mai and K. J. Kramer, Purification and char-
    acterization of two β-N-acetylhexosaminidases from the tobacco
    hornworm, Manduca sexta (L.) (Lepidoptera:Sphingidae), Arch. Biochem.
    Biophys. 212:546 (1981).

REGULATION OF <u>ARTEMIA</u> GLYCOGEN PHOSPHORYLASE: EFFECT OF DIGUANOSINE

NUCLEOTIDES

A. Sada, V. Carratora and M. G. Cacace

Istituto di Biochimica delle Proteine ed Enzimologia
CNR, Via Toiano 6
80072 Arco Felice, Italy

ABSTRACT

Dormant cysts of <u>Artemia</u> contain glycogen phosphorylase, a key enzyme in the utilization of energy reserves at resumption of metabolic activity. The enzyme is regulated through a phosphorylation/dephosphorylation mechanism similar to that operating in other higher organisms. This mechanism of interconversion is active during development and the relative ratio of the two forms changes significantly in the period preceding the emergence of the nauplius.

The enzyme was purified to homogeneity by a procedure involving ion-exchange and gel filtration chromatography and an affinity chromatography step on AMP-Sepharose.

Preliminary attempts to crystallize the protein were not successful due to difficulties in obtaining other than submicroscopic crystals.

The kinetic properties of the purified enzyme were determined. The activity is stimulated by AMP and is inhibited by diguanosine tetraphosphate (Gp_4G) one of the 5-5' dinucleoside polyphosphates present in large amounts in <u>Artemia</u> cysts.

Diguanosine triphosphate (Gp_3G) does not inhibit phosphorylase activity appreciably <u>in vitro</u>. The presence of these nucleotides at the first stages of development and their rapid degradation may contribute to the regulatory system affecting overall glycogen phosphorylase activity.

METALLOPROTEINS IN DEVELOPING <u>ARTEMIA</u>

Roger A. Acey[a], Benton N. Yoshida[b] and Martin E. Edep[c]

[a]Department of Chemistry, California State University, Long
Beach, California U.S.A.
[b]Department of Biology, California Institute of Technology
[c]University of California, San Francisco School of Medicine

INTRODUCTION

There is an increasing awareness of the importance of essential trace
metals, particularly zinc and copper, in normal cell growth and development [1]. Central to this theme has been the postulated direct involvement
of zinc in the control of gene expression [2]. A number of zinc metalloproteins have been reported to be involved in replication and transcription,
including RNA and DNA polymerases [3]. Thus, the question arises as to the
mechanism by which the activity of these enzymes might be directly regulated by cytosolic trace metals.

In conjunction with the idea of the essential interaction of trace
metals with the nuclear polymerizing enzymes involved in gene expression,
two nuclear regulatory transcription factors, TFIII and Spl, have also
been shown to require zinc for activity. TFIIIA, isolated from Xenopus,
is required for the accurate transcription of the 5s RNA genes [4]. This
40 Kd protein contains nine tandemly repeated sequences of approximately
30 amino acids; each of these repeated sequences is responsible for binding
a specific 5 bp DNA sequence [5]. These nucleic acid binding domains are
stabilized by zinc tetrahedrally coordinated to two cysteines and two
histidines. This nucleic acid binding motif is conveniently referred to
as a "zinc finger". It has been clearly demonstrated that removal of the
zinc prevents 5s RNA gene transcription. Similarily, transcription factor
Spl, which is responsible for activation of selective genes in animal cells,
contains three homologous DNA binding segments or "zinc fingers" [6]. In
this case, removal of zinc results in the loss of the sequence specific
DNA binding capability of Spl. Significant is the apparently ubiquitous
distribution of this nucleic acid binding motif. Similar sequences have
been detected in the yeast transcription factor ADR1[7] and several <u>Drosophila</u>
genes products, including those from the Kruppel [8], hunchback [9] and
serendipity [10] loci.

It is readily apparent that trace metals play a significant role in
gene expression. However, one is left with the enigma of how these
essential trace metals are stored intracellularly in a form which prevents
heavy metal toxicity while at the same time having the metal available for

transfer to specific nuclear/cytoplasmic apo-metalloproteins. Of particular
interest has been the involvement of metallothionein (MT). Metallothioneins
are a unique class of low molecular weight (ca. 6-8 Kd) metal binding
proteins found in a variety of organisms, including crustacea [11]. They
are characterized by a high cysteine content, lack of aromatic amino acids,
unusual heat stability, and broad metal binding specificity [12]. Moreover,
their primary structure is highly conserved, in particular, the positions
of the cysteine residues responsible for metal binding. In fact, in human,
mouse, and crab MT, the relative positions of the cysteines are invari-
ant [13]. Thus, MT's ubiquitous distribution and conserved primary sequence
suggest a vital biological role.

Metallothioneins' high affinity for and inducibility by heavy metals
such as cadmium and mercury suggest an involvement in metal detoxification.
The induction of MT by heavy metal is a direct result of changes in gene
transcription [12]. However, the expression of MT can also be influenced
by other seemingly unrelated factors, including glucocorticoids, inter-
ferons, and stress conditions [14]. The ability of this diverse group of
factors to influence the expression of MT is a consequence of unique DNA-
proteins interactions [15]. For example, the DNA sequences responsible
for glucocorticoid induction are distinct from those necessary to modulate
the expression of MT by heavy metal [16]. For these reasons, MT has been
used by numerous investigators to study the regulation of eukaryotic gene
transcription.

While its exact function is unknown, MT appears to play a general
"housekeeping" role in intracellular trace metal homeostasis in addition
to its ability to be induced by heavy metals. For example, it has been
shown that MT can activate apo-metalloenzymes by direct exchange of metal.
Incubation of apo-zinc metalloenzymes such as carbonic anhydrase and
alkaline phosphatase with zinc-thionein results in activation of the
enzyme [17]. In addition, Menkes' kinky hair disease and Wilson's disease,
both characterized by an accumulation of excess copper, are the result of
a loss in regulation of MT gene expression and synthesis of an abnormal
MT molecule, respectively [18,19].

The importance of MT in embryonic development is implied by studies
showing that MT is temporaly expressed in a tissue specific manner. In the
mid-gestation rodent embryo, MT is actively synthesized by parietal and
visceral cells of the extra-embryonic membranes [20]. In addition, expression of MT mRNA in hepatic tissue from developing rodents peaks twice
during development, once during the late fetal period and again in the
early neonate stage of development [21]. Similarly, expression of MT is
developmentally regulated in the sea urchin embryo; newly synthesized MT
mRNA begins to accumulate early in development, decreases during gastrulation, and increases again in the early pluteus stage of development.
Constitutive synthesis of MT mRNA is high in the ectoderm but occurs at low
levels in the mesoderm and endoderm [22]. While the factors responsible
for the differential induction of MT are unknown, it is clear that significant changes in MT gene expression occur during normal embryonic development.

We have been attempting to establish Artemia as a model eukaryotic
system for studying essential trace metal homeostasis and the role metal
binding proteins play in early embryonic development. Due to Artemia's
ease of handling, availability, and wealth of published biochemical data
on its development, it is an ideal organism with which to perform these
studies. In an earlier study, we reported the presence of MT-like metal
binding activities in developing Artemia [23]. Here we report on the
partial purification of these molecules and their developmentally regulated
expression.

204

MATERIALS AND METHODS

Shrimp Preparation

Metabolically dormant encysted embryos (San Francisco Bay Brand) were kindly donated by San Diego Brine Shrimp, Inc., Chula Vista, CA., and stored at -20°C. All manipulations were performed at 4°C. Cysts were sterilized by suspending them in antiformin with gentle stirring for five minutes [24]. The cysts were collected in a Buchner funnel lined with Miracloth and washed with 20 liters of water; the cysts were next resuspended in 0.01 M HCl for five minutes, collected, and again washed with 20 liters of water. The final 2 l of water was removed under vacuum. The Miracloth containing the embryos was removed from the funnel and placed on a stack of hand towels to remove residual liquid. The embryos were redistributed into five gram packets and stored at -20°C.

Shrimp Culture and Homogenization

Five gram lots of sterilized cysts were cultured at 30°C in 250 ml of sterile artifical sea water containing 5 mM sodium tetraborate, 25,000 units of penicillin and 25 mg of streptomycin sulfate [25]. The embryos were collected, washed with double distilled water and stored at -80°C. All reagents used in these experiments were reagent grade. To ensure high quality water, distilled water was first passed through a mixed-bed ion exchange resin, then charcoal filtered, and finally glass distilled a second time. Zinc and copper were below levels of detection as determined by atomic absorption spectrophotometry. Glassware used in the culture of shrimp and the preparation of soluble proteins was rinsed in 0.5 mM EDTA, acid washed with 10%(w/v) nitric acid, and finally rinsed thoroughly with double distilled water. Embryos were homogenized in 250 mM sucrose, 10 mM Tricine, 5 mM $MgCl_2$, 0.5 mMEDTA, 1 mM DTT, pH 8.0, supplemented with 1 mM PMSF and 10 µg/ml soybean trypsin inhibitor. Each gram of tissue was homogenized in a total volume of 2 ml. Unless otherwise specified, all experimental protocols were performed at 4°C. Using a minimum amount of buffer, embryos were ground to a thick paste in a porcelain mortar and pestle. The remaining buffer was added and the homogenate filtered through Miracloth into a Dounce homogenizer. The mixture was homogenized with ten strokes each of the loose and tight fitting pestle and then centrifuged at 15,000 x g for 30 minutes. The resulting post mitochondrial supernatant was centrifuged for an additional 90 minutes at 150,000 x g to obtain soluble proteins (150K sup). The lipid layer floating on the top of the solution was removed by vacuum aspiration. The 150K sup was next dialyzed overnight against nitrogen-saturated 50 mM Tris-HCl, 0.02% NaN_3(w/v) and stored under nitrogen at -80°C.

Fractionation of Soluble Proteins and Detection of Metal Binding Activity

Total protein concentrations were determined by the method of Lowry[26]. Metal binding activity was determined by incubating 150K sup with limiting amounts (1 to 5 pmoles) of [109]Cd (New England Nuclear; Sp. Act. = 1230 mCi/mg) for ten minutes and fractionating the mixture by gel filtration on either G-75-SF or G-50-SF (Sigma Chemical Company, St. Louis, MO). Unless otherwise indicated, columns were eluted with 50 mM Tris-HCl, 0.02% NaN_3 (w/v); the buffer was continually flushed with nitrogen to maintain reducing conditions. Column eluate was monitored simultaneously for 280 nm and 254 nm absorbance with an ISCO UA5 Monitor. (ISCO, Inc., Lincoln, NB). Five ml fractions were collected in tubes containing 5 µl of 1 M DTT and stored at -20°C. Radioactivity was determined using an Abbott Auto-Logic gamma counter. Endogenous zinc levels were determined by atomic absorption spectrophotometry using a Perkin-Elmer 5000 Atomic Absorption Spectrophotometer. When comparing protein mixtures from different periods of development by gel filtration, equal amounts of protein in a total volume of 25ml were applied to the colum.

Fractions showing metal binding activity were pooled. Subsequently, a 15 ml sample of the mixture was applied to a 1.0 x 15 cm column of A25-DEAE Sephadex (Sigma Chemical Co.) previously equilibrated with 200 mM Tris, 0.02% NaN_3(w/v), pH 8.0. The column was washed with one bed volume of equilibration buffer and eluted with a linear gradient of Tris (200 mM to 450 mM), 0.02% NaN_3(w/v), pH 8.0. Two ml fractions were collected and assayed for radioactivity and zinc as indicated above.

Determination of Total Metal Binding Capacity

Total metal binding capacity was determined by the cadmium saturation method of Eaton and Toal [27]. Fractions containing metal binding activity were pooled and a 200 ul aliquot transferred to a 1.5 ml microcentrifuge tube containing 200 μl of a $^{109}CdCl_2$ solution in 10 mM Tris, pH 7.4. Cadmium chloride was mixed with the $^{109}CdCl_2$ so that the total cadmium concentration of the solution was 2 μg/ml and the level of radio-activity 1.0 μCi/ml. Subsequently, 100 μl of 2% (w/v) bovine hemoglobin (Type II, Sigma Chemical Company) was added and the mixture placed in a boiling water bath for 2 minutes. This procedure was repeated a second time. Coagulated protein was pelleted by centrifugation. Subsequently, 200 μl of supernatant was removed and counted. To ensure that the metal binding ligand was the limiting factor in the assay, the procedure was repeated on samples which had been diluted two-and four-fold.

RESULTS AND DISCUSSION

Purified mammalian MT is normally associated with significant amounts of zinc and copper; a useful property of the molecule is its ability to instantaneously exchange endogenously bound zinc or copper with other exogenously added transition metals, most notably cadmium. Thus, independent of the degree of metal saturation, MT is easily detectable. Current methods for detection of MT or MT-like molecules in other organisms assumes similar metal exchange rates and involve incubating samples suspected of containing the protein with radioactive cadmium and fractionating the mixture by gel filtration. The presence of a low molecular weight cadmium binding protein is preliminary evidence for the presence of MT-like metal binding activities. Therefore, soluble proteins from metabolically dormant encysted embryos(cysts) were incubated with ^{109}Cd and fractionated on Sephadex G-75. As shown in Figure 1, the majority of radiolabeled metal is associated with a low molecular weight component, referred to a metal binding protein I(MBPI). For simplicity, we refer to all the metal binding activities described in this study as proteins, regardless of size. Proteins eluting in, or near, the void volume of the column are designated as high molecular weight proteins (HMWP). Both the HMWP and MBPI coelute with significant amounts of endogenous zinc (Fig. 1). While the LMWP are associated with minute amounts of copper, the metal is undectable in MBPI (data not shown). MBPI is heat stable and the bound cadmium cannot be removed by dialysis of the complex against 0.5 mM EDTA [23].

In most organisms, metallothionein exists in two isoforms, MTI and MTII [12]. The two isoforms are resolvable by anion exchange chromatography; the activity eluting first from the column is referred to as MTI and the latter MTII. It has been shown that the two isoforms are differentially regulated during development [28]. Mounting evidence suggests a tissue-specific expression of the isoforms [29]. With this in mind, we considered the possibility that MBPI is composed of multiple isoforms. Fractionation of MBPI on A-25 DEAE Sephadex resulted in the resolution of

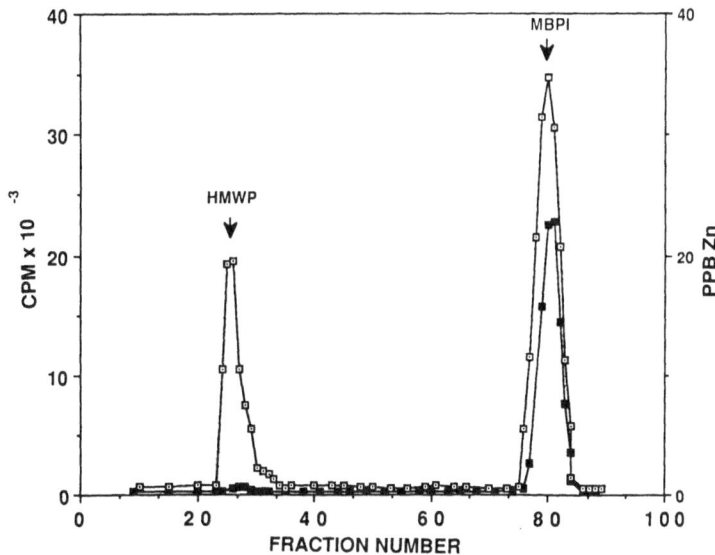

Fig. 1. Metal binding activity in metabolically arrested encysted embryos. Soluble proteins were prepared as
described in Materials and Methods. The mixture was
preincubated with [109]Cd for ten minutes and fractionated on a 2.5 x 100 cm column of Sephadex G-75. The
column was eluted with 50 mM Tris-HCl, 0.02% NaN$_3$,
pH 8.0. Radioactivity and zinc levels were determined
as described in Materials and Methods. The results
are expressed as CPM/fraction and PPB Zn/fraction,
respectively: (-■-) radioactivity; (-▫-) zinc; HMWP,
high molecular weight proteins; MPB1, metal binding
protein 1.

the endogenous zinc binding activity from the exogenous cadmium binding
activity, which we refer to as ZnBPI and CdBPI, respectively (Fig. 2).
The CdBPI elutes first from the column cleanly resolved from ZnBPI. Since
endogenously bound metal is undetectable in CdBPI, it is possible that
these "isoforms" are actually the same protein whose behavior on the ion
exchange column is influenced by it's metal content [30]; determination of
the amino acid composition for each of these two molecules will be required
to provide an unequivocal answer to the question of structural identity.

However, assuming true isoforms, the reason that exogenously added
cadmium associates exclusively with CdBPI is not clear. If one postulates
that ZnBPI, unlike MT, does not readily exchange endogenously bound zinc
with added cadmium, and if the metal binding sites of ZnBPI are completely
saturated, then one would not expect ZnBPI to be associated with the radiolabeled cadmium.

However, the situation is much more complicated if one assumes that
the metal binding sites of ZnBPI are not completely saturated; conceivably,
binding of cadmium to ZnBPI occurs at a rate too low to be observed under
our experimental conditions. On the other hand, assuming equal rates of
metal binding, the ability of CdBPI to exclusively sequester the trace

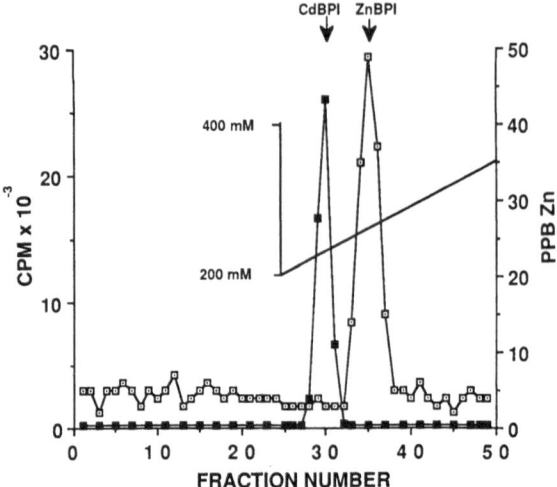

Fig. 2. Fractionation of MBP1 by ion-exchange chromatography.
MBP1 was applied directly to a 1.0 x 15 cm column
of DEAE A-25 Sephadex previously equilibrated with
200 mM Tris-HCl, 0.02% NaN$_3$, pH 8.0. The column
was eluted with a 200 mM to 450 mM Tris-HCl, pH 8.0
gradient. Radioactivity and zinc levels were determined as described in Materials and Methods. The
results are expressed as CPM/fraction and PPB Zn/
fraction, respectively: (-■-) radioactivity;
(-□-) zinc. The position of the salt gradient was
determined by measurement of refractive indicies.
CdBP1, cadmium binding protein 1; ZnBP1, zinc
binding protein 1.

amounts of exogenously added cadmium could be explained by CdBPI having a
significantly greater binding affinity for cadmium than ZnBPI. Thus, under
conditions which assume significantly greater rates of metal exchange and
binding affinity in CdBPI vs. ZnBPI, cadmium binding to CdBPI would occur
independent of the degree of metal saturation of ZnBPI. If one postulates
ZnBPI functioning as a store for intracellular essential trace metals and
CdBPI as playing á protective role by sequestering toxic heavy metals; it
would be logical for CdBPI to have a greater binding constant for toxic
heavy metals than ZnBPI. The obvious advantage to the embryo would be the
ability to cope with an exposure to toxic heavy metals while minimizing
the effects on intracellular trace metal homeostasis.

We next concerned ourselves with the possibility that one or both of
these metal binding proteins might be developmentally regulated. This was
prompted by the fact that the UV elution profiles of fractionated soluble
proteins changed dramatically during the first 48 hrs. of development
(Fig. 3). Of particular significance is the dramatic increase in the 254 nm
absorbance relative to the 280 nm absorbance of MBPI during the course of
development. Metallothioneins coordinate metal by thiolate clusters and
depending on the coordinated metal, the mercaptide bonds have a distinctive
absorption near 254 nm. A high 254/280 nm absorbance ratio is characteristic of metallothionein. Therefore, it is plausable that the temporally
increasing 254/280 nm absorption ratio observed for MBPI could be the

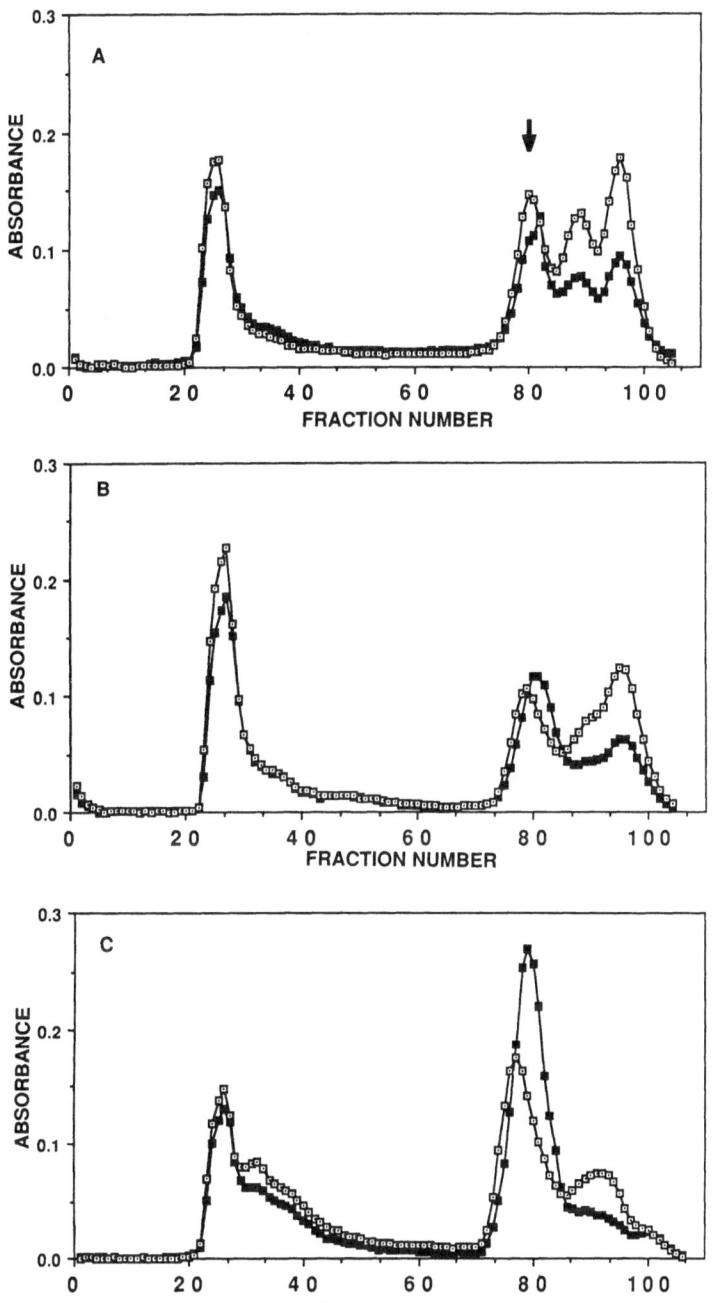

Fig. 3.   Absorption profiles of fractionated soluble proteins from
developing embryos.   Soluble proteins were prepared from encysted metabolically dormant embryos (0 hr.) and 24 and 48
hr. nauplii and as described in Materials and Methods. Equal
amounts of soluble protein (25 mg) from the three stages of
development were fractionated as described in Fig. 1. Column
eluate was monitored continuously at 254 nm and 280 nm
absorbance as described in Materials and Methods. Absorption
(Continued)

result of increased amounts of metal-saturated MT or an increase in metal flux onto preexisting apo-thionein. Increased levels of a specific protein with a characteristically high 254 nm absorbance would also account for the observation. Therefore, MBPI from cysts and 24 hr. and 48 hr. nauplii was collected and assayed for its total metal binding capacity using the cadmium saturation method of Eaton and Toal (Table 1) [27]. The total cadmium binding capacity of MBPI increases significantly during the first 48 hours of development, suggesting an increase in the level of a particular MT-like protein. Based on the fact that CdBPI will sequester limiting amounts of cadmium, the results of the binding capacity experiments could, in part, be explained by increased levels of CdBPI.

Subsequently, MBPI from the three stages of development was further resolved into CdBPI and ZnBPI. Using UV elution profiles as a measure of protein concentration, it is readily apparent that the level of CdBPI (or co-eluting protein) is developmentally regulated; the protein is barely detectable in metabolically dormant cysts, but increases dramatically during the next 48 hours of development (Fig. 4). During this same period of development there is little change in the level of ZnBPI. It should be noted that both proteins have high 254/280 nm absorbance ratios. However, since the estimated M_r of these molecules is approximately 2000 daltons [31], we are cautious in concluding that they are MT or MT-like. Alternatively, it may be that they are similar to the phytochelatins reported in plants [32]. The structure of these natural products has been established as (γ-glutamic acid- cysteine)$_n$ -glycine, where n = 3 to 7. They are inducible by heavy metals and appear to be the primary metal binding activity in plant cells. As expected, the metals are sequestered as thiolates. On the other hand, an endogenous 3600 dalton zinc-binding protein, closely resembling MT in amino acid composition, has been reported in unfertilized sea urchin eggs [33]. It is apparent that a more defined biochemical description of these molecules will be needed to answer questions posed by these experiments.

During the course of our experiments on the developmental regulation of MBPI, we detected a second larger molecular weight endogenous zinc metalloprotein which we refer to as ZnBPII. Like CdBPI, ZnBPII is un-detectable in dormant gastrula but increases significantly during the next 48 hrs of development. We estimate the M_r of the molecule to be 8000 daltons, comparable in size to mammalian MT [31].

Because of the possible similarity of ZnBPII and MT, and the involve-ment of MT in early development, soluble proteins from embryos between 3 and 24 hrs of development were analyzed for ZnBPII to determine the time course of its expression. In these experiments, mixtures of soluble proteins were fractionated on G-50 resin to increase the resolution of the fractionation procedure. These experiments revealed that significant levels of ZnBPII are detectable after 12 hours of development, coincident with embryonic hatching, and increase during the next 12 hours of development (Fig. 5). While it is possible that ZnBPII is comprised of more than one metal binding component, the fact that ZnBPII elutes as a sharp symmetrical peak of activity from the column, suggests that it is in fact a single protein. Whether the "appearance" of ZnBPII is the result of mobilization of zinc from cellular stores (ZnBPI?) onto a preexisting apo-protein or de novo synthesis of the metalloprotein is unknown. It is interesting to

_____

Fig.3
(Continued)   profiles from the G-75 column are presented for A) 0 hr
                cysts; B) 24 hr nauplii; C) and 48 hr nauplii (- ■ -) 254
                nm absorption; (- ▢ -) 280 nm absorption. The vertical
                arrow indicates the position of MBP1.

Table 1. Total Metal Binding Capacity of MBPI in Developing <u>Artemia</u>

Stage of development	Bound Cd (cpm)
0 Hr. (encysted embryos)	2727
24 Hr. nauplii	8115
48 Hr. nauplii	19080

Soluble proteins were obtained from cysts, 24 hr., and 48 hr. nauplii as described in Materials and Methods. For each developmental stage, an equal quantity of total soluble protein (25 mg) was applied to a 2.5 x 100 cm column of Sephadex G-75 and eluted with 50 mM Tris-HCl, 0.02% (w/v) NaN$_3$, pH 8.0. Fractions containing MBPI were collected and pooled. Total metal binding capacity was determined as described in Materials and Methods. Results are expressed as protein bound ^{109}Cd standardized to 220 nm absorbance. The same relative increase in metal binding activity was observed when the amount of bound radio-labeled cadmium was standardized to either 254 nm or 280 nm absorbance (data not shown).

note that in conjunction with the induction of ZnBPII, the level of zinc associated with MBPI remains constant for the first fifteen hours of development but increases two-fold within the next nine hours (Fig. 5).

Another significant aspect of these experiments involves the exogenously added ^{109}Cd used to detect the position of MBPI. Careful examination of the data shows that the label undergoes significant changes in distribution between the various protein pools, e.g., LMWP, MBPI, and ZnBPII, as development proceeds. During the first three to twelve hours of development, the majority of radiolabeled cadmium is distributed between LMWP and MBPI. However, in embryos between fifteen and twenty-four hours of age, the label is redistributed predominately between ZnBPII and MBPI (Fig. 5). This can be interpreted as resulting from the increasing levels of CdBPI and ZnBPII during this period of development. When one considers that the level of ZnBPII is very low (based on UV absorption) with respect to LMWP and MBPI, it would appear that ZnBPII has a greater affinity for trace metals than the LMWP and MBPI protein pools. On the other hand, since the composition of the LMWP is changing during this developmental period [34], loss of proteins which adventitiously bind metal from the LMWP pool could partially account for the observed decrease in LMWP cadmium binding activity. A decrease in competing metal binding proteins would also manifest itself as an increase in cadmium binding by MBPI and ZnBPII. However, one can not rule out the kinetics of metal binding (exchange) as a mechanism to explain the differential cadmium binding; it could be that these rates are much faster for ZnBPII than for MBPI.

With the onset of hatching, DNA synthesis and cell division are re-initiated [35]. In other systems, similar proliferative states are associated with increased amounts of zinc binding proteins [36]. With this in mind, we began to explore the possible involvement of ZnBPII in the hatching process and subsequent initiation of embryonic cell growth. In particular, we were interested in determining whether factors which delay hatching inhibit the expression of ZnBPII. Culture of <u>Artemia</u> embryos at elevated temperatures (heat shock) is known to inhibit hatching [37]. Therefore, embryos were cultured for 12 hrs at 30°C, then placed at 37°C for 6 hours. Cultures were either allowed to remain at the higher temperature for 6 or 12 hrs. or returned to 30°C for the same periods of time. As expected, the

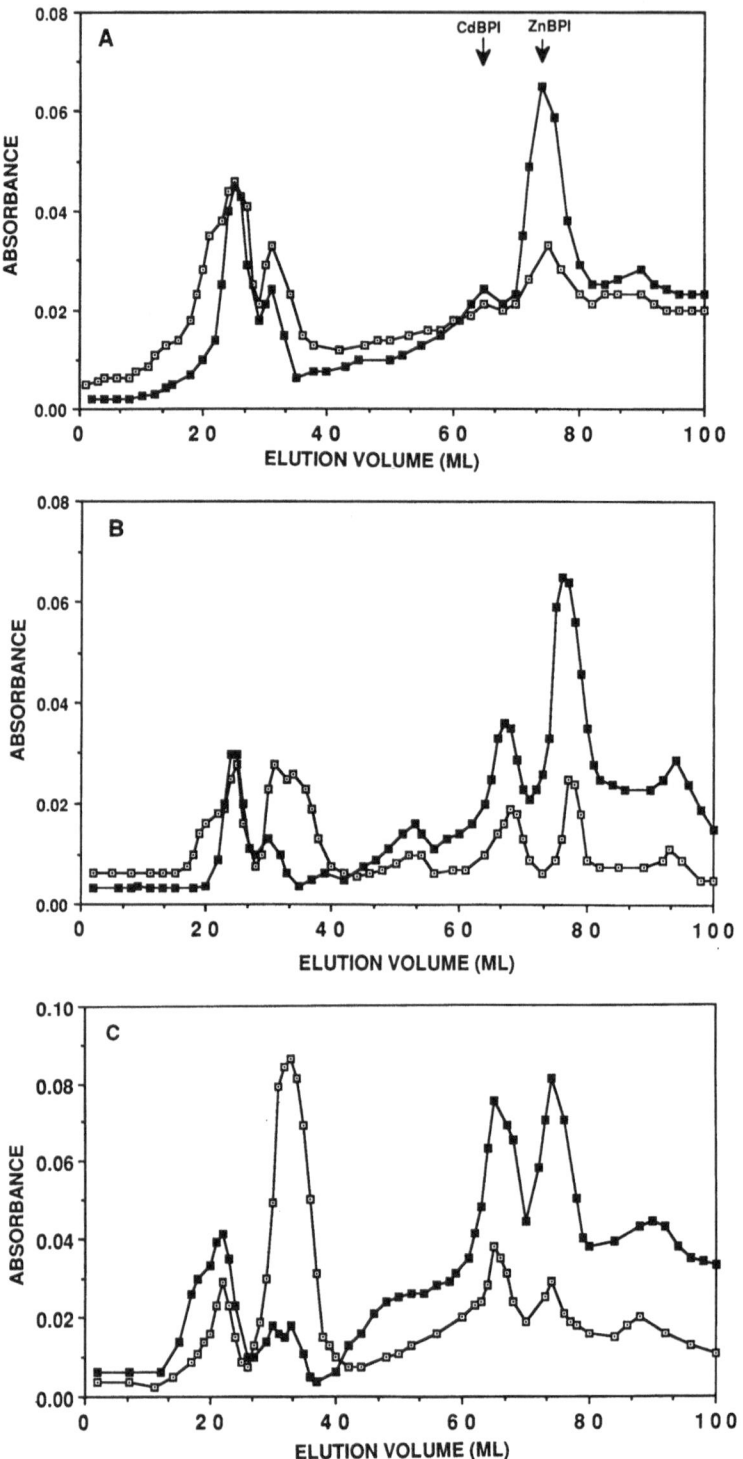

Fig. 4. Expression of CdBPI and ZnBPI in developing embryos.
Soluble proteins were prepared from the embryos as described in Materials and Methods and fractionated by gel

(Continued)

heat shock significantly diminished the degree of hatching. In cultures maintained at 37°C, those embryos which did not develop into free swimming nauplii either remained within their chitin-like shell or proceeded only to the E1 stage of development (data not shown). As shown in Fig. 6A, ZnBPII is undetectable in embryos arrested in development as a result of the initial 6 hour heat shock period. Moreover, the distribution of ^{109}Cd is remarkably similar to that observed for an earlier embryo (Fig. 5). Culture of these embryos at 37°C for an additional 6 or 12 hrs (24 hr and 30 hr old embryos, respectively) not only continues to supress ZnBPII expression but also inhibits the developmentally regulated increase in MBPI associated zinc, Fig. 6B and Fig. 6C, respectively. In contrast, embryos returned to 30°C after the initial heat shock, contain free swimming nauplii or embryos completely removed from their shell but still encased in their hatching membrane (data not shown). While ZnBPII is undetectable in embryos still encased in their hatching membrane (Fig. 7A), the increase in MBPI-associated zinc is readily evident. In addition, the ^{109}Cd binding profile is significantly different from that observed with embryos of the same age maintained at the higher temperature; the distribution of radiolabeled cadmium qualitatively represents that of an early embryo preparing for hatching, i.e., the radiolabeled cadmium undergoes a redistribution from the LMWP pool predominantly into MBPI and a small but significant level into fractions co-eluting with ZnBPII (Fig. 7B). It is apparent that we are looking at the embryo readjusting its metabolism to support normal growth and development after exposure to external stress and that exogenously added ^{109}Cd is a sensitive probe for determining embryonic stress as it relates to changes in metal binding molecules like ZnBPII. Figure 7C is the zinc and cadmium binding profiles for 30 hr. free swimming nauplii. The fact that ZnBPII is undetectable in embryos remaining within their hatching membrane suggests that, at a first approximation, the appearance of ZnBPII is associated with expulsion of the embryo from it's hatching membrane.

CONCLUSIONS

Results gleaned from a detailed time course study of ZnBPII expression in embryos arrested in development by heat shock should give further insights into the developmental program of Artemia. Moreover, from an ecological point of view, one or several of these proteins might be useful as a marker in determining the organism's level of exposure to environmental stress. Experiments are in progress to determine the effects of heavy metals and other physiological stress on expression of ZnBPII.

Fig. 4 (Continued)
filtration as described in Fig. 3. Total MBPI was collected from each stage of development and fractionated individually on DEAE A-25 Sephadex as described in Fig. 2. Column eluate was monitored continuously at 254 nm and 280 nm absorbance as described in Materials and Methods. Column fractions were assayed for radioactivity and zinc to determine the positions of CdBPI and ZnBPI, indicated by arrows. Absorption profiles are presented for the resolution of MBPI from A) 0 hr. cysts; B) 24 hr. nauplii; and C) 49 hr. nauplii. (- ■ -) 254 nm absorption; (- ▣ -) 280 nm absorption.

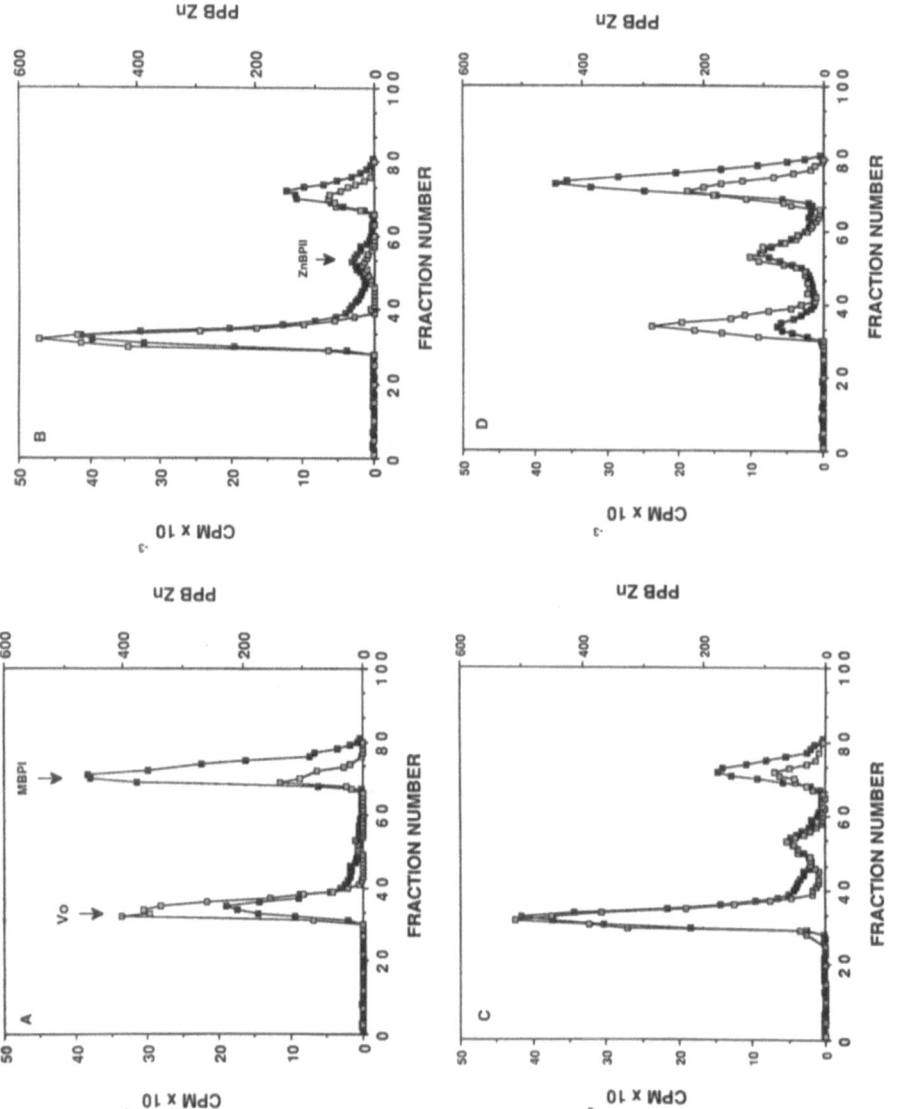

Fig. 5. Expression of ZnBP11 in developing embryos. Soluble proteins were prepared from the embryos as described in Materials and Methods. The mixture was preincubated with [109]Cd for ten minutes and fractionated on a 2.5 x 100 cm column of Sephadex G-50. The column was eluted with 50 mM Tris-HCl, 0.02% NaN_3, pH 8.0. Radioactivity and zinc levels were determined as described in Materials and Methods. The results are expressed as CPM/fraction and PPB Zn/fraction, respectively: (-■-) radioactivity; (-□-) zinc; V_0, void volume of the column; MPB1, metal binding protein 1; ZnBP11, zinc binding protein 2. Results are presented for; A) 3 hr cysts; B) 12 hr cysts; C) 15 hr cysts; and D) 24 hr nauplii.

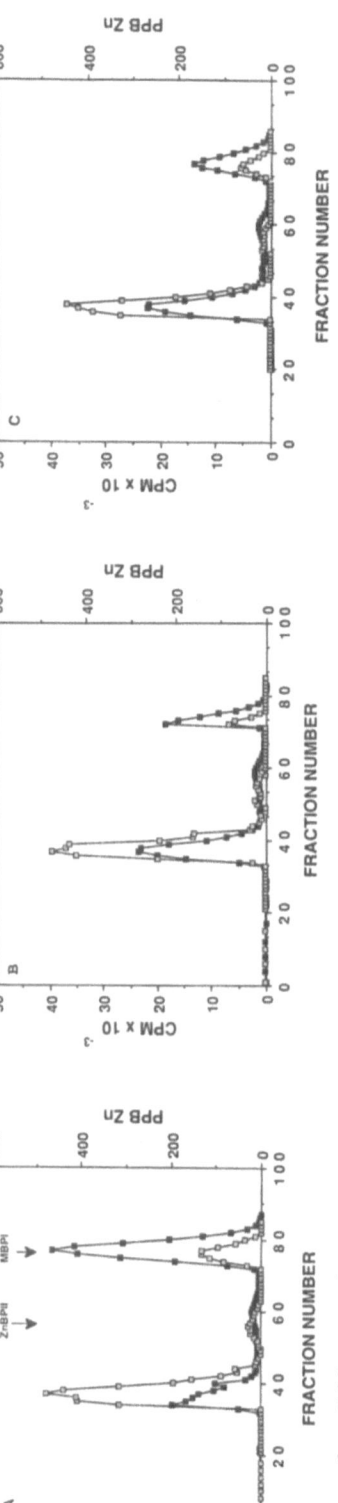

Fig. 6. Effects of heat shock on the expression of ZnBPII. Embryos were cultured for 12 hrs. at 30°C then placed at 37°C (heat shock) for an additional 6, 12, or 18 hrs. Soluble proteins were prepared as described in Materials and Methods. The mixture was preincubated with [109]Cd for ten minutes and fractionated on a 2.5 x 100 cm column of Sephadex G-50. The column was eluted with 50 mM Tris-HCl, 0.02% NaN_3, pH 8.0. Radioactivity and zinc levels were determined as described in Materials and Methods. The results are expressed as CPM/fraction and PPB Zn/fraction, respectively: (-■-) radioactivity; (-□-) zinc; V_0, void volume of the column; MPBI, metal binding protein 1, ZnBPII, zinc binding protein 2. Results are presented for; A) embryos heat shocked for 6 hrs; B) embryos heat shocked for 12 hours; and C) embryos heat shocked for 18 hours.

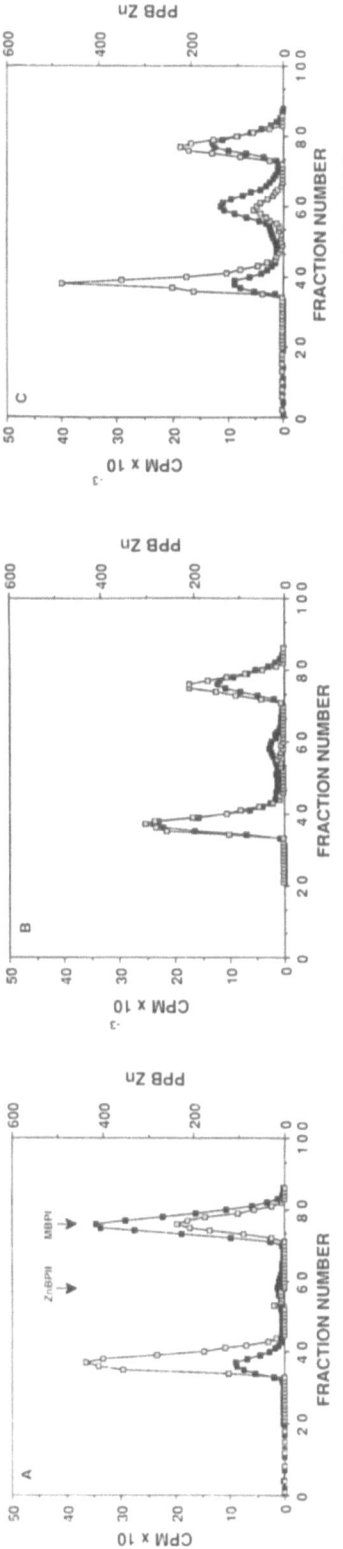

Fig. 7. Effects of initial heat shock and return to normal temperature on the expression of ZnBPII. Embryos were cultured for 12 hrs. at 30°C then placed at 37°C (heat shock) for another 6 hrs. The embryos were then returned to 30°C and cultured for an additional 6 or 12 hours. Free swimming nauplii were separated from embryos still encased in their hatching membrane. Soluble proteins were prepared as described in Materials and Methods. The mixture was preincubated with ^{109}Cd for ten minutes and fractionated on a 2.5 x 100 cm column of Sephadex G-50. The column was eluted with 50 mM Tris-HCl, 0.02% NaN$_3$, pH 8.0. Radioactivity and zinc levels were determined as described in Materials and Methods. The results are expressed as CPM/ fraction and PPB Zn/fraction, respectively: (-■-) radioactivity; (-□-) zinc; V$_0$, void volume of the column; MPBI, metal binding protein 1; ZnBPII, zinc binding protein 2. Results are presented for; A) developmentally arrested embryos returned to 30°C for 6 hrs after initial heat shock; B) developmentally arrested embryos returned to 30°C for 12 hrs after initial heat shock; and C) embryos (nauplii) cultured for 30 hrs. under normal conditions.

217

ACKNOWLEDGEMENTS

This study was supported by NIH AREA Grant 1 R15 GM36115-01. Martin Edep was a MARC Scholar (Minority Access to Research Careers) and supported by NIH Grant RR 08238-02. We would like to extend special thanks to Eric Itakura for his technical support during the course of these studies.

REFERENCES

1. S. M. Bolze, R. D. Reeves, F. E. Lindbeck and J. M. Elders, Influence of zinc on growth, somatomedin, and glycosaminoglycan metabolism in rats, Am. J. Physiol. 252:E21 (1987).
2. B. L. Vallee, A role for zinc in gene expression, J. Inher. Metab. Dis. 6:31 (1983).
3. D. P. Giedroc, K. M. Keating, C. T. Martin, K. R. Williams and J. E. Coleman, Zinc metalloproteins involved in replication and transcription, J. Inorganic Biochem. 28:155 (1986).
4. J. S. Hanas, D. J. Hazuda, D. R. Bogenhagen, F. Y. -H. Wu and C. -W. Wu, Xenopus transcription factor A requires a zinc for binding to the 5s RNA gene, J. Biol. Chem. 258:14120 (1983).
5. A. Klug and D. Rhodes, "Zinc fingers": a novel protein motif for nucleic acid recognition, Trends Biol. Sci. 12:464 (1987).
6. J. T. Kadonaga, K. R. Carner, F. R. Masiarz and R. Tjian, Isolation of cDNA encoding transcription factor Sp 1 and functional analysis of the DNA binding domain, Cell 51:1079 (1987).
7. T. A. Hartshorne, H. Blumberg and E. T. Young, Sequence homology of the yeast regulatory protein ADR1 with Xenopus transcription factor TFIIIA, Nature 320:283 (1986).
8. U. B. Rosenberg, C. Schroder, A. Preiss, A. Kielin, S. Cote, I. Riede and H. Jackel, Structural homology of the product of the Drosophila Krupple gene with Xenopus transcription factor IIIA, Nature 319:336 (1986).
9. D. Tautz, R. Lehmann, H. Schnurch, R. Schuh, E. Seifert, A. Kienlin, K. Jones and H. Jackel, Finger protein of novel structure encoded by Hunchback, a second member of the gap class of Drosophila segmentation genes, Nature 327:383 (1987).
10. A. Vincent, H. V. Colot and M. Rosbash, Sequence and structure of the Serendipity locus of Drosophila melanogaster, J. Mol. Biol. 186:149 (1985).
11. K. Yutaka and J. H. R. Kagi, Metallothionein, Trends Biochem. Sci. 3: 90 (1978).
12. D. H. Hamer, Metallothionein, Ann. Rev. Biochem. 55:913 (1986).
13. J. H. R. Kagi, M. Vasak, K. Lerch, D. E. O. Gilg, P. Hunziker, W. R. Bernhard and M. Good, Structure of mammalian metallothionein, Environ. Health. Perspect. 54:93 (1984).
14. M. Karin, M. Imagawa, R. J. Imbra, R. Chiu, A. Heguy, A. Haslinger, T. Cooke, S. Satabhama, C. Jonat and P. Herrlich, Hormonal and environmental control of metallothionein gene expression, in: "Transcriptional Control Mechanisms," D. Granner, G. Rosenfeld, S. Chang, eds., Alan R. Liss, New York (1987).
15. P. F. Searle, Metallothionein gene regulation, Biochem. Soc. Trans. 15:584 (1987).
16. K. E. Mayo and R. D. Palmiter, Glucocorticoid regulation of the mouse metallothionein 1 gene is selectively lost following amplification of the gene, J. Biol. Chem. 257:306 (1982).
17. A. O. Udom and F. O. Brady, Reactivation in vitro of zinc-requiring apo-enzymes by rat liver zinc-thionein, Biochem. J. 187:329 (1980).
18. D. H. Hamer, Metallothionein gene regulation in Menkes' Syndrome, Arch. Dermatol. 123:1384a (1987).
19. G. W. Evans, R. S. Dubois and K. M. Hambridge, Wilson's disease:

identification of an abnormal copper-binding protein, Science 181: 1175 (1973).

20. G. K. Andrews, E. D. Adamson and L. Gedamu, The ontogeny of expression of murine metallothionein: comparison with the α- fetogene, Dev. Biol. 103:294 (1984).

21. J. F. B. Mercer and A. Grimes, Variation in the amounts of hepatic copper, zinc and metallothionein mRNA during development of the rat, Biochem. J. 238:23 (1986).

22. M. Nemer, E. C. Travaglini, E. Rondinelli and J. D'Alonzo, Developmental regulation, induction and embryonic tissue specificity of sea urchin metallothionein gene expression, Dev. Biol. 102:471 (1984).

23. A. Thall and R. Acey, Cadmium binding proteins in developing Artemia salina, Fed. Proc. 44:1462 (1985).

24. Y. H. Nakanishi, T. Iwasaki, T. Okigaki and H. Kato, Cytological studies of Artemia salina. Embryonic development without cell multiplication after the blastula in encysted dry eggs, Annot. Zool. Japon. 35:223 (1962).

25. J. C. Bagshaw and R. Acey, Stage-specific gene expression in Artemia in: "Biochemistry of Artemia Development," J. C. Bagshaw and A. H. Warner, eds., University Microfilms International, Ann Arbor (1979).

26. O. H. Lowry, N. J. Rosebrough, L. A. Farr and R. J. Randall, Protein measurement with the Folin phenol reagent, J. Biol. Chem. 193:265 (1951).

27. D. L. Eaton and B. F. Toal, Evaluation of the Cd/Hemoglobin affinity assay for the rapid determination of metallothionein in biological tissues, Toxicol. Appl. Pharmacol. 66:134 (1982).

28. K. Suzuki, Y. Ebihara, H. Akitomi, M. Nishikawa and R. Kawamura, Change in ratio of the two hepatic isometallothioneins with development from prenatal to neonatal rats, Comp. Biochem. Physiol. 76C: 33 (1983).

29. D. G. Wilkerson and M. Nemer, Metallothionein genes MTa and MTb expressed under distinct quantitative and tissue specific regulation in sea urchin embryos, Mol. Cellular Biol. 7:48 (1987).

30. J. Pande, M. Vasak and J. H. R. Kagi, Interaction of lysine residues with the metal thiolate clusters in metallothionein, Biochemistry 24:6717 (1985).

31. R. Acey, Induction of a zinc binding protein during the early embryonic development of the brine shrimp Artemia salina, J. Cell. Biochem. Supplement D:354 (1988).

32. J. C. Steffens, D. F. Hunt and B. G. Williams, Accumulation of non-protein metal binding polypeptides (γ-glutamyl-cysteinyl)n-glycine in selected cadmium-resistant tomato cells, J. Biol. Chem. 261:13879 (1986).

33. H. Ohtake, T. Suyemitsu and M. Koga, Sea urchin (Anthocidaris crassispina) egg zinc-binding protein, Biochem. J. 211:109 (1983).

34. A. Cano, J. Crucus, I. Estepa, M. E. Gallego, M. A. G. Sillero, C. F. Heredia, P. Liorente, A. Olalla, C. Osuna, A. Pestana, J. Renart, A. Ruiz, L. Sastre, J. Sebastian and A. Sillero, Developmental changes of enzyme levels during Artemia salina differentiation, in: "Biochemistry of Artemia Development," J. C. Bagshaw and A. H. Warner, eds., University Microfilms International, Ann Arbor (1979).

35. D. K. McClean and A. H. Warner, Aspects of nucleic acid metabolism during development of the brine shrimp Artemia salina, Develop. Biol. 24:88 (1971).

36. A. J. Kraker and D. H. Petering, Tumor-host zinc metabolism: the central role of metallothionein, Biol. Trace Element Res. 5:363 (1983).

37. C. G. Vallejo, F. de Luchi, J. Laynez and R. Marco, The role of cytochrome oxidase in the resumption of the development of Artemia dormant cysts, in: "The Brine Shrimp Artemia," G. Persoone, P. Sorgeloos, O. Roels and E. Jaspers, eds., Universa Press, Wettren (1980).

A NEW METHOD FOR THE PURIFICATION OF NUCLEI FROM ARTEMIA

Sharon L. Squires and Roger A. Acey

Department of Chemistry
California State University
Long Beach, CA 90840

ABSTRACT

Yolk platelets outnumber nuclei about 1400 to 1. Since the densities
of these two organelles are comparable, it has been difficult to obtain
nuclear preparations free of contaminating yolk platelets using differential
centrifugation. We report here on a simple and reliable method for
obtaining Artemia nuclei free of yolk platelets using NaCl-Percoll
gradients. Free swimming nauplii were collected after 24 hrs. of culture
and homogenized in 0.01 M Tris-HCl, 0.15 M NaCl, 0.03 M $CaCl_2$, 0.024 M EDTA,
0.5% Triton X-100, pH 8.0 (SCE). All centrifugations were performed in a
Sorvall GSA-600 rotor. Shells and cellular debris were removed by a "quick
spin". The "quick spin" entails allowing the rotor to reach 1000 rpm followed by immediate deceleration. Nuclei were pelleted from the supernatant
by centrifugation at 2500 rpm for 10 minutes. This centrifugation was
repeated a second time. The nuclear pellet was next resuspended in SCE
buffer and centrifuged at 3000 rpm for 10 minutes. Subsequently, the nuclei
were resuspended in SCE not containing detergent, layered onto 0.15 M NaCl-
Percoll, and centrifuged at 12,500 rpm for 15 minutes. Nuclei, which form
a discrete layer just below the surface, were removed with a Pasteur pipette
and washed with SCE to remove the Percoll. Microscopic examination revealed
isolated, intact nuclei in a field free of yolk platelets. The preparation
is free of lactic dehydrogenase and isocitric dehydrogenase enzyme activities and incorporates ^{3}H-uridine triphosphate into trichloroacetic acid
precipitable material linearly over a two hour incubation period.

ACKNOWLEDGEMENT

Supported in part by NIH Grant 1 R15 GM36117-01.

ENZYMES OF DINUCLEOSIDE OLIGOPHOSPHATE METABOLISM IN ARTEMIA CYSTS AND LARVAE

Mark Prescott, Andrew D. Milne and Alexander G. McLennan

Department of Biochemistry
University of Liverpool
P.O. Box 147, Liverpool L69 3BX, U.K.

INTRODUCTION

After the discovery in 1963 of the abundant nucleotide store of P^1, P^4-diguanosine 5'-tetraphosphate, or P^1, P^4-bis (5'-guanosyl) tetraphosphate (Gp_4G) in Artemia embryos[1], it was logical to search for an enzyme capable of converting it into utilisable products. Such an enzyme, originally named diguanosine tetraphosphate asymmetrical-pyrophosphohydrolase (EC 3.6.1.17) was soon discovered which cleaved Gp_4G specifically to yield equimolar amounts of GTP and GMP[2]. Studies on the partially purified enzyme showed it to be primarily located in the soluble fraction of the cell, to have a molecular mass of 17,500 and to efficiently hydrolyse dinucleotides of the general structure Np_4N, including Gp_4G, Ap_4A, Xp_4X and Up_4U[3,4]; it had little activity towards Gp_3G which had also been found in significant quantities in embryos[2,5].

Two alternative schemes have been proposed for Gp_4G utilisation which require the participation of this pyrophosphohydrolase activity. The first involves a complex shuttle mechanism which generates cytosolic Gp_4G from the yolk platelet store; this is then cleaved to give GTP and GMP with further conversions of the latter leading to the adenine nucleotides[6-8]. The second scheme involves a base substitution reaction between adenine and Gp_4G to give guanine and Ap_4G which is then converted to ATP and GMP by the pyrophosphohydrolase[8,9]. A third possibility which does not involve this enzyme requires the pyrophosphorolysis of Gp_4G to two moles of GTP catalysed by the enzyme also responsible for its synthesis, Gp_4G synthetase[6-8].

As yet there is no overwhelming evidence to favour any one of these alternatives. However, the possible involvement of the asymmetrical-pyrophosphohydrolase in Gp_4G metabolism requires re-examination in view of the more recent isolation of a similar activity from a wide range of cell types which do not contain Gp_4G and the presence of an alternative substrate, P^1, P^4- diadenosine 5'-tetraphosphate, (P^1, P^4- bis(5'- adenosyl) tetraphosphate, Ap_4A) in all such cells. Asymmetrical-pyrophosphohydrolases (which we will now denote by Np_4Nase in view of their lack of base specificity) have been isolated from rat liver and other rat tissues[4,10-12], mouse liver[13], mouse Ehrlich ascites cells[14], human leukaemic cells[15] and lupin seed meal[16] while a symmetrical-Np_4Nase, which yields two moles of NDP from Np_4N has been purified from E. coli and other bacteria[17,18]

223

and <u>Physarum</u> <u>polycephalum</u>[19,20]. It is now widely assumed that the Np_4Nase is involved in the regulation of intracellular levels of Ap_4A. The precise function of this nucleotide has yet to be elucidated although roles in the initiation of DNA replication and, at least in prokaryotes, in the response to oxidative stress, have been proposed[21-24]. Ap_4A is believed to have a high turnover rate, its concentration is inversely proportional to cellular proliferation rate[23,25] and in some cases its level fluctuates by two to three orders of magnitude during the cell cycle[23,26]. We have monitored the level of Ap_4A during pre-emergence development of <u>Artemia</u> embryos and have shown a 130-fold increase from 25nM to 3.3μM during the redevelopment of brine-stored, decapsulated cysts up to the point of hatching followed by a more gradual decline during early larval development[27].

In contrast, the cytosolic concentration of Gp_4G has been estimated to be 1.2mM[6,7]. In addition to its function as purine ring store, Gp_4G has been assigned a more controversial role as a source of phosphate bond energy[8] and in the regulation of protein synthesis[6,7], and of certain enzymes including GMP reductase[28] and a trypsin-like protease[29]. At first sight it is difficult to imagine how the same enzyme could be responsible for the catabolism of two nucleotides of very different function and intracellular concentration without some form of regulation of its activity. In order to help solve this problem, we have purified the <u>Artemia</u> Np_4Nase to homogeneity and made a more detailed comparative study with the enzymes from other sources. In addition we have for the first time purified a separate enzyme from <u>Artemia</u> cysts which preferentially hydrolyses nucleotides of the form Np_3N, P^1, P^3 -dinucleoside 5'-triphosphatase, Np_3Nase (EC 3.6.1.29), and which could conceivably be involved in the degradation of Gp_3G, Gp_3A and/or Ap_3A, all of which have been detected in <u>Artemia</u>[5,-30,31].

MATERIALS AND METHODS

Great Salt Lake <u>Artemia</u> cysts were from the Sanders Brine Shrimp Co., Ogden, Utah, U.S.A. and were obtained in 1980. Nucleoside diphosphokinase, calf intestine alkaline phosphatase, pyruvate kinase and Ap_4A were from Boehringer. Ap_4G was a gift from G.E. Taylor. All other dinucleoside oligophosphates and protein molecular weight markers were from Sigma. Ap_4A-Sepharose was prepared by immobilising Ap_4A to epoxy-activated Sepharose 6B (Pharmacia) according to the manufacturer's instructions. $[\beta-^{32}P]Ap_3A$ was prepared by a two-step method involving the phosphorylation of AMP to $[\beta-^{32}P]ADP$ using $[\gamma-^{32}P]ATP$ (Amersham) and adenylate kinase (Boehringer) and the condensation of the labelled ADP with AMP using a water-soluble carbodiimide[32]. The final product was purified on a Mono-Q column. The carbodiimide condensation procedure was also used to prepare $[^3H]-Ap_3A$ and unlabelled Ap_4A in bulk for affinity chromatography. $[^3H]-Ap_4A$ (4.3Ci/-mmol) was from Amersham International.

<u>Assays</u>

Several different assay procedures were employed for the detection of both Np_3Nase and Np_4Nase activities. They were employed under different circumstances depending on the substrate or when either speed, lack of auxiliary enzymes or precise quantitation was most important.

<u>Method 1 - Luminescence assays.</u> These were based on previous procedures [33,34]. For Np_4Nase with Ap_4A as substrate, these contained 20mM Hepes-KOH, pH 7.8, 5mM Mg acetate, 100μM Ap_4A, and 25μl ATP-monitoring reagent (LKB) and for Np_3Nase, 23mM Hepes-KOH, pH 7.8, 4mM Mg acetate, 100μM Ap_3A, 2mM phosphoenolpyruvate, 2 units pyruvate kinase and 25μl ATP-monitoring

reagent in a total volume of 125µl. The increase in luminescence was monitored over a 2 min period at 25°C with an LKB 1250 Luminometer. With Gp_4G as substrate, Np_4Nase was assayed by a two-step procedure. First, Np_4Nase was incubated in 20mM Bicine-KOH, pH 8.4, 5mM Mg acetate, 50µM Gp_4G (25µl) for 10 min at 37°C followed by 5 min at 90°C to inactivate the Np_4Nase. Samples containing up to 10 pmol GTP were then added to an assay containing 20mM Hepes-KOH, pH 7.8, 5mM Mg acetate, 0.5µM ADP, 1.5U nucleoside diphosphokinase and 25µl ATP-monitoring reagent in 125µl and the luminescence recorded at 25°C.

Method 2-Radiolabel assays (DEAE). For Np_4Nase, these contained 20mM Bicine-KOH, pH 8.4, 2mM Mg acetate, 5 units alkaline phosphatase and 100µM $[^3H]-Ap_4A$ (40mCi/mmol) in a total volume of 50µl. After incubation for 10 min at 37°C, assays were chilled on ice and 300µl of a 25% (v/v) suspension of DEAE-Sephacel in 10mM Tris-HCl, pH 7.5 were then added followed by centrifugation at 16,000 x g for 5 min. 200µl of the supernatants were added to 4ml Optiphase-MP scintillant (LKB) and the radioactivity due to $[^3H]$-adenosine released from the products by the alkaline phosphatase was determined. Np_3Nase assays were similar except that the buffer was 20mM BisTris-HCl, pH 7.0, 100µg/ml BSA, 1mM Mg acetate and the incubation temperature was 30°C.

Method 3. As above except that the alkaline phosphatase was omitted from the primary incubation. After 10 min at 37°C, the assays were heated at 90°C for 10 min to inactivate the pyrophosphohydrolase then incubated for a further 10 min at 37°C with 5 units alkaline phosphatase.

Method 4 - Radiolabel assays (t.l.c.). Np_3Nase was assayed for 15 min at 30°C in a total volume of 25µl containing 20mM BisTris-HCl, pH 7.0, 100µg/ml BSA, 1mM Mg acetate and $[\beta-^{32}P] Ap_3A$ (20mCi/mmol). Samples 5µl were mixed with 0.2µl each of 10mM ADP and Ap_3A markers and spotted on to a PEI-cellulose plate which was then run its full length in 1.6M LiCl. The ADP spots were cut out and counted directly in Optiphase MP scintillant.

Method 5 - H.p.l.c. assays. 100 µM nucleotide substrates were incubated for 15 min at 30°C in 20mM BisTris-HCl pH 7.0, 100µg/ml BSA, 1mM Mg acetate (Np_3Nase) or 20mM Bicine-KOH, pH 8.4, 2mM Mg acetate, with or without 5 units alkaline phosphatase (Np_4Nase). Ammonium phosphate (70µl of 50mM), pH 5.2 was added and 50µl of the mixture injected on to a 4.6 x 250mm Partisil 10-SAX column. The column was developed isocratically for 10 min with 5% buffer B followed by a 44 min gradient from 5 to 80% buffer B where buffer A = 50mM ammonium phosphate, pH 5.2 and buffer B = 1M ammonium phosphate, pH 5.7[20].

Protein. This was assayed by the bicinchoninic acid[35] (Pierce) and silver-binding[36] methods.

Purification of Np_4Nase and Np_3Nase

Dry cysts (250g) were hydrated overnight at 4°C in 2.5 l seawater and then decapsulated[27]. The cysts were then filtered through nylon mesh, excess water removed by squeezing and the wet cake divided into three 250g portions which were ground separately for 20 min each at 4°C in a 7-inch unglazed mortar. The homogenate was then taken to a final volume of 1.25 l with 50mM Tris-HCl, pH 7.5, 0.25M sucrose, 10mM $MgCl_2$, 1mM β-mercaptoethanol, 0.1mM EDTA, left for 15 min and then centrifuged at 1400 x g for 15 min. The low speed supernatant was then centrifuged at 152,000 x g for 90 min to yield 670ml of crude extract from which the lipid pellicle was removed with cotton wool.

The crude extract was brought to 50% saturation by the addition of solid $(NH_4)_2SO_4$ and centrifuged at 20,000 x g for 30 min. The supernatant was then brought to 100% saturation and centrifuged as before. The pellet was redissolved in a final volume of 92ml Q-Sepharose loading buffer (20mM Tris-HCl, pH 8.0, 10% glycerol, 1mM β-mercaptoethanol) and dialysed overnight against 2 x 10 vol loading buffer, yielding a final volume of 146ml. This fraction was run onto a 5 x 16cm column of Q-Sepharose at 193ml/h and after elution of the unbound protein, the column was developed with a 2.5 l gradient of 0 to 0.6M NaCl in loading buffer. Active fractions were pooled and brought to 100% saturation with $(NH_4)_2SO_4$. The precipitate was recovered at 20,000 x g for 30 min and redissolved in a final volume of 31.5ml Ultrogel AcA44 running buffer (50mM potassium phosphate, pH 7.5, 10% glycerol, 1mM EDTA, 5mM β-mercaptoethanol).

This material was applied to a 5 x 95cm column of Ultrogel AcA44 and eluted with running buffer at 80ml/h. Fractions containing Np_3Nase and Np_4Nase activity were pooled separately and each dialysed overnight against 2 x 10 vol Mono-Q loading buffer (20mM Tris-acetate, pH 8.0, 10% glycerol, 1mM β-mercaptoethanol) to give 206ml Np_4Nase pool and 161ml Np_3Nase pool.

Further purification of the Np_4Nase. The Np_4Nase pool was applied at 4ml/min to an 8-ml Mono-Q HR10/10 column and the column eluted with a 160ml gradient of 0 to 0.75M sodium acetate in loading buffer. The fractions comprising peak 1 of Np_4Nase activity were combined and dialysed against 20 vol Ap_4A-Sepharose loading buffer (20mM MES, pH 6.0, 10% glycerol, 1mM β-mercaptoethanol, 0.1mM EDTA). This fraction (7.6ml) was applied to a 1ml column of Ap_4A-Sepharose and the column washed successively under gravity with 2ml loading buffer, 2ml loading buffer + 50mM NaCl and 8ml loading buffer + 100µM Ap_4A. The activity which eluted with Ap_4A was dialysed overnight against 2 x 1 liter HPHT equilibration buffer (10mM potassium phosphate, pH 6.8, 10µM $CaCl_2$) and 5.8ml applied to a 100 x 7.8mm Bio-Gel HPHT column (Bio-Rad) at 0.7ml/min. After eluting unbound protein, the column was eluted with a 31.5ml gradient of 10 to 500mM potassium phosphate, pH 6.8 containing 10µM $CaCl_2$. Fractions containing activity were re-run on the same HPHT column and finally concentrated on Mono-Q using a gradient of NaCl. Active fractions were brought to 50% glycerol, 100µg/ml BSA and stored in aliquots at -70°C.

Further purification of the Np_3Nase. The Np_3Nase pool from the AcA44 column was purified on Mono-Q as described for the Np_4Nase. The active fractions were dialysed against 50 vol HPHT equilibration buffer (final volume = 35.5ml) and chromatographed on Bio-Gel HPHT as described above. The active fractions were pooled and diluted with water to reduce the phosphate concentration to approx. 100mM and then concentrated to 0.55ml with an Amicon C-10 Centricon. Glycerol was added to 10% and β-mercaptoethanol to 1mM and the final preparation stored at 4°C.

RESULTS

Purification of Np_3Nase and Np_4Nase

Throughout the purification, fractions were assayed immediately with the luminescence assay. Pooled fractions were subsequently quantitated with the radiolabelled assay (DEAE method). The relatively mild extraction conditions employed appeared to avoid the extraction of non-specific phosphodiesterases which would interfere with the quantitation of the enzymes at early stages of the purification. Control experiments with French-pressed whole and decapsulated cysts showed that decapsulation with hypochlorite did not affect the yield of activity.

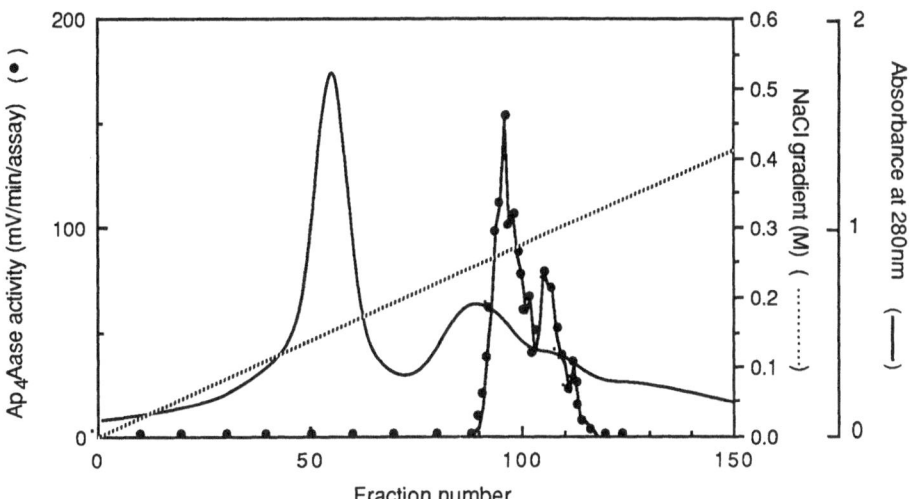

Fig. 1. Chromatography of dinucleoside tetraphosphatase and dinucleoside triphosphatase on Q-Sepharose. Fractions(19.5ml)were collected and 2.5μl portions assayed for Np_4Nase activity with the luminescence assay. Fractions were not assayed for Np_3Nase activity (see text). Fractions 92 to 116 which contained 97% of the recovered activity were pooled.

Np_4Nase activity eluted reproducibly from Q-Sepharose as two or three overlapping peaks (Fig.1). Such a charge heterogeneity has been noted before during chromatography on CM-cellulose[2] (see below). Np_3Nase activity co-eluted from Q-Sepharose with the Np_4Nase but could not be assayed conveniently with the luminescence assay owing to the presence of endogenous ADP-binding proteins hence its elution profile is not shown. The two pyrophosphohydrolases were separated completely by chromatography on Ultrogel AcA44 (Fig.2). From this point onwards, the two activities were purified separately.

The high resolving power of FPLC on Mono-Q clearly shows the heterogeneity of the Np_4Nase with two principal peaks eluting in the ratio 2:1 and a third minor activity (Fig.3). This does not appear to be the result of proteolysis since the pattern is reproducible and, although proteinase inhibitors were not employed in the bulk preparation, inclusion of PMSF (1mM), leupeptin (0.1mM), E-64 (10μM), EDTA (1mM), pepstatin (1μM), bestatin (10μM) and soybean trypsin inhibitor (100μg/ml) during small-scale extractions affected neither the yield nor the distribution of enzyme activity between peaks I and II (not shown). Peak I was further purified by affinity elution from Ap_4A-Sepharose (Fig.4). To our knowledge, this is the first time that such a resin has been used in the purification of an Ap_4A-binding protein. Final purification to homogeneity was achieved by chromatography on Bio-Gel HPHT with the activity eluting right at the start of the gradient (Fig.5). The large u.v.-absorbing peak eluting just after the Np_4Nase is residual Ap_4A from the affinity elution step. The final fraction was judged to be homogeneous by SDS-polyacrylamide gel electrophoresis and silver staining[37,38] (Fig.6). The overall yield was 2.7% with a purification of 11,000-fold over the crude extract (Table 1).

Np_3Nase also exhibited a degree of heterogeneity when purified further on Mono-Q with one major and one minor peak (Fig.7). The major peak had a higher affinity for Bio-Gel HPHT than did the Np_4Nase and coincided with a major protein peak (Fig.8). However the final fraction was not homogen-

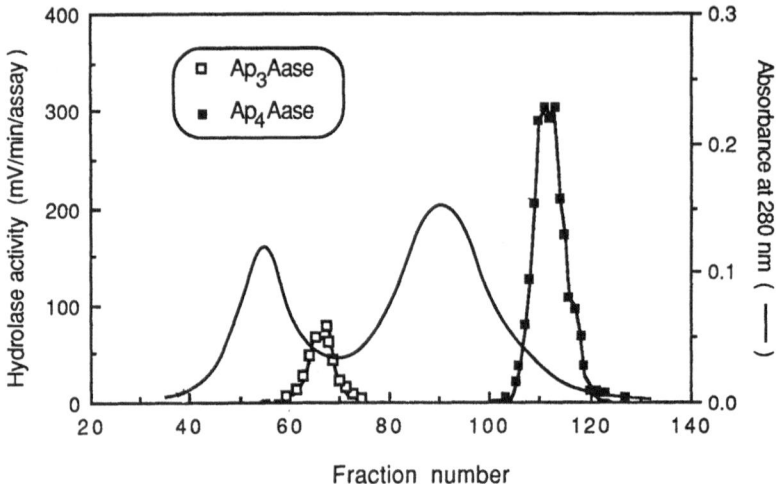

Fig. 2. Chromatography of dinucleoside tetrephosphatase and dinucleoside triphosphatase on Ultrogel AcA44. Fractions(12.4ml)were collected and 12.5μl portions assayed for Np_3Nase and Np_4Nase activity with the luminescence assays. Fractions containing Np_3Nase activity (59 to 71, 92% of total) and Np_4Nase activity (106 to 121, 97% of total) were pooled separately.

eous as shown by gel electrophoresis. The final purification factor was 215-fold with a yield of 2.7% (Table 2).

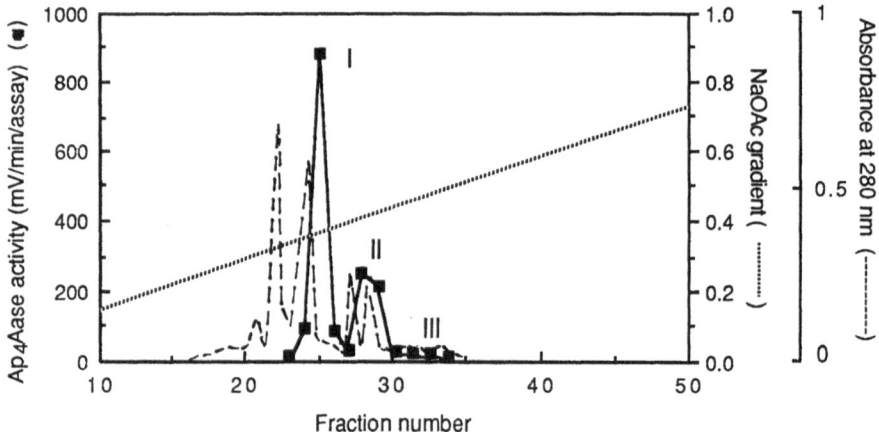

Fig. 3. Chromatography of dinucleoside tetraphosphatase on Mono-Q: Fractions(3ml)were collected and 2μl portions assayed for Np_4Nase activity with the luminescence assay. Fractions 25 and 26 (peak I, 60% of total) were combined for further purification and fractions 27 to 29 (peak II, 30%) and 30 to 32 (peak III, 4%) were pooled separately and stored at -70°C.

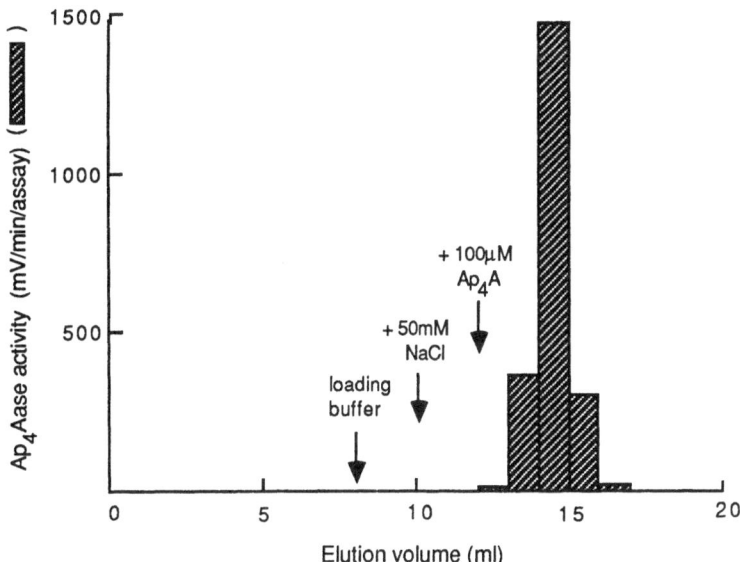

Fig. 4. Purification of dinucleoside tetraphosphatase on Ap_4A-Sepharose. Elution buffers were applied at the points indicated. Fractions (1ml) were collected and 2μl portions assayed for Np_4Nase activity with the luminescence assay. Fractions 14 to 17 comprising 99.6% of the recovered activity were pooled.

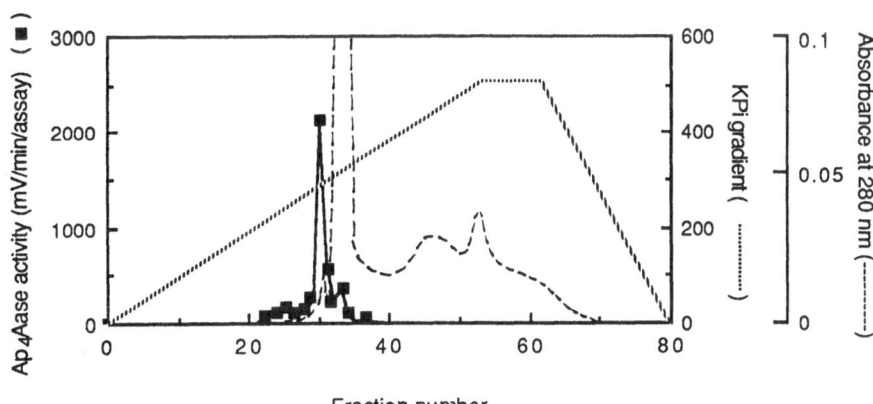

Fig. 5. Chromatography of dinucleoside tetraphosphatase on Bio-Gel HPHT: Fractions (0.5ml) were collected and 5μl portions assayed for Np_4Nase activity with the luminescence assay. 70% of the eluted activity was contained in fractions 28 to 31 which were retained.

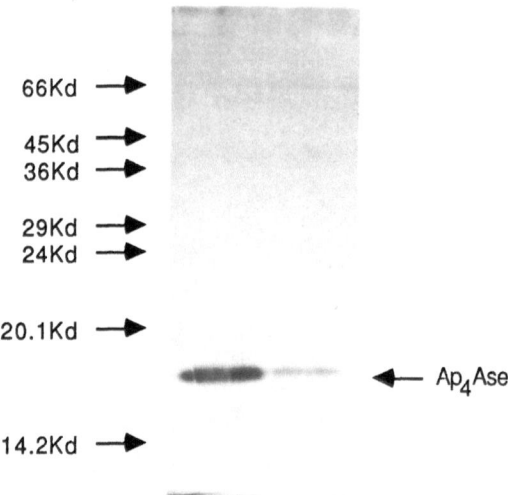

Fig. 6. SDS-polyacrylamide gel electrophoresis of purified Np_4Nase. 0.1μg each of the two peak fractions from the final Bio-Gel HPHT purification step were run on a 15% SDS-polyacrylamide gel and stained with silver. Molecular weight markers were lactalbumin (14,200), soybean trypsin inhibitor (20,100), trypsinogen (24,000), carbonic anhydrase (29,000), glyceraldehyde 3-phosphate dehydrogenase (36,000), ovalbumin (45,000) and bovine serum albumin (66,000).

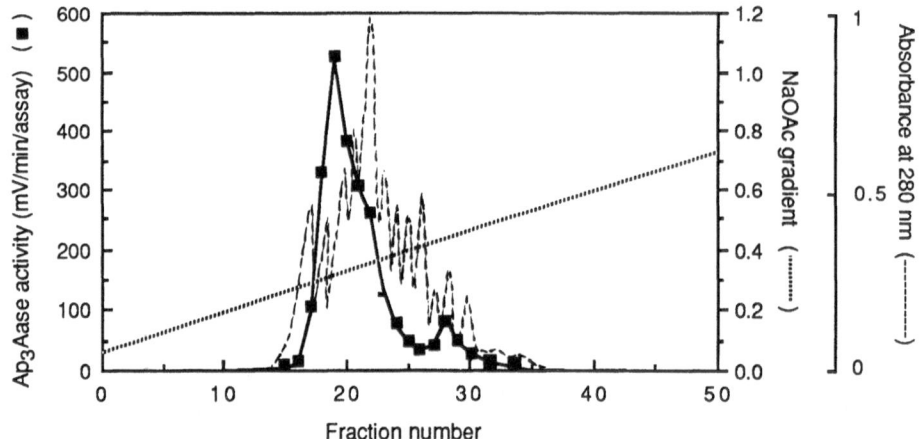

Fig. 7. Chromatography of dinucleoside triphosphatase on Mono-Q. Fractions (3ml) were collected and 12.5μl portions assayed for Np_3Nase activity with the luminescence assay. Fractions 17 to 24 comprising 87% of the recovered activity were pooled.

Table 1. Purification Scheme for Np$_4$Nase

Fraction	Volume (ml)	Protein (mg)	Activity (units)	Specific Activity (units/mg)	Yield (%)
Crude supernatant	670	7640	20,712	2.7	100
50-100% ammonium sulphate fraction	146	4220	22,075	5.2	107
Q-Sepharose	31.5	246	9323	38	45
Ultrogel AcA44	206	95	6773	71	33
Mono-Q	7.6	9.4	2468	262	12
Ap$_4$A-Sepharose	5.8	1.6	1276	798	6.2
Bio-Gel HPHT	2.0	0.019	569	29,947	2.7

Activities of each pooled fraction were determined by assay method 2 after dialysis of a sample to remove salts. 1 unit of Np$_4$Nase activity is defined as the amount of protein which degrades 1nmol of Ap$_4$A per min at 37°C.

Properties of Np$_3$Nase and Np$_4$Nase

Molecular weights. Calibration of a Sephacryl S-300 column and the Ultrogel AcA44 column with standard proteins gave native molecular masses of 117,000 for the Np$_3$Nase and 17,600 for Np$_4$Nase, respectively (not shown). The latter value agrees well with previous estimates for the Artemia enzyme [4]. Np$_4$Nases from other sources have molecular masses of 17,500 (human) [15], 18,500 (lupin)[16], 20-22,0000 (rat)[10,12] and 19,800 (Ehrlich ascites mouse cells)[14]. The polypeptide M$_r$ of 17,600 shows the native

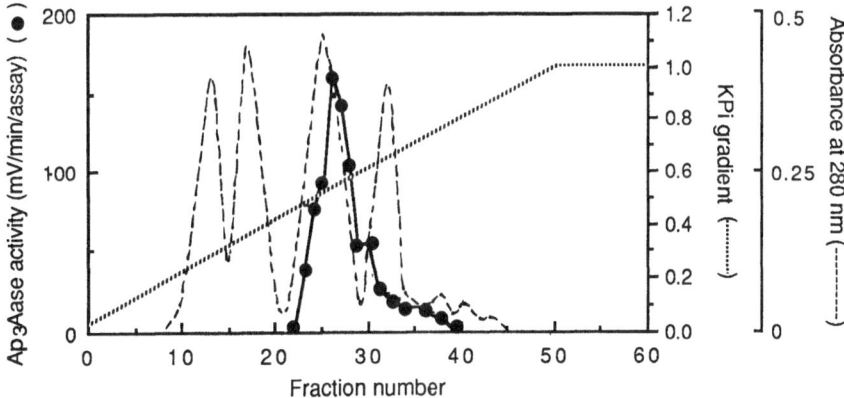

Fig. 8. Chromatography of dinucleoside triphosphatase on Bio-Gel HPHT. Fractions (0.5ml) were collected and 12.5µl portions assayed for Np$_3$Nase activity with the luminescence assay. 92% of the eluted activity was contained in fractions 23 to 33 which were retained.

Table 2. Purification Scheme for Np$_3$Nase

Fraction	Volume (ml)	Protein (mg)	Activity (units)	Specific Activity (units/mg)	Yield (%)
Crude supernatant	670	7640	271	0.035	100
50-100% ammonium sulphate fraction	146	4220	418	0.1	154
Q-Sepharose	31.5	246	93	0.38	34
Ultrogel AcA44	162	79	42	0.53	15
Mono-Q	36	12	9.7	0.81	3.6
Bio-Gel HPHT (concentrated)	0.6	0.97	7.3	7.5	2.7

Activities of each pooled fraction were determined by assay method 2 after dialysis of a sample to remove salts. 1 unit of Np$_3$Nase activity is defined as the amount of protein which degrades 1nmol of Ap$_3$A per min at 30°C.

enzyme to be a monomer (Fig.6). Since the Np$_3$Nase preparation is not yet homogeneous it has not been possible accurately to assign a particular polypeptide to this enzyme. However analysis of Bio-Gel HPHT fractions by gel electrophoresis (not shown) suggests that a polypeptide of M_r 50,000 most closely follows the elution profile of the Np$_3$Nase which would imply that the native enzyme may be a dimer. The lupin[16] and rat[39,40] enzymes have been shown to be monomers of molecular masses 41,000 and 35,000, respectively.

Temperature and pH optima. Optimal activity of Np$_3$Nase and Np$_4$Nase was observed at 30°C and 37°C, respectively although in both cases the dependence on temperature was not very critical within the range 25°C to 45°C. Both enzymes also had relatively broad pH optima with Np$_3$Nase being most active at neutral pH and Np$_4$Nase at a more alkaline pH around 8.5 (not shown). In these respects these enzymes are very similar to the other dinucleoside oligophosphate pyrophosphohydrolases which have been described. Neither enzyme was stable in Tris buffers, particularly the Np$_3$Nase which quickly lost activity at 4°C when stored in Tris-HCl.

Effect of divalent cations. The activity of the Np$_3$Nase is stimulated equally by Mg^{2+} ions at 0.5 to 2mM and Mn^{2+} ions at 0.1mM, while 1mM Ca^{2+} is 45% as effective (Fig.9a). Identical properties have been reported for the rat enzyme[39,41] although the lupin Np$_3$Nase is totally Mg^{2+}-dependent[16]. Interestingly, 14% activity is displayed in the absence of added divalent cations. This activity is abolished by pre-incubation with 1mM EDTA but only the basal level (12% maximum) can be restored by the addition of excess Mg^{2+} suggesting the involvement of a different endogenous ion in the activity of this enzyme.

The Artemia Np$_4$Nase is similar to all other asymmetrically-cleaving enzymes in its absolute requirement for a divalent cation. Mg^{2+} is the most effective in the range 1 to 5mM with Mn^{2+} 35% as effective at 0.1mM (Fig. 9b). In combination, these ions stimulate activity by a further 20% at 0.2mM Mn^{2+} and 2mM Mg^{2+}. Ca^{2+} ions did not support activity at any concentration as has also been reported for the rodent and lupin enzymes[11-

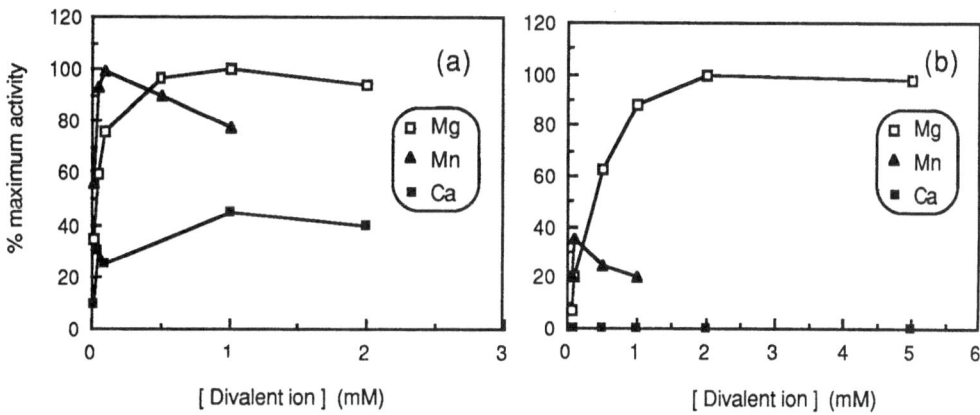

Fig. 9.   Dependence of (a) Np_3Nase and (b) Np_4Nase activities on divalent
cations. Np_3Nase was assayed by method 4 (100% activity = 42.4
pmol/min) and Np_4Nase was assayed by method 2 (100% activity =
120 pmol/min).

12,14,16]. In both the latter cases, as here, Ca^{2+} ions actually inhibit
the Mg^{2+}- dependent reaction (Fig. 10a); the inhibition would appear to be
competitive with Mg^{2+}.

The role of Zn^{2+} ions in the modulation of cellular Ap_4A levels has
been stressed as a result of their stimulation of the synthesis of Ap_4A by
aminoacyl-tRNA synthetases and their inhibition of the activity of certain
degradative Np_4Nases[23,42]. However, although rat liver Np_4Nase[43] and
Ap_4A-hydrolysing activity in crude extracts of BHK cells and sea urchins[42,
44] are inhibited by low concentrations of Zn^{2+} ions (e.g. the inhibitor
concentration required to achieve 50% inhibition of the reaction [IC_{50}] for
rat liver enzyme is 2μM), the <u>Artemia</u> enzyme is unaffected by Zn^{2+} ions up

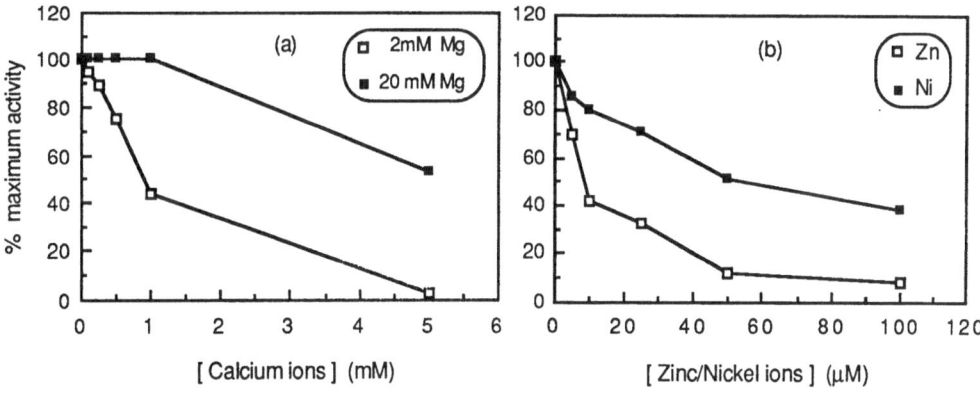

Fig. 10.   Inhibition of (a) Np_4Nase and (b) Np_3Nase activities by divalent
cations. Np_4Nase was assayed by method 2 (100% activity = 125pmol/
min) and Np_3Nase was assayed by method 4 (100% activity = 66.3pmol/
min). Control experiments showed that the excess alkaline phosphatase was still sufficiently active under these conditions not
to affect the assay.

to 100μM even when pre-incubated with them in the absence of Mg^{2+}. Zn^{2+} can in fact stimulate activity (7% of maximum at 2mM without Mg^{2+}). As the lupin Np_4Nase is only slightly inhibited by Zn^{2+} ions, the role of Zn^{2+} as a general regulator of Np_4N metabolism requires further investigation. On the other hand, Zn^{2+} ions are powerful inhibitors of the lupin[16], rat liver[39] and Artemia (Fig. 10b) Np_3Nases. The IC_{50} for the Artemia Np_3Nase is 25μM in the presence of 2mM Mg^{2+}. Ni^{2+} and Co^{2+} are also effective inhibitors of both the lupin[16] and Artemia Np_3Nases (Fig. 10b) but not the Np_4Nases (not shown). Zn^{2+} ions may therefore function in controlling the relative intracellular concentrations of Np_3N and Np_4N nucleotides through differential effects on their degradation. We have already suggested that the ratio of Ap_3A to Ap_4A within the cell may be more important than their absolute concentrations[31]. Furthermore the contrasting effects of Ca^{2+} ions on the Np_3N/Np_4N ratio. This area of potential control deserves more detailed investigation.

Substrate specificity and products. The Np_4Nase from lupin seeds has been shown to require a minimum of four bridging phosphates for activity and always produces an NTP as one of the products[16]. The Artemia enzyme shows the same specificity giving appreciable hydrolysis of Ap_5A, Ap_6A and Gp_5G in addition to the preferred substrates Ap_4A, Gp_4G and Ap_4G. However detectable activity was also found with Ap_3A and Gp_3G (0.5% and 2% of maximum, respectively, Table 3). An NTP was always one of the products. With substrates of the form Ap_nA, the Artemia Np_3Nase has a very strong preference for Ap_3A with no activity against Ap_2A and 8% or less when n>3 (Table 3). However with Gp_nG, both Gp_4G and Gp_5G are roughly half as effective as substrates as Gp_3G which itself is 70% as effective as Ap_3A. Where similar determinations have been made with the rat and lupin enzymes, there is good agreement[16,39] although the marked lack of chain length specificity with Gp_nG has not been examined or observed before. One major difference between the lupin and Artemia enzymes is in the nature of the products. The lupin Np_3Nase always produces an NDP, however for each substrate examined by h.p.l.c, the Artemia enzyme always produces an NMP as one of the products e.g. Gp_3G gives GMP + GDP, Gp_4G gives GMP + GTP and Gp_5G gives GMP + Gp_4. The products of hydrolysis by the rat or E. coli enzymes have not been determined with substrates containing more than three phosphate groups[39,-45].

Kinetic parameters for Np_4N substrates. These were calculated with the HYPER computer program and from Eadie-Hofstee plots. When the Np_4Nase was assayed by method 2 (alkaline phosphatase present) a K_m of 33μM and rate constant of $12.7s^{-1}$ were determined for Ap_4A. This K_m value is much higher than previous estimates of 2μM for the partially purified enzyme[4] and, from our own previous work using the luminometer assay (without alkaline phosphatase), of 4.2μM[46]. Values for other asymmetrical Np_4Nases are all reported within the range 0.5 to 5μM. When K_m and k_{cat} were redetermined by assay method 3 (alkaline phosphatase omitted) values of 4.4μM and $1.5s^{-1}$, respectively were obtained which represent 7.5- and 8.5- fold reductions. The latter value for the rate constant is in close agreement with that calculated for the homogeneous lupin enzyme of $1.2 s^{-1}$[16]. The reason for this discrepancy appears to be the presence of an alkaline phosphatase-sensitive inhibitor in the commercial Ap_4A, possibly adenosine 5'-tetraphosphate which is a known contaminant of some adenine nucleotide preparations [40,43] and is a powerful competitive inhibitor of the enzyme[3,10,14]. Pretreatment of the Ap_4A with alkaline phosphatase results in higher values of K_m and k_{cat}. This finding is reported here to indicate that some previous determinations of the kinetic parameters of Np_4Nases may not have allowed for this factor, particularly when commercial preparations of Ap_4A have been used directly or even when repurified material has been used. The co-incubation with alkaline phosphatase employed here compensates for any spontaneous generation of Ap_4 which might occur upon storage. Assuming a K_i of 50nM

Table 3. Substrate Specificity of Np_3Nase and Np_4Nase

Substrate	% relative hydrolysis	
	Np_3Nase	Np_4Nase
Ap_2A	0	0
Ap_3A	100	0.5
Ap_4A	8	100
Ap_5A	9	25
Ap_6A	1.4	18
Gp_2G	0	0
Gp_3G	71	2
Gp_4G	30	48
Gp_5G	33	10
Ap_4G	n.d.	60

Substrate specificity was determined by h.p.l.c (assay method 5, without alkaline phosphatase). Rates were determined by quantitation of substrates and products (with alkaline phosphatase in the case of Np_4Nase) by u.v.-absorbance peak area integration taking into account the hyperchromicity of the products relative to substrates. Rates are expressed relative to Ap_3A for Np_3Nase and relative to Ap_4A for Np_4Nase.

for Ap_4, a 0.2% contamination would be sufficient to reduce k_{cat} by a factor of 8.5.

The K_m for Gp_4G (pretreated with alkaline phosphatase) was 5µM as previously shown[4] and k_{cat} = 6.2s^{-1}. Specificity constants (k_{cat}/K_m) for Ap_4A and Gp_4G show that the enzyme has an approximately 3-fold preference for Gp_4G over Ap_4A as a substrate (Table 4).

Ap_4G has been proposed as a substrate for the Artemia Np_4Nase in vivo[9]. We have therefore investigated its behaviour as a substrate for this enzyme in vitro. k_{cat} was 7.6s^{-1} and, when the products were investigated by h.p.l.c. after digestion to completion, a marked asymmetry in the cleavage pattern was found: AMP and GTP were present in a 4.5-fold excess over GMP and ATP. This ratio, and the ratio of the rate constants for Ap_4A, Gp_4G and Ap_4G imply that, for nucleotides of the form Np_4N, the rate of cleavage is determined by the nature of the 'Nppp' portion of the molecule in the binding site. Therefore Ap_4G is 4.5-fold more likely to bind with the high affinity 'Gppp' end in the binding site and be cleaved at the rate of 6.2s^{-1} than it is to bind with the low affinity 'Appp' end in the binding site when it is cleaved at the rate of 12.7s^{-1}. The net rate is therefore 7.6s^{-1}. This pattern of cleavage appears to be inconsistent with the model of Van Denbos and Finamore for Gp_4G catabolism which, in its simplest form, requires preferential production of ATP and GMP from an Ap_4G intermediate[9].

Table 4. Kinetic Parameters of Np$_4$Nase

Substrate	Km (μM)	k_{cat} (s^{-1})	Specificity Constant (M^{-1}s^{-1})
Ap$_4$A	33	12.7	3.84 x 10^5
Gp$_4$G	5	6.2	1.22 x 10^6
Ap$_4$G	n.d.	7.6	n.d.

The K_m of the Np$_3$Nase for Ap$_3$A was determined to be 15μM by assay method 1. This compares with values of 12μM for the rat[39,40] and E.coli [45] enzymes and 1.2μM for the lupin Np$_3$Nase[16].

Changes in Np$_4$Nase and Np$_3$Nase Activities during Embryonic and Early Larval Development

One possible way of determining the relative importance of the Artemia Np$_4$Nase and Np$_3$Nase to the metabolism of adenine- and guanine-containing dinucleoside oligophosphates is to correlate their activities with the pool sizes of these nucleotides in cysts and embryos. Gp$_4$G pools show a slow decline from the onset of redevelopment at a rate of 0.2pmol/h/embryo increasing to 1.5pmol/h/embryo after hatching while Gp$_3$G is degraded much more slowly, at an average rate of 0.06pmol/h/embryo after hatching[1,6,7, 47]. In contrast, Ap$_4$A and Ap$_3$A accumulate rapidly up to the point of hatching to reach maximum levels of 15fmol/embryo (3.7pmol/10^6 cells) and 32fmol/embryo (7.6pmol/10^6 cells), respectively. Thereafter they decline at average rates of about 0.35fmol/h/embryo and 0.63fmol/h/embryo, respectively[27] (see also Figs. 11a,c).

It has previously been calculated that there is more than sufficient Np$_4$Nase activity in the cysts to account for the above rate of Gp$_4$G degradation. Our own data would support this with the theoretical available rate being 12pmol Gp$_4$G/h/embryo. With the ability to degrade Ap$_4$A at a possible rate of 24pmol/h/embryo (70,000 x the observed net rate) it may be argued that this is an unnecessary excess for an enzyme intended for Ap$_4$A metabolism (however, see Discussion). Similar calculations for the Np$_3$Nase show that while the extracted activity is 30 to 50-fold less than the Np$_4$Nase, there is still enough of this enzyme to account for net degradation rates of Gp$_3$G and Ap$_3$A of 0.34 and 0.48pmol/h/embryo, respectively, again well in excess of those required. These figures may be underestimates since the rat Np$_3$Nase is at least 50% membrane bound and requires detergent for full extraction[39,48]. Therefore, in general terms, the activities of both enzymes seem to relate more to the levels of the guanine nuceotides, however this does not take into account the turnover rates of Ap$_3$A and Ap$_4$A which are believed to be high[25].

When Np$_4$Nase activity was determined in crude extracts of decapsulated cysts and larvae at various times after the reinitiation of development little change was observed until hatching; however a gradual 1.8-fold increase in activity occurred between hatching and 72h (Fig. 11b). This is in good agreement with original observations on this enzyme[49]. However

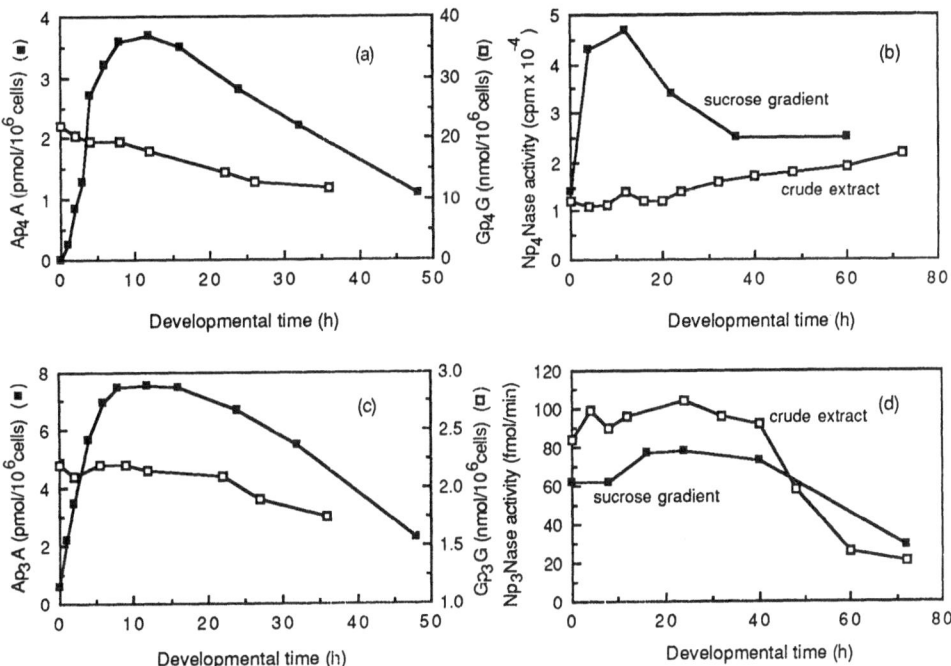

Fig. 11. Changes in dinucleoside oligophosphates and pyrophosphohydrolase activities during embryonic and early larval development of Artemia. (a) Data for Ap$_4$A and Gp$_4$G pool sizes are reproduced from references 27 and 47, respectively. (b) Crude extracts were prepared from decapsulated cysts or larvae at various developmental stages by homogenizing 1g portions per ml of 10mM potassium phosphate, pH 7.5, 10% glycerol, 5mM β-mercaptoethanol, 1mM EDTA, 1mM PMSF, 100µg/ml soybean trypsin inhibitor with a Potter-Elvehjem homogenizer. After high speed centrifugation and dialysis, 200µl samples were layered on 5-20% sucrose gradients in homogenizing buffer and centrifuged at 182,000 x g for 21h. 0.5µl of each crude extract and of all gradient fractions were assayed for Np$_4$Nase activity by method 2. For each gradient, the activities of those fractions which contained Np$_4$Nase were summed and the total activity obtained plotted against time of development. (c) Data for Ap$_3$A and Gp$_3$G pool sizes are reproduced from references 31 and 47, respectively. (d) Np$_3$Nase was prepared and assayed as in (b).

a very different picture emerged when the extracts were subjected to sucrose gradient sedimentation analysis and the activities of the fractions containing Np$_4$Nase summed. In this case activity was seen to increase 3.2-fold up to hatching and thereafter fall again to the level characteristic of 72h larvae (Fig. 11b). This pattern is remarkably similar to the changes in the Ap$_4$A pool (but not Gp$_4$G) over the same period. A rise in the activity of Np$_4$Nase is not necessarily in conflict with a concomitant increase in the Ap$_4$A pool as this nucleotide has a high turnover rate[25] and a rise in its concentration may be accompanied by unequal increases in both its rates of synthesis and degradation. The reasons why this pattern of activity change is apparently masked in crude extracts is not clear but could reflect the separation of the Np_4Nase from an inhibitor or regulatory factor. Further work will be required to clarify this point.

An analogous study of the Np_3Nase showed that, in contrast to the Np_4Nase, the activity in crude extracts and sucrose gradient fractions behaved in a similar manner, both rising slightly upon hatching then declining to about 20-30% of maximum between 40 and 72h (Fig 11d).

DISCUSSION

Now that asymmetrical-Np_4Nases have been purified from a variety of sources and their properties described in some detail, it seemed appropriate to extend the original observations made with the partially purified Artemia activity and so allow it to be compared more fully with these other enzymes. Sillero and co-workers showed the similarity between the Artemia and rat enzymes in 1976[4] and the additional data which we have provided here confirms that the Artemia Np_4Nase is indeed a member of a widespread family of enzymes of very similar properties (although the mouse liver Ap_4Aase does appear quite different)[13]. Therefore as this enzyme cannot be regarded as unique to Artemia, its role in the metabolism of Gp_4G in this organism must be given more careful consideration than before. Although Gp_4G and several other non-adenylated dinucleoside oligophosphates have recently been detected at low concentrations in yeast and E. coli[50], Artemia is unique in possessing levels of this nucleotide which if accessible to the Np_4Nase, would seriously compromise Ap_4A degradation (which must be regarded as the likliest function of this enzyme in other organisms). Therefore, if this enzyme is responsible for the catabolism of Ap_4A or indeed both nucleotides, how can Ap_4A degradation occur in the presence of Gp_4G and be separately controlled? And what about the catabolism of Ap_3A in the presence of Gp_3G? There are several possibilities:

Compartmentalization

Gp_4G and Gp_3G are located predominantly in the yolk platelets[51,52]. A cytoplasmic concentration of Gp_4G as high as 1.2mM has been reported on the basis of subcellular fractionation studies[6,7] but it is not known to what degree this concentration is the result of leakage from the platelets during cellular disruption. Such a concentration would lead to strong inhibition of mRNA cap-binding by the cap-binding protein[53] and permanent activation of GMP reductase[28] and so it has been suggested that in vivo Gp_4G may be confined to the yolk platelets[8].

Ap_4A is probably synthesised in the cytoplasm[21,23] although it may be subsequently translocated to the nucleus[54], while Np_4Nase is predominantly cytoplasmic[2,3] (or possibly nucleoplasmic if free to leak out during aqueous fractionation). If Gp_4G were located and metabolised exclusively within the yolk platelets, possibly by the reversal of the Gp_4G synthetase reaction, the cytoplasmic/nuclear Np_4Nase would have uninhibited access to cytoplasmic/nuclear Ap_4A. Even if Gp_4G is present in kinetically significant quantities in the cytoplasm, access to the active site of the Np_4Nase may be restricted in some way, perhaps by metabolic channeling. This is presumably the case for ATP which is also present at millimolar levels in the cytoplasm and which competitively inhibits the Artemia Np_4Nase in vitro with a K_i of 30μM[3]. A more detailed study of the intracellular location of Ap_4A, Gp_4G and the Np_4Nase should prove useful. An anti-Np_4Nase antibody which we have recently raised should be of value here. The subcellular location of the Np_3Nase may be more suited to a dual role with 50% particulate in rat liver[39,48].

Multiple Enzymes

The distinct heterogeneity of the Artemia Np_4Nase upon ion-exchange chromatography has not been reported for other such enzymes. It is tempting therefore to suggest that Artemia has evolved two (or more) distinct but related enzymes, one appropriate for Ap_4A metabolism and the other for Gp_4G. Unfortunately, although we have not purified the peak II activity from Mono-Q any further, we have been unable to detect any significant differences between the kinetic properties and assay requirements of peaks I and II. Nevertheless a more detailed investigation of the peak II Np_4Nase and of any truly platelet-associated activity must still be performed before this possibility can be completely eliminated.

One enzyme that could be considered as an alternative Gp_4G pyrophosphohydrolase is the Np_3Nase which degrades Gp_4G at a rate much higher than Ap_4A. However the calculated maximum rate in vivo of 0.14pmol Gp_4G/h/embryo would appear to be too slow, unless of course we have under-estimated the amount of this enzyme as previously indicated.

Control of Ap_4A Concentration at the Level of Synthesis and Control of Substrate Recognition

In Artemia, and perhaps other organisms, changes in the concentration of Ap_4A may be regulated solely at the level of synthesis, the degradation rate remaining constant under all conditions. Given the apparent excess of Np_4Nase activity, a constant rate of Ap_4A degradation of 0.35fmol/h/embryo could still be achieved in competition with a 500-fold higher concentration of Gp_4G. Indeed we have previously noted the stability of Ap_4A during the harvesting of cysts in contrast to its reported lability during the handling of mammalian cells[25] and we have suggested that this may be due to the 'protective' effect of the excess Gp_4G[27]. It may even be possible to regulate activity specifically towards Ap_4A in the presence of Gp_4G by mechanisms which alter K_m or k_{cat} only for this nucleotide. For example, reversible modification of a single enzyme either covalently or by association with regulatory factors might allow the necessary degree of discrimination between Ap_4A and Gp_4G to occur at the appropriate times during development. Two observations may be relevant here.

First, although we have for the moment pointed to adenosine 5'-tetra-phosphate as the factor probably responsible for the apparent 7.5- and 8.5-fold reductions in the values of K_m and k_{cat} for the Np_4Nase when assayed in the absence of alkaline phosphatase, we have not ruled out an additional effect of alkaline phosphatase on the enzyme itself. Preliminary experiments show that pre-treatment of the Np_4Nase with alkaline phosphatase followed by complete separation of the two enzymes by high-performance gel permeation chromatography results in an Np_4Nase with approximately 3-fold higher activity. Interestingly, the symmetrical-Np_4Nase from Physarum has been shown to exhibit biphasic kinetics with Ap_4A as substrate, with two K_m values of 2.6µM and 37µM and two V_{max} values which also differ by an order of magnitude suggesting the existence of two enzyme species[19], while the Np_4Nase from rat intestinal mucosa has a K_m for Gp_4G of 50µM, some 10-fold higher than the K_m for the corresponding enzymes from other rat tissues[12]. Since alkaline phosphatase is known to be able to dephosphorylate protein phosphotyrosine residue[55], it is tempting to speculate that reversible phosphorylation of the Np_4Nase might specifically alter K_m and/or k_{cat} for only one of the two major nucleotide substrates. Even if such covalent modifications do not in fact take place, other regulatory factors may bind to and alter the kinetic properties of the enzyme. The increase in Np_4Nase activity during pre-emergence development which is observed only after sucrose density gradient sedimentation may indicate the existence of such factors. Although a similar increase was not found for the Np_4Nase, many of the same arguments may be applied to the role of this enzyme in the catabolism of Ap$_3$A and Gp$_3$G.

In conclusion, the dilemma with which one is initially confronted regarding the dual roles of the Np$_4$Nase and Np$_3$Nase in Ap$_n$A and Gp$_n$G metabolism may be more apparent than real. There are several ways in which the same enzymes could perform these seemingly mutually exclusive roles. Nevertheless, much more research into this interesting metabolic problem is required before any final conclusions can be drawn.

ACKNOWLEDGEMENTS

The financial support of the Science and Engineering Research Council (Grant GR/E/26358), The Wellcome Trust (Grant 17218/1.5) and the Royal Society is gratefully acknowledged.

REFERENCES

1. F. J. Finamore and A. H. Warner, The occurrence of P^1, P^4-diguanosine 5'-tetraphosphate in brine shrimp eggs, J. Biol. Chem. 238:344 (1963).
2. A. H. Warner and F. J. Finamore, Isolation, purification and characterization of P^1, P^4-diguanosine 5'-tetraphosphate asymmetrical-pyrophosphohydrolase from brine shrimp eggs, Biochemistry, 4:1568 (1965).
3. C. G. Vallejo, M. A. G. Sillero and A. Sillero, Diguanosine tetraphosphate guanylohydrolase in Artemia salina, Biochim. Biophys. Acta 358:117 (1974).
4. C. G. Vallejo, C. D. Lobaton, M. Quintanilla, A. Sillero and M. A. G. Sillero, Dinucleosidetetraphosphatase in rat liver and Artemia salina, Biochim Biophys Acta 438:304 (1976).
5. A. H. Warner and F. J. Finamore, Isolation, purification and characterization of P^1, P^3-diguanosine 5'-triphosphate from brine shrimp eggs, Biochim Biophys Acta 108:525 (1965).
6. A. H. Warner, Studies on the biosynthesis and function of dinucleoside polyphosphates in Artemia embryos, in: "Regulation of Macromolecular Synthesis by Low Molecular Weight Mediators," G. Koch and D. Richter, eds., Academic Press, New York (1969).
7. A. H. Warner, The biosynthesis, metabolism and function of dinucleoside polyphosphates in Artemia embryos: a compendium, in: "The Brine Shrimp Artemia Vol. 2, Physiology, Biochemistry, Molecular Biology," G. Persoone, P. Sorgeloos, O. Roels and E. Jaspers, eds., Universa Press, Wetteren, Belgium (1980).
8. A. Sillero and M. A. Günther Sillero, Interconversion of purine nucleotides in Artemia: a review, in: "Artemia Research and its Applications Vol. 2, Physiology, Biochemistry, Molecular Biology," W. Decleir, L. Moens, H. Slegers, E. Jaspers and P. Sorgeloos, eds., Universa Press, Wetteren, Belgium (1987).
9. G. Van Denbos and F. J. Finamore, An unusual pathway for the synthesis of adenosine triphosphate by the purine-requiring organism Artemia salina, J. Biol. Chem. 249:2816 (1974).
10. C. D. Lobaton, C. G. Vallejo, A. Sillero and M. A. G. Sillero, Diguanosine tetraphosphatase from rat liver: Activity on diadenosine tetraphosphate and inhibition by adenosine tetraphosphate, Eur. J. Biochem. 50:495 (1975).
11. J. C. Cameselle, M. J. Costas, M. A. Günther Sillero and A. Sillero, Two low K_m hydrolytic activities on dinucleoside 5', 5'''-P^1, P^4-tetraphosphates in rat liver, J. Biol. Chem. 259:2879 (1984).
12. J. C. Cameselle, M. J. Costas, M. A. Günther Sillero and A. Sillero, Bis(5'-guanosyl) tetraphosphatase in rat tissues, Biochem J. 201:405 (1982).

13. M. Höhn, W. Albert and F. Grummt, Diadenosine tetraphosphate hydrolase from mouse liver: Purification to homogeneity and partial characterization, J. Biol. Chem. 257:3003 (1982).

14. A. Moreno, C. D. Lobaton, M. A. Günther Sillero and A. Sillero, Dinucleoside tetraphosphatase from Ehrlich ascites tumour cells: inhibition by adenosine, guanosine and uridine 5'-tetraphosphates, Int. J. Biochem. 14:629 (1982).

15. A. Ogilvie and W. Antl, Diadenosine tetraphosphatase from human leukemia cells: Purification to homogeneity and partial characterization, J. Biol. Chem. 258:4105 (1983).

16. H. Jakubowski and A. Guranowski, Enzymes hydrolyzing ApppA and/or AppppA in higher plants: purification and some properties of diadenosine triphosphatase, diadenosine tetraphosphatase and phosphodiesterase from yellow lupin (Lupinus luteus) seeds, J. Biol. Chem. 258:9982 (1983).

17. P. Plateau, M. Fromant, A. Brevet, A. Gesquière and S. Blanquet, Catabolism of bis(5'-nucleosidyl) oligophosphates in Escherichia coli: metal requirements and substrate specificity of homogeneous diadenosine-5',5'''-P^1, P^4-tetraphosphate pyrophosphohydrolase, Biochemistry 24:914 (1985).

18. A. Guranowski, H. Jakubowski and E. Holler, Catabolism of diadenosine -5',5'''-P^1, P^4-tetraphosphate in procaryotes: purification and properties of diadenosine-5',5'''-P^1, P^4-tetraphosphate symmetrical pyrophosphohydrolase from Escherichia coli K12, J. Biol. Chem. 258:14784 (1983).

19. L. D. Barnes and C. A. Culver, Isolation and characterization of diadenosine- 5',5'''-P^1, P^4-tetraphosphate pyrophosphohydrolase from Physarum polycephalum, Biochemistry 21:6123 (1982).

20. P. N. Garrison, G. M. Robberson, C. A. Culver and L. D. Barnes, Diadenosine-5',5'''-P^1, P^4-tetraphosphate pyrophosphohydrolase from Physarum polycephalum: substrate specificity, Biochemistry 21:6129 (1982).

21. P. C. Zamecnik, Diadenosine-5',5'''-P^1, P^4-tetraphosphate (Ap_4A): its role in cellular metabolism, Anal. Biochem. 134:1 (1983).

22. E. F. Baril, S. A. Coughlin and P. C. Zamecnik, 5',5'''-P^1, P^4 diadenosine tetraphosphate (Ap_4A): a putative initiator of DNA replication, Cancer investigation 3:465 (1985).

23. F. Grummt, Diadenosine tetraphosphate as a putative intracellular signal of eukaryotic cell cycle control, in: "Modern Cell Biology", Vol. 6, B. H. Satir, ed., Alan R. Liss Inc., New York (1988).

24. B. R. Bochner, P. C. Lee, S. W. Wilson, C. W. Cutler and B. N. Ames, AppppA and related adenylated nucleotides are synthesised as a consequence of oxidation stress, Cell 37:225 (1984).

25. E. Rapaport and P. C. Zamecnik, Presence of diadenosine- 5',5'''-P^1, P^4-tetraphosphate (Ap_4A) in mammalian cells in levels varying widely with proliferative activity of the tissue: a possible "pleiotypic activator", Proc. Natl. Acad. Sci. USA 73:3984 (1976).

26. C. Weinmann-Dorsch, A. Hedl, I. Grummt, W. Albert, F. J. Ferdinand, R. R. Friis, G. Pierron W. Moll and F. Grummt, Drastic rise of intracellular adenosine(5') tetraphospho(5')adenosine correlates with onset of DNA synthesis in eukaryotic cells, Eur. J. Biochem. 138:179 (1984).

27. A. G. McLennan and M. Prescott, Diadenosine-5',5'''-P^1, P^4-tetraphosphate in developing embryos of Artemia, Nucleic Acids Res. 12:1609 (1984).

28. M. F. Renart, J. Renart, M. A. G. Sillero and A. Sillero, Guanosine monophosphate reductase from Artemia salina: inhibition by xanthosine monophosphate and activation by diguanosine tetraphosphate, Biochemistry 15:4962 (1976).

29. B. Ezquieta and C. G. Vallejo, Diguanosine-5',5'''-P^1,P^4-tetraphosphate causes specific inhibition and desensitization of Artemia trypsin-like protease in the hydrolysis of a high-affinity, arginine-rich

substrate, Biochim. Biophys Acta 883:380 (1986).

30. S. J. Gilmour and A. H. Warner, The presence of guanosine 5'-diphospho
   -5'-guanosine and guanosine 5'-triphospho-5'-adenosine in brine
   shrimp embryos, J. Biol. Chem. 253:4960 (1978).

31. D. Miller and A. G. McLennan, Changes in intracellular levels of Ap_3A
   and Ap_4A in cysts and larvae of Artemia do not correlate with changes
   in protein synthesis after heat shock, Nucleic Acids Res. 14:6031
   (1986).

32. K. E. Ng and L. E. Orgel, The action of a water soluble carbodiimide on
   adenosine 5'-polyphosphates, Nucleic Acids Res. 15:3573 (1987).

33. A. Ogilvie, Determination of diadenosine tetraphosphate (Ap_4A) levels
   in subpicomole quantities by a phosphodiesterase luciferin-luciferase
   coupled assay: application as a specific assay for diadenosine
   tetraphosphatase, Anal. Biochem. 115:302 (1981).

34. A. Ogilvie and P. Jakob, Diadenosine-5',5'''-P^1, P^3-triphosphate in
   eukaryotic cells: identification and quantitation, Anal. Biochem.
   134:382 (1983).

35. P. K. Smith, R. I. Krohn, G. T. Hermanson, A. K. Mallia, F. H. Gartner,
   M. D. Provenzano, E. K. Fujimoto, N. M. Goeke, B. J. Olson and
   D. C. Klenk, Measurement of protein using bicinchoninic acid, Anal.
   Biochem. 150:786 (1985).

36. G. Krystal , A silver binding assay for measuring nanogram amounts of
   protein in solution, Anal. Biochem. 167-86 (1987).

37. U. K. Laemmli, Cleavage of structural proteins during the assembly of
   the head of bacteriophage T4, Nature 227:680 (1970).

38. W. Wray, T. Boulikas, V. P. Wray and R. Hancock Silver staining of
   proteins in polyacrylamide gels, Anal. Biochem. 118:197 (1981).

39. M. J. Costas, J. C. Cameselle, M. A. Günther Sillero and A. Sillero,
   Occurence of dinucleoside triphosphatase in the cytosol and
   particulate fractions from rat liver, Int. J. Biochem. 17:903 (1985).

40. M. J. Costas, J. M. Montero, J. C. Cameselle, M. A. Günther Sillero and
   A. Sillero, Dinucleoside triphosphatase from rat brain, Int. J.
   Biochem. 16:757 (1984).

41. M. A. G. Sillero, R. Villalba, A. Moreno, M. Quintanilla, C. D. Lobaton
   and A. Sillero, Dinucleoside triphosphatase from rat liver:purifica-
   tion and properties, Eur. J. Biochem. 76:331 (1977).

42. F. Grummt, C. Weinmann-Dorsch, J. Schneider-Schaulies and A. Lux, Zinc
   as a second messenger of mitogenic induction: effects on diadenosine
   tetraphosphate (Ap_4A) and DNA synthesis, Exp. Cell Res. 163:191
   (1986).

43. J. C. Cameselle, M. J. Costas, M. A. Günther Sillero and A. Sillero,
   Dinucleosidetetraphosphatase inhibition by Zn(II), Biochem. Biophys.
   Res. Commun. 113:717 (1983).

44. M. Morioka and H. Shimada, Ap_4A-hydrolysing activity in sea urchin
   embryos, Exp. Cell Res. 169:57 (1987).

45. C. Hurtado, A. Ruiz, A. Sillero and M. A. Günther Sillero, Specific
   magnesium-dependent diadenosine-5',5'''-P^1, P^3-triphosphate
   pyrophosphohydrolase in Escherichia coli,J. Bacteriol. 169:1718
   (1987).

46. G. M. Blackburn, G. E. Taylor, G. R. J. Thatcher, M. Prescott and
   A. G. McLennan, Synthesis and resistance to enzymic hydrolysis of
   stereochemically-defined phosphonate and thiophosphate analogues of
   P^1, P^4-bis(5'-adenosyl) tetraphosphate, Nucleic Acids Res. 15:6991
   (1987).

47. A. H. Warner and F. J. Finamore, Nucleotide metabolism during brine
   shrimp embryogenesis, J. Biol. Chem. 242:1933 (1967).

48. M. J. Costas, J. C. Cameselle and A. Sillero, Mitochondrial location of
   rat liver dinucleoside triphosphatase, J. Biol. Chem. 261:2064 (1986).

49. P. C. Beers, Diguanosine tetraphosphate pyrophosphohydrolase in the
   development of the brine shrimp, Artemia salina, Dissertation,
   University of Windsor, Ontario (1971).

50. H. Coste, A. Brevet, P. Plateau and S. Blanquet, Non-adenylated bis(5'-nucleosidyll) tetraphosphates occur in _Saccharomyces_ _cerevisiae_ and in _Escherichia_ _coli_ and accumulate upon temperature shift or exposure to cadmium, J. Biol. Chem. 262:12096 (1987).

51. A. H. Warner, J. G. Puodziukas and F. J. Finamore, Yolk platelets in brine shrimp embryos; site of biosynthesis and storage of the diguanosine nucleotides, Exp. Cell Res. 70:365 (1972).

52. C. G. Vallejo, R. Perona, R. Garesse and R. Marco, The stability of the yolk granules of _Artemia_. An improved method for their isolation and study, Cell Differentiation 10:343 (1981).

53. W. Filipowicz, Y. Furiuchi, J. M. Sierra, S. Muthukrishnan, A. J. Shatkin and S. Ochoa, A protein binding the methylated 5'-terminal sequence m^{7}GpppN of eukaryotic messenger RNA, Proc. Natl. Acad. Sci. USA 73:1559 (1976).

54. C. Weinmann-Dorsch and F. Grummt, Diadenosine tetraphosphate is compartmentalized in nuclei of mammalian cells, Exp. Cell Res. 165:550 (1986).

55. G. Swarup, S. Cohen and D. L. Garbers, Selective dephosphorylation of proteins containing phosphotyrosine by alkaline phosphatases, J. Biol. Chem. 256:8197 (1981).

COMPARISON OF DIGUANOSINE POLYPHOSPHATE LEVELS IN VARIOUS POPULATIONS OF

ARTEMIA CYSTS

M. J. Sonnenfeld and A. H. Warner

University of Windsor
Windsor, Ontario
Canada   N9B 3P4

ABSTRACT

Thirteen populations of encysted embryos of Artemia obtained from the Artemia Reference Centre in Belgium were analyzed for their guanosine nucleotide content using high performance liguid chromatography of neutralized perchloric acid extracts of the cysts. Results showed that the guanosine nucleotide content of Artemia cysts varied from 44 pmol/cyst (Thailand) to 125 pmol/cyst (Yugoslavia), while the diguanosine tetraphosphate content varied between 27 pmol/cyst (Thailand) and 78 pmol/cyst (Yugoslavia). These differences were found to reflect mainly the size of the cysts; cysts from Yugoslavia were the heaviest (5.95 ± 0.56 μg/cyst) while those from Thailand were the lightest (2.46 ± 0.14 μg/cyst). Cyst viability may also be an important factor. We also determined the energy charge of each population of cysts based on their guanosine and diguanosine nucleotide content which collectively represent about 90% of the total nucleotide pool in Artemia. The guanylate-based energy charge (ECg) was found to vary from 0.803 (Canada) to 1.192 (Philippines). Comparison between ECg and population viability using regression analysis showed a small linear correlation (r = 0.5) between these two parameters. However, no linear correlation was found between ECg and cyst mass (r = -0.2).

ACKNOWLEDGEMENT

Supported by NSERC of Canada.

PHOSPHONATE ANALOGUES OF DIADENOSINE TETRAPHOSPHATE AS SUBSTRATES AND

INHIBITORS OF <u>ARTEMIA</u> DINUCLEOSIDE TETRAPHOSPHATASE

Alexander G. McLennan, Graham E. Taylor*, Mark Prescott and
G. Michael Blackburn*

Department of Biochemistry, University of Liverpool,
P.O. Box 147, Liverpool L69 3BX, and
*Department of Chemistry, University of Sheffield,
Sheffield S3 7HF, U.K.

INTRODUCTION

P^1,P^4-bis(5'-nucleosidyl) tetraphosphate pyrophosphohydrolase (dinucleoside tetraphosphatase, Np_4Nase, E.C.3.6.1.17) is believed to be responsible for the catabolism of the proposed signal nucleotide P^1,P^4-bis(5'-adenosyl) tetraphosphate (diadenosine tetraphosphate, Ap_4A) and related compounds in a wide variety of cell types. It is also a candidate for the enzyme involved in the degradation of the purine ring store Gp_4G during pre-emergence and early larval development in <u>Artemia</u> (see accompanying paper[1] for detailed discussion and references).

We have recently begun to study synthetic analogues of Ap_4A with two aims in mind. First, to assist in the elucidation of Ap_4A function through their application to microinjected and permealysed cells and subcellular systems, and secondly to use them as probes of the structures of the binding sites of Ap_4A-binding proteins including DNA polymerase-α and the Np_4-Nase and Ap_4A-binding protein kinase from <u>Artemia</u>[2]. For example, the three diastereomers of diadenosine 5',5'''-P^1,P^4-(P^1,P^4-dithio-P^2,P^3-methylene) - tetraphosphate ($Ap_spCH_2pp_sA$) are extremely resistant to hydrolysis by the <u>Artemia</u> Np_4Nase and are also effective competitive inhibitors of the enzyme (K_i = 1 to 1.5 μM)[3]. They should therefore be of use in the analysis of Ap_4A function. These analogues were designed to resist symmetrical cleavage through the introduction of a $P\beta$ -C-$P\beta'$, linkage and also asymmetrical cleavage through the substitution of oxygens on $P\alpha$ and $P\alpha'$ by sulphur. However, recent studies with the <u>asymmetrical</u> - Np_4Nase from lupin seeds has shown that a $P\beta$ -C-$P\beta'$ linkage on its own is sufficient to confer resistance to asymmetrical hydrolysis[4]. Furthermore, the more electronegative $\beta\beta'$-CHBr- substituted analogue was a slightly better substrate than the $\beta\beta'$-CH_2 counterpart.

In order to investigate the substrate requirements of the active site of the <u>Artemia</u> Np_4Nase in more detail, we have determined the ability of a series of $\beta\beta'$-substituted and $\alpha\beta\alpha'\beta'$-disubstituted phosphonate analogues of Ap_4A of varying electronegativities and substituent size to act as substrates and inhibitors of this enzyme. By exploring the nature of the active site, we should eventually arrive at an understanding of how the recognition of alternative substrates, such as Ap_4A and Gp_4G, might be regulated[1].

MATERIALS AND METHODS

$H_2^{18}O$ (97.66 atom %) and [^{3}H] Ap_4A (4.3 Ci/mmol) were from Amersham International. Ap_4A and calf intestine alkaline phosphatase were from Boehringer.

Synthesis of Phosphonate Analogues

Details of the syntheses will be published elsewhere. Briefly, $\alpha\beta$- substituted ADP analogues were first prepared by the reaction of the tris- tetrabutylammonium salt of the appropriate substituted bisphosphonate with 5'-O-tosyl-2',3'-isopropylideneadenosine. After removal of excess bisphos- phonate by precipitation, the products were purified by anion-exchange chromatography on DEAE-Sephadex with yields of 50-70%. The ADP analogues were then self-condensed with dicyclohexylcarbodiimide to form the $\alpha\beta$, $\alpha'\beta'$-disubstituted Ap_4A derivatives (yield = 40%). $\beta\beta'$- substituted Ap_4A analogues were prepared by the reaction of the appropriate substituted bisphosphonate with AMP -morpholidate in pyridine. The $\beta\gamma$- substituted ATP which slowly formed at room temperature for three weeks was converted to the $\beta\beta'$- substituted Ap_4A by heating at 60°C for 24h and the product purified with a final yield of 40-50% by anion-exchange chromatography. For those analogues where diastereomers exist, these were not resolved.

Determination of Inhibition Constants

Assays contained 20mM Bicine-KOH pH 8.4, 2mM Mg acetate, 5 U alkaline phosphatase, 10 or 50µM [^{3}H]-Ap_4A (200mCi/mmol), 45mU homogeneous <u>Artemia</u> Np_4Nase[1] and various concentrations of analogues in a total volume of 50µl. After incubation of 10 min at 37°C, assays were chilled on ice and 300µl of a 25% (v/v) suspension of DEAE-Sephacel in 10mM Tris-HCl, pH 7.5 were then added followed by centrifugation at 16,000 x g for 5 min. 200 µl of the supernatants were added to 4ml Optiphase-MP scintillant (LKB) and the radioactivity in the [^{3}H]-adenosine released by alkaline phosphatase from the [^{3}H]-AMP and [^{3}H]-ATP products determined.

Determination of Degradation Rates

Analogues (400µM) were incubated with Np_4Nase (0.045 to 35 U) for between 10 and 60 min at 37°C in 20mM bicine-KOH pH 8.4, 2mM Mg acetate, 5 units alkaline phosphatase in a volume of 50µl. 10µl of 0.3M ammonium phosphate, pH 5.2 were added and 50µl of the mixture injected on to a 4.6 x 250mm Partisil 10-SAX column. The column was developed at 1ml/min with a 20 min gradient of 5 to 60% buffer B where buffer A = 50mM ammonium phos- phate pH 5.2 and buffer B = 1M ammonium phosphate pH 5.7[5].

Analysis of Degradation Products of Ap_4A by Mass Spectrometry

Reaction mixtures (150µl) containing 20mM Bicine-KOH pH8.4, 2mM Mg^{2+} and 200µM Ap_4A were freeze-dried and reconstituted in the original volume of $H_2^{16}O$ or $H_2^{18}O$. The Ap_4A was digested to completion for 2h at 37°C with Np_4Nase and the products separated by chromatography on a Mono-Q HR5/5 column using a 20ml gradient of 50mM to 0.7M NH_4HCO_3, pH 8.6 at 1ml/min. Peaks corresponding to AMP and ATP were collected, freeze-dried and analy- sed by FAB (negative ion) mass spectrometry.

RESULTS AND DISCUSSION

Ap_4A is Cleaved by Attack at P

A complete interpretation of the interaction between phosphonate

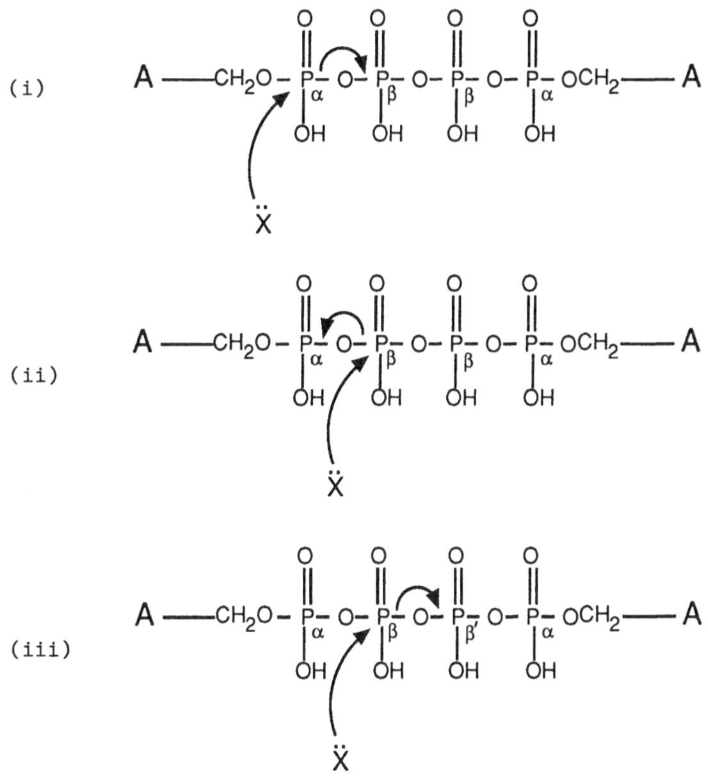

(i)

(ii)

(iii)

Fig. 1. Possible cleavage mechanisms for asymmetrical (i,ii) and symmetrical (iii) dinucleoside tetraphosphatases.

analogues of Ap_4A and the Np_4Nase requires knowledge of the cleavage mechanism of Ap_4A itself. Asymmetric cleavage of Ap_4A con conceivably proceed
by attack at either $P\alpha$ or $P\beta$. In Fig. 1, X represents either an enzyme-
associated nucleophile which could hydrolyse Ap_4A via a nucleotidyl-enzyme
intermediate or water itself. The nature of the nucleophile is presently
unknown. Attack on $P\alpha$ followed by the elimination of the $P\alpha$ - O bond (Fig.
l(i)) or attack at $P\beta$ followed by elinination of the $P\beta$ - O bond adjacent
to $P\alpha$ (Fig. l(ii)) would both lead to AMP and ATP as products. In the case
of symmetrical Np_4Nases, cleavage must occur through attack at $P\beta$ with the
elimination of the $P\beta$ - O bond adjacent to $P\beta'$ (Fig. l(iii)).

In order to distinguish between mechanisms (i) and (ii) for the Artemia
Np_4Nase, Ap_4A was hydrolysed to completion in $H_2{}^{18}O$ and the distribution
of the ^{18}O between AMP and ATP determined by FAB (negative ion) mass
spectrometry. This experiment clearly shows that the ^{18}O is incorporated
exclusively into the AMP. The [M-H]$^-$ pseudo-molecular ion of AMP is increased by 2 mass units from M/Z = 346 to M/Z = 348 when the hydrolysis
is carried out in $H_2{}^{18}O$ while the [M-H]$^-$ ion of ATP has M/Z = 506 in both
normal and heavy solvents (Fig. 2). This indicates that the site of attack
is $P\alpha$ [mechanism (i)].

Phosphonate Analogues as Substrates and Inhibitors of Np_4Nase

Isosteric phosphonate analogues of nucleotides containing a P - CH_2-P
bridge tend strongly to resist enzymic hydrolysis owing to the reduced

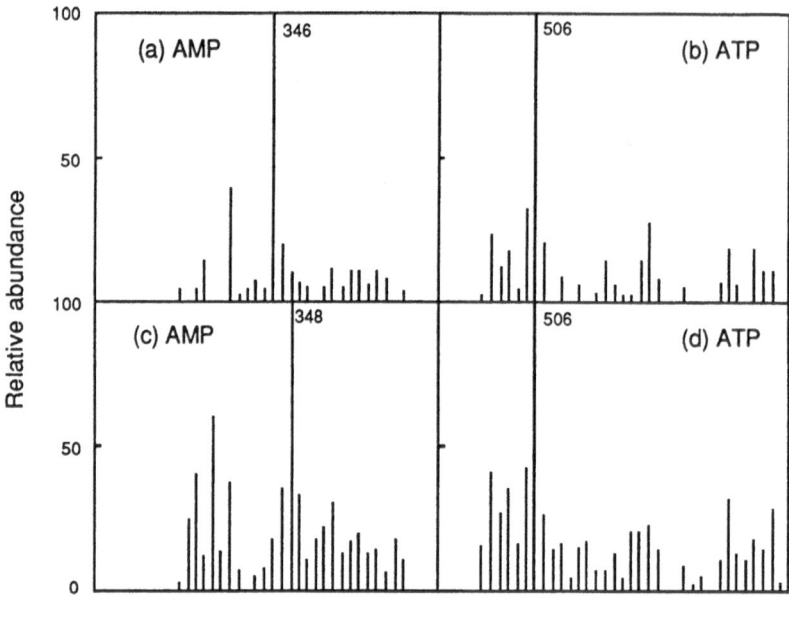

Fig. 2. Analysis of degradation products of Ap_4A by mass spectrometry. Samples (2.5µg) of the purified AMP (a,c) and ATP (b,d) products from reactions performed in the presence of $H_2^{16}O$ (a,b) or $H_2^{18}O$ (c,d) were subjected to negative ion FAB mass spectrometry in a polyethylene glycol matrix. Limited range scanning was performed with a VG Micromass 7070F mass spectrometer.

electrophilicity of the phosphorus atoms[6]. Replacement of the methylene hydrogens by more electronegative halogen atoms can improve substrate efficiency[7,8]. A series of analogues in which the O of the $P\beta$ -O-$P\beta'$ bridge was replaced by CF_2, CHF, CCl_2, CHCl or CH_2 were tested as substrates of the Artemia Np_4Nase. Their degradation was determined using an h.p.l.c. system which gave excellent resolution of all substrates and possible products (Table 1). In each case they were hydrolysed asymmetrically to yield AMP (shown in the absence of alkaline phosphatase) and the corresponding β,γ -substituted ATP analogue at a rate which precisely reflected the electronegativities of the substituents (Table 2). The isopolar and isosteric $AppCF_2ppA$ was a very good substrate (V_{rel}=0.7) while the isosteric $AppCH_2ppA$ proved to be highly resistant to hydrolysis (V_{rel}=0.025) as was shown for the lupin enzyme[4]. As the site of hydrolytic attack is $P\alpha$, this may mean that catalysis requires the considerable transmission of electron withdrawing power through the oligophosphate chain.

As expected, these analogues were also competitive inhibitors of Ap_4A degradation. K_i values were determined at two substrate concentrations according to Cheng and Prusoff[9]. Analysis of the K_i values of the 'simple' $\beta\beta'$ -substituted compounds i.e. $\beta\beta'$ -CF_2, CHF, CCl_2, CHCl and CH_2, shows that, except for $AppCCl_2ppA$ where the size of the Cl atoms may reduce affinity, binding is inversely related to the electronegativities of these groups (Table 2). Assuming the K_m = K_s, this group of analogues binds more tightly to the enzyme than Ap_4A itself which may indicate that these compounds bear some resemblance to the transition state for Ap_4A. If so, the $P\beta$ -O-$P\beta'$ oxygen atom must participate in some unfavourable binding interaction which is gradually relieved by substituting it with groups of decreasing electronegativity. The relative specificities for each analogue,

Table 1. H.P.L.C. Retention Times of Nucleotides and Nucleotide Analogues

Compound	R.T. (min)	Compound	R.T. (min)	Compound	R.T. (min)
Ap_4A	10.54	ATP	14.24	$ApCF_2ppCF_2pA$	10.13
$AppCF_2ppA$	10.06	$AppCF_2p$	14.27	$ApCCl_2ppCCl_2pA$	9.60
$AppCCl_2ppA$	9.97	$AppCCl_2p$	13.02	$ApCHFppCHFpA$	9.81
$AppCHFppA$	10.07	$AppCHFp$	11.49	$ApCHClppCHClpA$	10.00
$AppCHClppA$	10.21	$AppCHClp$	10.70	$ApCH_2ppCH_2pA$	8.81
$AppCH_2ppA$	9.50	$AppCH_2p$	7.82	$ApCH_2CH_2ppCH_2CH_2pA$	6.45
$AppCH=CHppA$	8.36	$AppCH=CHp$	8.36	ADP	8.40
$AppCH_2CH_2ppA$	7.55	$AppCH_2CH_2p$	6.21	$ApCF_2p$	8.91
AMP	4.50	Adenosine	2.65	$ApCCl_2p$	7.94
$ApCHFp$	6.68	$ApCHClp$	6.20	$ApCH_2p$	5.25
$ApCH_2CH_2p$	4.32				

i.e. the values of k_{cat}/K_i relative to k_{cat}/K_m for Ap_4A, are roughly 2-fold higher (except again for $AppCCl_2ppA$) (Table 2) which may also indicate a similarity to the transition state[10]; however, these increases may be insignificant within the approximations for the equvalence of kinetic constants.

Therefore an alternative explanation for the reductions in both k_{cat} and $K_i(K_m)$ is that these analogues can bind non-productively in a different mode within the active site. Studies on the oligophosphate chain length specificity of the Np_4Nase indicate that the enzyme probably has a binding pocket which recognises an 'Nppp' portion of the substrate[1]. In this model, loss of an important binding interaction involving the oxygen of the $P\beta$-O-$P\beta'$ bridge which has been replaced (either directly or through its electron-withdrawing ability) may result in a compensatory interaction with one of the $P\alpha$-O-$P\beta$ oxygens causing a 'frameshift' binding mode with an 'Npp' now in the binding site. The juxtaposition of the $P\beta$-C bond with the catalytic groups which results from this alternative binding mode would explain the somewhat unexpected resistance of these analogues to hydrolysis. The probability of non-productive binding and hence reduction in k_{cat} and K_i would increase with decreasing electronegativity.

Not unexpectedly, increasing the distance between the β and β' phosphorous atoms severely affects the recognition of the molecule since neither of the 'complex' $\beta\beta'$- substituted analogues, $AppCH=CHppA$ and $AppCH_2CH_2ppA$ is an effective substrate or inhibitor (Table 2). The 4-fold higher rate with the trans-ethylene analogue may reflect the greater polarity of this substituent. Alterations in the metal-chelating properties of these two compounds may also contribute to their poor recognition.

Further studies with the corresponding $\alpha\beta,\alpha'\beta'$-analogues where both $P\alpha$-O-$P\beta$ bridges are substituted showed that binding as reflected in the K_i

Table 2. Efficiency of Phosphonate Analogues of Ap$_4$A (AppppA) as Substrates and Inhibitors of the _Artemia_ Np$_4$Nase

Compound	V_{rel}	$K_i (\mu M)$	Relative Specificity
AppppA	1.0	33*	1.0
AppCF$_2$ppA	0.7	12	1.9
AppCHFppA	0.20	3.5	1.9
AppCCl$_2$ppA	0.18	23	0.26
AppCHClppA	0.11	2.3	1.6
AppCH$_2$ppA	0.025	0.4	1.8
AppCH=CHppA	0.012	120	3.4×10^{-3}
AppCH$_2$CH$_2$ppA	0.003	120	8.2×10^{-4}
ApCHFppCHFpA	0.030	50	0.02
ApCH$_2$ppCH$_2$pA	0.004	23	5.8×10^{-3}
ApCH$_2$CH$_2$ppCH$_2$CH$_2$pA	0.004	>1500	$< 8 \times 10^{-5}$
ApCHClppCHClpA	0.003	40	2.6×10^{-3}
ApCF$_2$ppCF$_2$pA	0.002	300	2.2×10^{-4}
ApCCl$_2$ppCCl$_2$pA	0.001	1200	1.3×10^{-4}

*K_m for Ap$_4$A. Assays were performed as described in Materials and Methods. Alkaline phosphatase was included in the incubations to degrade contaminating traces of inhibitors, assumed to be phosphonate analogues of adenosine 5'-tetraphosphate. Rates were determined by integration of the product peaks (adenosine and β,γ-ATP analogue for the $\beta\beta'$-substituted compounds and the α,β-ADP analogue for the $\alpha\beta,\alpha'\beta'$-substituted compounds). k_{cat} for Ap$_4$A = 12.7s^{-1}.

values is again much as expected with the least bulky CH$_2$, CHF and CHCl compounds having an affinity similar to Ap$_4$A, although less than the corresponding $\beta\beta'$-analogues, while the CF$_2$, CCl$_2$ and CH$_2$CH$_2$ compounds bind very poorly. By virtue of having P$_\alpha$-C-P$_\beta$ bridges, these compounds would be expected to be very resistant to hydrolysis and this is indeed the case with hydrolysis rates between 0.1 and 0.4% of that for Ap$_4$A except for ApCHFppCHFpA which is cleaved at the appreciable relative rate of 3% (Table 2). Analysis of the products of this unexpected reaction reveals that the P - C bonds have in fact remained intact and that the cleavage has occurred symmetrically, the sole product of the reaction being ApCHFp (Fig. 3). Cleavage of the other $\alpha\beta, \alpha'\beta'$ -substituted analogues is also symmetrical. This reaction is unlikely to be due to a contaminating symmetrical - Np$_4$Nase activity since

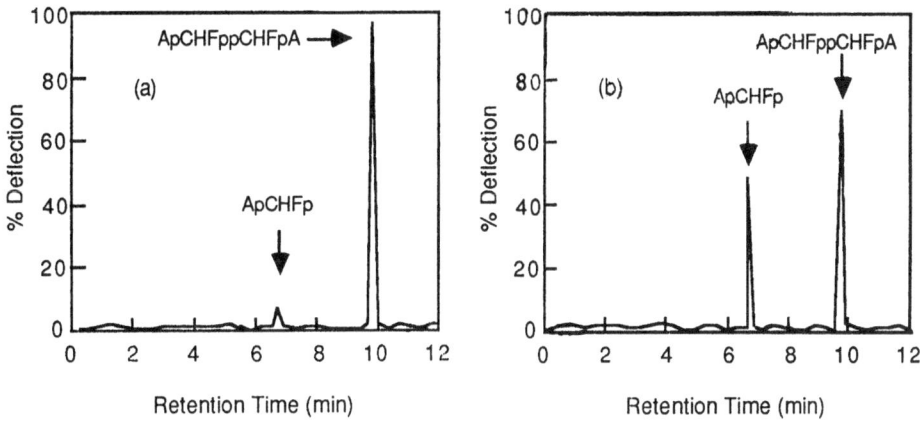

Fig 3. Symmetrical hydrolysis of ApCHFppCHFpA by the Artemia asymmetrical - Np_4Nase. ApCHFppCHFpA (400μm) was incubated (a) without and (b) with 3.6U of Np_4Nase for 60 min as described in Materials and Methods and the products determined by h.p.l.c.

such an activity has never been observed during the course of our studies with Artemia and the enzyme preparation used is homogeneous on a silver-stained SDS-polyacrylamide gel[1].

As symmetrical cleavage can only occur by attack at P (Fig. 1iii), this result reinforces the non-productive binding model for the behavior of the αβ-substituted analogues. In this case, binding of ApCHFppCHFpA with an 'Npp' moiety (i.e. ApCHFp) in the binding site is the productive mode. This is presumably the way in which Ap_3A (V_{rel} =0.005) and Gp_3G (V_{rel} = 0.02) bind to the enzyme[1]. We have previously observed the symmetrical cleavage of the mixed diastereomers of diadenosine 5', 5'''-P^1, P^4 -(P^1, P^4 - dithio) - tetraphosphate (Ap_sppp_sA) by the Artemia enzyme[3] and the slow, asymmetrical cleavage of the $\beta\beta'$-CF_2, CHF and CH_2 analogues by the symmetrical - Np_4Nase activity from E.coli (G.E. Taylor, unpublished observations). The apparent ability of the binding site to accommodate the substrate in alternative ways may therefore be a property of Np_4Nases in general. $ApCH_2ppCH_2pA$ has been reported by Guranowski et al. [4] to be totally resistant to hydrolysis by the lupin asymmetrical - Np_4Nase, but it is not clear whether $ApCH_2p$ was looked for as a possible product in their thin layer chromatography assay or whether the assay was sufficiently sensitive to detect it. Our h.p.l.c. assay detects all possible products with high sensitivity.

Analogues of Ap_4A have now been used as aids towards understanding the mechanisms and binding site geometry of a number of enzymes for which Ap_4A is a substrate, product or target such as Ap_4A pyrophosphohydrolase and Ap_4A phosphorylase[3,4,11], aminoacyl-tRNA synthetase[12,13] and poly (ADP-ribose) polymerase[14] as well as enzymes for which it is thought to behave as a transition state analogue. The latter include adenosine kinase[15,16] and adenylate kinase[15,17,18]. The series of analogues which we have studied here has been particularly valuable in delineating some aspects of the binding site of the Artemia Np_4Nase. When combined with data obtained with base and sugar substituents, a detailed picture of the catalytic centre of this enzyme should emerge.

ACKNOWLEDGEMENTS

We thank the Science and Engineering Research Council for financial support through a CASE studentship with Amersham International to G.E.T. and Research Grants GR/D/35438 (to G.M.B.) and GR/E/26358 (to A.G.McL.). The general support of the North West Cancer Research Fund and the Royal Society (to A.G.McL.) is also acknowledged. Mass spectrometric analysis was performed by R. Evershed and M.C. Prescott.

REFERENCES

1. M. Prescott, A. D. Milne and A. G. McLennan, Enzymes of dinucleoside oligophosphate metabolism in _Artemia_ cysts and larvae, accompanying article this volume.
2. M. Prescott and A. G. McLennan, The protein kinase activity associated with the major bis(5'-adenosyl) tetraphosphate-binding protein of _Artemia_, in "_Artemia_ Research and its Applications, Vol. 2, Physiology, Biochemistry, Molecular Biology," W. Decleir, L. Moens, H. Slegers, E. Jaspers and P. Sorgeloos, eds., Universa Press, Wetteren, Belgium (1987).
3. G. M. Blackburn, G. E. Taylor, G. R. J. Thatcher, M. Prescott and A. G. McLennan, Synthesis and resistance to enzymic hydrolysis of stereochemically-defined phosphonate and thiophosphate analogues of P^1, P^4-bis(5'-adenosyl) tetraphosphate, _Nucleic Acids Res._ 15:6991 (1987).
4. A. Guranowski, A. Biryukov, N. B. Tarussova, R. M. Khomutov and H. Jakubowski, Phosphonate analogues of diadenosine 5', 5''' -P^1, P^4-tetraphosphate as substrates and inhibitors of procaryotic and eucaryotic enzymes degrading dinucleoside tetraphosphates, _Biochemistry_ 26:3425 (1987).
5. L. D. Barnes and C. A. Culver, Isolation and characterization of diadenosine-5',5''' -P^1, P^4-tetraphosphate pyrophosphohydrolase from _Physarum polycephalum_, _Biochemistry_ 21:6123 (1982).
6. K. -H. Scheit, "Nucleotide Analogs: Synthesis and Biological Function," J. Wiley and Sons, New York (1980).
7. G. M. Blackburn, T. D. Perree, A. Rashid, C. Bisbal and B. Lebleu, _Chemica Scripta_ 26:21 (1986).
8. G. M. Blackburn, G. E. Taylor, R. H. Tattershall, G. R. J. Thatcher and A. G. McLennan, Phosphonate analogues of biological phosphates, _in_: "Biophosphates and Their Analogues - Synthesis, Structure, Metabolism and Activity," K. S. Bruzik and W. J. Stec, eds., Elsevier, Amsterdam (1987).
9. Y. C. Cheng and W. H. Prusoff, Relationship between the inhibition constant (K_i) and the concentration of inhibitor which causes 50 per cent inhibition (I_{50}) of an enzymatic reaction, _Biochem. Pharmacol._ 22:3099 (1973).
10. A. R. Fersht, Catalysis, binding and enzyme-substrate complementarity, _Proc. R. Soc. London B_ 187:397 (1974).
11. A. J. Guranowski, Specific enzymes degrading diadenosine tetraphosphate, _Nucleosides and Nucleotides_ 6:307 (1987).
12. T. I. Merkulova, M. K. Nurbekov, N. B. Tarussova and G. K. Kovaleva, Studies of ATP- and PPi-binding sites of bovine tryptophanyl-tRNA synthetase with the use of Ap_4A- phosphonate analogues, _Biopolymers and Cell_ 2:179 (1986).

13. N. B. Tarussova, T. I. Osipova, T. V. Tyrtysh and A. I. Biryukov, Phosphonate analogues of P^1, P^4 - bis(5'-adenosyl) tetraphosphate (Ap_4A), in: "Biophosphates and Their Analogues - Synthesis, Structure, Metabolism and Activity, "K. S. Bruzik and W. J. Stec, eds., Elsevier, Amsterdam (1987).

14. H. Suzuki, Y. Tanaka, D. T. Buonamassa, B. Farina and E. Leone, inhibition of ADP-ribosylation of histone H1 by analogs of diadenosine 5', 5'''-p^1, p^4-tetraphosphate, Mol. Cell. Biochem. 74:17 (1987).

15. M. Prescott and A. G. McLennan, unpublished results.

16. R. Bone, Y. C. Cheng and R. Wolfenden, Inhibition of adenosine and thymidylate kinases by bisubstrate analogues, J. Biol. Chem. 261:16410 (1986).

17. P. Feldhaus, T. Fröhlich, R. S. Goody, M. Isakov and R. H. Schirmer, Synthetic inhibitors of adenylate kinases in the assays for ATPases and phosphokinases, Eur. J. Biochem. 57:197 (1975).

PURIFICATION AND CHARACTERIZATION OF A PHTHALATE ESTER HYDROLYZING ENZYME

FROM ARTEMIA

David S. Miller, Patricia A. Healy and Roger A. Acey

Department of Chemistry
California State University
Long Beach, CA 90840

ABSTRACT

Previous studies have established the embryotoxicity of various dialkyl phthalate esters on developing Artemia, most notably di-n-butylphthalate (DBP). Hudson and coworkers [1] have reported the partial purification of an enzyme (DBPase) responsible for the hydrolysis of DBP and suggested that the observed embryotoxicity might be the result of DBP preventing DBPase from performing its normal biological function. We now report on the purification of this enzyme to near homogeneity and its partial biochemical characterization. The assay for DBPase involves incubation of appropriate samples with radiolabeled ^{14}C-carbonyl-DBP and quantifying the amount of monobutyl ester found during a 2 hr incubation period. Free swimming nauplii were collected and homogenized in 50 mM Tris, 0.2 mM EDTA, 0.1 mM DTT, pH 7.5 (TED). Cytosol was made 50% (w/v) in ammonium sulfate and stored at -20°C to precipitate the enzyme. The pellet was dissolved in TED, desalted, and fractionated on Bio-Gel P-100, $M_r=51,000$. Peak fractions were pooled, applied to Bio-Gel DEAE, and eluted with a linear salt gradient (0.0-1.2 M KCl) in TED. Subsequently the eluted DBPase activity was fractionated by high performance liquid chromatography (HPLC) on a TSK-GW 3000 gel permeation column. Electrophoretic analysis on a 12% sodium dodecylsulfate polyacrylamide gel of the preparation revealed two polypeptides, $M_r=51,000$ and 98,000, respectively. The enzyme has a neutral pH optimum. In addition, the enzyme is unaffected by soybean trypsin inhibitor but its activity is almost completely inhibited by phenylmethyl sulfonylfluoride. Furthermore, when either EDTA or dithiothreitol are excluded from the purification protocol, DBPase activity is significantly diminished.

ACKNOWLEDGEMENT

Supported by a grant from the Office of University Research, CSULB.

REFERENCE

1. R. A. Hudson, T. Giancarlo II, C. F. Austerberry and J. C. Bagshaw, Isolation and partial purification of phthalate ester hydrolyzing enzyme(s) from the brine shrimp, Artemia, Toxicol. Letters 10:389 (1982).

MESSENGER RIBONUCLEOPROTEINS OF CRYPTOBIOTIC GASTRULAE OF <u>ARTEMIA</u>:

MECHANISMS OF ACTIVATION AND REPRESSION OF NON-POLYSOMAL MESSENGER RIBO-

NUCLEOPROTEINS

H. Slegers and M. Aerden

Department of Biochemistry
Universitaire Instelling Antwerpen
Universiteitsplein 1 - B 2610 Antwerpen-Wilrijk, Belgium

INTRODUCTION

The mechanisms involved in the termination of dormancy of cryptobiotic gastrulae of <u>Artemia</u> as well as the sequential events of pre-emergence are largely unknown. Key phenomena seem to be the increase in the intracellular pH from 6.2 to > 7.9 [1], degradation of trehalose affecting the stability of intracellular membranes [2], yolk degradation by proteolytic enzymes [3,4], a shift of poly $(A)_1$ polymerase from the cytosol to the cytoskeleton [5,6] and an increase in P^1, P^4 -bis(5' -adenosyl) tetraphosphate (Ap_4A) formed by aminoacyl tRNA synthetases from activated aminoacyladenylate and ATP and proposed to be involved in the stimulation of DNA synthesis [7]. After resumption of development one or more of these phenomena trigger the activation of stored mRNP.

Our research on stored messenger ribonucleoproteins demonstrated that at least two apparently independent mechanisms regulate the translational activity of stored mRNP [8]. One mechanism is phosphorylation-dephosphorylation of mRNP proteins by an mRNP-associated protein kinase and two cytoplasmic protein phosphatases. The protein kinase is a casein kinase II with a Mr of 136000 and an $\alpha_2 \beta_2$ subunit structure. The α subunit is the catalytic subunit which is easily converted to α' by limited proteolysis without loss of activity. The function of the β subunit is unknown. The subunits have Mr of 36500 (α), 33000 (α') and 28000 (β). The enzyme is cAMP independent, requires Mg^{2+} or Mn^{2+} and is inhibited by monovalent ions, heparin, caffeine and poly(L) glutamic acid. The enzyme is active between pH 6.8 and 8.5 with an optimum at pH 8.0. The initial pH increase observed after termination of dormancy probably results in the activation of this enzyme. However, the detection of at least two compartments with a different pH indicates that termination of dormancy involves more than a simple increase in pH [2]. As phosphorylation of mRNP proteins results in the inhibition of their translation, the activation of the protein kinase has to be counteracted by protein phosphatases and/or inhibition by specific effectors. The phosphorylation of mRNP proteins is counteracted by two cytoplasmic polycation stimulated protein phosphatases (PCS) [8]. PCS-X and PCS-Y have a Mr of 225000 and 346000 respectively. PCS-X has an $\alpha_2 \beta_2$ structure with subunit Mr of 75000 (α) and 40000 (β). The subunit is the catalytic subunit. PCS-Y has a more complex structure but is composed of the same core subunits. Both enzymes have an optimum activity at pH 7.2, are inhibited by NaF, ATP and PPi and are activated by heparin, protamine, poly(L) lysine and histone H_1.

A second mechanism of translational regulation is due to the action of an inhibitor ribonucleic acid associated with stored mRNP. The latter inhibitor abolishes the ternary complex formation between eIF2, Met-tRNA$_f$ and GTP. The association of eIF2 with the translatable mRNP fraction suggests that the inhibitor interferes with the binding of eIF2 to mRNP [9].

In this communication both mechanisms of translational regulation have been further studied. Ap_4A has been identified as an inhibitor of the mRNP-associated protein kinase and is involved in the activation of stored mRNP. The inhibitor ribonucleic acid associated with repressed mRNP has been further characterized.

RESULTS AND DISCUSSION

The Regulation of the Messenger Ribonucleoprotein Associated Protein Kinase

A discrete number of proteins is associated with non-polysomal mRNP of cryptobiotic gastrulae. Several mRNP proteins have been identified and characterized e.g. the poly(A)-binding proteins, eIF2 and a protein kinase [8]. Our identification studies support the involvement of mRNP proteins in the activation of stored mRNP. Evidence for the regulation of mRNP translation by phosphorylation-dephosphorylation comes from the observed correlation of mRNP protein phosphorylation and inhibition of translation. The translational activity of mRNP is abolished after incubation with ATP or GTP [10]. Although exogeneous eIF2 is very efficiently phosphorylated by the mRNP-associated protein kinase, in situ phosphorylation of eIF2 is not observed. The main phosphorylated mRNP proteins are the poly(A)-binding proteins [11]. Phosphorylation of the poly(A)-binding Mr 38000 protein increases its interaction with the poly(A)-sequence of mRNA and results in the inhibition of mRNP translation [12]. From these results we may conclude that activation of stored mRNP has to be preceeded by an inhibition of the mRNP associated protein kinase. Two potential protein kinase effectors have been compared for inhibition of the mRNP associated protein kinase viz. 2,3-bisphosphoglycerate and Ap_4A.

In erythroid cells 2,3-bisphosphoglycerate has been identified as a regulator of casein kinase II activity [13]. Thirty minutes after activation of cryptobiotic gastrulae, approximately 75% of added inorganic phosphate is incorporated in the glycolytic intermediate 2,3-bisphosphoglycerate [14]. The majority of this intermediate is formed in yolk platelets. As yolk degradation is one of the initial events observed after resumption of development, 2,3-bisphosphoglycerate is released in the cytoplasm and may inhibit the mRNP associated protein kinase.

In the initial hours of <u>Artemia</u> development an increase in Ap_4A is observed. This dinucleoside tetraphosphate is formed in the back reaction of aminoacyl tRNA synthetase and attains a maximal concentration just before hatching. Although Ap_4A is often proposed as a regulator of the initiation of transcription it seemed worthwhile to test Ap_4A as a possible inhibitor of the mRNP-associated protein kinase in order to link activation of stored mRNP with the formation of aminoacyl tRNA. In <u>Artemia</u> Ap_4A has been shown to bind to a protein kinase unrelated to the mRNP-associated protein kinase [15]. The binding of Ap_4A did not affect the activity of this enzyme.

In Fig. 1, the inhibition of the mRNP-associated protein kinase by

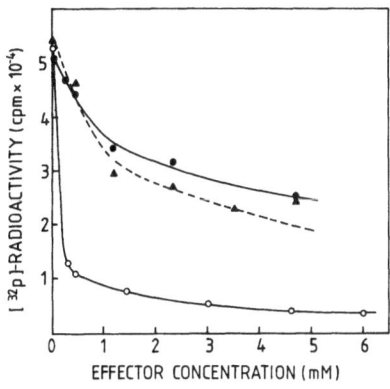

Fig. 1. Inhibition of mRNP-associated protein kinase by 2,3-
bisphosphoglycerate and Ap$_4$A. The mRNP-associated
protein kinase was prepared from a 1.5 M KCl wash of
oligo(dT)-cellulose bound mRNP by ion-exchange chromatography on phosphocellulose P11 and affinity chromatography on casein-agarose as described [10]. The
kinase assay contained 10 mM Tris-HCl (pH 8.2), 12.5
mM MgCl$_2$, 50 mM KCl, 7 mM 2-mercaptoethanol, 70 μg
casein or 7 μg poly(A)-binding protein, 1 μg protein
kinase, 1 μM (γ-^{32}P)-ATP (50 Ci/mmol) and various
concentrations of effector. The reaction mixture of
200 μl was incubated for 30 min at 37°C and the
reaction terminated by addition of trichloroacetic
acid to a final concentration of 10% (w/v). Effect
of 2,3-bisphosphoglycerate on the phosphorylation of
casein (▲---▲) or the Mr 38000 poly(A)-binding
protein (●-●) and the effect of Ap$_4$A on the phosphorylation of the Mr 38000 poly(A)-binding protein
(O-O).

2,3-bisphosphoglycerate and Ap$_4$A was compared using casein and the Mr 38000
poly(A)-binding protein as substrates. Ap$_4$A is a far better inhibitor of
the protein kinase than 2,3-bisphosphoglycerate. Fifty percent inhibition
was obtained at a concentration of 0.15 mM Ap$_4$A and 5 mM 2,3-bisphospho-
glycerate respectively. The inhibition by Ap$_4$A is noncompetitive with
respect to ATP (Table 1). The inhibition of casein kinase II by 2,3-bis-
phosphoglycerate is reported to be noncompetitive with ATP but competitive
with the protein substrate [13].

For the first time evidence is presented in favour of a function for
Ap$_4$A in the regulation of translation of stored mRNP in cryptobiotic
gastrulae of Artemia. Further experiments are in progress to determine
the mechanism of the protein kinase inhibition by Ap$_4$A.

The Inhibitor Ribonucleic Acid Associated with Messenger Ribonucleoprotein

An inhibitor of protein synthesis has been isolated from translationally repressed non-polysomal mRNP. mRNP purified by affinity chromatography
on oligo(dT)-cellulose was separated into a 17S and 7S fraction. Previously
an inhibitor ribonucleoprotein was purified from the 7S RNP fraction [16].
Recently we demonstrated that this inhibitor is also associated with the
17S RNP [9]. The inhibitor RNA (iRNA) isolated from the repressed 17S RNP

261

Table 1. Inhibition by Ap$_4$A is Non-competitive with ATP

ATP (mM)	Incorporated radioactivity	
	casein	poly(A)-binding protein
1.9	2670	2020
3.7	3020	1660
5.6	2950	3360

The kinase assay described in the legend of Fig. 1 was supplemented with 0.6 mM Ap$_4$A and various concentrations of (γ-^{32}P) ATP (0.1 Ci/mmol). In the control assay with 1.25 mM ATP but without added Ap$_4$A, 12080 cpm were incorporated into casein.

is as active as the iRNA purified from the 7S RNP fraction (Fig. 2). Fifty percent inhibition is obtained with 10 ng iRNA. Ninety percent inhibition is observed with 110 ng and 170 ng iRNA purified from the 17S RNP and 7S RNP fractions respectively.

The iRNA preparations were 3'-end labelled with (5'-^{32}p)-pCp and analyzed on a 8% (w/v) polyacrylamide gel (Fig. 3). The main RNA component has a length of 85 ± 4 nucleotides (RNA$_{85}$). Minor RNAs have a length of 112 (RNA$_{112}$), 100 (RNA$_{100}$) and 95 (RNA$_{95}$) nucleotides. The preparations also contained oligonucleotides with a length below n = 20 which probably resulted from RNA degradation. The iRNA preparations from the 7S and 17S RNA fractions have approximately the same RNA composition (Fig. 3).

In an attempt to assign the inhibitory activity to a particular RNA component of the iRNA preparation, several separation methods have been tested without much success. Finally we subjected a 3'-end (5'-^{32}p)pCp labelled iRNA preparation from the 7S RNP fraction to ion-exchange chromatography on a benzoylated DEAE- cellulose column (Fig. 4A). Bound RNA was eluted with a linear salt gradient. The inhibitor activity eluted in a salt concentration range from 600 mM up to 1050 mM NaCl with an optimum at 800 mM NaCl. The inhibitory activity was slightly shifted from the majority of the RNA, which eluted in a salt concentration range from 400 mM up to 1200 mM NaCl with an optimum at 700 mM NaCl. Pooled fractions were analyzed by polyacrylamide gel electrophoresis (Fig. 4B). The inhibition of globin mRNA translation is not coincident with RNA$_{112}$, RNA$_{100}$ and RNA$_{95}$. The fractions of optimum inhibiton contained RNA$_{85}$ and oligonucleotides generated by RNA degradation, but the inhibition is not coincident with the maximum amount of RNA$_{85}$. This may be explained by: i) the existence of a sequence heterogeneous population of RNAs with a length of 85 nucleotides, of which only a fraction is active in the inhibition of globin mRNA translation, ii) a more complex inhibition mechanism which involves iRNA$_{85}$ and/or other oligonucleotides, iii) inhibition by a molecule not labelled with (5'-^{32}P)pCp. This is however very unlikely as dephosphorylation of the 3'-end before labelling did not indicate the existence of other RNA components. Molecules other than RNA are also very unlikely as no differences in inhibition were observed after phenolization and protease degradation.

In an attempt to elucidate the mechanism of inhibition, we tested the binding of iRNA preparations to rabbit reticulocyte lysate proteins.

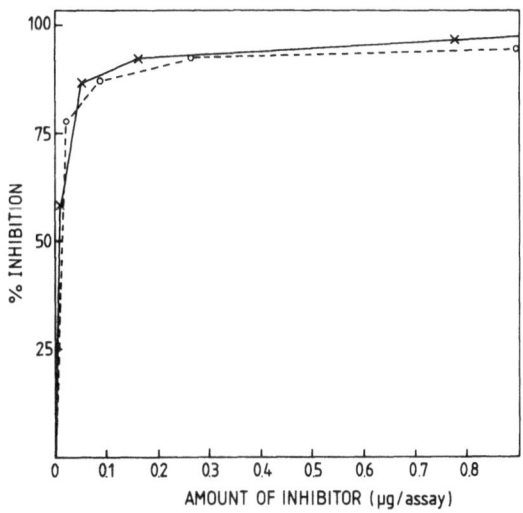

Fig. 2.   Comparison of the inhibition of globin mRNA
          translation by the inhibitor RNAs purified
          from the 17S RNP and 7S RNP.
          The purification of the 7S RNP-associated
          inhibitor has been described previously [16].
          The 17S RNP associated inhibitor was dissociated
          from RNP by centrifugation on a 10-30% (w/v)
          sucrose density gradient in 20 mM Tris-HCl
          (pH 7.2), 500 mM KCl, 0.1 mM DTT and 0.1 mM
          PMSF. Centrifugation was in the SW28Ti rotor
          at 27000 rpm for 47 h. Inhibitor-containing
          fractions were filtered through oligo(T)-
          cellulose, the unbound material diluted to
          20 mM Tris-HCl (pH 7.2), 100 mM KCl and 0.1 mM
          DTT and filtered through poly(A)-Sepharose 4B.
          The unbound material was concentrated by ultra-
          filtration, desalted on Sephadex G-25 and
          lyophilized. Increasing amounts of purified
          inhibitor RNA were mixed with 25 ng globin
          mRNA before addition to rabbit reticulocyte
          lysate. Inhibitor RNAs from 17S RNAs (x-x)
          and 7S RNP (o--o).

Twenty five to thirty percent of the $(5'-^{32}P)pCp$-labelled RNA was able to
bind to nitrocellulose filters after mixing with lysate proteins in 20 mM
Tris-HCl (pH 7.2), 100 mM KCl. Analysis by polyacrylamide gel electrophoresis demonstrated the heterogeneous nature of the bound RNA lacking
any specificity of binding. The majority of the bound RNA were small
oligonucleotides generated by degradation.

So far we have only been able to show that iRNA preparations inhibited
the ternary complex formation between eIF2, Met $tRNA_f$ and GTP [8]. Further
experiments are necessary to evaluate the observed inhibition by iRNA
preparations.

ACKNOWLEDGEMENT

This investigation is supported by grants from the Fund for Joint Basic

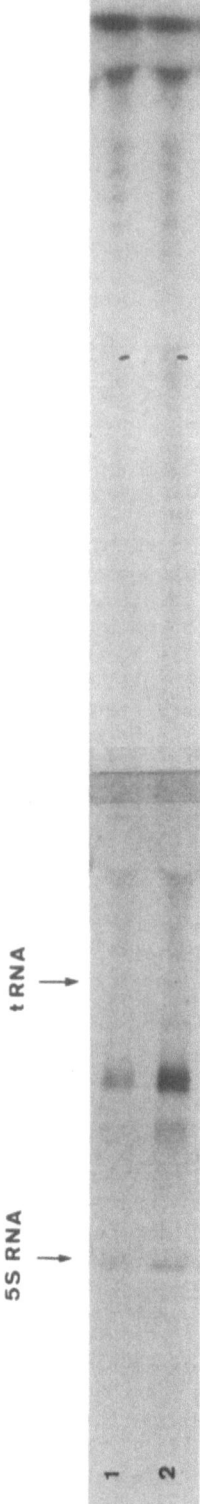

Fig. 3. Analysis of iRNA by polyacrylamide gel electrophoresis. iRNA purified from the 7S RNP (lane 1) and from the 17S RNP (lane 2) were labelled with $(5'-^{32}P)pCp$, precipitated with ethanol and dissolved in 20 mM Tris–HCl (pH 7.5), 8 M urea, 1 mM EDTA, 0.05% (w/v) xylene cyanol FF and 0.05% (w/v) bromophenolblue. After heating at 60°C for 30 s, RNA was electrophoresed on a 8% (w/v) polyacrylamide gel (0.38 x 800 mm). Gels were prerun overnight at 700 V. Electrophoresis was at 1700 V. The RNA length was determined using 5S rRNA (n = 120) and tRNA (n = 75) as standards.

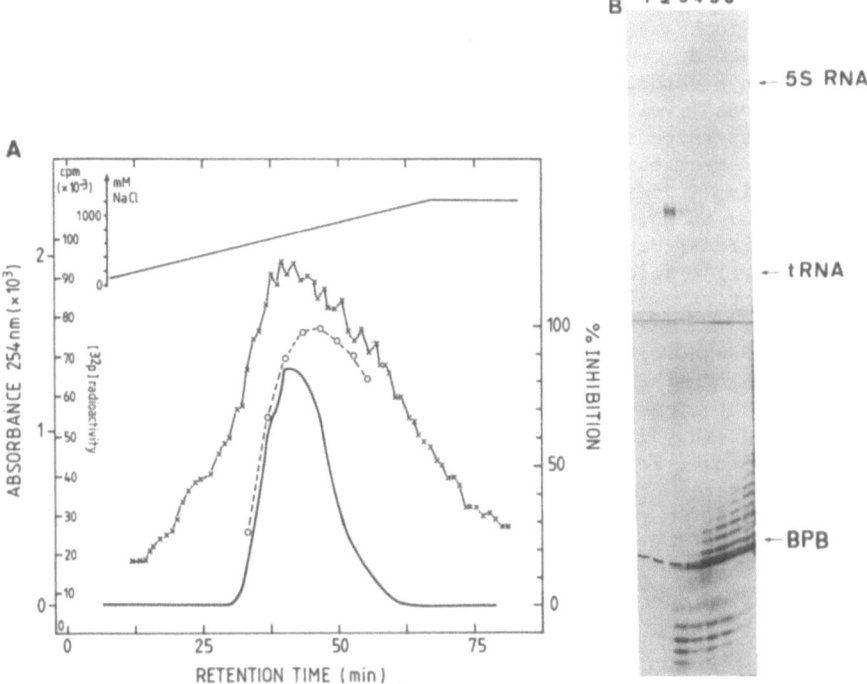

Fig. 4. A) Ion-exchange chromatography of iRNA on benzoylated
DEAE-cellulose. iRNA purified from the 7S RNP
fraction was bound to the column (0.6 x 10 cm) in
20 mM sodium acetate (pH 7.0), 100 mM NaCl and 10
mM MgCl$_2$. Bound RNA was eluted with a linear
gradient from 100 mM to 1200 mM NaCl (2 x 15 ml) at
a flow rate of 0.5 ml/min. Absorbance at 254 nm
(----), (^{32}P)pCp radioactivity (X——X), inhibition
of translation of globin mRNA in rabbit reticulocyte
lysate (0----0).

B) Analysis of gradient fractions by polyacrylamide
gel electrophoresis.
RNA of pooled fractions was desalted, precipitated
with ethanol and dissolved in 20 mM Tris-HCl (pH
7.5), 8 M urea, 1 mM EDTA, 0.05% (w/v) xylene cyanol
FF and 0.05% (w/v) bromophenol blue. After heating
at 60°C for 30 s, RNA was electrophoresed on a 8%
(w/v) polyacrylamide gel (0.38 x 800 mm).
Lane 1: 35-37 min, lane 2: 38-40 min, lane 3: 41-43
min, lane 4: 44-46 min, lane 5: 47-49 min, lane 6:
50-52 min.

Research of the Belgian National Fund for Scientific Research. M.A. is a
fellow of the I.W.O.N.L.

REFERENCES

1. W. B. Busa, J. H. Crowe and G. B. Matson, Intracellular pH and the

metabolic status of dormant and developing Artemia embryos, Arch. Biochem. Biophys. 216:711 (1982).

2. J. H. Crowe, L. M. Crowe, L. Drinkwater and W. B. Busa, Intracellular pH and anhydrobiosis in Artemia cysts, in "Artemia Research and its Applications", Vol. 2, W. Decleir, L. Moens, H. Slegers, P. Sorgeloos and E. Jaspers, eds., Universa Press, Wetteren (1987).

3. R. Perona, B. Ezquieta and C. G. Vallejo, The degradation of yolk in Artemia, in "Artemia Research and its Applications," Vol. 2, W. Decleir, L. Moens, H. Slegers, P. Sorgeloos and E. Jaspers, eds., Universa Press, Wetteren (1987).

4. A. H. Warner, The role of proteases and their control in Artemia development, in "Artemia Research and its Applications", Vol. 2, W. Decleir, L. Moens, H. Slegers, P. Sorgeloos and E. Jaspers, eds., Universa Press, Wetteren (1987).

5. N. Jeyaraj, S. Talib, J. Louis, C. Susheela and K. Jayaraman, Occurrence of poly(A) polymerase in particles rich in poly(A) RNA in the developing embryos of Artemia, in "The Brine Shrimp Artemia", Vol. 2, G. Persoone, P. Sorgeloos, O. Roels and E. Jaspers, eds., Universa Press, Wetteren (1980).

6. L. Sastre and J. Sebastian, Developmental changes in poly(A) polymerase activity in Artemia, Eur. J. Biochem. 135:69 (1983).

7. A. G. McLennan and M. Prescott, Diadenosine 5',5'''-P[1], P[4]-tetraphosphate in developing embryos of Artemia, Nucleic Acids Res. 12:1609 (1984).

8. H. Slegers, E. De Herdt, E. Piot, H. Backhovens, C. Thoen, L. Van Hove, E. Roggen and M. Aerden, Activation of stored messenger ribonucleo-proteins: Identification and function of the proteins associated with non-polysomal poly(A)-containing messenger ribonucleoproteins of cryptobiotic gastrulae of Artemia sp., in "Biochemistry and Cell Biology of Artemia", T. H. MacRae, J. C. Bagshaw and A. H. Warner, eds., CRC Press, Boca Raton (1988).

9. E. De Herdt, E. Piot, A. Wahba and H. Slegers, Initiation factor eIF2 associated with non-polysomal poly(A)-containing messenger ribo-nucleoproteins of cryptobiotic gastrulae of Artemia salina, Eur. J. Biochem. 151:455 (1985).

10. C. Thoen, L. Van Hove, E. Piot and H. Slegers, Purification and char-acterization of the messenger ribonucleoprotein-associated casein kinase II of Artemia salina cryptobiotic gastrulae, Biochim. Biophys. Acta 783:105 (1984).

11. C. Thoen, L. Van Hove and H. Slegers, Identification of the substrates of the casein kinase II associated with non-polysomal messenger ribonucleoproteins of A. salina cryptobiotic embryos, Mol. Biol. Rep. 11:69 (1986).

12. L. Van Hove, C. Thoen, P. Cohen and H. Slegers, Dephosphorylation of cytoplasmic non-polysomal messenger ribonucleoproteins from crypto-biotic gastrulae of Artemia salina, Biochem. Biophys. Res. Commun. 131:1241 (1985).

13. G. M. Hathaway and J. A. Traugh, Regulation of casein kinase II by 2, 3 bisphosphoglycerate in erythroid cells, J. Biol. Chem. 259:2850 (1984).

14. L. M. V. Raja, J. Louis and K. Jayaraman, Demonstration of the formation of 1,3 diphosphoglyceric acid as an intermediate in the high energy phosphate metabolism during reinitiation of development in Artemia, Biochim. Biophys. Acta 723:410 (1983).

15. M. Prescott and A. G. McLennan, The protein kinase activity associated with the major bis(5'-adenosyl) tetraphosphate-binding protein of Artemia, in "Artemia Research and its Applications", Vol. 2, W. Decleir, L. Moens, H. Slegers, P. Sorgeloos and E. Jaspers, eds., Universa Press, Wetteren (1987).

16. E. Piot, H. Backhovens and H. Slegers, The inhibitor ribonucleoprotein of poly(A)-containing non-polysomal messenger ribonucleoprotein of A. salina cryptobiotic embryos, FEBS Letters 175:16 (1984).

KINETIC AND THERMODYNAMIC PARAMETERS OF <u>ARTEMIA</u> RIBOSOME

SUBUNIT INTERACTIONS: THE EFFECTS OF POLYAMINES

Dixie J. Goss, Thomas C. Becker and Donna J. Rounds

Department of Chemistry
Hunter College of the City University of New York
New York, N.Y.

INTRODUCTION

Virtually all eucaryotic cells contain significant amounts of the polyamines-- spermine, spermidine, and putrescine. The physiological function of these amines is not well understood at the molecular level, although recent studies have demonstrated that polyamines are required for cellular growth and differentiation and that the concentration within the cell is highly regulated[1,2]. In quiescent cells polyamines are present in very small amounts, although polyamine activity can be dramatically increased within a few hours of exposure to trophic stimuli[3]. The pathway for polyamine biosynthesis has been determined, detailed descriptions are given in several recent reviews[4,5].

Studies of the effects of polyamines on protein synthesis and ribosomes have yielded a variety of results. Spermidine reduces aggregation of ribosomal subunits in low concentrations[6] but is also a powerful promoter of association[7]. Polyamines at high concentrations inhibit protein synthesis, but stimulate it at low concentrations[8]. Polyamines effectively compete for Mg^{2+} binding sites on ribosomes and, like Mg^{2+}, bind to the phosphate groups of the rRNA in a non-covalent charge neutralization reaction. However, the effects of polyamines differ from those of Mg^{2+}; rRNA must retain some Mg^{2+} in order to maintain its structural and functional integrity[9]. Further, the total charge neutralization during subunit ribosomal association by Mg^{2+} and polyamines combined is much less than the charge neutralization by Mg^{2+} alone[9].

Dry cysts of <u>Artemia</u> are metabolically inactive, and as such, are nearly devoid of protein initiation factors, elongation factors, and endogenous mRNA. The ribosomes exist almost exclusively as uncomplexed 80S monomeric particles[10,11]. Furthermore, as quiescent cells, <u>Artemia</u> ribosomes are low in polyamines compared to ribosomes from other sources. Data showed that salt washed <u>Artemia</u> ribosomes contain less than half the amount of polyamines per mg ribosomes compared to reticulocyte salt-washed ribosomes[3]. For these reasons, ribosomes and subunits from the dry cysts of <u>Artemia</u> are particularly useful for studying partial reactions of protein synthesis.

A primary event of initiation of protein synthesis in eucaryotes and procaryotes is the Mg^{2+}-dependent reversible association and dissociation of the ribosomal subunits. The contribution of cations in controlling the position of equilibrium of ribosomal subunits has been well established. Divalent cations such as Mg^{2+} and polyvalent cations such as spermine and spermidine facilitate association; on the other hand, monovalent cations such as K^+ facilitate dissociation. Furthermore, eucaryotic polypeptide chain initiation factor 3 (eIF3) is believed to shift the equilibrium towards dissociation as well as aid in the formation of the preinitiation complex with the 40S subunit, Met-tRNA, and mRNA. But the details of these cationic contributions and the role of eIF3 in initiation remain to be clarified. Elucidation of the possible translational control mechanisms and the pathway of initiation requires a detailed knowledge of the kinetic parameters for the various reactions. In recent studies we have used laser light scattering to examine the kinetics and thermodynamics of ribosomal subunit interactions. We reported the effects of protein initiation factor eIF3 on the equilibrium and kinetics of eucaryotic ribosomal subunit association[12]. We also reported the first rate constants obtained for the binding of unmodified eIF3 to the 40S ribosomal subunit and the 80S monosome using ribosomes and initiation factors from wheat germ[13]. In the present study, we examine the effect of polyamines on Artemia ribosomal subunit interactions.

MATERIALS AND METHODS

Artemia 80S ribosomes were prepared as described by MacRae et al.[14]. Ribosomes were stored under liquid nitrogen in buffer consisting of 20 mM tris (hydroxymethyl)-aminomethane hydrochloride (Tris-HCl), pH 7.6, 25% glycerol, 100 mM KCl, 0.1 mM EDTA and 9 mM $MgCl_2$. For most experiments, ribosomes were used within 48 hours of preparation. Buffer B consisting of 20 mM HEPES, pH 7.6, 1 mM dithiothreitol, 75 mM KCl with $MgCl_2$ and polyamines as indicated was used for kinetic experiments. Tris (Trizma) base was a product of Sigma Chemical Co. All other chemicals were reagent grade.

Determination of Ribosome and Subunit Concentration

Ribosomes were assumed to have an absorbance of 0.121 at 260 nm for a 0.001% solution[15]. The molecular weights of the 80S ribosome and the 60S and 40S subunits were assumed to be 3.8×10^6, 2.4×10^6 and 1.4×10^6, respectively, as reported by Nieuwenhuysen et al.[16].

Kinetic Measurements

Association and dissociation reactions for Artemia ribosomal subunits were monitored by 90° light scattering. Kinetic measurements were performed using a Hi-Tech SF-51 stopped-flow device interfaced to a Zenith Z-100 computer with 768K memory and dual disk drives. The light source for the stopped-flow was a Liconix 4210 NB Helium-Cadmium laser with either 325 or 441 nm optics.

The ribosome association reaction was measured by flowing ribosomal subunits in Buffer B containing 0.5 mm Mg^{2+} with Buffer B containing 0.5 mM Mg^{2+} and either 0.25, 0.5, 1.0 or 2.0 mM spermine, spermidine or putrescine. Control reactions without polyamines were monitored. The polyamine was then added to the flow syringe, allowed to equilibrate, and the association reaction again monitored. Association kinetics were also measured as a function of temperature. Experiments were carried out in the stopped-flow device maintained at the indicated temperature by a Fisher Scientific Model 900 circulating waterbath. Values of pH were determined at the appropriate temperature prior to each experiment. Apparent rate constants and equilibrium constants were obtained under these conditions for four different temperatures and constant pH.

The ribosome dissociation reaction was measured in the stopped-flow utilizing unequal driving syringes to give a 1:6 dilution. According to Le Chatelier's principle, such a dilution will cause dissociation of some of the complex.

Data Fitting

The values obtained for ΔV, the normalized voltage change for a light-scattering experiment, are directly proportional to the fraction of 80S ribosomes[7]. Equilibrium constants were determined both by the ratio of the rate constants ($K_1 = K_1/K_{-1}$), and from the amplitude of the kinetic curves. The amplitude of the kinetic curves is related to the concentration of species (see)[12]. The values of ΔG were calculated from the standard equation $\Delta G = -RT\ln K_{eq}$. For each sample a minimum of 8 kinetic runs was averaged by computer, thus providing 900 raw data points to generate a curve suitable for data fitting. The minimum of the response function for theoretical models was found by the Fletcher-Powell nonlinear minimization algorithm[17]. For each apparent minimum, an estimate of the standard errors in the parameters was obtained by a statistical routine, which gives the variance-covariance matrix of the parameters[18].

RESULTS

Polyamine Effects on the Association of the 60S and 40S Ribosomal Subunits

The effects of varying the polyamine concentration for the association reaction between the ribosomal subunits have been determined. Figure 1 shows the spermidine-induced association reaction monitored by light-scattering. Table 1 depicts the rate constants obtained for various spermine and spermidine concentrations. The rate constants for association were calculated at 0.25, 0.5, 1.0, 2.0 mM polyamine. As the ion concentration of spermidine was increased from 0.25 to 2.0 mM in the presence of 75 mM KCl and 0.5 mM $MgCl_2$ at 29°C the association rate was relatively unchanged. For spermine, the rates increased slightly with increasing polyamine concentration.

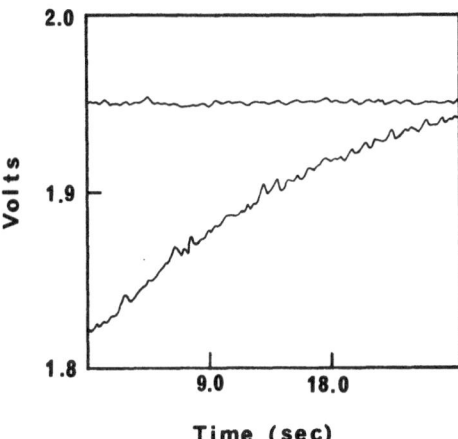

Fig. 1. Association of 25 nM 40S and 60S ribosome subunits with 1.0 mM spermidine at 29°C. Rate constants are given in Table 1.

Table 1. Rate Constants for <u>Artemia</u> Ribosomal Subunit Association[a]

| Concentration | K_1 ($\mu M^{-1} s^{-1}$) | |
	Spermine	Spermidine
0.25 mM	2.61	4.10
0.5 mM	3.31	4.22
1.0 mM	4.22	4.45
2.0 mM	--	3.50

[a]Association was induced by a shift in polyamine concentration from 0 to the indicated values. Temperature of the reaction was 29°C. Uncertainties in parameters are approximately 15% of the value reported, calculated from fitting an average of eight kinetic runs.

The extent of association of ribosomal subunits was determined from the amplitude of the light-scattering signal and previous ultracentrifugation experiments[6]. The theoretical maximum for the light-scattering signal change can be calculated as described elsewhere[7]. The molecular weights reported by Nieuwenhuysen et al.[16] were used to estimate the change in light-scattering signal of the ribosome solutions. The theoretical change in scattering intensity after the background is subtracted for completely dissociated ribosomes going to 100% associated 80S ribosomes is $I_{80S} = 1.87 \ I_{40S + 60S}$.

Thermodynamic Parameters

These formal thermodynamic parameters were not derived from calorimetric measurements, but rather through the use of kinetic techniques under the assumption that the rate-determining step of the observed reactions was the actual association or dissociation of the ribosomal subunits. The rate constants for association and dissociation reactions were calculated at 1 mM spermidine with 75 mM K^+ and 0.5 mM Mg^{2+}, for four temperatures. Table 2 shows the association rate constants obtained for the temperature range of 2.5°C to 30°C at pH 7.5. Table 3 shows the dissociation rate constants obtained for the same temperature range and pH. An Arrhenius plot of ln k vs 1/T for the association and dissociation rate constants was used to calculate the activation energies. Figure 2 compares the activation energies for the association reactions with spermidine and Mg^{2+}. Figure 3 shows an Arrhenius plot for the dissociation reaction. ΔH and ΔS were calculated from the slope and intercept of the Arrhenius plots where: slope = $-E_a/R$ and $\Delta H = E_{aon} - E_{aoff}$; $\Delta S/R$ is given by the difference of the intercept of a plot of ln k_{on} vs. 1/T minus the intercept of a plot of ln k_{off} vs. 1/T. These equations gave values of approximately 5.60 kcal/mol and 52.0 cal/(mol deg) for ΔH and ΔS, respectively. The equilibrium constants, determined by the ratio of the rate constants, ranged from 9.1 x 10^6 M^{-1} at 2.5°C to 21.9 x 10^6 M^{-1} at 30°C. The values obtained for K_{obsd} were used to calculate $\Delta G°$ for complex association. A relatively low activation energy for association and the counterion dependence are consistent with a diffusion controlled reaction governed mainly by electrostatic interactions.

DISCUSSION

Ribosomal RNAs are polyanions which develop a strong electrostatic

Table 2. Temperature Dependence of Spermidine-Induced Ribosomal Subunit Association[a]

T(°C)	k_{+1} (μM^{-1} s^{-1})
2.5°C	0.39
13°C	0.59
23°C	0.83
31°C	3.50

[a]Association was induced by a shift in spermidine concentration from 0 to 2 mM. Uncertainties in parameters are approximately 15% of the value reported, calculated from fitting an average of eight kinetic runs.

potential. Interactions with counterions modulate this electrostatic potential according to the polyelectrolyte theory[19,20]. The mechanism of association of ribosomal subunits has been postulated to be a charge neutralization reaction. Our data suggest that the interaction of the subunits with Mg^{2+} and polyamines is predominantly electrostatic, due to charge neutralization, based on the signs and magnitude of the thermodynamic terms, and the association/dissociation rates constants. In previous work we examined the Mg^{2+}-induced association and dissociation of Artemia ribosomal subunits[6]; in the present work we examined the polyamine-induced effects on Artemia ribosomal subunits.

Artemia ribosomal subunits associated with Mg^{2+} only are relatively unstable; however, with a small addition of polyamine the reaction is stabilized. Spermine was the most effective polyamine in stabilizing the ribosomes, spermidine had a moderate effect while putrescine was inactive. The greater effectiveness of spermine than spermidine suggests that spermine may bind with a higher affinity than spermidine to ribosomes.

Equilibrium between the ribosome and its subunits is sensitive to small increases in counterion concentration. Polyamine-associated subunits take longer to achieve equilibrium than Mg^{2+}-associated subunits. (The dissociation reactions showed no concentration dependence for either Mg^{2+} or spermidine as expected for the reaction going to completion). Anticooperative behavior among bound counterions may account for the extra time

Table 3. Temperature Dependence for Ribosome Dissociation[a]

T(°C)	k_{-1} (s^{-1})
2.5°C	0.043
13°C	0.060
23°C	0.043
31°C	0.047

[a]Dissociation was induced by a 1:6 dilution. Uncertainties in the parameters are approximately 25% of the value reported, calculated from fitting an average of eight kinetic runs.

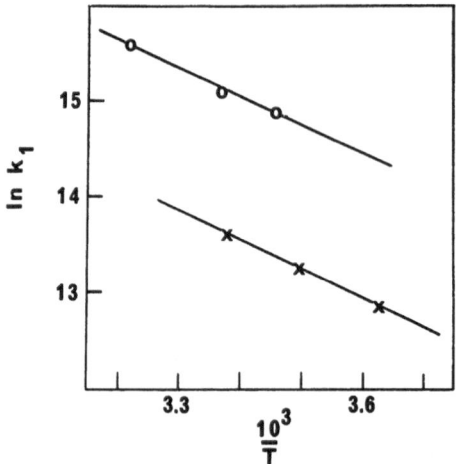

Fig. 2. Arrhenius plot of ln k_1 vs. 1/T for 40S and 60S
ribosome subunits associated with, (x) 2 mM
spermidine, and (o) 9 mM Mg^{2+}. Activation energies were 6.0 and 5.5 kcal/mol for spermidine
and Mg^{2+}, respectively.

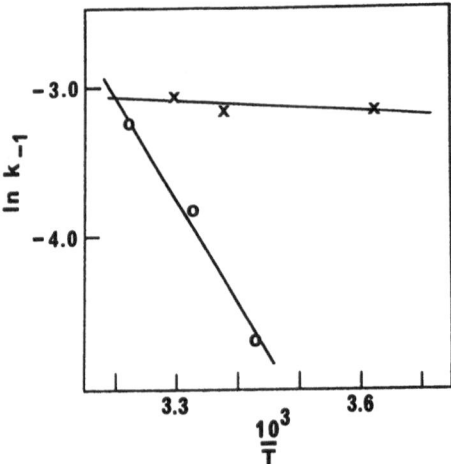

Fig. 3. Arrhenius plot of ln k_{-1} vs. 1/T for 80S ribosome
particles. Dissociation was induced by a shift
in [spermidine] from 3 mM to 0.5 mM (x), and by
a shift in [Mg^{2+}] from 6mM to 0.2 mM (o). Activation energies were 0.4 and 14 kcal/mol for
spermidine and Mg^{2+}, respectively.

to achieve equilibrium and the reduced rate constants compared with Mg^{2+}-associated subunits.

With Mg^{2+}, the association of subunits is exothermic with an enthalpy term that contributes more to the free energy than the corresponding entropy term. With spermidine, ΔH decreases in magnitude and, in fact, changes sign; association is endothermic. In this case, the entropy term is the primary contributor to the free energy term. Entropy contributions may come from either the solvent, the ribosomal proteins, or the rRNA and may depend upon the specificity of the interactions. Pure electrostatic interactions should not participate in the selectivity of recognition between these polymer groups since an interaction between a positive and negative charge should not depend on the nature of the base. However, long-range interactions may depend on the propagation of charge through the nucleic acid, protein or solvent matrix, consequently electrostatic interactions may be involved in the formation of hydrogen bonds which are expected to be directional.

The model used for fitting these data does not distinguish between a site-specific model, where the cation binds to specific sites on the ribosome, or an electrostatic model. However, bound Mg^{2+} exists in nearly the same environment as Mg^{2+} free in solution: the bound state is almost completely hydrated and has free translational and rotational mobility.

One can visualize the ribosomal subunit association reaction as a hypothetical two step process, A $->$ B $->$ C. Step A represents a hydrophobic interaction, which involves the mutual penetration of hydration layers. This hydrophobic process is a result of the tendency of water to form more ordered structure around nonpolar hydrocarbon groups and would involve mainly ribosomal proteins. Estimates of the thermodynamic parameters for this process would suggest ΔG negative, ΔH positive and ΔS positive. In the second step, B $->$ C, the hydrophobically associated species participate in further interactions which yield a fully associated complex. This second step could be considered to involve mainly ionic interactions and hydrogen bonds. These interactions are characterized by negative ΔS and ΔH values for hydrogen bonds and relatively small values of either sign for H resulting from electrostatic interactions.

Polyamines, as compared to Mg^{2+}, cause shifts in the ΔH from negative to positive. The reaction goes from being enthalpy- to entropy-controlled; consequently, the thermodynamic driving force of association must change. Low levels of spermidine and Mg^{2+} may cause a specific association governed partially by hydrogen bonds and electrostatic interactions. These are accounted for by the A $->$ B $->$ C model where the final step dominates the thermodynamic parameters. At higher spermidine concentrations, governed by hydrophobic interactions, the reaction may be less specific and hence translational misincorporation increases. In this case, because of the reduced electrostatic and possibly hydrogen bonding caused by polyamines, the first step, A $->$ B, dominates the thermodynamic parameters. This also accounts for reduced stability of ribosomes associated with polyamines[9]. Such a process was described by Ross and Subramanian[21] for protein-ligand interactions.

Acknowledgements

We wish to thank Patricia Reinhold for her expert technical assistance. This work was supported by PSC-CUNY faculty research award, by an American Heart Association-NYC Affiliate investigator award (to D.J.G.), by NSF Grant 8607070, and by the RCMI Program of NIH at Hunter College (Grant RR03037).

REFERENCES

1. U. Bachrach, Effect on growth processes, in: "Function of Naturally Occurring Polyamines", Academic Press, New York (1973).
2. D. H. Russell, Polyamines in growth - normal and neoplastic, in: "Polyamines in Normal and Neoplastic Growth", Raven Press, New York (1973).
3. G. Kramer and B. Hardesty, Ribosomes from Artemia cysts in cell-free translation of eukaryotic mRNA, in: "The Brine Shrimp Artemia", G. Persoone, P. Sorgeloos, O. Roels, and E. Jaspers, eds., Universa Press, Wetteren, Belgium (1980).
4. A. E. Pegg and P. P. McCann, Polyamine metabolism and functions, Am.J. Physiol. 243:212 (1982).
5. C. W. Tabor and H. Tabor, Polyamines, Annu. Rev. Biochem. 53:749 (1984).
6. D. J. Goss and T. Harrigan, Magnesium ion dependent equilibria, kinetics, and thermodynamic parameters of Artemia ribosome dissociation and subunit association, Biochemistry 25:3690 (1986).
7. H. Gorisch, D. J. Goss and L. J. Parkhurst, Kinetics of ribosomes dissociation and subunit association studied in a light-scattering stopped-flow apparatus, Biochemistry 15:5743 (1976).
8. C. L. Woodley and A. J. Wahba, The development of a translation system to examine mRNA and messenger ribonucleoproteins from Artemia, in: "The Brine Shrimp Artemia," G. Persoone, P. Sorgeloos, O. Roels and E. Jaspers, eds., Universa Press, Wetteren (1980).
9. J. M. Sperrazza and L. L. Spremulli, Quantitation of cation binding to wheat germ ribosomes; influences on subunit association equilibria and ribosome activity, Nucleic Acids Res. 11:2655 (1983).
10. T. Hultin and J. E. Morris, The ribosomes of encysted embryos of Artemia salina during cryptobiosis and resumption of development, Devel. Biol. 17:143 (1968).
11. A. Golub and J. S. Clegg, Protein synthesis in Artemia salina embryos. l. Studies on polysomes, Devel. Biol. 17:644 (1968).
12. D. J. Goss, D. J. Rounds, T. Harrigan, C. L. Woodley and A. J. Wahba, Effects of eucaryotic initiation factor 3 on eucaryotic ribosomal subunit equilibrium and kinetics, Biochemistry 27:1489 (1988).
13. D. J. Goss and D. J. Rounds, A kinetic light-scattering study of wheat germ protein synthesis initiation factor 3 to 40S ribosomal subunits and 80S ribosomes, Biochemistry 27:3610 (1988).
14. T. H. MacRae, M. Roychowdury, K. J. Houston, C. L. Woodley and A. J. Wahba, Protein synthesis in brine shrimp embryos: dormant and developing embryos of Artemia salina contain equivalent amounts of chain initiation factor 2, Eur. J. Biochem. 100:67 (1979).
15. P. Nieuwenhuysen and J. Clauwaert, Physicochemical characterization of ribosomal particles from the eukaryote Artemia, J. Biol. Chem. 256:9623 (1981).
16. P. Nieuwenhuysen, F. Devoeght and J. Clauwaert, The molecular weight of Artemia ribosomes, as determined from their refractive-index increment and light-scattering intensity, Biochem. J. 197:689 (1981).
17. R. Fletcher and M. J. D. Powell, A rapidly convergent descent method for minimization, Comput. J. 6:163 (1963).
18. N. A. Draper and H. Smith, "Applied Regression Analysis", Wiley, New York (1966).
19. J. S. Kliber, G. Hui Bon Hui, P. Douzou, M. Graffe and M. Grunberg-Manago, Implications of electrostatic potentials on ribosomal proteins, Nucleic Acids Res. 3:3423 (1976).
20. G. S. Manning, The molecular theory of polyelectrolyte solutions with applications to the electrostatic properties of polynucleotides, Q. Rev. Biophys. 11:179 (1978).
21. P. D. Ross and S. Subramanian, Thermodynamics of protein association reactions: forces contributing to stability, Biochemistry 20:3096 (1981).

STUDY OF POLY(ADP-RIBOSYL)ATION IN ARTEMIA EMBRYOS AT VARIOUS STAGES OF

DEVELOPMENT

M. M. Gorski and A. H. Warner

University of Windsor
Windsor, Ontario
Canada N9B 3P4

ABSTRACT

Artemia embryos at different stages of development were analyzed for
chromatin associated poly(ADP-ribose) synthetase activity. The enzyme was
partially purified and characterized to determine the average polymer
length and acceptor proteins. Using (^{3}H)-NAD as substrate, the amount of
acid insoluble radio-activity bound to protein during incubation was determined. The reaction was dependent on divalent cations and Mg^{++}, Ca^{++} and
Zn^{++} showed similar activities. However, Mn^{++} did not support the reaction.
The newly synthesized polymers were stripped from their protein acceptors
by alkali treatment and purified on a dihydroxyboronate column. The polymers were hydrolyzed with snake venom phosphodiesterase and alkaline phos-
phomonoesterase to yield adenosine, ribosyladenosine and diribosyladenosine.
These products were analyzed by reversed phase high performance liquid
chromotography (HPLC) to determine chain lengths and the degree of
branching. The chain lengths obtained were similar for all developmental
stages studied and this was found to be less than one unit long, suggesting
a high rate of polymer breakdown during the synthesis. Acceptor proteins
were determined by SDS-polyacrylamide gel electrophoresis and fluorography
of the resulting gel. Our results show that during the first 12 hours of
development many proteins serve as acceptors of ADP-ribose, while at 24
and 48 hour stages of development, the main acceptor proteins are histones
and the polymerase itself.

ACKNOWLEDGEMENT

Research supported by NSERC of Canada

ADVENTURES AND MISADVENTURES IN THE GENOME OF ARTEMIA

Joseph C. Bagshaw

Worcester Polytechnic Institute

Worcester, MA 01609

INTRODUCTION

As this and other recent publications clearly show, Artemia is an exceptionally interesting and useful biological system for a wide variety of experimental studies [1,2]. Among the areas of active research interest are biochemistry, cell biology, molecular biology, developmental biology, morphology, ecology, and population biology. Much of the research involving Artemia, especially in the areas of biochemistry, cell and molecular biology, focusses on developmental aspects. This is a natural outgrowth of the fact that dormant encysted gastrulae are readily available commercially at any time and in large quantity and can be easily raised in the laboratory to produce swimming nauplis larvae. This easily accessible period of development has attracted the attention of quite a few investigators. Despite its short generation time, Artemia has been relatively little studied by classical or genetic means, perhaps owing to the difficulty of generating and isolating mutant strains in comparison with more familiar organisms such as Drosophila and Caenorhabditis. Over the past decade, an increasing number of research groups has begun to explore the molecular genetics of Artemia. As in other areas of investigation, many of these studies are also developmental in nature, aimed at understanding the mechanisms that regulate gene expression during Artemia development.

Applications of recombinant DNA methods, have created powerful new tools for studying phylogeny and evolution, provided that the biological samples are available in sufficient quantity and diversity. Restriction endonucleases can be used to examine specific gene structure in comparisons of individuals, populations, species or higher taxa. DNA sequencing allows a direct comparison of gene structure at its most fundamental level, and sequence data can be used to establish relationships and estimate evolutionary distances. Artemia is well suited for molecular studies of phylogeny and evolution. Natural and transplanted populations exist on every continent except Antarctica, and many of these are available either commercially or through the Artemia Reference Center. Genomic DNA is easily isolated in relatively large quantity. A number of cloned DNAs isolated in my laboratory and others provide the molecular tools needed for systematic study of gene structure in various species and populations of Artemia.

Our adventures in the genome of Artemia began with studies of the structure and expression of Histone genes. These studies were prompted by the observation of McClean & Warner [3] that DNA synthesis does not occur between the resumption of development in post-gastrula embryos and hatching of pre-nauplius larvae. In somatic cells, histone synthesis is usually coupled to DNA replication [4]. However, this situation is not always found during early stages of development. Thus it was necessary to demonstrate that in Artemia histone synthesis does not occur during pre-emergence development when DNA replication is also not occuring, and that histone synthesis accompanies the resumption of DNA replication in hatched larvae. Radiolabelling studies of DNA and histone showed that this is indeed the case [5]. This led us to the cloning and characterization of the histone genes of Artemia franciscana and to the unexpected discovery that the 5S ribosomal RNA genes are interspersed with histone genes in Artemia. A series of fortuitous accidents led us to our current studies of the polymorphism of histone genes between different populations and species of Artemia, and to studies of repeated DNA sequences in Artemia.

HISTONE GENES AND GENE EXPRESSION

The first step in the isolation of the Artemia histone genes was the construction of a library of randomly generated genomic DNA fragments in the bacteriophage vector lambda Charon 4A. For this construction we used genomic DNA isolated from the Great Salt Lake strain of Artemia franciscana. We screened the library, using as a probe cloned Drosophila histone gene, cDm500, a generous gift of D. Hogness. In a screen of about 10,000 recombinant phage, we isolated and purified four separate recombinants containing histone genes. Even at this early point we realized there must be polymorphism of the histone genes either within the individual genome or within the population, because the four recombinant phage DNAs exhibited three distinct restriction fragment patterns upon digestion with the restriction enzyme Eco RI [6].

In the next phase of our work, we wished to determine whether the histone messengers that were being translated in hatched larvae were stored in dormant gastrulae or arose de novo during the course of development. By using one of our recombinant phage as a probe against Northern blots of whole cell RNA isolated from different stages of development, we showed that in fact histone messenger RNAs are not present in the dormant gastrulae of Artemia, and therefore must be transcribed at some time during post-gastrula development [5]. In these same experiments, we discovered that the recombinant phage containing histone genes also hybridized to a small, highly abundant RNA which we subsequently proved was the 5S ribosomal RNA. Using end-labelled 5S RNA as a probe, we found that each of our histone gene recombinant phage also contained a sequence homologous with 5S ribosomal RNA. Solution hybridization kinetic experiments conducted in collaboration with Dr. Jack Vaughn showed that the histone genes and the putative 5S ribosomal RNA genes are present in equal number in the Artemia genome [6]. This strongly suggested that the 5S-related sequences we had found were the true 5S RNA genes of Artemia and were indeed interspersed with histone genes.

In order to characterize the 5S RNA gene more fully, we determined the DNA sequence of the gene and its surrounding region (Fig. 1). This region contains a sequence of 120 base pairs synonymous with the previously published sequence of Artemia 5S ribosomal RNA [7]. Immediately flanking this sequence on the 3' side is a series of thymidine residues which is characteristic of termination signals for RNA polymerase III genes. Within the gene sequence are several pairs of guanine residues that have been proposed to be binding sites for the 5S- specific transcription factor IIIA [8]. These findings, together with the hybridization kinetic data, make it virtually certain that this is an authentic 5S ribosomal RNA gene,

278

```
          10          20          30          40          50          60
TCTAG AGGGA GAAGG TGAGG AGGTG GGTAC TTTAA AATAC CTTCC CGGGC ATACT TTAGC
          70          80          90         100         110         120
CTGTA GACCC ATCCC TGAAA GTTTC ATTTT TCTAA CCTAA ACCCT TTCTG AGATA TCAAG
         130         140         150         160         170         180
AAGTC AATTC ACTAG AATTT TACCA AAACA ATTAA TAAAC ATATT TTGAT TCTTT TTAAA
         190         200         210         220         230         240
TATTT GTCTA CTAAA AACCT TGGTT CACGA CTCTA AAACT AGGCA TTTTA TTTCT TTTCA
         250         260         270         280         290         300
GTTTG TGGAC CAACA ACTAT TTTGT TACAA AATTT CTCAG ACCAA CGGCC ATACC ACGTT
         310         320         330         340         350         360
GAAAG TACCC AGTCT CGTCA GATCC TGGAA GTCAC ACAAC GTCGG GCCCG GTCAG TACTT
         370         380         390         400         410         420
GGATG GGTGA CCGCC TGGGA ACACC GGGTG CTGTT GGCAT TTTTT TTGTT TTTGT TTTTT
         430         440         450         460         470         480
TTTCT TATAT ATATA TTTAT ATATT ATAAT GAATT ATTTT CAAGT AAAAT AATAT TCTTT
         490         500         510         520         530         540
AACAA ACTGC TTTCC TTATT CTAAT AGCGC CAAGC TGATT ACATC TTTTA TAAGA AAAAA
         550         560         570         580         590         600
GAAAA ACAGG AAGAG ACTAA GTTAC ATCTC C
```

Fig. 1. Sequence of the region including the 5S rRNA gene. The sequence synonymous with 5S rRNA is underlined. Dots and arrows indicate direct and inverted repeats, respectively. Open circles with dashed lines indicate G residues implicated in transcription factor binding.

and that these are the only 5S RNA genes in A. franciscana.

Both histone and 5S RNA genes are repeated in the genome of all eukaryotic cells studied to date, but in other organisms these genes are either scattered or present in separate tandemly repeated structures. Thus our discovery of 5S RNA genes interspersed with histone genes was completely novel at the time. Our publication of this finding was quickly followed by a report from Doolittle's group that in the copepod Calanus, 5S ribosomal RNA genes are interspersed with the nucleolar ribosomal RNA genes [9]. Although two observations certainly do not make a case, it may be that in the crustacea the location of 5S ribosomal RNA genes is unlike that in any other phylum.

A question of great interest is how this curious juxtaposition of 5S and histone genes came about. The most likely explanation is that the 5S RNA gene was already adjacent to histone genes or was transposed to that position prior to the amplification of the entire unit. The alternative explanation-that a 5S RNA gene was transposed to the exact same position in each of 100 copies of histone genes is far less plausible. Does the DNA sequence reveal any lingering evidence of an ancient transposition event? We can rule out the possibility of an RNA-mediated retrotransposition. The 5S gene is flanked on the downstream side by a T-rich region, not a A-rich region as would be expected in the case of retrotransposition. Both direct and inverted repeat sequences are found in the region flanking the 5S RNA gene. As shown in Fig. 1, a perfect seven base pair repeat (TTTTTCT) is found with one copy 194 base pairs upstream from the 5S gene and two tandem copies beginning 132 base pairs downstream from the gene. However, six of the seven base pairs are AT, and given the high AT content of the Artemia genome, it is not certain that this sequence is significant. An imperfect direct repeat (14 of 16 base pairs) is also present, the downstream copy of which includes the last five base pairs of the coding sequence. This sequence is also T-rich, but being more extensive may be more significant. These clues are tantalizing but inconclusive. A comparison of sequences flanking the 5S RNA genes in a variety of populations and species of Artemia might provide more definitive evidence regarding the evolution of this gene structure.

POPULATION POLYMORPHISM OF HISTONE AND RIBOSOMAL RNA GENES

Restriction fragment length polymorphisms (RFLPs) are valuable molecular tools for phylogenetic and evolutionary studies. Point mutations can lead to the loss or acquisition of restriction enzyme sites and insertions or deletions alter the spacing between restriction sites. When these mutations become fixed in a population, the result is changes in the pattern observed when restriction digests of genomic DNA are reacted with specific gene probes (RFLPs).

Our entry into the field of RFLP analysis was unintentional. About five years ago we noticed that the restriction pattern of the histone gene repeat unit was no longer the same as the structure we had originally determined. Our original cloned histone genes were derived from the Great Salt Lake strain of Artemia franciscana. Because of hatching problems with the Great Salt Lake strain, we had switched to using cysts distributed by San Francisco Bay Brand, Inc. These cysts were presumably harvested from artificial salterns in the vicinity of San Francisco Bay. It is well documented that the Great Salt Lake and San Francisco strains are fully cross fertile and are classified as the same species, Artemia franciscana [10]. We subsequently learned that cysts sold as San Francisco Bay Brand are not always from the San Francisco area, and that some lots had been imported from China. We confirmed this rumor by obtaining samples from several populations, including Tientsin, PRC, from the <u>Artemia</u> Reference Center. Examples of RFLP analysis using histone and ribosomal RNA gene probes are shown in Fig. 2. From these studies it was immediately evident that San Francisco Bay Brand lot #1113 (Lanes 3 and 8) is very different from the two samples of <u>Artemia franciscana</u>, and at least for histone genes this lot is identical to the sample from Tientsin, which we believe to be <u>Artemia parthenogentica</u>. Examination of the ribosomal RNA genes however (Panel B) shows that the San Francisco lot #1113 and Tientsin samples are not identical after all. It appears likely that this particular lot of San Francisco Bay Brand cysts was imported from some region in China but not from Tientsin. Thus our RFLP analysis can be used to characterize and identify populations. This is equally true in <u>Artemia franciscana</u> where clear differences can be seen in both histone and ribosomal RNA genes between the Great Salt Lake strain and the San Francisco Bay strain. The sample from the Philippines appears in all respects identical to the authentic San Francisco population. This is not surprising since <u>Artemia</u> raised commercially in the Phillippines were transplanted from San

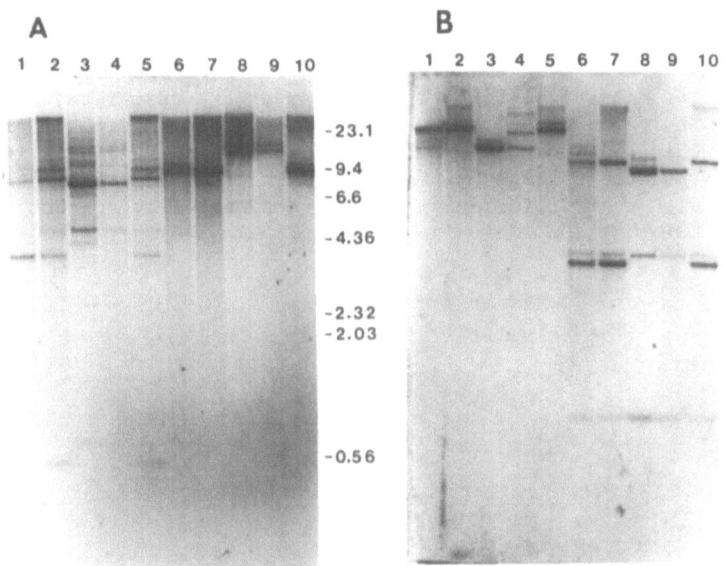

Fig. 2. Southern blot analysis of histone and ribosomal RNA genes. Genomic DNAs were isolated from 24 hr nau-plius larvae. Cysts were obtained from the follow-ing sources: lanes 1 and 6, Great Salt Lake, Utah, Sanders Brine Shrimp Co., lot #12715; lanes 2 and 7, San Francisco Bay Brand, lot #1808; lanes 3 and 8, San Francisco Bay Brand, lot #1113; lanes 4 and 9, Tientsin, China, from the <u>Artemia</u> Reference Center; lanes 5 and 10, Philippines, unspecified site, from the A.R.C. Genomic DNAs were digested with Eco RI (lanes 1-5) or Bam HI (lanes 6-10). Restriction fragments were separated by agarose gel electrophoresis, blotted onto a charged nylon mem-brane, and hybridized separately with recombinant plasmid DNAs containing either histone genes (panel A) or part of the ribosomal RNA gene (panel B). Sizes in kilobase pairs (kb) of marker DNA fragments (Hind III-digested lambda DNA) are shown to the right of panel A.

Francisco Bay (Patrick Sorgeloos, personal communication). This kind of analysis can obviously be used to establish relationships between populations. For example, the question arises, where did the San Francisco strain originate? Results shown in Fig. 2 indicate that the San Francisco population was not originally derived from the Great Salt Lake. Another likely source would be salterns in Mexico, Central or South America, from which Artemia cysts might have been transported by migrating birds. RFLP analysis of specific genes from these populations should enable us to answer this question.

REPEATED SEQUENCES IN THE ARTEMIA GENOME

The genomes of all eukaryotic organisms contain DNA sequences that are repeated anywhere from a few tens to millions of times. The existance and distribution of repeated sequences in the Artemia genome was first demonstrated by the hybridization studies of Dr. Jack Vaughn [11]. Sometimes highly repeated DNA sequences can be seen as discreet bands in restriction enzyme digests of genomic DNA. This requires either that the sequence be tandemly repeated, or that the sequence if dispersed, contains two uniformly spaced sites for a given restriction enzyme. A well known example of this phenomenon in Artemia is the satellite DNA that has been independently described by two different research groups [12,13]. This satellite is easily seen by digestion of genomic DNA with the appropriate restriction endonuclease. Another example of this phenomenon is shown in Fig. 3. In this experiment genomic DNA from the Great Salt Lake strain of Artemia franciscana and San Francisco Bay Brand lots 1808 (authentic) and 1113 (China) were digested with restriction enzymes and fractionated by agarose gel

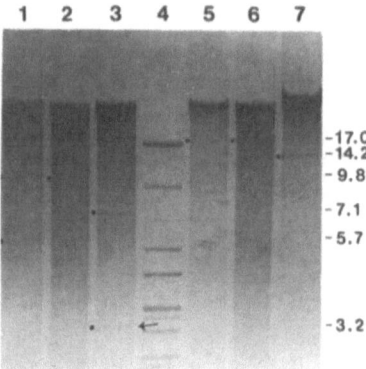

Fig. 3. Repeated sequences in the Artemia genome. Genomic
         DNA from the Great Salt Lake (lanes 1 and 5) and
         San Francisco (lanes 2 and 6) strains of A. franci-
         scana and an unspecified strain of A. partheno-
         genetica (lanes 3 and 7) were digested with Bam HI
         (lanes 1-3) or Eag I (lanes 5-7) and fractionated
         by electrophoresis in a 0.8% agarose gel. The gel
         was stained with ethidium bromide and photographed.
         Markers are lambda DNA digested with Eco RI and Bam
         HI (lane 4). Dots on the left side of each lane
         indicate bands that appear to be unique to one
         species or the other. The approximate sizes (kb)
         of these bands are shown to the right. The 3.2 kb
         band in lane three indicated by the arrow was cloned
         in plasmid pMB 48 (see Fig. 4A).

electrophoresis. Several characteristic bands can be seen in each prepara
tion, and in fact we now consider these bands to be diagnostic of both the
quality of the DNA preparation and of the enzyme digestion. Some of the
bands, which must be derived from highly repeated DNA sequences, are
unique to one or the other species. A particularly striking example of
this is the 14.2 kb band derived from A. parthenogenetica DNA by digestion
with the enzyme Eag I. These discrete bands are also forms of RFLPs that
can be used to characterize species, but we have never seen differences
in these bands between isolated populations of the same species. In order
to learn more about the nature of these repeated DNA sequences, we decided
to clone and study them, beginning with the 3.2 kb Bam HI fragment from
A. parthenogenetica shown in Fig. 3 (arrow, lane 3). We cloned the 3.2
kb Bam HI fragment simply by inserting fragments of this size range into
the plasmid vector pBR322 and selecting recombinants that contained repeated DNA sequences. We anticipated that most or all of these selected
recombinants would contain the 3.2 kb band of interest. To our surprise,
every isolate we examined contained a different inserted DNA fragment.
We chose several of these isolates to use as probes against Southern blots
of digested genomic DNA, and the results shown in Fig. 4 illustrate the
three general kinds of observations we have made. The plasmid used as
probe in panel A, pMB48, contains the 3.2 kb band that we originally set
out to clone (see lane 8). The fact that this clone hybridizes to a broad
spectrum of fragments in the Eco RI digest of the same DNA (lane 3) tells
us that this sequence is widely dispersed in the genome and not tandemly
repeated. Moreover, we can see that the same or a closely related sequence is also found in A. franciscana, but that its structure must be
significantly different in that it does not produce the characteristic
3.2 kb band fragment. Note also that lanes 2 and 7 contain ten times more
DNA than other lanes, thus the number of copies of this sequence in A.
franciscana is clearly much lower than in A. parthenogenetica, from which
it was originally cloned. Panel B shows a different kind of result for
plasmid pMB14. This sequence is also dispersed in the genome of A. parthenogenetica and hybridizes to a large number of genomic fragments. The
homologous sequence in A. franciscana is much lower in copy number and
present primarily as several discrete bands. Indeed, this sequence can
only be seen in A. franciscana by greatly increasing the quantity of DNA
on the blot (compare lanes 1 and 2). Panel C shows a particularly interesting result with plasmid pMB25. The sequence cloned in this plasmid is
also widely dispersed as shown by its hybridization to a large number of
Bam HI fragments. However hybridization to Eco RI fragments is almost
entirely limited to three discrete bands (lanes 3 and 4). Again, the sequence is present in A. franciscana but in a much reduced copy number, and
there is no evidence of the discrete structure seen in A. parthenogenetica.
The restriction patterns shown here suggest the possibility that these
Eco RI fragments represent one or more transposable element. If A. parthenogenetica is a decendent of A. franciscana, these sequences have been
greatly amplified during the evolution of the genus Artemia.

Figure 5 shows another completely unanticipated observation. In
collaboration with Dr. Tom MacRae, we isolated recombinant phage containing putative alpha and beta tubulin genes. At the time these recom-
were constructed we did not realize that we were working with DNA from A.
parthenogenetica. As a step toward repeating the cloning with A. franciscana DNA, we performed Southern blot analysis of genomic DNAs using our
tubulin clones as probes. To our great surprise, we discovered that these
clones hybridized to a broad spectrum of DNA fragments, indicating that
they also contain sequences repeated in the Artemia genome. Indeed, in
other experiments not shown here, we have directly demonstrated the presence
of repeated sequences in these clones, as well as tubulin genes. A particularly fascinating result is that phage t-beta-3 hybridizes to two discrete
and highly abundant fragments in the Eco RI digest. One of these genomic

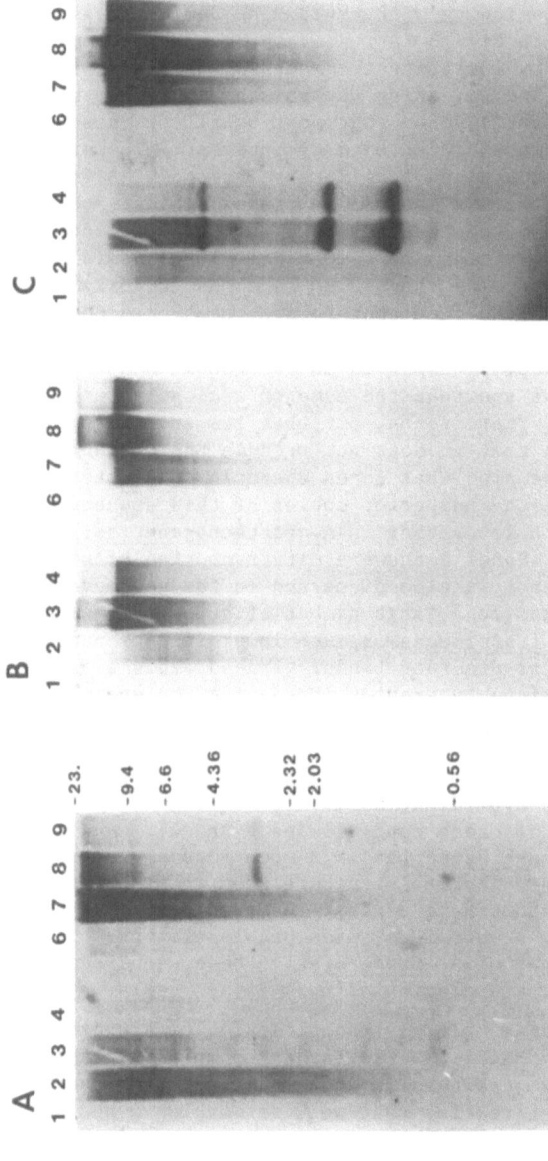

Fig. 4. Southern blot analysis of repeated DNA sequences. Sources of DNA were the same as in Fig. 2 except that lanes 2 and 7 contain approximately 10-fold more DNA than other lanes. A single blot was probed successively with recombinant plasmids pMB48 (panel A), pMB14 (panel B), and pMB25 (panel C). Marker fragment sizes (Hind III-digested lambda DNA) are shown to the right of panel A. DNA in lanes 1-4 was digested with Eco RI and in lanes 6-9 with Bam HI. (no lane 5 in any panel).

A

1 2 3

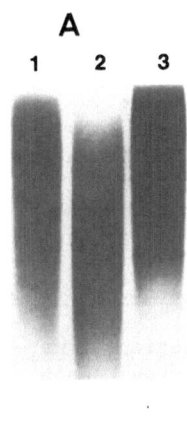

B

1 2 3

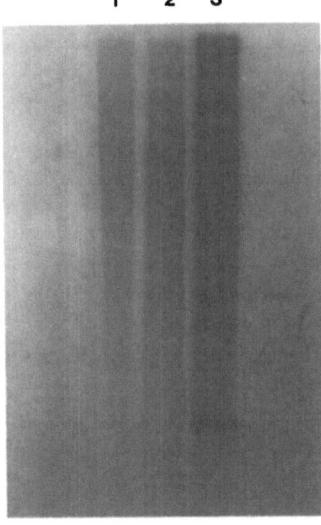

Fig. 5. Southern blot analysis of tubulin genes. Sources
of genomic DNA are the same as in Fig. 3. Eco RI
digests were fractionated on 1% agarose gels and
blotted onto charged nylon membranes. Panel A
shows a blot probed with pMLGα225, a plasmid containing a α-tubulin gene from A. parthenogenetica.
Panel B shows a different blot probed with tβ3,
a lambda phage containing a β-tubulin gene from
A. parthenogenetica. Marker fragment sizes (Hind
III-digested lambda DNA) are shown to the right
of each panel.

DNA bands also hybridizes with plasmid pMB25 (Fig. 4C, lanes 3 and 4). We
are not yet certain that recombinant beta 3 contains a true tubulin gene,
but we do know that it contains a protein-coding sequence because it
hybridizes with a cDNA probe. Thus we have an example in Artemia parthenogenetica of a highly repeated DNA sequence suggestive of a transposable
element, one copy of which is found in the vicinity of a protein coding
gene. Certainly this and other aspects of the repeated sequences in the
Artemia genome will be of great interest for further investigation.

ACKNOWLEDGEMENTS

The work reported here includes contributions by many students and
colleagues both past and present. The author gratefully acknowledges the
efforts of the following: Matt Andrews, Brian Perry, Paula Smolenski,
Mike Casasanta, Mitchell Sanders, Maryanne Lemaire, Lynda Gryzb, Gerry
Farley and Mike Buckholt. Drs. Jack Vaughn and Tom MacRae have been involved in several of the cloning and characterization studies. This
research was supported by NSF Grants PCM8111485, PCM8442443 and DCB8510471.

REFERENCES

1. T. H. MacRae, J. C. Bagshaw and A. H. Warner, in: "Biochemistry and Cell Biology of Artemia," CRC Press, Boca Raton (1988).
2. P. Sorgeloos, D. A. Bengtson, W. Decleir and E. Jaspers, in: "Artemia Research and its Applications," Vols 1, 2 and 3, Universa Press, Wetteren (1987).
3. D. K. McClean and A. H. Warner, Aspects of nucleic acid metabolism during development of the brine shrimp Artemia salina, Dev. Biol. 24:88 (1971).
4. L. H. Kedes, Histone genes and histone messengers, Annu. Rev. Biochem. 48:837 (1979).
5. J. C. Bagshaw, Developmentally regulated gene expression in Artemia: histone gene expression in newly hatched larvae, Dev. Genet. 3:41 (1982).
6. M. T. Andrews, J. C. Vaughn, B. A. Perry and J. C. Bagshaw, Interspersion of histone and 5S RNA genes in Artemia, Gene 51:61 (1987).
7. L. Diels, R. De Baere, A. Vandenberghe and R. DeWachter, The sequence of the 5S ribosomal RNA of the crustacean Artemia salina, Nucl. Acids Res. 9:5141 (1981).
8. D. Rhodes and A. Klug, An underlying repeat in some transcriptional control sequences corresponding to half a double helical turn of DNA, Cell 46:123 (1986).
9. G. Drouin, J. D. Hofman and W. F. Doolittle, Unusual ribosomal RNA gene organization in copepods of the genus Calanus, J. Mol. Biol. 196: 943 (1987).
10. L. S. Clark and S. T. Bowen, The genetics of Artemia. VII. Reproductive isolation, J. Heredity 67:385 (1976).
11. J. C. Vaughn, DNA reassociation kinetics analysis of the brine shrimp Artemia salina, Biochem. Biophys. Res. Commun. 79:525 (1977).
12. J. Cruces, M. L. G. Wonenburger, M. Diaz-Guerra, J. Sebastian and J. Renart, Satellite DNA in the crustacean Artemia, Gene 44:341 (1986).
13. G. Badaracco, L. Baratelli, E. Ginelli, R. Meneveri, P. Plevani, P. Valsanini and C. Barigiozzi, Variations in repetitive DNA and heterochromatin in the genus Artemia, Chromosoma 95:71 (1987).

STRUCTURE AND EXPRESSION OF RIBOSOMAL RNA GENES FROM <u>ARTEMIA</u>

Jaime Renart, Jesús Cruces, Margarita Diaz-Guerra, Elvira
Domínguez, Enrique Franco, Inés Gil, María Teresa Macías
Ignacio Palmero and Leandro Sastre

Instituto de Investigaciones Biomédicas del C.S1.C. and
Departamento de Bioquimica de la Universidad Autónoma
Arzobispo Morcillo, 4. 28029-Madrid, Spain

INTRODUCTION

Stable RNAs in the eukaryotic cell comprise several species all involved in the protein synthesis machinery. They are the 28S, 5.8S and 5S
in the large ribosomal subunit, the 18S in the small ribosomal subunit,
the 16S and 12S ribosomal RNAs in the mitochondrial ribosomes, and the
large families of tRNAs. The study of these RNAs is useful for several
reasons. One of them is the construction of evolutionary trees, based on
sequence comparisons [1]. A second important point regards their gene
structure, at least for the cytoplasmic ribosomal RNA genes. These genes
form repeated families, and so the question arises as the whole family is
maintained homogeneous, and how mutations are eliminated or fixed at a
rate not compatible with selective pressure mechanisms [2]. The structure
of the intergenic spacer, made up from small repetitions, also poses
questions about the initiation of transcription, in a system where transcription is almost species-specific [3]. It is in this direction that we
are approaching the study of these genes in <u>Artemia</u>. Studies on this
crustacean could help to fill a gap in molecular evolution, where
"arthropod" almost always means "insect".

MATERIALS AND METHODS

Cysts used came from San Francisco Bay Brand, and were purchased in
1987. For the type II rRNA experiments, cysts were of Chinese origin,
although also obtained from the same supplier as San Francisco Bay cysts
in 1982. Standard techniques of DNA manipulation were as described by
Maniatis et al. [4].

TYPE II RIBOSOMAL RNA GENES

We have already described in <u>Artemia</u> two types of rDNA repeats, of
16.5 and 12.2 kb respectively [5]. The larger one, type I, has been characterized more extensively and the promoter and initiation sites studied
by us and others [6,7].

Type II rDNA is found in Eurasian parthenogenetic populations; the

size of the repetition is only 12.2 kb, the main difference, being the
size of the intergenic spacer, of only 5.2 kb. All restriction sites
found in coding sequences are maintained in both types of repeats, whereas
sites in non-coding regions vary between them [5]. Restriction analysis
failed to detect any repetitive pattern in the intergenic spacer of type II
genes. However, the sequence reveals subrepeats, as shown in the scheme
of Fig. 1. These subrepeats are of different length, of 611, 460 and 364
bp for repetitions I', II' and III' respectively. Subrepeats II' and III'
are truncated forms of I'. Upstream of subrepeat III', there are several
50 bp repeats (we have detected six up to now), that are homologous to the
boundaries between subrepeats I' and II' and III'. The large repeat of
type II genes is very similar to those found in type I genes.

    In order to identify the initiation site of transcription, we have
done S1 analysis of type II repeats, as shown if Fig. 2. The initiation
site is located 710 bp 5' to the start of 17S RNA, inside suprepeat I',
as happens in type I genes [6]; the sequence of the initiation site is also
homologous to that of type I genes.

STRUCTURE OF THE 5S-HISTONE GENE COMPLEX

    In Artemia, 5S rRNA genes are found in a complex repeat unit with the
histone genes, as shown for the first time by Bagshaw and coworkers [8,9,
22]. We have focused on the genomic organization of these genes and also
on the sequences around the 5S rRNA [10]. We have found that 5S-histone
genes in Artemia are arranged in two different organizations, shown in
Fig. 3. One of them is 9 kb, whereas the other is 8.5 kb. All histones
account for 3.2 kb and are located around 500 bp upstream from the 5S gene
which spans 300 bp. The spacer of this complex unit is therefore 4.5-5 kb,
unusually large for these types of genes.

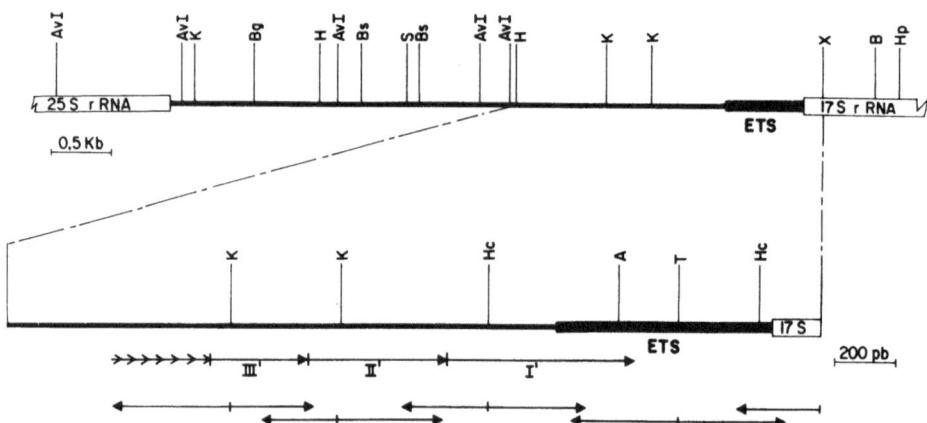

Fig. 1. Restriction map of the promoter region of type II rDNA genes.
         Double arrowheads indicate the strategy for sequencing. Sub-
         repeats I', II' and III' are also indicated, as well as the
         external transcribed spacer (ETS). Restriction endonucleases
         are; Av, Aval; K, Kpnl; Bg, Bgll; H, Hindlll; Bss, BstEll; S,
         Sall; X, Xbal; Hp, Hpal; Hc, Hincll; A, Accl; T, Taql; B, BamHl.

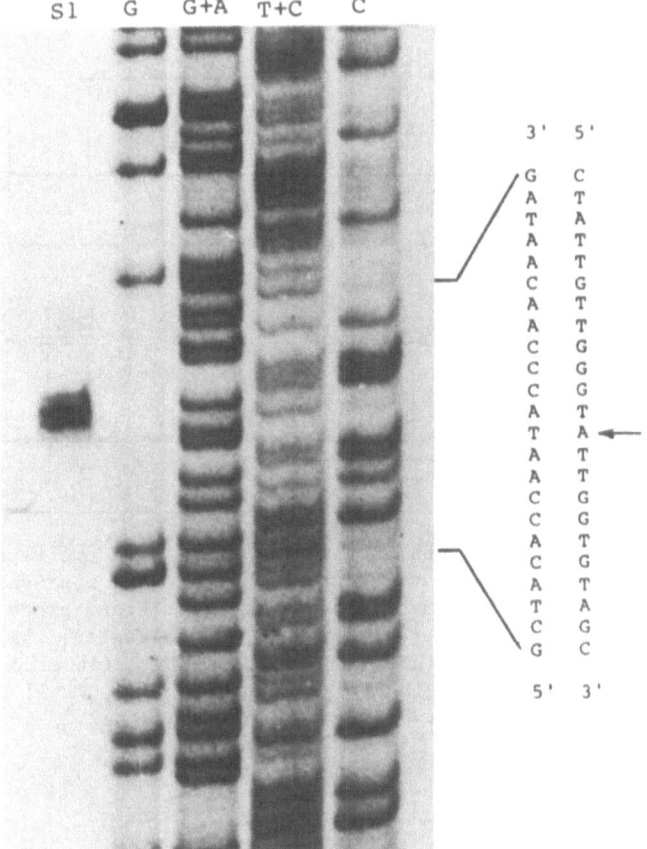

Fig. 2. S1 analysis of type II rDNA. The Acc1/HincII
band that spans subrepeat I' was labeled with
polynucleotide kinase and used for S1 protection
experiments with total RNA. The gel also shows
the sequencing reactions [20].

The organization of 5S-histone genes is distinct among different
Artemia populations. We always found two repeat units of different size
in every population, but they differ from population to population. Taking
these differences, Artemia populations can be classified into three groups:
bisexuals from North America, bisexuals from Eurasia and parthenogenetic
populations (always from Eurasia).

To study the organization around the 5S gene, we have sequenced three
clones isolated by hybridization with this RNA. The organization of these
clones can be seen in Fig. 4. pArt5H-a comes from a population of San
Francisco Bay; the two others, pArt5H-b and pArt5-b come from a parthenogenetic population of China. One of them, pArt5b, does not contain histones
and will be discussed later.

In both pArt5H-a and pArt5H-b, the histone closest to 5S RNA is H2A
(not shown in the figure for pArt5H-b). Downstream from the H2A coding

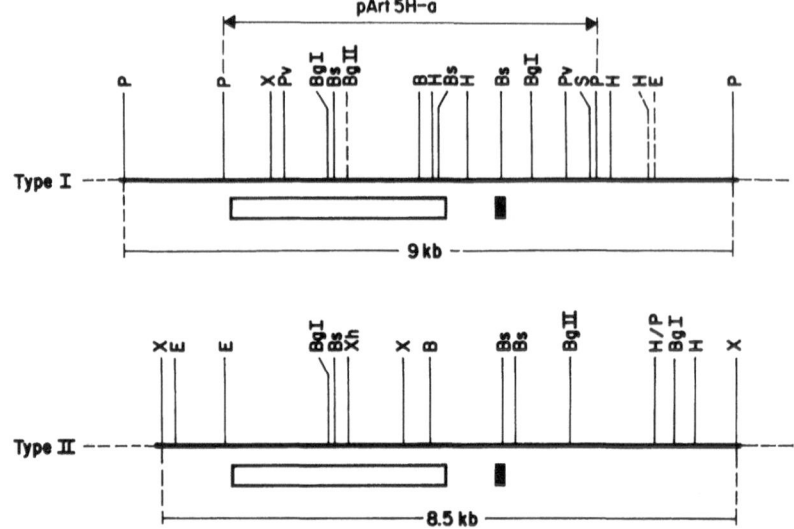

Fig. 3. Genomic map of the 5S-histone genes. The two repeats, of 9 and 8.5 kb are shown. The white box represents the region that hybridizes with the histone probe cDm500 [21]. The solid box is the 5S region. pArt5H-a is indicated in the type I map. Restriction endonucleases are as in Fig. 1 except for : P, PstI; Pv, PvuII; Bgl, BglI; Bgll, BglII; E, EcoR1; Xh, XhoI.

sequence, we find the 3' non translated consensus sequences described by Birnstiel et al. [11]. The sequences flanking 5S DNA are similar for the three clones we have characterized, both in 5' and 3' direction (boxes 'c' and 'd'), although box 'c' is truncated in pArt5-b. Further downstream, pArt5H-b and pArt5-b differ from pArt5H-a, and again, pArt5-b has a truncated box 'a'. We think pArt5-b is an orphon, an isolated gene from a repeated family, as defined by Childs et al. [12].

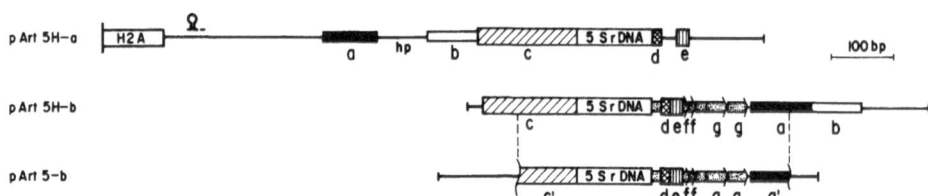

Fig. 4. Schematic representation of the sequences surrounding the 5S gene in the plasmids pArt5H-a, pArt5H-b and pArt5-b. Letters represent repeated sequences found in all plasmids, as shown. Thin lines represent unique sequences in each plasmid.

Comparison between pArt5H-a (from bisexual populations) and pArt5H-b (from parthenogenetic populations) shows that the 5S gene (from box 'c' to box 'e') seems to have been transposed to different positions in this region of DNA. As there is no 5S or histone genes alone (except for orphons which are not very abundant), we think that these genes were joined together before amplification of the two families. We are now looking for this arrangement in other crustaceans, to see whether this is a peculiarity of Artemia or is a more general feature. So far, we have not found any link between 5S and histone genes in lobster and some species of crabs.

MITOCHONDRIAL 16S rRNA

When looking for developmentally regulated genes by differential screening of cDNA libraries [13], we isolated a very abundant clone that is highly homologous to the 3' half of mitochondrial 16S rRNA. This surprising finding is due to the fact that mitochondrial 16S rRNA is polyadenylated and therefore oligo (dT) can prime cDNA synthesis efficiently. The restriction map of the complete mitochondrial 16S rRNA is shown in Fig. 5. pArLSU-1 was obtained from a cDNA library in λgt11 [13], and comprises half of the molecule, up to the EcoR1 site. pArLSU-2 comprises 95% of the mitochondrial 16S rRNA sequence, and was obtained from a library made in pUC18 with BamH1 linkers, using the RNaseH method of Gubler and Hoffman [14].

Table 1 shows the molecular properties of Artemia mitochondrial 16S rRNA in comparison to that of insects and mammals. The size of the RNA is smaller in Artemia than in insects or mammals. The G+C content of Artemia mitochondrial 16S rRNA is close to that of mammals, unlike the G+C content of the nuclear genome. This is especially worth noting in the 5' half of the molecules where the difference between Artemia (or mouse) and mosquito is three-fold; in the 3' half the difference is two-fold. These differences in G+C content do not disturb the secondary structure of the RNAs [15]. In fact, in the structure proposed for Artemia mitochondrial 16S rRNA, there appears G-C pairs in positions where there are A-U pairs in the mosquito structure. The similarity of these molecules is more pronounced in the 3' half, with 67% for insects and 62% for mammals, whereas it is only 34% and 50% in the 5' half. The higher similarity of Artemia mitochondrial 16S rRNA to the mouse molecule at the 5' end than to the mosquito one probably is just a consequence of the more similar base composition of Artemia and mouse molecules in this region.

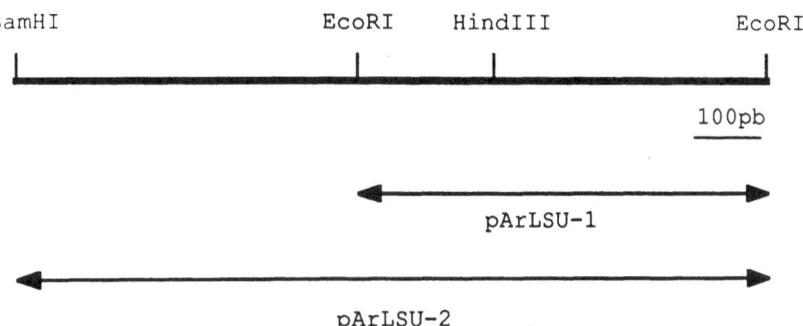

Fig. 5. Restriction map of the two plasmids coding for mitochondrial 16S rRNA.

Table 1. Properties of Mitochondrial 16S rRNA from _Artemia_

	Artemia	Mosquito[a]	Mouse[b]
Size	1250[c]	1335	1582
G+C content			
Total	33.2	16.7	35.3
5' half[d]	28.6	9.1	31.4
3' half	41.6	25.8	38.1
% similarity			
Total		53	56
5' half		34	50
3' half		67	62

a) HsuChen et al., [18]
b) Bibb et al., [19]
c) The size of _Artemia_ mitochondrial 16S rRNA has been estimated from the nucleotide sequence. Alignment of this sequence to other mitochondrial rRNAs has shown that there are 83 nucleotides missing at the 5' end of our cDNA clones which have been included in the predicted size.
d) The halves of the molecules are with relation to the _EcoR_ 1 site present in all species.

PERSPECTIVES

The work we have done with structural RNAs of _Artemia_ will allow us to study the genetic relationships between different populations of this crustacean. For instance, the presence of satellite 1 [16,17] can distinquish between American and Eurasian populations. The results presented in this communication with the 5S-histone gene complex extend the possibility to distinguish between bisexual and parthenogenetic populations from Eurasia. The mitochondrial 16S rRNA could be used for the same purposes. We are now developing _in vitro_ transcription systems for the ribosomal RNA genes, to understand the specificity of RNA polymerase 1 transcription.

ACKNOWLEDGEMENTS

The work done in our laboratory was supported by grants from the Comision Asesora para la investigacion Científica y Técnica and Fondo de Investigaciones Sanitarias de la Seguridad Social.

REFERENCES

1. J. A. Lake, Origin of the eukaryotic nucleus determined by rate-invariant analysis of rRNA sequences, _Nature_ 331:184 (1988).
2. G. A. Dover, Molecular drive: a cohesive mode of species evolution, _Nature_ 299:111 (1982).
3. B. Sollner-Webb and J. Tower, Transcription of cloned eukaryotic

ribosomal RNA genes, Ann. Rev. Biochem, 55:801 (1986).

4. T. M. Maniatis, E. F. Fritsch and J. Sambrook, in "Molecular Cloning, a Laboratory Manual", Cold Spring Harbor Laboratory, Cold Spring Harbor, New York (1982).

5. M. E. Gallego, M. Diaz-Guerra, J. Cruces, J. Sebastian and J. Renart, Characterization of two types of rRNA gene repeat units from the crustacean Artemia, Gene 48:175 (1986).

6. I. Gil, M. E. Gasllego, J. Renart and J. Cruces, Identification of the transcriptional initiation site of ribosomal RNA genes in the crustacean Artemia, Nucleic Acids Res. 15:6007 (1987).

7. H. T. Koller, K. A. Frondorf, P. D. Maschner and J. C. Vaughn, In vivo transcription from multiple promoters during early development and evolution on the intergenic spacer in the brine shrimp Artemia, Nucleic Acids Res. 15:5391 (1987).

8. J. C. Bagshaw, H. B. Skinner, T. C. Burn and B. A. Perry, Nucleotide sequence of the 5S RNA gene and flanking regions interspersed with histone genes, Nucleic Acids Res. 15:3628 (1987).

9. M. T. Andrews, J. C. Vaughn, B. A. Perry and J. C. Bagshaw, Interspersion of histone and 5S RNA genes in Artemia, Gene 51:61 (1987).

10. J. Cruces, M. Diaz-Guerra, J. Sebastian and J. Renart, Organization and sequence of 5S rRNA genes in Artemia, in "Artemia Research and its Applications," Vol. 2, W. Decleir, L. Moens, H. Slegers, E. Jaspers and P. Sorgeloos, eds., Universa Press, Wetteren, Belgium (1987).

11. M. L. Birnstiel, M. Busslinger and K. Strub, Transcription termination and 3' processing: the end is in site, Cell 41:349 (1985).

12. G. Childs, R. Maxson, R. H. Cohn and L. Kedes, Orphons: dispersed genetic elements derived from tandem repetitive genes of eukaryotes, Cell 23:651 (1981).

13. I. Palmero, J. Renart and L. Sastre, Isolation of cDNA clones coding for mitochondrial 16S ribosomal RNA from the crustacean Artemia, Gene in the press (1988).

14. U. Gubler and B. J. Hoffman, A simple and very efficient method for generating cDNA libraries, Gene 25:263 (1983).

15. R. G. Gutell and G. E. Fox, A compilation of large subunit RNA sequences presented in a structural format, Nucleic Acids Res. 16:r175 (1988).

16. J. Cruces, M. L. G. Wonenburger, M. Diaz-Guerra, J. Sebastian and J. Renart, Satellite DNA in the crustacean Artemia, Gene 44:341 (1986).

17. G. Badaracco, L. Barabelli, E. Ginelli, R. Meneveri, P. Plevani, P. Valsasnini and C. Barigozzi, Variations in repetitive DNA and heterochromatin in the genus Artemia, Chromosoma 95:71 (1987).

18. C. C. HsuChen, R. M. Kotin and D. T. Dubin, Sequences of the coding and flanking regions of the large ribosomal subunit RNA gene of mosquito mitochondria, Nucleic Acids Res. 12:7771 (1984).

19. M. J. Bibb, R. A. Van Etten, C. T. Wright, M. W. Walberg and D. A. Clayton, Sequence and gene organization of mouse mitochondrial DNA, Cell 26:167 (1981).

20. A. Maxam and W. Gilbert, Sequencing end-labeled DNA with base-specific chemical cleavages, Methods Enzymol. 65:499 (1980).

21. R. P. Lifton, M. L. Goldberg, R. W. Karp and D. S. Hogness, The organization of the histone genes in Drosophila melanogaster: functional and evolutionary implications, Cold Spring Harbor Symp. Quant. Biol. 42:1047 (1978).

22. J. C. Bagshaw, M. T. Andrews and B. A. Perry, Organization and expression of histone genes in the brine shrimp Artemia, in "Histone Genes, Structure, Organization and Regulation", G. S. Stein, J. L. Stein and W. F. Marzluff, eds., John Wiley and Sons, New York (1984).

RIBOSOMAL RNA GENES DURING DEVELOPMENT IN THE BRINE SHRIMP ARTEMIA:

STRUCTURE, EXPRESSION AND EVOLUTION

Jack C. Vaughn, Hannelore T. Koller, Kathleen A. Frondorf,
Patricia D. Maschner and Walter H. Swanson

Department of Zoology
Miami University
Oxford, Ohio 45056

INTRODUCTION

Artemia is not yet a very useful organism in terms of its genetics,
and certainly does not compare with Drosophila or yeast in this regard.
However, the combined use of molecular and biochemical approaches can be
very powerful in elucidating mechanisms of gene control during early development. To make effective use of the potential strength of this combined
approach, it is important to select a suitable gene for in-depth study. A
number of developmentally regulated genes have been identified in Artemia,
and several excellent examples are included in this Symposium volume. The
particular gene set in which we are interested is the one encoding ribosomal RNAs. Ribosomal RNA genes were the first eukaryotic genes to be
isolated in pure form, and these genes are among the most extensively
studied. Despite this, we do not yet have a detailed molecular understanding as to the molecular mode of regulation for this gene class. The brine
shrimp Artemia offers some unusual advantages in this regard, an apprecia-
tion of which can be gleaned from reviews by Hentschel and Tata [1] and
Warner et al. [2]. We believe that gene regulatory molecules, predicted
to exist based on molecular data, can be isolated from the appropriate
developmental stage(s) for further in-depth characterization of their
molecular mode of operation.

Eukaryotic promoters all operate by first establishing stable transcription complexes, a process involving interaction with various protein
factors. These stable transcription complexes are then recognized and re-
peatedly read by the corresponding polymerases. Among eukaryotic organisms,
the greater than 10,000 different chromosomal genes of a given species are
all believed to be transcribed by one of only three different classes of
RNA polymerases. Taking these in reverse order, Pol III is largely used
to transcribe transfer RNAs and also the 5S class of ribosomal RNAs, while
Pol II transcribes genes which produce messenger RNAs. The promoters of
these two gene classes have been relatively well characterized. Pol I
transcribes only the major (18S + 28S) ribosomal RNA gene class, and its
promoter is the least well understood of the three classes (reviewed by
Sollner-Webb and Tower, [3]). Early progress in characterizing a model
Pol I gene's expression was slowed, in part, by the general lack of conserved promoter DNA tracts between distantly related species [4], whose
recognition could have signaled their potential significance, and by the

lack of reproducible in vitro transcription systems. In the well studied Xenopus system, the existence of conserved sequence tracts surrounding transcription initiation sites at the gene promoter and within duplicated spacer promoters [5,6] did eventually correlate very nicely with short DNA tracts identified from in vitro promoter mapping assays [7]. We have made extensive use of the comparative sequence approach in several of our experimental designs. This paper presents recent work done in our laboratory on the structure, expression and evolution of ribosomal RNA (rRNA) genes during development in the brine shrimp Artemia franciscana. These organisms have been obtained from the population living in salterns near San Francisco Bay, California.

ORGANIZATION AND STRUCTURE OF rRNA GENES AND rRNAs IN ARTEMIA

The organization of rRNA genes in Artemia is diagrammed in Fig. 1. These genes are organized into long tandemly repeated units, as is typical for eukaryotes, in which these genes are localized at the nucleolar organizer locus. It is believed that virtually all the Artemia rRNA genes are organized into one cluster, or possibly a number of identical such clusters, because there are examples in which cutting genomic DNA with only a single restriction enzyme, such as PstI or SalI (each of which only occurs once per repeat unit), followed by electrophoretic size fractionation, results in all the rRNA genes localizing in a single band. As is generally the case in eukaryotes, the coding regions in Artemia rRNA gene repeat units (rDNA units) are separated from those in adjacent repeats by long so-called "nontranscribed" or intergenic spacers (IGS). Virtually every repeat unit is very nearly identical in length to its neighbors, for length heterogeneity owing to differing numbers of sub-repeat structures in the IGS, which is common in many other organisms, is rare in Artemia [8]. It has been reported [9] that in Artemia from the San Francisco, Calif. population, two types of rRNA gene repeat units are present differing from one another in repeat unit length and in restriction sites in the IGS and ETS. It was subsequently learned (J. C. Bagshaw, personal communications; see also J. C. Bagshaw, this Symposium volume) that Artemia packaged from "San Francisco Bay, Calif." have recently in fact been obtained from China prior to packaging! We have found only one major repeat unit length class in genomic rDNA maps prepared from the legitimate San Francisco population (H. T. Koller and J. C. Vaughn, unpublished data). The rDNA gene promoter in Artemia franciscana is located about 800 base pairs upstream of the 18S rRNA coding region, at the ETS/IGS junction. Transcription is initiated within the promoter near its 3'-end, proceeds through the ETS, and continues through the 18S, 5.8S and 26S rRNA coding regions to produce pre-rRNA molecules. Our present preliminary results indicate that transcription terminates about 150 nucleotides downstream of the 26S rRNA coding sequence (D. S. Russell, C. J. Lee and J. C. Vaughn, unpublished data).

The 26S rRNA coding region contains a short intervening sequence (IVS), which is probably transcribed but is excised in an event which does not include splicing, so that the mature 26S rRNA is composed of two half molecules of similar size which are associated by weak chemical bonding. This feature is common to the 26S rRNAs of most protostome animals [10]. The nucleotide sequence of Artemia 26S rRNA immediately 5' to this nick has been reported [11]. Additional processing events cleave off the 18S and 5.8S rRNAs, and the latter is found associated with the 5'-end of the 26S rRNA molecule by complementary base pair interaction in ribosomes. These three rRNAs are joined by the 5S rRNAs coded for by genes located elsewhere in the genome, and are then assembled into ribosomal subunits for transport to the cytoplasm, where they function in protein synthesis.

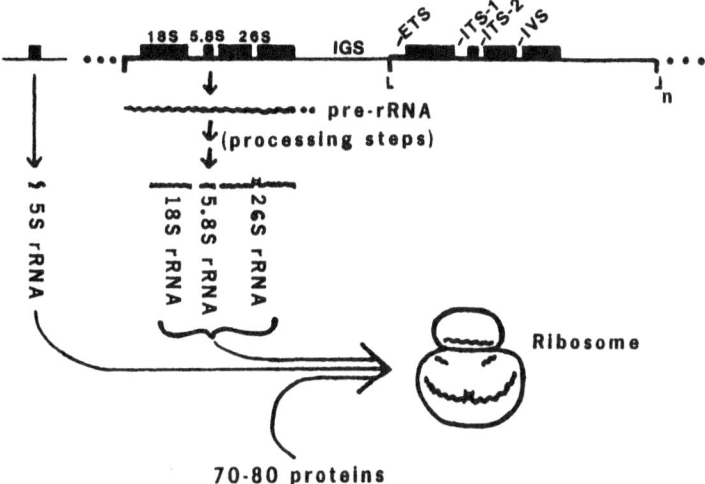

Fig. 1. Organization of Artemia ribosomal RNA genes and
relationship of their transcription products to
ribosome structure. A typical repeat unit is
represented by n number of tandem gene copies.
Each repeat unit contains coding regions (elevated boxes); the 26S region is interrupted by
an intervening sequence (IVS). Spacers include
the "nontranscribed" intergenic region (IGS) and
three transcribed spacers: external transcribed
spacer (ETS) and two internal transcribed spacers
(ITS-1 and ITS-2). The gene promoter is at the
ETS/IGS interface.

In Artemia, the 5S rRNA genes have been shown to have an unusual and
thus far unique organization, in which one gene is present in each histone
repeat unit [12]. This arrangement differs from the more common eukaryotic
pattern, in which 5S rRNA genes are tandemly organized. The nucleotide
sequence of Artemia 5S rRNA has been reported [13], and is identical to
that for the 5S DNA sequence [14]. An Artemia 5.8S rRNA sequence has been
published [15], which is virtually identical to that reported for the DNA
sequence [16], showing that an intervening sequence is not present here
despite its existence in some insect 5.8S genes [17,18]. The Artemia 18S
rRNA coding sequence has been reported [19], and these authors have also
advanced a secondary structure model for its corresponding RNA transcript.

The number of rDNA units per haploid Artemia genome has been estimated
at about 300 copies, based on saturation plateau curves, in which filter-
immobilized genomic DNA was hybridized with labeled rRNA molecules [20].
We have also conducted slave mini-cot experiments (Fig. 2), in which a vast
excess of total genomic DNA is permitted to renature in the presence of
trace levels of particular radiolabeled cloned genes, which preferentially
hybridize to the driver genomic DNA under such conditions. Comparison of
the hybridization kinetics between Artemia single-copy DNA sequences and
those of cloned tracer 5S or (18S + 26S) rRNA genes permits estimation of
the copy number for these genes. In these studies, labeled 5S rRNA genes
hybridize to their genomic counterpart 100 times faster than single-copy
sequences hybridize to each other, from which it can be concluded that the
genomic 5S genes are present in about 100 copies per haploid genome [12].

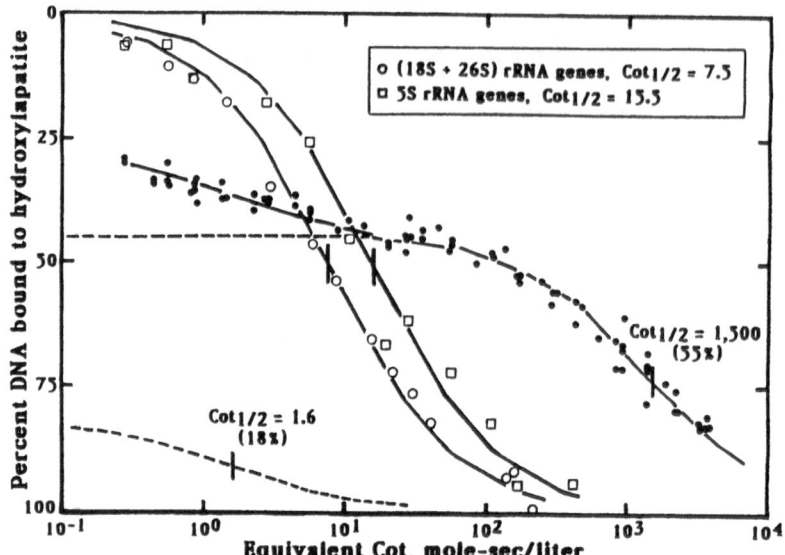

Fig. 2. Slave mini-cot experimental estimation of copy
numbers for Artemia 5S and (18S + 26S) rRNA
genes. Sheared, unlabeled total genomic DNA in
vast excess was hybridized with tracer amounts
of cloned, labeled gene fragments in solution,
followed by separation of single- from double-
stranded DNA molecules on hydroxylapatite
columns. The comparative hybridization kinetics
of single-copy driver genomic DNA (-●-) and
tracer slave DNAs (-O-, -□-) permit estimation of gene copy numbers. In this experiment,
55% of the total genomic DNA follows the single-
copy reaction kinetic curve and 18% falls into
an intermediately-repetitive component, within
which the ribosomal RNA genes are located.

The (18S + 26S) rRNA genes hybridize 200 times faster than single-copy
sequences, and are therefore present in about 200 copies per haploid genome.
This latter number is in approximate agreement with our earlier estimate
of 300 copies.

Several years ago, J. C. Bagshaw's laboratory constructed the first
recombinant DNA gene library for Artemia, utilizing size-selected 15-25 kb
DNA fragments following partial EcoRI digestion of genomic DNA. We colla-
borated to screen this phage library for clones containing rRNA genes. One
of the primary clones, which we named lambda Ch4A BSr1 [21], was subcloned
into pBR322 for convenience in handling. The three resulting recombinant
subclones, whose structures are shown in Fig. 3, were named pBSr3, pBSr5
and pBSr6. Subclone pBSr5 contains the entire 18S rRNA coding region and
also the ETS and rDNA gene promoter, while subclones pBSr3 and pBSr6 collectively contain the 5.8S and 26S rRNA coding regions. These subclones
have been extensively utilized in the work described in this paper.

It is now well established that 5.8S rRNA molecules interact with the
5'-end of 26S rRNA molecules by complementary base pair interactions. The
primary structures of 5.8S rRNAs were early shown to be remarkably similar

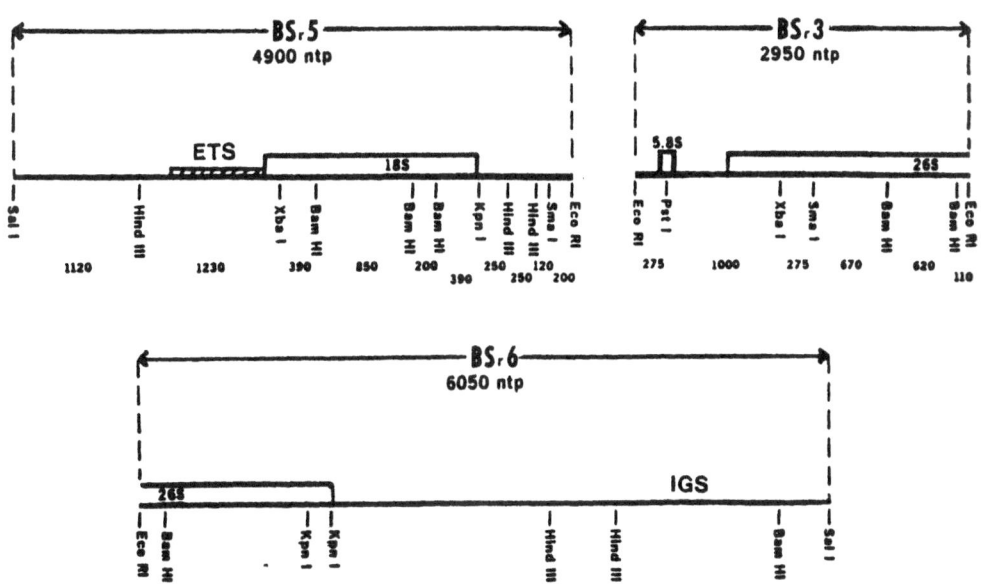

Fig. 3. Subclones constructed from one complete *Artemia* rDNA repeat unit, which had been originally cloned in a lambda Ch4A phage vector. Positions of restriction enzyme recognition sites are shown, and fragment sizes are given in nucleotide pairs.

to that of the 5'-terminus of E.coli 23S rRNA [22,23], and the two sequences are clearly evolutionarily homologous. We have shown that these homologous polynucleotide tracts have virtually identical secondary structures [24], which strongly implies a common functional role within ribosomes. The homologous structure in E.coli plays a major role in the biogenesis of the large ribosomal subunit, by serving as the recognition site for the early binding ribosomal protein L-24 [25], which is the assembly initiator protein [26]. Study of the homologous region in eukaryotes may lead to an understanding of the ribosome assembly process in higher organisms. This will require detailed knowledge of the secondary structure of the ribosomal RNA molecules involved.

We have sequenced the Artemia 5.8S rRNA coding region, as mentioned previously, and have proposed a universal secondary structure model applicable to all 5.8S rRNA molecules thus far sequenced [24]. Recently, we have completed a similar project for the 5'-end of the Artemia 26S rRNA molecule, and its sites of interaction with 5.8S rRNA [27,50,51]. Our experimental plan for determination of the 5'-end of the Artemia 26S rRNA coding sequence involved excision of the appropriate fragment from recombinant plasmid pBSr3, and preparation of a detailed restriction map of the critical region. A sequencing strategy was then devised, from which the structure of the 5'-end of the 26S rRNA coding sequence was obtained. The resulting sequence was compared to those available for the corresponding region from over 30 different species, derived from eukaryotes, eubacteria, and archaebacteria, as well as from mitochondria and chloroplasts [28]. We utilized the comparative phylogenetic approach of Fox and Woese [29] to derive a general secondary structure model for this region. Our model, as illustrated for Artemia, is shown in Fig. 4. There are 19 helices in structural domain I,

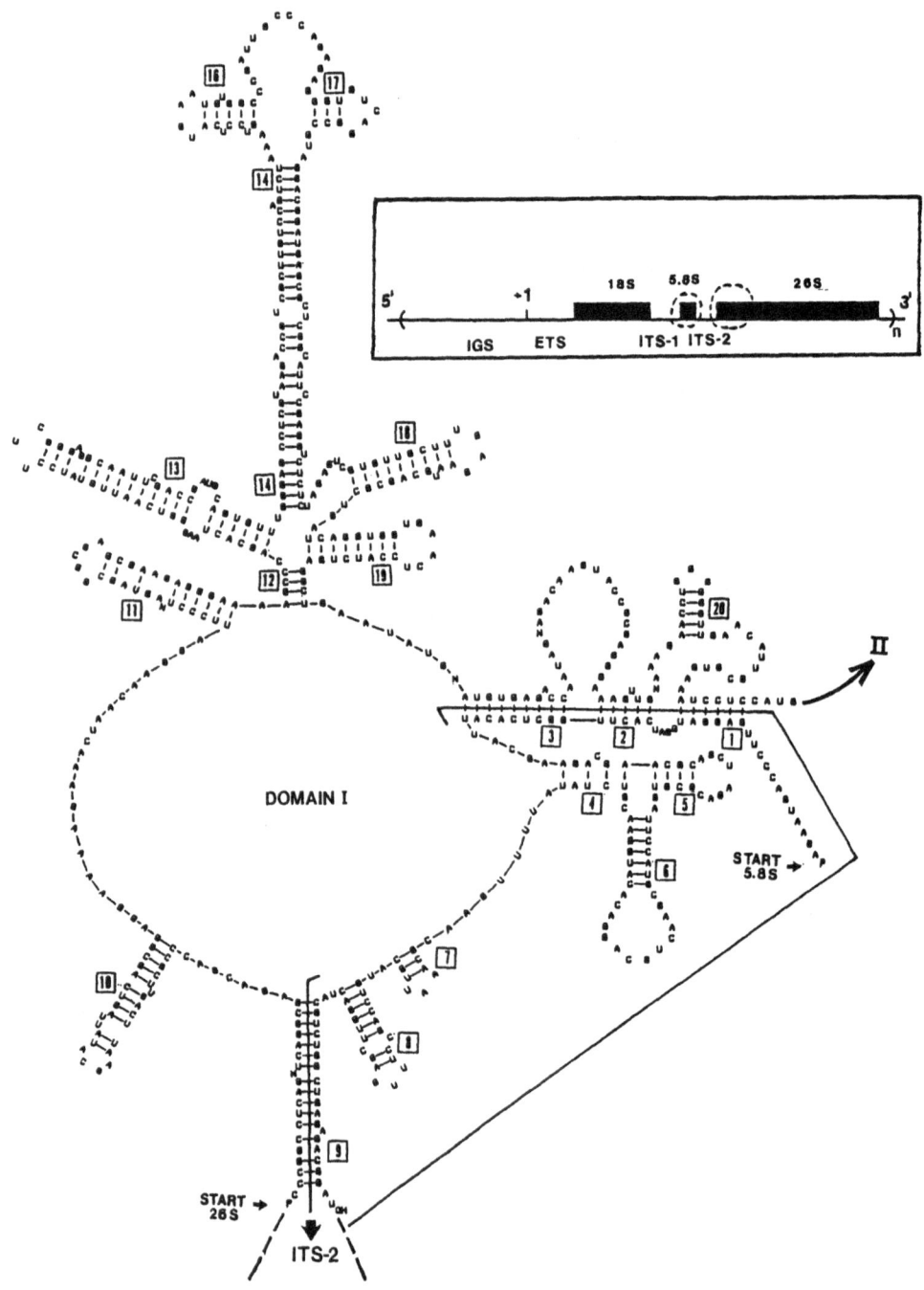

Fig. 4. Secondary structure model for the Artemia large subunit
rRNA domain I, involving base pair interaction between
5.8S rRNA and the 5'-end of the 26S rRNA molecule. The
helices were derived using the comparative phylogenetic
approach. The four proposed regions of contact between
the two molecules are universal.

of which five are entirely within the 5.8S molecule and four others are located at sites of interaction between the 5.8S and 26S rRNA molecules. The helices in this model are each superimposable with those for all the other more than 30 species and organellar sequences we have compared. The existence of each helix in the model has been proven by the demonstration of compensatory base changes in every case.

EXPRESSION OF rRNA GENES IN <u>ARTEMIA</u> AND rDNA PROMOTER STRUCTURE

A key paper for anyone studying the expression of rRNA genes during development in <u>Artemia</u> was published a number of years ago by McClean and Warner [30]. They showed that at about the hatching stage of development, during which time a developing embryo emerges from its shell, there is a burst of RNA synthesis, much of which is probably rRNA. We re-examined that point, using cloned rRNA genes to identify the transcripts during development. Our results were like McClean and Warner's for the initial parts of the study (Fig. 5<u>A</u>); there is a burst of total embryo RNA and also

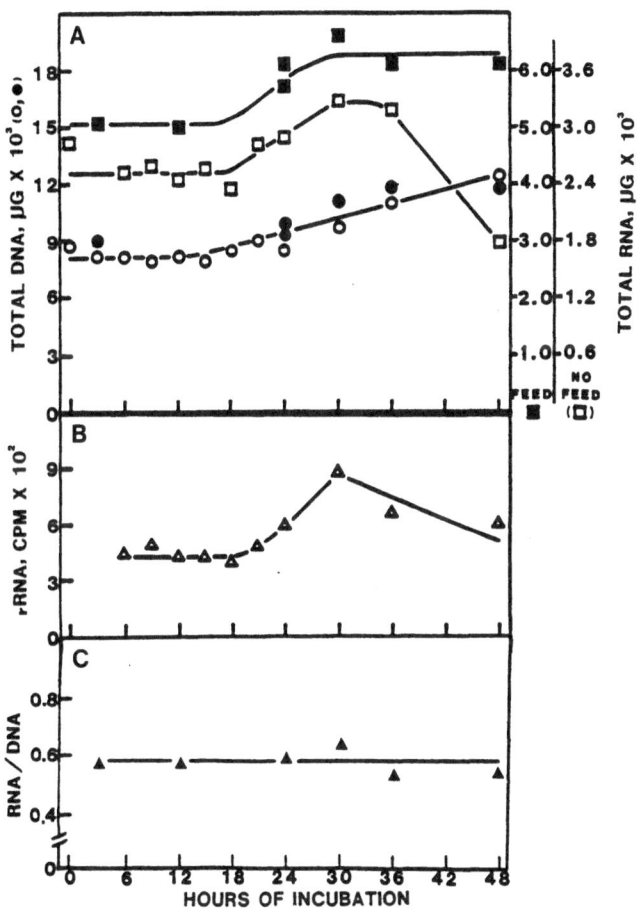

Fig. 5. Expression of rRNA genes during <u>Artemia</u> development. In <u>A</u> total embryo RNA and DNA content are measured in both fed (-●-, -■-) and starved (-○-, -□-) embryos. In <u>B</u> total RNA (unfed series) has been probed for rRNAs, using a dot-blot method. In <u>C</u> the ratio of total embryo RNA (fed series) which is largely rRNA, to total DNA is plotted.

DNA synthesis at about the time of hatching (18 hrs of incubation at 30 C). It is common in these short term experiments not to feed the animals, and it has been suggested that declining RNA levels reported by some workers might be due to starving the larvae [1], which has been reported to result in a dramatic decline in the activity of RNA polymerase I in these organisms [31]. We examined the effects of feeding vs not feeding the developing organisms, and found that there is a dramatic decline in total RNA synthesis in unfed animals. Starvation obviously represents an unnatural additional stress factor which could easily lead to altered expression of various physiological and biochemical parameters, so that results obtained in the same or different laboratories could easily differ in unexpected ways if this factor is not controlled.

Total embryo RNA samples were filter-immobilized and hybridized to labeled, cloned Xenopus rRNA genes using a quantitative "dot-blot" method (Fig. 5B). It was found that the curve for rRNA accumulation (unfed animals) follows that for total embryo RNA. Since more than 80% of a cell's total RNA is typically rRNA in eukaryotes, we conclude that much of the burst in RNA synthesis at hatching is due to rRNA. Since this is a stage-specific event, we also conclude that rRNA synthesis in Artemia is under developmental regulation. It is known that the total cell number remains constant in developing Artemia from the gastrula stage until hatching, then begins to increase [32]. Since three parameters (cell number, total embryo rRNA, and total embryo DNA) follow a parallel trend of increasing following hatching, we suspected that the burst in rRNA levels at hatching reflected a need for additional ribosome synthesis, to keep up with the additional cells. To test this idea, we calculated the ratio of total cell RNA (which is largely rRNA) to total cell DNA, looking for a parallel relationship. This ratio (Fig. 5C) remains constant during early development. If total cell DNA content is proportional to total cell number during these developmental stages, then we would conclude that the burst in rRNA synthesis observed at hatching is in response to the need for additional rRNA required for ribosome synthesis and assembly. This is not a very surprising result, since it has been estimated that a single eukaryotic cell must produce more than a million new ribosomes per generation, which are used to support protein synthesis in the daughter cells.

We have utilized this information on rRNA synthesis to tell us from which developmental stages we should isolate RNAs which will contain the greatest enrichment for pre-rRNA molecules, which we have then utilized in experiments designed to locate the rRNA gene promoter.

Fig. 6A shows our experimental design for localization of the sites to which the 5'-ends of pre-rRNAs map. In this experiment, we had expected to find only one such site and to find that this would be in the promoter region. This result was, however, not the one we obtained. A cloned restriction fragment spanning the entire IGS, within which the promoter was expected to be localized, was 5'-end labeled on its coding strand. This probe was then hybridized to total cell RNA isolated from various developmental stages. The resulting pre-rRNA/DNA hybrids were then treated with S1 nuclease to remove unpaired regions, and electrophoretically sized to determine the distance from the protected labeled end to the pre-rRNA 5'-terminus expected to lie within the promoter. To our surprise, we found that some 14 sites exist within the spacer (autoradiograph: Fig. 6B). The greatest number of such sites occurs at about the hatching stage (18 hr) of development, which is the stage at which appearance of new cellular rRNA is first detected during development (Fig. 5B). It is puzzling that new cellular rRNA does not begin to appear until the ca. 18 hr hatching stage (Fig. 5B), since between 0-18 hr we detect appreciable levels of pre-rRNAs (Fig. 6B). Perhaps these precursor molecules are left over from pre-desiccation stage embryos, and are detected due to the absence of

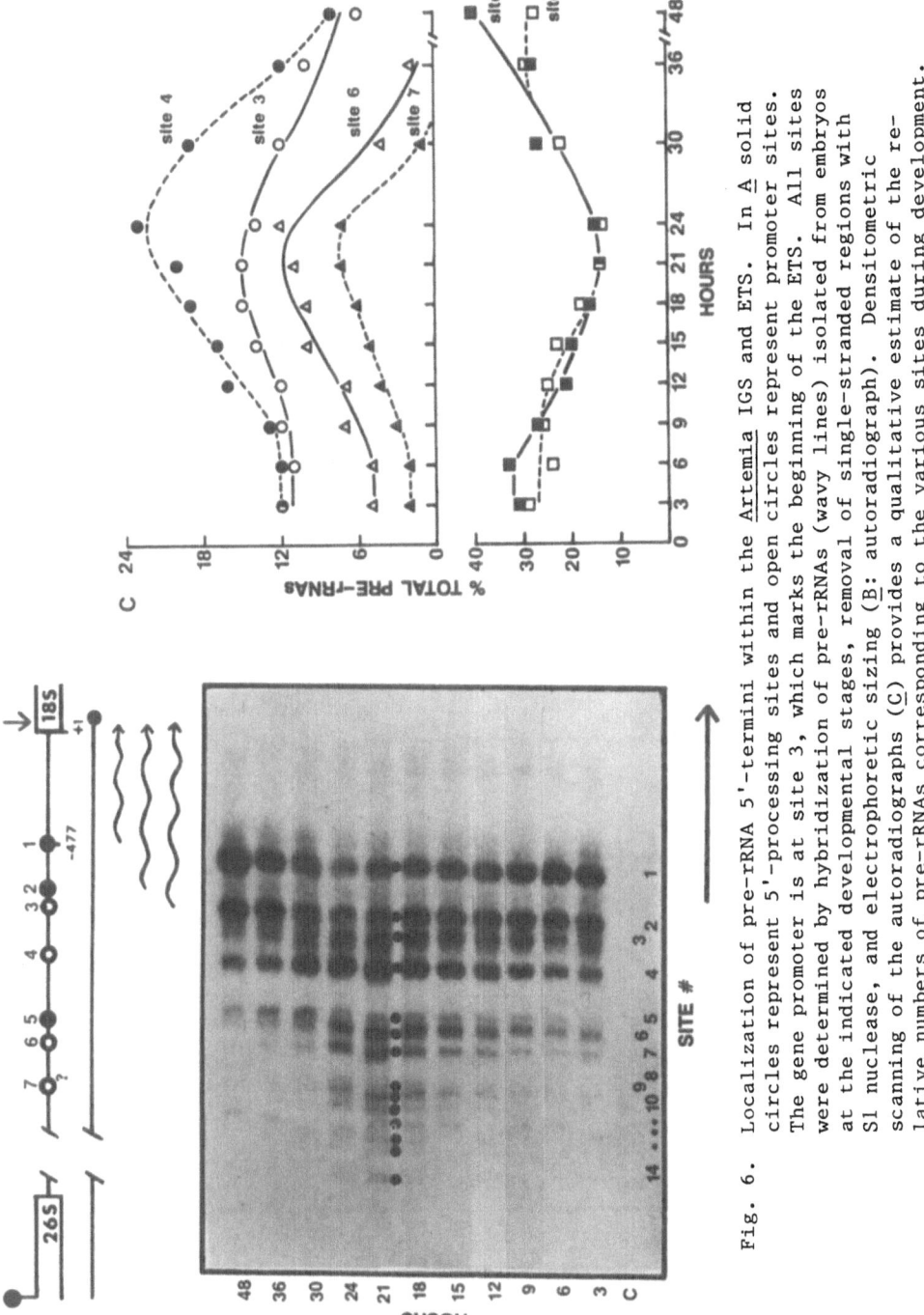

Fig. 6. Localization of pre-rRNA 5'-termini within the Artemia IGS and ETS. In A solid circles represent 5'-processing sites and open circles represent promoter sites. The gene promoter is at site 3, which marks the beginning of the ETS. All sites were determined by hybridization of pre-rRNAs (wavy lines) isolated from embryos at the indicated developmental stages, removal of single-stranded regions with S1 nuclease, and electrophoretic sizing (B: autoradiograph). Densitometric scanning of the autoradiographs (C) provides a qualitative estimate of the relative numbers of pre-rRNAs corresponding to the various sites during development.

processing enzymes prior to the hatching stage of development.

    To begin characterization of these many pre-rRNA 5'-termini, we
designed the experiment shown in Fig. 7. First, a restriction fragment
containing several of these 5'-termini sites was cut from pBSr5. A detailed
restriction map was then prepared for the spacer region containing these
sites. Short restriction fragments spanning several of these sites were
isolated, 5'-end labeled on the coding strands, and split into two portions.
One portion of each was mixed with total cell RNA isolated from developmental stages shown to be synthesizing maximal levels of rRNA (i.e. 21-24
hr embryos), denatured, hybridized, and unpaired regions were removed by
S1 nuclease treatment. The resulting protected pre-rRNA/DNA hybrid molecules were then denatured and run alongside a Maxam/Gilbert sequencing
ladder derived from the other portion of the same labeled DNA fragment.
Comparison of the base sequence revealed in the sequencing ladder to the
positions of the S1-protected DNA fragments told us the precise nucleotide
encoding the first position of the pre-rRNA corresponding to each site,
while the sequence of the surrounding DNA was also read from the same gel.

    When the results obtained for several of these pre-rRNA 5'-termini
were compared (Fig. 8), we found that the sequences fell into two structural
classes. The sequences surrounding sites 1,2,5 and 8 were similar to one
another, and alignment revealed the existence of a conserved consensus
sequence: 5'...Pu-G-T-Pu-T-T-G...3'. The final "G" is coincident with the
first pre-rRNA nucleotide in this class.

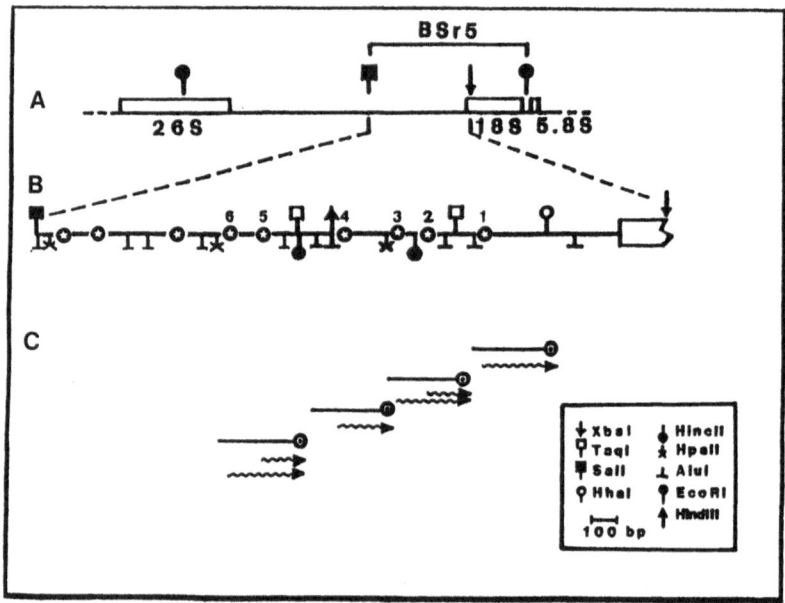

Fig. 7. Experimental design for determination of precise
         nucleotide sequences encoding each of the first six
         Artemia pre-rRNA 5'-termini (stars). Restriction
         fragments spanning these sites (B) were isolated,
         5'-end labeled on their coding strand (circle with
         dot in center), hybridized with pre-rRNA isolated
         from 21-24 hr embryos (C), treated with S1 nuclease,
         and the resulting protected end-labeled fragments
         were sequenced (see text).

A

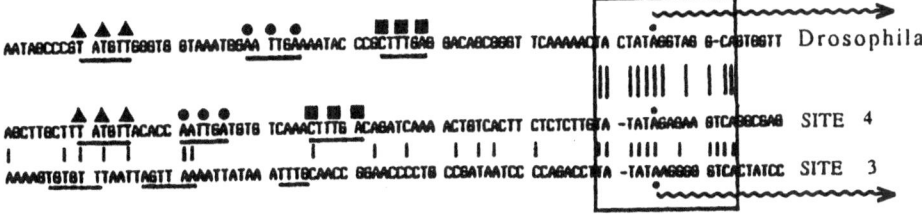

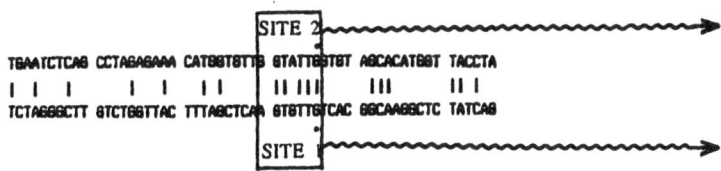

B

Fig. 8. Identification of two different classes of conserved nucleotide tracts among DNAs corresponding to Artemia pre-rRNA 5'-termini. Following alignment of non-coding strands, identical nucleotides are indicated by vertical bars. Transcription begins with the nucleotides indicated at the end of the wavy lines. In A, the Drosophila rDNA gene promoter is also shown. Three identical short upstream conserved tracts are identified when the Artemia site 4 promoter is compared with the Drosophila rDNA promoter. Similar but not identical tracts also exist for the Artemia site 3 promoter upstream region (underlined). In (B), a similar exercise reveals conserved tracts at sites believed to serve as 5'-processing enzyme recognition sites in Artemia.

The sequences surrounding sites 3,4,6 and 9 were also determined by the S1 nuclease protection approach, and found to be similar to one another but different from the other class just described. Alignment showed that the conserved consensus sequence here is: 5' ...T-A-T-A-T-Pu-Pu-Pu-G-Pu-Pu-G-T-C-A ...3'. The underlined "Pu" (purine) is coincident with the first pre-rRNA nucleotide in this class.

Comparison of the conserved consensus sequence surrounding the 5'-ends of these two classes of pre-rRNAs with rDNA sequence tracts from other arthropod species has led to the tentative assignment of their functional significance. Alignment of the site 3 and 4 consensus sequence class to that obtained by other workers for rDNA promoters reveals some striking similarities (Fig. 9). It is possible to write out a consensus sequence for all these arthropod rDNA promoters, including the Artemia site 3 + site 4 class, and we have concluded that this site class represents a component of the Artemia rDNA promoter(s).

```
-10                    +10
  GACCTTA-TATAAGGGGGTCA       ARTEMIA, site 3 (gene)

  TCTTGTA-TATAGAGAAGTCA       ARTEMIA, site 4 (spacer)

  CTACGTA-TGTATGCATG-CG       B. MORI, gene

  AAACCTA-TACATGGTGAGCA       D. HYDEI, gene

  AAACCTA-TTCATGGTGAGCA       D. VIRILIS, gene

  AAACCTA-TATAGGGAGTGGT       D. VIRILIS, spacer

  AAAACTACTATAGGTAGG-CA       D. MELANOGASTER, gene

  AAAACTACTATAGGTAGG-CA       D. MELANOGASTER, spacer

  AAAACTACTATAGGTAGG-CA       D. ORENA, gene

  AAAACTACTATAGGTAGG-CA       D. ORENA, spacer

  AAAAGTACTATAGGAGGTATG       G. MORSITANS, gene

  RAA  TA TATA G RGR CR       CONSENSUS
```

Fig. 9. Alignment comparison of non-coding DNA tracts corresponding to sequences surrounding arthropod rDNA transcription initiation sites and <u>Artemia</u> pre-rRNA 5'
-termini. The experimentally determined points of
initiation are indicated by <u>black</u> <u>dots</u> except for the
last entry, where the initiating nucleotide is shown
by an <u>open</u> <u>circle</u>. Sources of sequences are: <u>B. mori</u>
[45]; <u>D. hydei</u> and <u>G. morsitans</u> [46]; <u>D. virilis</u> [39];
<u>D. melanogaster</u> gene [47]; <u>D. melanogaster</u> spacer [48];
<u>D. orena</u> [49].

     A similar comparison has been made for the other <u>Artemia</u> pre-rRNA
class [33], and striking similarities have again been found between this
sequence class and known rDNA 5'-processing sites in other arthropod
species, from which we have concluded that this site class may be involved
in recognition by 5'-processing enzymes in <u>Artemia</u>.

     The location and size of the 15 nucleotide long conserved putative
rDNA promoter consensus sequence in <u>Artemia</u> is very similar to the "proximal
core promoter element" described in other species, including the frog
<u>Xenopus</u> [7] and various mammals [34]. However, rDNA promoters in most well-
studied systems, including the fruit fly <u>Drosphila</u>, contain a relatively
large "distal core promoter element," extending upstream from the transcription start site for about 60 nucleotides (reviewed in [3]). The comparative
sequence approach utilized by us did not reveal a candidate for the critical
nucleotides in this region when <u>Artemia</u> site 3 was compared to <u>Artemia</u>
site 4. This may possibly be because the distal element of this promoter
domain is concerned with transcriptional efficiency, as is the case for
rDNA promoters in other species. We know that <u>Artemia</u> site 3 is not used
as efficiently as site 4 (see Fig. 6<u>C</u>). We therefore explored the possibility that the comparative approach could perhaps be used to reveal potenti-
ally important tracts in the upstream region if site 4 alone was compared
to the rDNA gene promoter of another arthropod species. This exercise
revealed the existence of three additional short upstream sequence tracts
in <u>Artemia</u> (Fig. 8<u>A</u>). Each is six nucleotides long and is 100% identical
to the corresponding tract in the authenticated <u>Drosophila</u> <u>melanogaster</u>

rDNA promoter. Three very similar sequence tracts are also present in the _Artemia_ site 3 promoter. When these tracts are projected onto a DNA helix in the β-configuration (Fig. 10), an interesting picture emerges. These short tracts all fall on the same face of the helix in these two species, suggesting that they would be "seen" by the transcriptional machinery in the same way. Since this region falls inside the minus 60 region, which corresponds to the distal core promoter element of the rDNA promoter in other species, it is conceivable that these short tracts could play a critical role in the function of the _Artemia_ rDNA promoter.

Eukaryotic promoters have been defined as DNA sequences whose function is concerned with the accurate and efficient initiation of transcription [35]. Our present working hypothesis (Fig. 10), derived by comparison with other, better-studied systems, is that the conserved sequence tract surrounding the _Artemia_ transcription start site is functionally concerned with the accuracy of transcription initiation and probably represents the minimal requirement for accurate transcription initiation to occur. The functional significance of the upstream elements may revolve around the efficiency of transcription, as is the case in other systems. Verification of these interpretations should be attainable from an _in vitro_ transcription system for the study of _Artemia_ gene expression. Such a system is yet to be devised.

EVOLUTION OF THE IGS AND rDNA CONTROL REGION IN _ARTEMIA_

We have thus far shown that in _Artemia_, pre-rRNA 5'-termini fall into two structural/functional classes: one is believed to be concerned with transcription initiation and the other with 5'-processing of the resultant primary transcripts. Why are there so _many_ pre-rRNA 5'-termini in _Artemia_? How did these multiple functional sites arise, and what is the functional

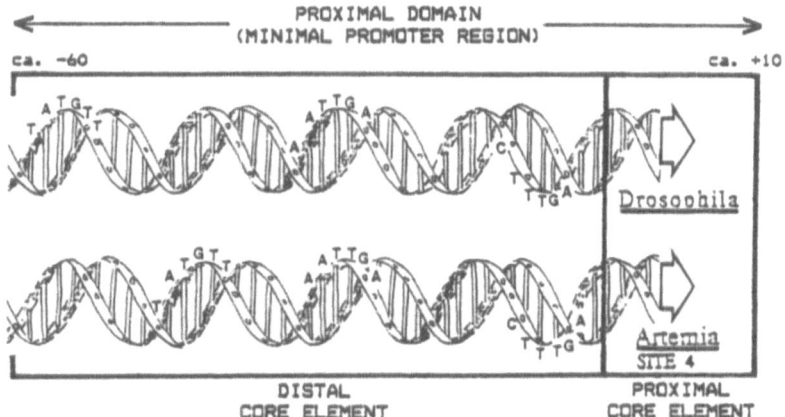

Fig. 10. Model for proximal and distal core element structure within the _Artemia_ rDNA minimal promoter, as deduced from comparative intra- and inter-species sequence analysis. The correspondense in conserved sequence tract positions between _Drosophila_ and _Artemia_ implies that the transcription machinery may recognize these promoters in a comparable fashion, i.e. from the same face of the DNA helix in both species.

significance of having multiple sites? To begin the process of understanding these and related questions, we have sequenced much of the spacer
upstream of the 18S rRNA coding region, for a distance of about 2,200
nucleotides [33]. This sequence was then analyzed with the assistance of
computer programs. Homology matrix analysis revealed the sequence to
consist of nearly identical sub-repeat units, each measuring about 600
base pairs in length (Fig. 11). Sub-repeat A, which was completely sequenced, is 617 nucleotides long and is 90% homologous to the sequenced
region in sub-repeat B. Both A and B are about 90% homologous to sub-repeat
C. Within each sub-repeat, there is an identical pattern of pre-rRNA 5'
-termini, consisting of one transcription start site and two 5'-processing
sites. It is clear that a series of ancestral duplication events or saltatory replications could easily explain the origin of these 600 base pair
sub-repeats, and thereby explain much of the multiplicity observed in sites
encoding pre-rRNA 5'-termini.

What of the origin of the original 600 base pair sub-repeat unit
itself? Each of these units contains two 5'-processing sites, which have
very similar sequences. Was there an even earlier round of duplication,
which produced the original 600 base pair unit? We have compared the

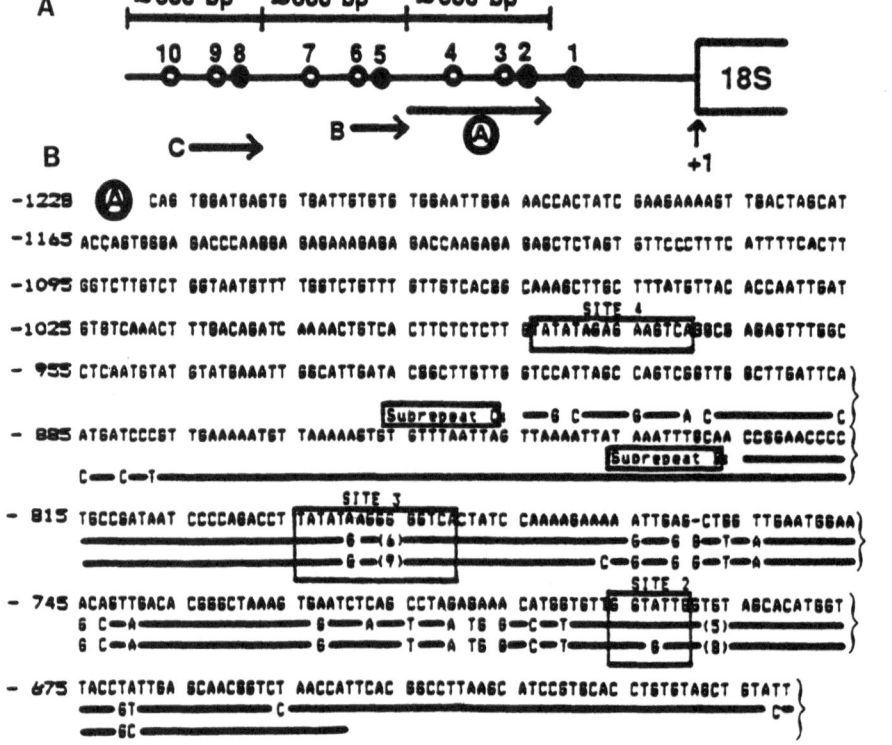

Fig. 11. Alignment comparison of Artemia spacer DNA sequence tracts
          (heavy arrows) within ca. 600 base pair subrepeats. In A,
          orientation diagram is provided showing positions of sites
          corresponding to rDNA promoters (open circles) and 5'-
          processing sites (solid circles). In B, the complete non-
          coding strand DNA sequence for subrepeat A (upper continu-
          ous line) is aligned with partial sequences obtained for
          subrepeats B and C (lower two lines). Only nucleotides
          which differ from subrepeat A are shown in the two lower
          lines, and gaps introduced to facilitate alignment are
          indicated by dashes. Identical nucleotides are indicated
          by thick black lines.

sequences of regions surrounding DNAs encoding various pre-rRNA 5'-termini in sub-repeat A. The degree of homology observed [33] is less than that seen when the 600 base pair sub-repeats are compared, as would be expected for an even earlier duplication event. For example, a 110 base pair region surrounding site 1 is 72% identical in sequence to a tract located near the 5'-end of the 600 base pair subrepeats. In the alignment, the site 1 conserved processing consensus sequence box lies opposite a very similar sequence, which may once have served a similar function, prior to the accumulation of mutational base changes. We have designated such sequence modified tracts M-sites, and note that this one occurs about 100 base pairs upstream of site 4. A roughly 45% sequence identity can be demonstrated when the ca. 90 base pair regions surrounding and upstream of homologous sites 1 and 2 are compared. Site 2, which we consider to be another M-site, is still fully functional as a 5'-processing site. A similar result is observed when the regions surrounding and upstream of sites 3 and 4 are compared, where a roughly 50% sequence identity is seen. Site 4 is believed to represent another functional M-site.

These observations, and the regular pattern of spacing which is apparent between sites within each subrepeat (ca. 200-200-100-100 base pairs), have led us to propose a model for the evolution of the Artemia IGS within the rDNA gene control region (Fig. 12). We picture that the ancestral element was about 200 base pairs in length, containing one transcription start site and one 5'-processing site. A series of saltatory replications produced an alternating pattern of such sites, each originally separated from the next by about 100 base pairs. This sequence decayed, with the subsequent introduction of mutational changes, and some of these sites lost their original functions. Some are still recognizable as M-sites today. An additional, later round of saltatory replications produced the present-day pattern of 600 base pair sub-repeats, which are only recognizable because of the highly conservative nature of base changes in ribosomal RNA repeat units.

CONCLUDING REMARKS AND DIRECTIONS FOR FUTURE RESEARCH

What is the functional significance of having multiple spacer promoters for rDNA expression? This situation has also been observed in several other species, including Xenopus [36] and Drosophila [37], and its significance may lie in the increased efficiency of trapping components which are essential for construction of transcription complexes [38,39]. This function would confer some selective advantage, thereby providing for their maintenance. This idea awaits experimental testing in Artemia. Activation of upstream spacer rDNA promoters plays a major role during Artemia development at stages when there is a relatively sudden need for quantities of additional rRNA in rapidly growing and dividing somatic cells. This phenomenon may be an important somatic counterpart to the amplification of rRNA genes utilized during oogenesis in germ line cells of many animal species.

Upstream promoters have virtually identical sequence to downstream ones, yet are only expressed during the burst of rRNA synthesis observed at the hatching stage of development in Artemia. What keeps the upstream rDNA promoters turned off at other times? Conformational changes within the rDNA chromatin at upstream sites may play a key role here. Differential supercoiling of chromatin could be important, but again this idea awaits experimental testing in Artemia. Multiple upstream promoters are functional in Xenopus embryos (Pruitt and Reeder, Ms in preparation), where differential supercoiling has been shown to play an important role in their expression [40].

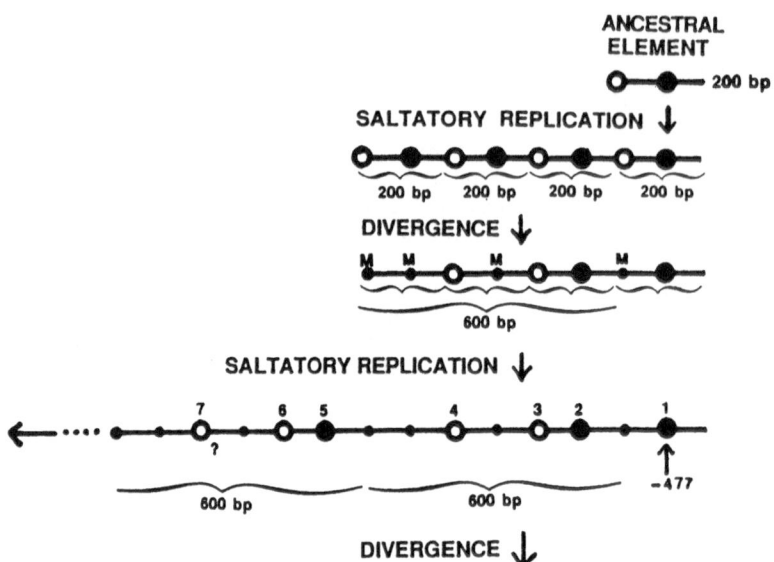

Fig. 12. Model for evolutionary history of Artemia rDNA control
          region within the intergenic spacer. On the lower line,
          the organization of the present-day spacer shows several
          regions corresponding to promoters (open circles) and
          5'-processing sites (solid circles). Site 1 is located
          at -477 with respect to the start of the 18S rRNA coding
          sequence. The ca. 600 base pair subrepeats have very
          similar, but not identical, sequences and arose by
          saltatory replication from an ancestral ca. 600 base
          pair element. This ancestral element arose by an
          earlier saltatory replication of an ca. 200 base pair
          element to produce an originally alternating array of
          start and processing sites, some of which have since
          accumulated base substitutions to become "M-sites."
          The original spacer ca. 200 base pair element contained
          one promoter and one 5'-processing site.

Is transcriptional read-through between adjacent rDNA repeat units an
important component in the Artemia system, as it is in Xenopus [41,42],
Drosophila [43], and (in modified form) in mouse [44]? Preliminary
indications are that it is not, and that Artemia lacks a fail-safe terminator [33], but this again awaits careful experimental testing. Are the
conserved sequence tracts we have identified within the putative Artemia
rDNA promoter region and 5'-processing sites really functionally significant? DNase footprinting experiments may provide additional insight into
this question. The development of an in vitro transcription system for
study of Artemia rRNA gene transcription would surely be of great benefit
in answering such questions in the future.

Much is now known about the Artemia ribosomal RNA gene system, but
the most interesting questions await elucidation by future investigators.
The demonstrated advantages of having synchronously developing stages in
early development, and of having virtually unlimited quantities of experimental material available at any desired developmental stage, should greatly
aid in the future elucidation of the molecular mechanisms which govern the

control of expression of ribosomal RNA genes, as well as other developmentally regulated gene systems, in this important experimental organism.

REFERENCES

1. C. C. Hentschel and J. R. Tata, The molecular embryology of the brine shrimp, Trends Biochem. Sci. 1:97 (1976).
2. A. H. Warner, T. H. MacRae and A. J. Wahba, The use of Artemia salina for developmental studies: preparation of embryos, tRNA, ribosomes and initiation factor 2, Methods Enzymol. 60:298 (1979).
3. B. Sollner-Webb and J. Tower, Transcription of cloned eukaryotic ribosomal RNA genes, Annu. Rev. Biochem. 55:801 (1986).
4. I. Grummt, E. Roth and M. Paule, Ribosomal RNA transcription in vitro is species specific, Nature 296:173 (1982).
5. B. Sollner-Webb and R. H. Reeder, The nucleotide sequence of the initiation and termination sites for ribosomal RNA transcription in X. laevis, Cell 18:485 (1979).
6. R. Bach, B. Allet and M. Crippa, Sequence organization of the spacer in the ribosomal genes of Xenopus clivii and Xenopus borealis, Nucl. Acids Res. 9:5311 (1981).
7. B. Sollner-Webb, J. Wilkinson, J. Roan and R. H. Reeder, Nested control regions promote Xenopus ribosomal RNA synthesis by RNA polymerase I, Cell 35:199 (1983).
8. J. Cruces, J. Sebastian and J. Renart, Restriction mapping of the rRNA genes from Artemia larvae, Biochem. Biophys. Res. Commun. 98:404 (1981).
9. M. E. Gallego, M. Diaz-Guerra, J. Cruces, J. Sebastian and J. Renart, Characterization of two types of rRNA gene repeat units from the crustacean Artemia, Gene 48:175 (1986).
10. H. Ishikawa and R. W. Newburgh, Studies of the thermal conversion of 28S RNA of Galleria mellonella to an 18S product, J. Mol. Biol. 64:135 (1977).
11. L. Nelles, C. VanBroeckhoven, R. DeWachter and A. Vandenberghe, Location of the hidden break in large subunit ribosomal RNA of Artemia salina, Naturwissenschaften 71:634 (1984a).
12. M. T. Andrews, J. C. Vaughn, B. A. Perry and J. C. Bagshaw, Interspersion of histone and 5S RNA genes in Artemia, Gene 51:61 (1987).
13. L. Diels, R. DeBaere, A. Vandenberghe and R. DeWachter, The sequence of the 5S ribosomal RNA of the crustacean Artemia salina, Nucl. Acids Res. 9:5141 (1981).
14. J. C. Bagshaw, H. B. Skinner, T. C. Burn and B. A. Perry, Nucleotide sequence of the 5S RNA gene and flanking regions interspersed with histone genes in Artemia, Nucl. Acids Res. 15:3628 (1987).
15. D. Ursi, A. Vandenberghe and R. DeWachter, The sequence of the 5.8S ribosomal RNA of the crustacean Artemia salina. With a proposal for a general secondary structure model for 5.8S ribosomal RNA, Nucl. Acids Res. 10:3517 (1982).
16. J. C. Vaughn, S. J. Sperbeck and M. J. Hughes, Molecular cloning and characterization of ribosomal RNA genes from the brine shrimp. Nucleotide sequence analysis and evolution of the 5.8S rRNA gene region and its flanking nucleotides, Biochim. Biophys. Acta 783:144 (1984a).
17. G. N. Pavlakis, B. R. Jordan, R. M. Wurst and J. N. Vournakis, Sequence and secondary structure of Drosophila melanogaster 5.8S and 2S rRNAs and of the processing site between them, Nucl. Acids Res. 7:2213 (1979).
18. B. R. Jordan, M. Latil-Damotte and R. Jourdan, Coding and spacer sequences in the 5.8S-2S region of Sciara coprophila ribosomal DNA, Nucl. Acids Res. 8:3565 (1980).
19. L. Nelles, B.-L. Fang, G. Volckaert, A. Vandenberghe and R. DeWachter,

Nucleotide sequence of a crustacean 18S ribosomal RNA gene and secondary structure of eukaryotic small subunit ribosomal RNAs, Nucl. Acids Res. 12:8749 (1984b).

20. M. P. Roberts and J. C. Vaughn, Ribosomal RNA sequence conservation and gene number in the larval brine shrimp, Biochim. Biophys. Acta 697:148 (1982).

21. J. C. Vaughn, D. J. Whitman, J. C. Bagshaw and J. C. Helder, Molecular cloning and characterization of ribosomal RNA genes from the brine shrimp, Biochim. Biophys. Acta 697:156 (1982).

22. R. N. Nazar, A 5.8S rRNA-like sequence in prokaryotic 23S rRNA, Febs Lett. 119:212 (1980).

23. B. Jacq, Sequence homologies between eukaryotic 5.8S rRNA and the 5' end of prokaryotic 23S rRNA: evidences for a common evolutionary origin, Nucl. Acids Res. 9:2913 (1981).

24. J. C. Vaughn, S. J. Sperbeck, W. J. Ramsey and C. B. Lawrence, A universal model for the secondary structure of 5.8S ribosomal RNA molecules, their contact sites with 28S ribosomal RNAs, and their prokaryotic equivalent, Nucl. Acids Res. 12:7479 (1984b).

25. R. A. Zimmermann, Interactions among protein and RNA components of the ribosome, in: "Ribosomes," G. Chambliss, G. R. Craven, J. Davies, K. Davis, L. Kahan and M. Nomura, eds., Univ. Park Press, Baltimore (1980).

26. V. Nowotny and K. H. Nierhaus, Initiator proteins for the assembly of the 50S subunit from Escherichia coli ribosomes, Proc. Natl. Acad. Sci. USA 79:7238 (1982).

27. J. C. Vaughn and W. H. Swanson, Secondary structure of the putative large subunit assembly initiator domain at the 5'-end of 23-28S rRNAs: a general model for the ancestral molecule and its relevance to mitochondrial origins, J. Mol. Evol. (In Press, 1988b).

28. R. R. Gutell and G. E. Fox, A compilation of large subunit RNA sequences presented in a structural format, Nucl. Acids Res. 16(suppl.): r175 (1988).

29. G. E. Fox and C. R. Woese, 5S RNA secondary structure, Nature 256:505 (1975).

30. D. K. McClean and A. H. Warner, Aspects of nucleic acid metabolism during development of the brine shrimp Artemia salina, Dev. Biol. 24:88 (1971).

31. H. C. Birndorf, J. D'Alessio and J. C. Bagshaw, DNA-dependent RNA polymerases from Artemia embryos. Characterization of polymerases I and II from nauplius larvae, Dev. Biol. 45:34 (1975).

32. Y. H. Nakanishi, T. Iwasaki, T. Okigaki and H. Kato, Cytological studies of Artemia salina. I. Embryonic development without cell multiplication after the blastula stage in encysted dry eggs, Annot. Zool. Jap. 35:223 (1962).

33. H. T. Koller, K. A. Frondorf, P. D. Maschner and J. C. Vaughn, In vivo transcription from multiple spacer rRNA gene promoters during early development and evolution of the intergenic spacer in the brine shrimp Artemia, Nucl. Acids Res. 15:5391 (1987).

34. I. Financsek, K. Mizumoto, Y. Mishima and M. Muramatsu, Human ribosomal RNA gene: nucleotide sequence of the transcription initiation region and comparison of three mammalian genes, Proc. Natl. Acad. Sci. USA 79:3092 (1982).

35. J. Windle and B. Sollner-Webb, Upstream domains of the Xenopus laevis rRNA promoter are revealed in microinjected oocytes, Mol. Cell. Biol. 6:1228 (1986).

36. P. Boseley, T. Moss, M. Machler, R. Portmann and M. Birnstiel, Sequence organization of the spacer DNA in a ribosomal gene unit of Xenopus laevis, Cell 17:19 (1979).

37. E. S. Coen and G. A. Dover, Multiple Pol I initiation sequences in rDNA spacers of Drosophila melanogaster, Nucl. Acids Res. 10:7017 (1982).

38. R. H. Reeder, J. G. Roan and M. Dunaway, Spacer regulation of Xenopus

ribosomal gene transcription: competition in oocytes, Cell 35:449 (1983).

39. V. L. Murtif and P. M. M. Rae, In vivo transcription of rDNA spacers in Drosophila, Nucl. Acids Res. 13:3221 (1985).

40. S. C. Pruitt and R. H. Reeder, Effect of intercalating agents on RNA polymerase I promoter selection in Xenopus laevis, Mol. Cell. Biol. 4:2851 (1984).

41. R. DeWinter and T. Moss, The ribosomal spacer in Xenopus laevis is transcribed as part of the primary ribosomal RNA, Nucl. Acids Res. 14:6041 (1986).

42. B. McStay and R. H. Reeder, A termination site for Xenopus RNA polymerase I also acts as an element of an adjacent promoter, Cell 47: 913 (1986).

43. D. Tautz and G. A. Dover, Transcription of the tandem array of ribosomal DNA in Drosophila melanogaster does not terminate at any fixed point, EMBO J. 5:1267 (1986).

44. I. Grummt, A. Kuhn, I. Bartsch and H. Rosenbauer, A transcription terminator located upstream of the mouse rDNA initiation site affects rRNA synthesis, Cell 47:901 (1986).

45. H. Fujiwara and H. Ishikawa, Structure of the Bombyx mori rDNA: initiation site for its transcription, Nucl. Acids Res. 15:1245 (1987).

46. N. Cross and G. A. Dover, Tse-tse fly rDNA: an analysis of structure and sequence, Nucl. Acids Res. 15:15 (1987).

47. E. O. Long, M. L. Rebbert and J. B. Dawid, Nucleotide sequence of the initiation site for ribosomal RNA transcription in Drosophila melanogaster: comparison of genes with and without insertions, Proc. Natl. Acad. Sci. USA 78:1513 (1981).

48. B. D. Kohorn and P. M. M. Rae, Nontranscribed spacer sequences promote in vitro transcription of Drosophila ribosomal DNA, Nucl. Acids Res. 10:6879 (1982).

49. D. Tautz, C. Tautz, D. Webb and G. A. Dover, Evolutionary divergence of promoters and spacers in the rDNA family of four Drosophila species, Implications for molecular coevolution in multigene families. J. Mol. Biol. 195:525 (1987).

50. W. H. Swanson and J. C. Vaughn, Nucleotide sequence of the putative large subunit assembly initiator domain at the 5'-end of 26S ribosomal RNA in the brine shrimp Artemia, Nucl. Acids Res. (In Press, 1988).

51. J. C. Vaughn and W. H. Swanson, A general secondary structure model for domain I in the ancestral 23-28S rRNA molecule relevance to mitochondrial origins, J. Cell Biol. (In Press, 1988a).

SPECIES AND POPULATION POLYMORPHISM OF THE HISTONE AND 5S GENE REPEAT UNIT
IN ARTEMIA

J. C. Bagshaw

Department of Biology and Biotechnology
Worcester Polytechnic Institute
Worcester, Massachusettes 01609, U.S.A.

ABSTRACT

The genes encoding histones and 5S rRNA are interspersed in a single
tandemly repeated unit in the genome of the brine shrimp, Artemia [1].
In both the Great Salt Lake and San Francisco populations of A. franciscana
there are three variants of the repeat unit with lengths of 8.6, 9.1 and
9.9 kb. The total number of repeat units is about 100 and the distribution
among the three variants, defined by digestion of genomic DNA with BamHI,
is the same in both populations. However, digestion with other restriction
enzymes reveals subtle differences in the genomic restriction patterns,
indicating that some sequence divergence has occurred since the San
Francisco population became established in man-made salterns. In A.
parthenogenetica from Tiensin, PRC, there are two variants of the repeat
unit with lengths of 12.5 and 14.1 kb. These two variants are present in
approximately equal number in this species and appear to be more numerous
than in A. franciscana. Except for the unique BamHI site in each repeat
unit, there is little restriction fragment length similarity between these
two species.

ACKNOWLEDGEMENTS

This research was sponsored by NSF Grant DCB85-10471.

REFERENCE

1. M. T. Andrews, J. C. Vaughn, B. A. Perry and J. C. Bagshaw, Interspersion of histone and 5S RNA genes in Artemia, Gene 51:61 (1987).

ORGANIZATION OF 5S-HISTONE GENES IN DIFFERENT ARTEMIA POPULATIONS

J. Cruces, M. Díaz-Guerra, I. Gil and J. Renart

Instituto de Investigagiones Biomédicas
Facultad de Medicina de la Universidad Autónoma
Arzobisop Morcillo, 4, 28029 Madrid, Spain

ABSTRACT

We have been studying 5S rRNA genes in different populations of Artemia. We have found in all cases that 5S genes are linked to histone genes. In all of the populations studied, there are two types of 5S-histone genes, differing both in size and restriction map. According to the organization of these genes, the populations we have studied can be grouped into three categories: bisexuals from North America, bisexuals from Eurasia and parthenogenetic populations from Eurasia.

We have studied in detail three recombinant clones that hybridize with 5S rRNA, one from a North American bisexual and the other two from a Chinese parthenogenetic population. Histone genes are clustered in 3.5 kb segments, separated by 500 bp from the 5S rRNA gene; one is of around 350 bp, including the spacer regions. The histone closest to the 5S gene is H2A. Comparison of the distribution of flanking sequences around the 5S gene, as well as the fact that the majority of 5S and histone genes are linked together suggest that 5S rRNA gene sequences invaded the histone repeat before amplification of this complex unit.

We are currently studying whether this organization of 5S-histone genes is of wider occurrence, or whether it is just a peculiarity of the genus Artemia.

ACKNOWLEDGEMENT

This research was supported by grants from CAICYT and FISSS.

cDNA CLONING OF DEVELOPMENTALLY REGULATED ARTEMIA GENES

Leandro Sastre, Ignacio Palmero, Maria-Teresa Macias, Ines Gil, Enrique Franco, Elvira Dominguez, Margarita Diaz-Guerra, Miguel Quintanilla, Jesus Cruces and Jaime Renart

Instituto de Investigaciones Biomedicas del C.S.I.C. and Departamento de Bioquimica de la Universidad Autonoma de Madrid Arzobispo Morcillo, 4. 28029 - Madrid. Spain

INTRODUCTION

Artemia cryptobiotic embryos are characterized by their total absence of metabolic activity, including DNA, RNA and protein synthesis. Many efforts have been devoted to the study of the mechanisms that regulate the interruption of these processes in the cyst and their activation following cyst rehydration. In the case of protein synthesis, all the necessary components are present in the cyst (mRNA, ribosomes, initiation and elongation factors, etc,) and reactivation of translation seems to depend on some modifications, such as protein phosphorylation or mRNP disruption that make these components available for protein synthesis (reviewed by Wahba and Woodley [1]).

The situation seems somewhat similar for the process of RNA synthesis. Both chromatin and RNA polymerases are present in the cyst and must undergo some modifications in order to activate RNA synthesis. The biochemical properties of RNA polymerases have been extensively studied in vitro and no change during cyst activation has been observed [2]. Little is known about the changes that could occur in the other component of the transcription system, chromatin, during this process. One possibility is that there is no transcription in Artemia cysts because promoter regions are not available for RNA polymerase binding. This situation could be the consequence of general DNA modification (for example, hypermethylation of the DNA), the existence of a particular chromatin structure, the presence of DNA-binding proteins that would block initiation of transcription, the lack of activating DNA-binding proteins, etc.

Once the cyst has been activated, embryonic development is continued from the gastrula stage. Embryonic development is regulated by the differential induction of the expression of a number of genes so that specific activation of gene expression takes place in Artemia at the same time as or immediately after general activation of transcription. The existence of both general and specific activation of gene expression makes Artemia a very interesting system to study the mechanisms involved in these processes. The first step in this study would be the isolation of a number of genes whose expression is induced in this period of development. These genes would be used to study their genomic structure, characterize their regulatory elements (promoter, enhancer and silencer regions) and the possible modifications that accompany their activation, including possible changes in their chromatin structure, DNA methylation, DNA-binding proteins associated with their regulatory regions, etc.

We have started this project by focussing on a number of genes whose encoded proteins are known to be induced during _Artemia_ development, such as P-type ATPases or actins, as well as on genes that could play a regulatory role during development, such as the _ras_ proto-oncogene (involved in intracellular signalling) or ubiquitin (involved in gene expression and protein degradation). The isolation of cDNA clones coding for these proteins and the characterization of their expression during _Artemia_ cyst activation and embryonic development will be described in this paper.

RESULTS AND DISCUSSION

Artemia cDNA clones were isolated from cDNA libraries synthesized from mRNA isolated at several different stages of development. cDNA was synthesized according to Gubler and Hoffman [3] and inserted into the lambda vector λgt11 through _EcoRI_ linkers or into the plasmid vector pUC18, through _Bam_ HI linkers.

Probes from other species were used for the isolation of some of the _Artemia_ cDNA clones. A 0.46 kb fragment containing most of the coding sequence of the viral Ha-ras gene [4] was used to isolate a 0.9 kb cDNA clone from an _Artemia_ adult cDNA library. Nucleotide sequencing of this clone showed the existence of an open reading frame whose predicted amino acid sequence is 80% similar to human and _Drosophila_ ras genes. _Drosophila melanogaster_ genomic clones coding for ubiquitin [5] and Actin (DmA2, [6]) were used to isolate the homologous _Artemia_ cDNA clones. Using ubiquitin, a 0.87 kb cDNA clone was isolated that codes for 2.7 repeats of the ubiquitin monomer, 14 additional amino acids of the carboxy terminus of the polyprotein and 245 nucleotides of 3' untranslated sequences. The amino acid sequence deduced from this clone is 99% similar to insect and mammalian ubiquitin genes. Two distinct cDNA clones were isolated using the _Drosophila_ actin probe. _Artemia_ actin cDNA clones are 1.35 and 1.45 kb long and contain the complete coding sequence for actin, 50-90 nucleotides of 5' untranslated region and 160-250 nucleotides of 3' untranslated region, respectively. These two clones differ by five positions in their predicted amino acid sequences, which are over 95% similar to mammalian β and γ-actin and to insect actins.

The isolation of P-type ATPase cDNA clones was approached using an oligonucleotide probe designed for the highly conserved autophosphorylation region of these enzymes [7]. Two different cDNA clones were isolated from _Artemia_ cDNA libraries. Nucleotide sequencing of oligonucleotide-hybridizing fragments from these clones showed that both code for peptides very similar to the autophosphorylation region of P-type ATPases (Fig. 1a). Actually, the similarity extends beyond the heptapeptide used for the probe. One of the clones was identical in this region to rabbit sarcoplasmic reticulum Ca^{++}/Mg^{++} ATPases [8], while the second clone was more similar to sheep Na^+/K^+ ATPase α subunit [9]. Complete nucleotide sequencing of these clones showed that the amino acid sequence predicted from the former clone is 70% similar to rabbit sarcoplasmic reticulum Ca^{++}/Mg^{++} ATPases. This clone is 3.1 kb long and contains the complete coding sequence for this protein, 79 nucleotides of 5' untranslated sequence and 200 nucleotides of 3' untranslated sequence. The predicted amino acid sequence of the 5' end of the second clone is identical to the previously reported N-terminal sequence of the α subunit of _Artemia_ Na^+/K^+ ATPase [10] (Fig. 1b). The Na^+/K^+ ATPase clone is 1.2 kb long and codes for amino acids 7 to 405 of

A

```
Oligonucleotide probe:                  CSDKTGT
                                        *******
Rabbit.SR Ca++/Mg++ATPase   KKNAIVRSLPSVETLGCTSVICSDKTGTLTTNQMSV
                            ************************************
pArATCa151                  KKNAIVRSLPSVETLGCTSVICSDKTGTLTTNQMSV
                             **  *  *  ***** **  ********* * * *
pArATNa10                   SKNCLVKNLEAVETLGSTSTICSDKTGTLTQNAMTV
                            ********************************* ***
SheepNa+/K+ATPase           RKNCLVKNLEAVETLGSTSTICSDKTGTLTQNRMTV
```

B

Artemia Na+/K+ATPase Protein sequence: AKKKQKKGKDLNELKKELDIDFHKIP
pArATNa10 Predicted aminoacid sequence: KGKDLNELKKELDIDFHKIP

Fig. 1. Verification of Artemia P-type ATPase cDNA clones. A. The nucleotide sequence of oligonucleotide hybridizing fragments from Artemia cDNA clones pArATCa151 and pArATNa10 was determined and the deduced amino acid sequence compared to that reported for the autophosphorylation region of sheep Na^+/K^+ ATPase and rabbit sarcoplamic reticulum Ca^{++}/Mg^{++} ATPases. B. Terminal sequence of Artemia cDNA clone pArATNa10 was determined. The amino acid sequence predicted from this clone is compared to the amino acid sequence determined for the purified α subunit of Artemia Na^+/K^+ ATPase.

this protein. The predicted amino acid sequence is 69% similar to either α, α+ or αIII forms of the mammalian protein [11]. When the amino acid sequences predicted for Artemia Na^+/K^+ and Ca^{++}/Mg^{++} ATPases are compared, some similar regions are evident (Figure 2). Most of these regions are conserved in all P-type ATPases compared to date (Regions a to f in Figure 2, named according to Serrano [12]). An additional region of homology is present at the amino-terminal region of the protein (Region 1, Figure 2). Artemia Na^+/K^+ ATPAse is as similar to Ca^{++}/Mg^{++} ATPase in this region as to other Na^+/K^+ ATPases which might be relevant since this region has been implicated in cation translocation [9].

We have examined the mRNA levels during development for the different genes we have cloned to determine when newly transcribed mRNAs start to accumulate in the cyst. These data are very important in Artemia since it is not possible to measure mRNA synthesis due to the impermeability of the cyst to radioactive precursors, and it is not known when RNA synthesis resumes after cyst activation.

RNA levels have been determined both by Northern- and Dot-blotting experiments for every cDNA clone and the results are summarized in Figure 3. The analyzed genes can be distributed in two groups according to their mRNA levels during development. The first group of genes (ras and ubiquitin) have large amounts of mRNA stored in the cyst and their mRNA levels decrease more than ten-fold during development, especially after hatching

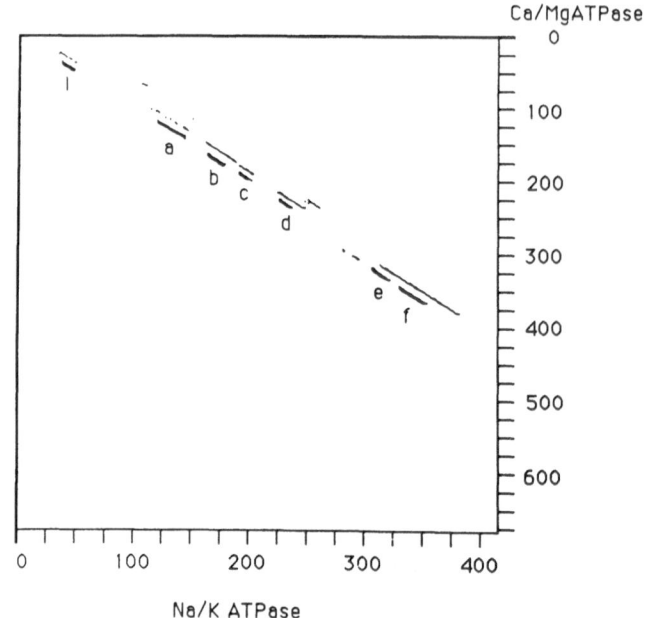

Homology region I:

Ca/MgATPase: PERGLALEQUKKNQEKYGPNEL
               ** **   * * * *** *
Na/K ATPase: PETGLTNAQARSNIERDGPNCL

Fig. 2. Protein homology between _Artemia_ Na$^+$/K$^+$ and
         Ca^{++}/Mg^{++} ATPases. The amino acid sequence
         deduced from nucleotide sequencing of _Artemia_
         Na$^+$/K$^+$ and Ca^{++}/Mg^{++} ATPase cDNA clones was
         compared by dot matrix analyses. Regions that
         are similar between all known P-type ATPases
         are underlined with a dark line and named
         according to Serrano [12]. An additional
         region of homology between _Artemia_ ATPases is
         underlined and named as Homology region I.
         Amino acid aligment for homology region I is
         shown in the lower part of the figure.

of the nauplii. _ras_ mRNA levels increases twice during the first five hours
of development and it markedly decreases after this stage. The second
group of genes (Na$^+$/K$^+$ ATPase, Ca^{++}/Mg^{++} ATPase and actins) have less mRNA
stored in the cyst. These levels increase two-to four-fold after activation of the cyst and stay high later during the period of development
analysed (25 to 120% of the levels present at the hatching time). Accumulation of mRNAs over cyst levels starts between five and ten hours of
development for these genes, which can reach maximal levels of expression
as early as 10 or 15 hours after activation of the cyst. The increase in
mRNA levels precedes the reported increase in protein levels. For example,
maximal levels for protein and enzymatic activity of the Na$^+$/K$^+$ ATPase are
reached by 32 hours of development [13], and maximal mRNA levels by 10 hours

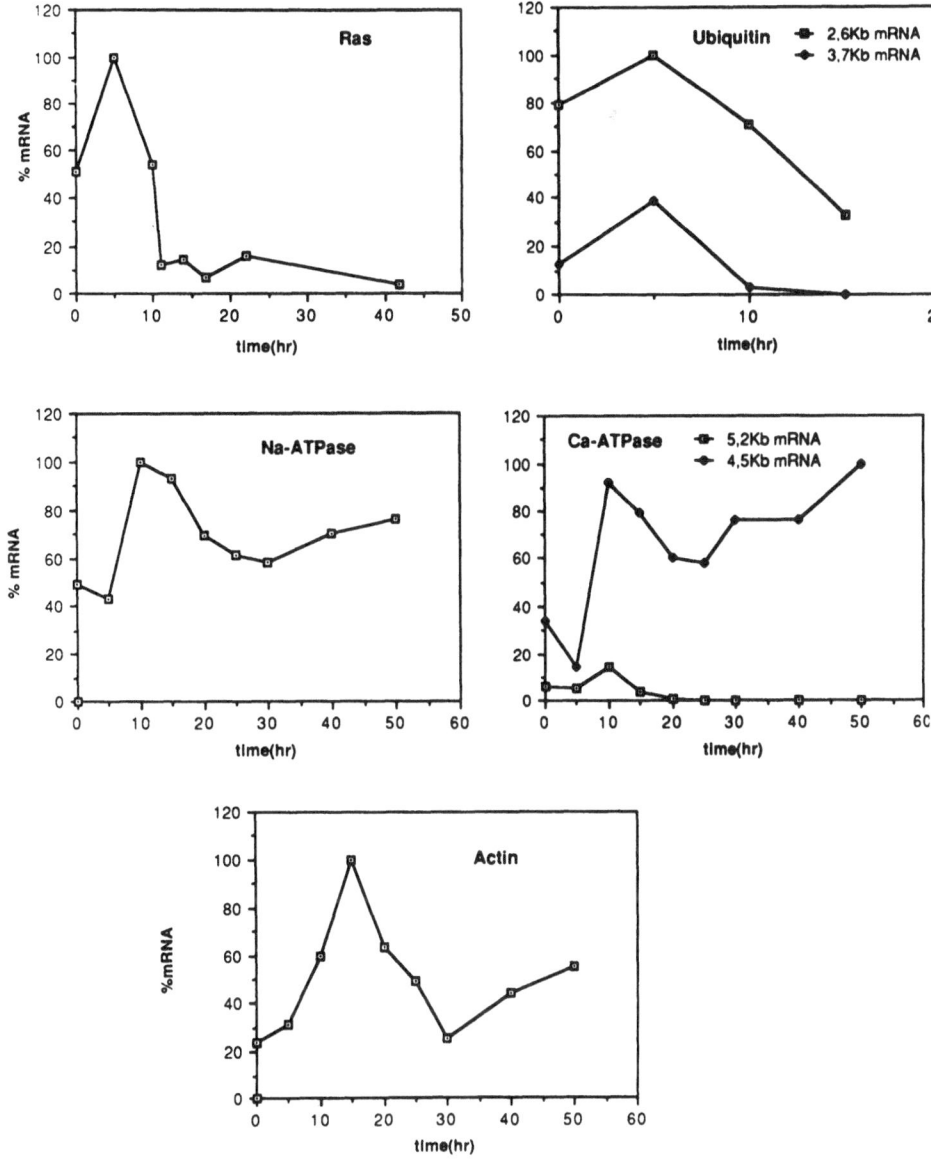

Fig. 3. Expression of cloned cDNA genes during <u>Artemia</u> development.
Total or Poly(A)$^+$ RNA was isolated at different stages of
<u>Artemia</u> development and the relative levels of expression
of each cDNA clone determined by northern- and dot-blotting. The hybridization of each RNA was quantified by
densitometry and corrected for the amount of RNA applied
to each sample, as determined by EtBr staining of RNA gels
or hybridization to rDNA probes for the dot-blots. Relative hybridization values are represented where we have
assigned the value 100% to the developmental period where
expression of the tested cDNA is maximal. Hatching of the
cysts occured at 14 hours of culture; over 80% of the cysts
were viable in these experiments.

of development. For actin, a three-fold increase in protein level is observed only after hatching of the nauplii [14], whereas maximal mRNA levels are reached by 15 hours of development just around hatching time.

Two mRNA species of 4.5 and 5.2 kb hybridize to the Ca^{++}/Mg^{++} ATPase cDNA clone. The larger mRNA is only present in the first hours of development and is no longer detectable after hatching. This mRNA could represent a precursor of the 4.5 kb mRNA which is stored in the cyst. An alternative hypothesis would be that the 5.2 kb mRNA codes for a Ca^{++}/Mg^{++} ATPase isoform that is expressed only during prenaupliar development of Artemia. Ubiquitin cDNA also hybridizes to two mRNAs of 3.5 and 2.7 kb but, in this case, the relative levels of both mRNAs are similar during development; both are present at high levels in the cyst and decrease thereafter.

The experiments described provide direct evidence for the presence in the cyst of stored mRNA for all genes studied. The relative levels of stored mRNA depend on the gene: ras and ubiquitin have very high levels of stored mRNA while the levels of stored mRNA is lower for the other genes. We can not estimate how long the stored mRNA persists during development, since it is not possible to measure RNA synthesis in the cysts. We can estimate from these data the RNA synthesis starts very soon after activation of the cyst since after five hours of development there is already a two-fold increase in steady state levels of ras mRNA. Accumulation of mRNA begins between five and ten hours of development for other genes, such as actins and ATPases.

Sequencing of cDNA clones has allowed us to establish the frequency of codon usage in Artemia (Figure 4). Our interest in this study comes from the fact that the Artemia genome has a very low G+C content (32%, [15]) in comparison to most other organisms. We wanted to know if this difference was reflected in the frequency of use of each codon. The codon frequency has been calculated from the deduced amino acid sequences of the cDNA clones described in this paper, the cDNA clone coding for the elongation factor 1a [16] and the last 30 amino acids deduced from the genomic sequences for histone H2A described in the accompanying paper (Renart et al., this volume). A total of 2766 codons were analysed.

The frequency of Artemia codon usage is compared in Figure 4 with that of Drosophila [17] and humans [18], whose DNA has a G+C content of 40 and 42%, respectively. Artemia codon usage frequency markedly differs from that of Drosophila and humans. For example, the codon CGU accounts for 52% of the arginine codons in Artemia, for 9% in humans and for 17% in Drosophila. Glycine is coded by GGU in 52% of the cases in Artemia, 15% in human and 22% in Drosophila.

In order to see if there was a preference for any base at the third codon position, we have calculated the frequency of each nucleotide at the third position in those cases were all four nucleotides in the third position code for the same amino acid (leucine, valine, serine, proline, threonine, alanine, arginine and glycine). The results are presented in Table 1. There is a marked preference for thymidine at the third position in Artemia when compared with human or Drosophila (46% in Artemia, 21% Human, 14% Drosophila). On the contrary, guanine is present at a very low frequency, 8%, in Artemia compared with 26% in humans and 30% in Drosophila. The frequency of adenine at the third codon position is similar in the three organisms compared, while the frequency of cytidine is slightly higher in human and Drosophila than in Artemia. There is a good correlation between the frequency of G+C at the third codon position and the G+C content of the DNA for each organism (Table 1), although the G+C frequency at the third codon position is always higher than the G+C content of the DNA.

324

aa	codon	Artemia	Human	Drosoph.	aa	codon	Artemia	Human	Drosoph.
F	UUU	0.57	0.35	0.30	Y	UAU	0.53	0.47	0.30
	UUC	0.43	0.65	0.70		UAC	0.47	0.53	0.70
L	UUA	0.10	0.05	0.02	H	CAU	0.53	0.42	0.36
	UUG	0.23	0.09	0.17		CAC	0.47	0.58	0.64
	CUU	0.32	0.11	0.06					
	CUC	0.14	0.22	0.14	Q	CAA	0.58	0.26	0.23
	CUA	0.08	0.07	0.06		CAG	0.42	0.74	0.77
	CUG	0.13	0.46	0.55					
					N	AAU	0.42	0.34	0.41
I	AUU	0.58	0.23	0.29		AAC	0.58	0.66	0.59
	AUC	0.33	0.64	0.60					
	AUA	0.09	0.13	0.11	K	AAA	0.55	0.45	0.18
						AAG	0.45	0.55	0.82
M	AUG	1.00	1.00	1.00					
					D	GAU	0.49	0.38	0.47
V	GUU	0.38	0.13	0.15		GAC	0.51	0.62	0.53
	GUC	0.35	0.27	0.30					
	GUA	0.16	0.09	0.06	E	GAA	0.78	0.40	0.20
	GUG	0.11	0.50	0.49		GAG	0.22	0.60	0.80
S	UCU	0.28	0.17	0.06	C	UGU	0.61	0.30	0.24
	UCC	0.22	0.26	0.29		UGC	0.39	0.70	0.76
	UCA	0.25	0.11	0.06					
	UCG	0.07	0.07	0.23	W	UGG	1.00	1.00	1.00
	AGU	0.07	0.11	0.12					
	AGC	0.10	0.29	0.24	R	CGU	0.52	0.09	0.17
						CGC	0.08	0.19	0.42
P	CCU	0.37	0.24	0.08		CGA	0.10	0.10	0.11
	CCC	0.14	0.41	0.40		CGG	0.04	0.15	0.12
	CCA	0.43	0.24	0.17		AGA	0.17	0.24	0.07
	CCG	0.06	0.11	0.35		AGG	0.09	0.23	0.11
T	ACU	0.37	0.20	0.09	G	GGU	0.52	0.15	0.22
	ACC	0.31	0.47	0.52		GGC	0.22	0.44	0.26
	ACA	0.25	0.21	0.14		GGA	0.22	0.17	0.47
	ACG	0.07	0.12	0.25		GGG	0.04	0.24	0.05
A	GCU	0.48	0.31	0.18	st	UAA	0.83	0.33	0.57
	GCC	0.34	0.40	0.54		UAG	0	0.15	0.23
	GCA	0.13	0.17	0.12		UGA	0.17	0.52	0.20
	GCG	0.05	0.12	0.16					

Fig. 4. Frequency of codon usage in _Artemia_. The frequency of use of each codon for protein synthesis was calculated in _Artemia_ from the deduced amino acid sequence of cloned genes and compared to the frequencies described for human and _Drosophila_.

SUMMARY AND PERSPECTIVES

We have used probes from other species and oligonucleotide probes to isolate the following genes from _Artemia_: ras, ubiquitin, Ca^{++}/Mg^{++} ATPase, Na^+/K^+ ATPase α subunit and two actin isoform genes. Nucleotide sequencing of these genes showed a significant similarity to homologous genes from other species (70-99%). The cloned genes have been used to measure the steady state level of their mRNA during _Artemia_ development. ras and ubiquitin mRNAs are stored in large amounts in the cyst and their levels decrease markedly during development. Actin and ATPase mRNAs, although

Table 1. Nucleotide Frequency at the Third Codon Position in Artemia[a]

Nucleotide	Artemia	Human	Drosophila
A	0.18	0.18	0.14
C	0.27	0.35	0.42
G	0.08	0.26	0.30
T	0.46	0.21	0.14
G + C	0.35	0.61	0.72
G+C content	32%	42%	40%

[a] Codons where all possible nucleotides at the third codon position code for the same amino acid have been analyzed and the frequency of each nucleotide at the third position of these codons determined for Artemia, human and Drosophila. The G+C content of genomic DNA for these species is indicated.

present in the cyst, increase their level during development, reaching high levels of expression by 10-15 hours of development. Accumulation of newly synthesized mRNA has been detected after five hours of development for ras mRNA and 10-15 hours for actin and ATPases mRNAs. Analyses of coding sequences for these genes has shown a high frequency of thymidine and low frequency of guanine at the third codon position when compared with other organisms; these data are in correlation with the low G+C content of Artemia DNA.

The isolation of Artemia cDNA clones for several genes and the characterization of their level of expression during development makes it possible to study mechanisms involved in the control of their expression. Genomic clones for these genes are being isolated in our laboratory in order to study their promoter elements. Structural and functional changes in the promoters and the protein factors that interact with them at the moment of their activation during development would be studied to determine how gene expression is regulated during the activation of Artemia cysts and their embryonic development.

ACKNOWLEDGMENTS

We want to thank Amador Ovejero and Antonio Fernandez for the preparation of the figures. This project was supported by grants from the Comision Asesora para la Investigacion Cientifica y Tecnica and Fondo de Investigaciones Sanitarias de la Seguridad Social.

REFERENCES

1. A. J. Wahba and C. L. Woodley, Molecular aspects of development in the brine shrimp Artemia, Prog. Nucl. Acid Res. Mol. Biol. 31:221 (1984).
2. C. Osuna and J. Sebastian, Levels of RNA polymerases during early larval development of Artemia, Eur. J. Biochem. 109:383 (1980).
3. U. Gubler and B. J. Hoffman, A simple and very efficient method of

generating cDNA libraries, <u>Gene</u> 25:263 (1983).

4. R. W. Ellis, D. DeFeo, J. M. Maryak, H. A. Young, T. Y. Shih, E. H. Chang, D. R. Lowy and E. M. Scoinick, Dual evolutionary origin for the rat genetic sequences of Harvey murine sarcoma virus, <u>J. Virol</u> 36:408 (1980).

5. M. Izquierdo, C. Arribas, J. Galcera, J. Burke and V. M. Cabrera, Characterization of a <u>Drosophila</u> repeat mapping at the early-ecdysone puff 63F and present in many eucaryotic genomes, <u>Biochem. Biophys. Acta</u> 783:114 (1984).

6. E. A. Fyrberg, B. J. Bondy, N. D. Hershey, K. S. Mixter and N. Davidson, The actin genes of <u>Drosophila</u>: Protein coding regions are highly conserved but intron positions are not, <u>Cell</u> 24:107 (1981).

7. G. E. Shull and J. B. Lingrel, Molecular cloning of the rat stomach $(H^+ + K^+)$ ATPase, <u>J. Biol. Chem.</u> 261:16788 (1986).

8. C. J. Brandl, M. Green, B. Korczak and D. H. Mac Lennan, Two $Ca2^+$ ATPase genes: Homologies and mechanistic implications of deduced amino acid sequences, <u>Cell</u> 44:597 (1986).

9. G. E. Shull, A. Schwartz and J. B. Lingrel, Amino acid sequence of the catalytic subunit of the $(Na^+ + K^+)$ ATPase deduced from a complementary DNA, <u>Nature</u> 316:691 (1985).

10. M. Morohashi and M. Kawamura, Solubilization and purification of <u>Artemia salina</u> (Na,K)-activated ATPase and NH_2-terminal amino acid sequence of its larger subunit, <u>J. Biol. Chem.</u> 259:14928 (1984).

11. G. E. Shull, J. Greeb and J. B. Lingrel, Molecular cloning of three distinct forms of the Na^+, K^+-ATPase α-subunit from rat brain, <u>Biochem.</u> 25:8125 (1986).

12. R. Serrano, Structure and function of proton translocating ATPase in plasma membranes of plant and fungi. <u>Biochem. Biophys. Acta</u> 947:1 (1988).

13. G. L. Peterson, L. Churchill, J. A. Fisher and L. E. Hokin, Structural and biosynthetic studies of the two molecular forms of the $(Na^+ + K^+)$-activated adenosine triphosphatase large subunit in <u>Artemia salina</u> nauplii, <u>J. Exp. Zool.</u> 221:295 (1982).

14. H. Grosfeld and U. Z. Littauer, The translation <u>in vitro</u> of mRNA from developing cysts of <u>Artemia salina</u>, <u>Eur. J. Biochem.</u> 70:589 (1976).

15. J. Cruces, M. -L. Wonenburger, M. Diaz-Guerra, J. Sebastian and J. Renart, Satellite DNA in the crustacean <u>Artemia</u>, <u>Gene</u> 44:341 (1986).

16. J. A. Lenstra, A. V. Vliet, A. C. Arnberg, F. V. Hemert and W. Moller, Genes coding for the elongation factor EF-1a in <u>Artemia</u>, <u>Eur. J. Biochem.</u> 155:475 (1986).

17. S. Aota, T. Gojobori, F. Ishibashi, T. Maruyama and T. Ikemura, Codon usage tabulated from the GenBank genetic sequence data, <u>Nucl. Acid Res.</u> 16:r315 (1988).

18. R. Lathe, Synthetic oligonucleotide probes deduced from amino acid sequence data. Theoretical and practical considerations, <u>J. Mol. Biol.</u> 183:1 (1985).

ARTEMIA MITOCHONDRIAL DNA

Roberto Marco, Beatriz Batuecas, Carmen Moratilla, Jesús Diez-
Sebastián, Ruben Gómez-Rioja, Fernando Díaz-Otero, Alfonso
Valencia, Manuel Calleja, José Ramón Valverde and Rafael
Garesse

Departmento de Bioquímica
de la UAM e Instituto de Investigaciones Biomédicas del CSIC
Facultad de Medicina, Universidad Autónoma
Arzobispo Morcillo, 4 Madrid 28029 Spain

INTRODUCTION

Previous work by members of our group [1,2,3] suggested that in Artemia
early development a considerable proportion of the mitochondria were stored
in an inactive state in the yolk granules from which they were released
during the developmental process, becoming active as power generators in
the nauplii. In order to test conclusively this idea, we turned to the
mitochondrial DNA as we thought that it could provide the best marker for
the presence and quantitation of these organelles, independent of their
functional status. Very little was known about this interesting informa-
tional molecule in Artemia [4], but in other systems as distant as sea
urchin, amphibians and mammals [5,6,7] it is known that it is basically
synthesized during oogenesis, the mature oocyte containing all the mitochondria necessary for early embryogenesis.

In this paper we review the current state of our analysis of Artemia
mt-DNA, the first crustacean mt-DNA well characterized to date. We have
determined the overall main genomic organization of Artemia mt-DNA, and
we discuss it in terms of its evolutionary implications. The relationships
among the New and Old World species of Artemia has begun to be established
using restriction site analysis of their mt-DNAs. Finally, the sequences
of three different genes contiguously mapping in the Artemia mt-DNA have
been obtained: apocytochrome b, and NAD dehydrogenase subunits 6 and 4L.
Using several approaches at the level of comparison between the primary
and secondary structures of these genes, it is possible to start outlining
some of the main evolutionary factors behind the conservation of sequence
in mt-DNAs.

THE GENOMIC ORGANIZATION OF ARTEMIA mt-DNA

We started by cloning the mt-DNA purified from a sample of Artemia
franciscana, obtained from Metaframe Corporation. San Francisco Bay [8].
Artemia mt-DNA is one of the smallest mt-DNAs studied to date. It is only
15 kilobase pairs (kb) kbs in length. Only the 14,284 bp of the nematode

worm _Ascaris suum_, studied by Wolstenholme and coworkers [9], is smaller than the _Artemia_ mt-DNA. The physical map of the _Artemia_ mt-DNA was established by ordering the DNA fragments obtained with the following restriction enzymes: BamHI, EcoRI, Hind III, Pst I, Pvu II, and Ava I. In Figure 1, we show the physical organization of these restriction sites in _Artemia franciscana_ mt-DNA, as well as the parts of the molecule that we have sequenced so far.

The genomic organization of the mt-DNA major genes, i.e., the ribosomal RNAs and 11 of the 13 protein genes, has been determined by a sequencing strategy [10]. We had to turn to this approach since the more rapid approach of using hybridization with _Drosophila_ mt-DNA probes for known genes [11,12], which allowed mapping the more conserved genes for 16 S rRNA and cytochrome oxidase subunit I [8] failed in detecting the rest of the genes, due very likely to little conservation of the sequence (see below). We subcloned the mt-DNA (see Figure 1 for the location of the subclones used), and by sequencing the ends of the subclones, translating the sequences and comparing them to _Drosophila_ and vertebrate sequences, the genes shown in Figure 2 were mapped. As mentioned above, we had initially mapped only 11 of the 13 genes [10]. Recently (unpublished results), by extending the same strategy we have localized an additional gene, the cytochrome oxidase subunit II gene, which is now localized in Figure 2. Among the main genes only one protein gene, ATPase subunit 8, has not been located. There is enough space in the same position as in _Drosophila_ mt-DNA to encode a gene of similar size to ATPase 8 in other sequenced animal mt-DNAs, and in fact we have already identified there an open reading frame (unpublished results). However, the available partial sequence does not show sufficient homology to the corresponding gene of _Drosophila_ to confirm the identification.

These results provide strong support for our conclusion [10] that the overall organization of the main genes in _Artemia_ mt-DNA is exactly the same as the organization found in insects, as exemplified by the mt-DNAs of _Drosophila_ _yakuba_ [13] and _Drosophila_ _melanogaster_ [11,12]. This similar organization holds even for the direction of transcription, i.e., the mt-DNA strand on which the genes are encoded. Starting from the putative origin of replication and transcription, located in the AT rich region of the _Drosophila_ mt-DNA [14], we have the 16 S rRNA and ND-1 in the minus strand, then apocytochrome b and ND6 in the plus strand, ND4L, ND-4 and ND-5 in the minus strand and finally ND-3, cytochrome oxidase subunit III, ATPase subunit 6, ATPase subunit 8 (?, not yet completely identified), cytochrome oxidase subunits II and I, and ND-2 in the plus strand (Figure 2). This order is exactly the same in all the arthropod mt-DNAs studied to date, in insect (_Drosophila_) and crustacean classes (_Artemia_).

THE EVOLUTIONARY DISTANCE BETWEEN THE DIFFERENT _ARTEMIA_ SPECIES AS INDICATED BY THE VARIATION IN THEIR mt-DNA

Having cloned the mt-DNA from _Artemia franciscana_, we have started studying the mt-DNA organization in different species of _Artemia_, mainly those located in Spain. As shown in Figure 3, preliminary analysis by restriction enzyme site location indicates a great deal of divergence between New and Old World bisexual species and among Old World bisexual and parthenogenetic species. This is important since there is very little biological information on which to base the distances among parthenogenetic animals, as they are by definition not amenable to interbreeding. Our preliminary analysis indicates that there are very few restriction enzyme

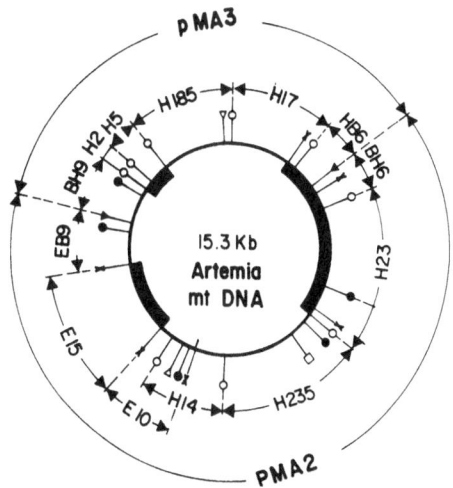

Fig. 1.   Physical map of <u>Artemia</u> <u>franciscana</u> mt-DNA showing
the subclones of which the ends have been sequenced
to establish the gene organization as well as the
part of the molecule which we have sequenced (dark
lines). Restriction sites: (▼) Bam H-1, (♀) Hind
III, (⚡)Eco R-I, (Ɣ) Pst I, (⚐) Pvu II and (●)
Ava I. The subclones are named by the initial(s) of
the restriction enzyme(s) sites bordering the fragment
and a number which indicates its length. With the
exception of single digit numbers which correspond to
hundreds of base pairs, the first digit corresponds to
kilobases.

sites that have been conserved among Old World bisexual and parthenogenetic
species, and even between diploid and tetraploid parthenogenetic strains
there are also many changes. To date we have examined these DNAs using 3
restriction enzymes, EcoR 1, Bam H1 and Hind III, which together produce
more than 40 different fragments in these DNAs. Only two coincide, which
indicates a very high degree of divergence at the sequence level among
these DNAs. Using Upholt's statistical analysis [15] and the molecular
estimation by Brown et al.[16] for the <u>Artemia</u> mt-DNAs which share at
least one fragment, it can be deduced that Old World bisexuals diverged
from tetraploid parthenogenetic species 9±2 million years (MY) ago. Diploid
and tetraploid parthenogentic strains found in Spain diverged 7±1.5 million
years (MY) ago. It would be interesting to have samples of the bisexual
<u>Artemia</u> <u>urmiana</u>, the proposed ancestor of the current parthenogenetic
strains, to verify this relationship and the actual time at which the
emergence of parthenogenesis occured. Unfortunately, the current samples
of <u>Artemia</u> from Lake Urmia (Iran), seem to be of the parthenogenetic type
(F. Amat, R. Browne, personal communication). Old World bisexuals diverged
from New World bisexuals even longer ago, since they do not share any
restriction site fragment. These data, which have to be extended in the
future using more enzymes and sequencing data, largely agree with the
genetic distances obtained by Abreu-Grobois using allozymic data [18]. The
main discrepancy, which should be clarified when more molecular information
becomes available, is on the timing of the branching of different ploidy
populations of <u>Artemia</u> <u>parthenogenetica</u>, which was of much more recent
origin using Abreu-Grobois's data.

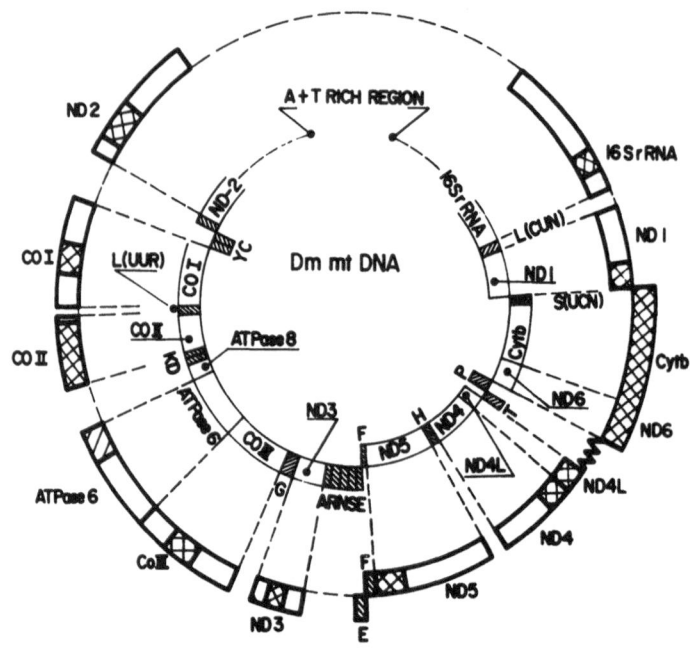

Fig. 2. Gene organization of <u>Artemia</u> mt-DNA (outside)
compared to that of <u>Drosophila</u> (inside).
The coding strand is indicated by the side
of the circle on which the box is drawn.
tRNAs are indicated by single letter amino
acid code.

EVOLUTIONARY IMPLICATIONS OF THE MAIN GENE ORGANIZATION OF mt-DNAS

In Figure 4, the classical evolutionary tree of the metazoan kingdom
is represented. This tree is the result of morphological and embryological
considerations. In the animals showing bilateral symmetry, two main
branches are recognized: the protostome and deuterostome branchs of evolution. Protostomian animals are those animals in which the larval or the
definitive mouth appears to correspond to the initial embryonic mouth,
marked by the initial blastopore in the gastrula. In deuterostomes, the
definitive mouth is formed in a secondary process at some distance anter-
iorly without having any dependence upon the blastopore [17]. Additionally,
the animals in each branch show high analogy in their initial cleavage
plan. Deuterostomes undergo a radial mode of cleavage by which the different blastomeres come to lie one directly on top of another, as seen in the
sea urchin and primitive vertebrates. Cleavage in most protostomes follows
a spiral mode, in which the blastomeres are cleaved and displaced clock
or counter clockwise in the blastula, so that they come to lie in between
the blastomeres in the next row of the blastula. Mt-DNAs have been studied
in the animals shown in frames in Figure 4. In the deuterostomia these
include echinoderms, the sea urchins <u>Paracentrotus</u> <u>lividus</u> and <u>Strongylo-
centrotus</u> <u>purpuratus</u> [19,20], one primitive vertebrate, the amphibian
<u>Xenopus</u> <u>laevis</u> [21] and several mammals, mouse [22], cow [23] and humans
[24]. In the protostomian branch, the gene organization has been studied
in the nematode worm, <u>Ascaris</u> <u>suum</u> [9], and in the more highly evolved
animals of this evolutionary branch, the arthropods, both in <u>Drosophila</u> [11,

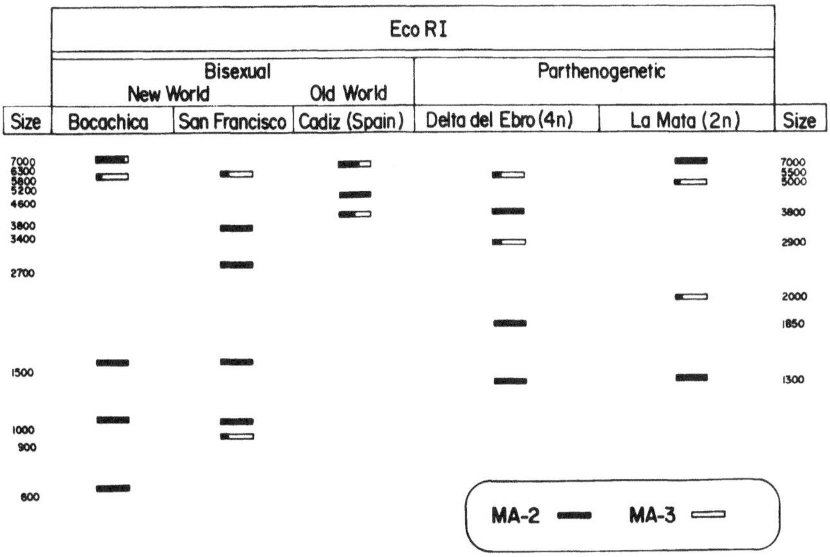

Fig. 3. Restriction site polymorphism of mt-DNAs of
Artemia parthenogenetica and Old and New World bisexuals. The fragments hybridizing to the two Bam Hl
subclones from Artemia franciscana (Fig. 1) are indicated (stippled vs non-stippled). New World bisexuals:
San Francisco and Bocachica (Venezuela), Old World
bisexual (Cadiz, Spain), tetraploid parthenogenetic
Artemia (Delta del Ebro, Spain) and diploid parthenogenetic Artemia (La Mata, Spain).

12,13] and in Artemia [10]. Therefore, the evolutionary relationships can
be studied at the molecular level using mt-DNA, which is believed to have
a conserved rate of molecular evolution, behaving as a reasonably good
molecular clock [16]. This can be studied at the level of protein evolution, by comparison of sequences (see next section of this paper) or at
the level of the gene organization. In fact, it has been proposed that
gene rearrangements rather than point mutations drive morphological evolution [25]. Therefore, we think it is interesting to discuss the differences
so far encountered in the gene organization of mt-DNAs of these species,
distributed across the evolutionary tree.

In the deuterostomian branch, the agreement between the different mt-
DNAs known is quite good. Vertebrates maintain exactly the same organization of main genes and t-RNA genes. The sea urchin mt-DNAs show differences, but the basic mt-DNA plan can still be recognized. Two main genes
differ in location (Fig. 5): 16 S rRNA, which is located between the ND-2
and cytochrome oxidase subunit I genes, and ND-4L, which is located between
ND-3 and ND-4 in vertebrates and between cytochrome oxidase subunit II and
ATPase 8 in sea urchins. tRNA genes are extensively rearranged in sea
urchin mt-DNA, appearing in a cluster close to the replication and transcription origin in which 15 of the 22 tRNAs genes are located. Nevertheless, 6 out of 7 tRNAs that remain interspersed with the main genes are in
the same relative position as the corresponding genes in mammalian mt-DNA.
Only S (UCN) tRNA differs in location but maintains its direction of tran-

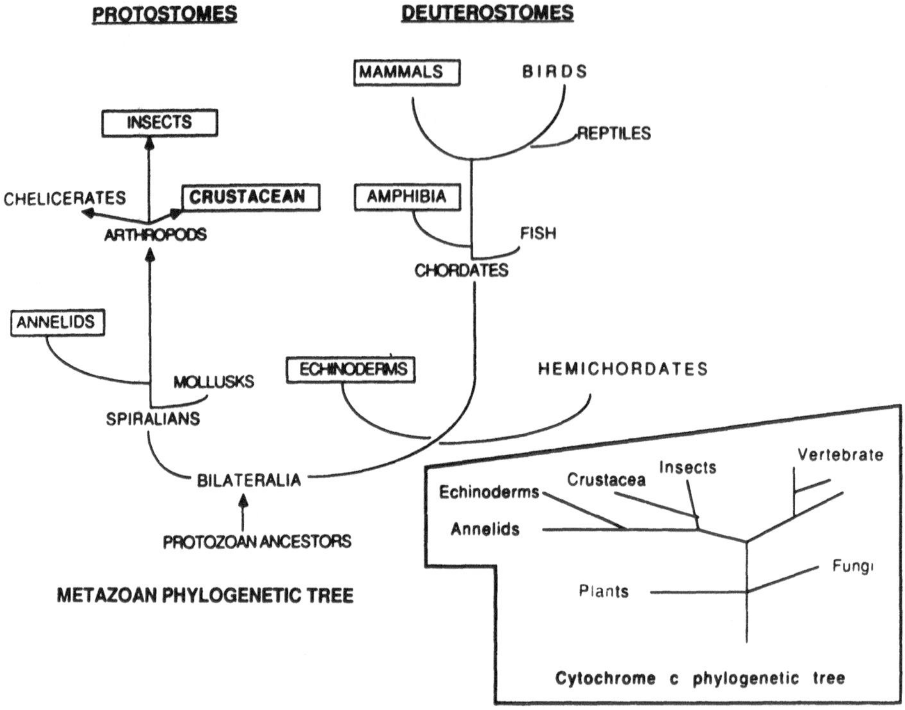

Fig. 4. Classical phylogenetic tree based on embryological and morphological criteria and tree based on cytochrome c sequences [17]. Enclosed in frames are the classes of animals from which the mt-DNA organization is known.

scription. In the case of the main genes the direction of transcription, i.e., the sense strand, is also conserved. All main genes but ND-6 are coded by the plus strand; ND-6 is coded by the opposite minus strand. Some of the tRNAs which are in a cluster in P. lividus differ in direction of transcription, i.e, E, P and C tRNAs are coded in the plus strand of the P. lividus mt-DNA and A, V and D tRNAs are coded in the minus strand, unlike vertebrate mt-DNA.

If we consider now the organization of arthropod mt-DNAs, it is remarkable to see how exactly the overall organization of the main genes has been conserved, both in organization and in direction of transcription, i.e., sense strand (Figure 2). In fact, as already indicated by Wolstenholme [26] for Drosophila, it is not very difficult with a few transpositions to switch from the vertebrate to the arthopod main gene organization. The gene segment from ND-1 through 16 S rRNA and 12 S rRNA to the intergenic region and the one from ND-4L to ND-5 seem to have undergone inversions in the arthropods, with respect to the deuterostomian mt-DNAs (Fig. 5). This type of inversion has also been described for chloroplast DNA [27]. On the other hand, tRNA genes have been altered more profoundly. We have not yet completed the sequence of the Artemia mt-DNA and we do not know where all of the tRNA genes are located, but some of them do not maintain the positions found in Drosophila mt-tRNAs. For instance K, D, P, T and S (UCN) tRNAs have not been found in the corresponding positions

334

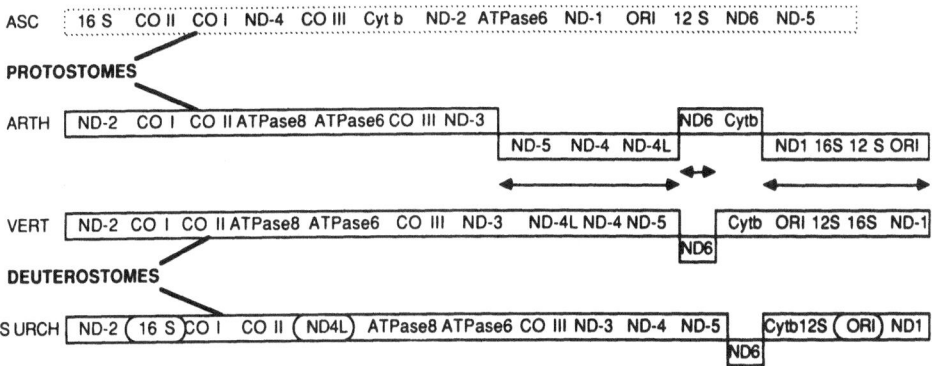

Fig. 5. Comparative gene organization of animal mitochondrial
         DNAs. Asc = _Ascaris_ (nematodes), Arth=Arthropods, Vert=
         Vertebrates and S Urch=sea urchin (echinoderms).
         ←→ Possible Inversions   ⊂⊃ Possible Translocations.

that they occupy in _Drosophila_ (Fig. 2 and unpublished results). This
difficulty in identifying tRNAs in _Artemia_ mt-DNA, in spite of having
already sequenced close to 50% of the whole molecule, suggests that there
may be one or two clusters of tRNAs in _Artemia_.

While all the mt-DNAs discussed so far tend to follow the general
evolutionary trend, a more primitive protostome mt-DNA has also been
sequenced and the gene organization reported [9]. Surprisingly, as can be
estimated from the published data [9], _Ascaris_ mt-DNA is quite unrelated
both to the deuterostome and the arthropod mt-DNAs. From all the combina-
tions that can be adopted by the 12 main genes in the case of _Ascaris_ mt-
DNA (since no ATPase 8 gene has been identified) only two features are
conserved between arthropod and nematode DNA: the orientation of transcription of the ND-6 gene, which is in the same direction as the apocytochrome
b and not opposite as in the deuterostome mt-DNA, and the close association
between the cytochrome oxidase subunits I and II. It is also in this region
that some slight conservation of the tRNA genes is found: the C tRNA is
close to the cytochrome oxidase subunit I and the D tRNA is close to the
cytochrome oxidase subunit II. The 20 additional tRNA genes are located
in completely different positions in insect and nematode mt-DNAs.

These differences are somewhat difficult to reconcile with the "classical" evolutionary tree, since it seems that the arthropod mt-DNA is closer
to the deuterostome mt-DNA than to the annelid mt-DNA. Interestingly, this
closer relationship is also visible in phylogenetic trees based on other
molecular characteristics, such as cytochrome c protein sequences. As
shown in Figure 4, the arthropod cytochrome c sequences [28] are closer to
the vertebrates than to the annelids. In this tree the echinoderms do not
appear at the base of the chordate evolution, but appear together with the
mollusks (not shown in the Figure). Therefore, it may be possible to
question based on all these data the traditional way of drawing the evolutionary tree for higher eucaryotes. In fact, it is not clear at all,
whether the pattern of cleavage of the arthropods is really of the spiral
type and whether they can be truly classified as protostomes, since at
least in higher arthropods, the pattern of gastrulation has little resemblance with that of any other group. If relationships based on molecular

data were to be held as more important and confirmed by their extension to other types of animals, they would suggest a different branching of the arthropod phyla, separately from the true spiralians like the annelids. Although there is no doubt that arthropods derive from a segmented ancestor, their relationship to annelids has been a matter of debate [29]. A recent analysis [30] of cytoplasmic 18 S rRNA sequences is also unable to demonstrate at the molecular level the existence of two main branches, protostome and deuterestome, of metazoan evolution. A rapid radiation of four coelomate groups is suggested, namely: chordata, echinodermata, arthropoda and eucoelomata protostomes (annelida, molusca, etc.), more in agreement with the mt-DNA data. An alternative explanation, one that cannot yet be ruled out for lack of information on the mt-DNA of species at intermediate evolutionary distances, is that this mt-DNA organization is the result of convergent evolution rather than of phylogenetic relationship. Nevertheless, the data available in the arthropod phylum is in agreement with the idea that the main gene organization reported here for Artemia and Drosophila mt-DNAs is fixed in this phylum, as the corresponding one has become fixed in vertebrates, independently of the evolutionary mechanism involved.

THE CONSERVATIVE FEATURES OF THE mt-DNA GENES AT THE SEQUENCE LEVEL

We have completely sequenced so far (unpublished data) 3 main Artemia mt-DNA genes: apocytochrome b, ND-6 and ND-4L. In Figure 6 we present the sequences of the 3 genes at the amino acid level, using the genetic coding for insect mt-DNA, which seems also to be applicable to crustacean mt-DNA. Since there is already considerable information across the animal kingdom on these genes, we have started to compare them at the level of the primary sequences and also at the level of secondary structures. Secondary structures can be deduced from the primary sequence by using computational methods that calculate several important parameters, such as hydrophobicity, alpha and beta sheet potential, and beta turns [31,32, Gomez-Rioja et al., in preparation]. Here we review the preliminary results of this analysis.

The three sequenced genes from the brine shrimp mt-DNAs correspond to three different categories of genes: apocytochrome b is a gene whose primary sequence is relatively well conserved all along the gene (Fig. 7), ND-6 is only conserved in the central region, especially if we compare evolutionarily unrelated animals (Figure 8) and ND-4L is a poorly conserved gene, in terms of primary sequence (Figure 9). The low primary sequence conservation of most of ND-6 and ND-4L is observed when we compare mt-DNAs from animals of different classes, but not in the closely related animals. For instance, Drosophila yakuba and melanogaster still share much primary sequence similarity in the case of their ND-4L genes (data not shown). It is interesting to note that the situation is quite different when the hydrophobicity predictions are taken into account, as can be seen in Figures 7 to 9, since the conservation at this level of analysis is quite extraordinary, even between the more separated animals. Although this conclusion is general, it is interesting to report in this book on Artemia, that in several aspects the Artemia mt-DNA genes are relatively more changed among the different animals compared. For example, in the case of apocytochrome b, the hydrophobic regions are somewhat altered. The amino terminal part of the molecule, in particular the second loop, is missing, although the conservation of sequence is similar in the rest of the molecule. On the other hand, from residue 200 to 270 in which the conservation of the sequence is lower then in the rest of the molecule, the conservation of this structural parameter is striking (Fig. 7). A similar parallelism is found at the level of beta potential, at least for apocytochrome b (data not shown).

APOCYTOCHOME b

	10	20	30	40
	MKCYLYVSRN	LPLKIINSAL	VDLPVPANIS	IWWFKISPG
	50	60	70	80
	PMSSYLNCYW	TFLAMHITAS	VDLAFSSVAH	ICADVNYGWL
	90	100	110	120
	LATVHANGAS	FFFICIYFHI	GRGMYYGSFH	YFETWMTGIA
	130	140	150	160
	LLFLVMAAAF	LGYVLPWQM	SFWGATVITN	LVSAVPYIGN
	170	180	190	200
	DVVQWLLWGF	AVDNPTLTRF	FTFHFLIPFL	VAGLTMIHLL
	210	220	230	240
	FLHQSGSNNP	LGINANLDKL	PFHPYFTIKD	TVGFMVLIFF
	250	260	270	280
	LVTLSLTSPY	LLGDPDNFIP	ANPLVTPAHI	QPEWYFLFAY
	290	300	310	320
	AILRSIPNKL	GGVIALVSSI	LILVSLPFTF	QPQFRGLEFY
	330	340	350	360
	SVAQPLFWSW	VSVFLLLTVI	GARPVEDPYI	FLGQILTCAY
	370			
	FSYFVFTPIV	INLNDKIV		

NAD DEHYDROGENASE-6

	10	20	30	40
	IKLPFDILGS	IVVISMFMLL	MNHPLAFTLS	LFVKTLLICV
	50	60	70	80
	MLKNVSLWIS	LILFLIFLGG	ILVMFIIVSS	LSANEKFAVD
	90	100	110	120
	LTSFMWVVET	IVLSFLVLNK	NFMFMSPSSG	YLYPTDFVII
	130	140	150	160
	NFNVNSLTML	AYSFMVVYLF	LALLLVIDFL	NSNKKPLASM

I

NAD DEHYDROGENASE -4L

	10	20	30	40
	IMMIYLSLSL	GLLIFSSSNK	HLLVTLLSLE	FLILLLFSLL
	50	60	70	80
	VYSNYMSMIN	AFIFLSVTVC	EGALGLSVLV	SLVRSSGSDQ

VQFLNE

Fig. 6. Derived amino acid sequences of Artemia mitochondrial
apocytochrome b, ND-6 and ND-4L.

The conclusion of this comparative study of three mt-DNA genes is
clear. These proteins may be basically conserved at the level of their
secondary structures. Hydrophobic loops alternate with beta turns as
would be expected for membrane proteins. The number, size and location
of the hydrophobic regions as well as the beta turns are well conserved
independent of whether this conservation also extends to the level of the
primary sequence. The secondary structure is low in alpha chains and high
in beta sheet formation, both in the hydrophobic loops and in the intervening shorter hydrophylic regions (data not shown). One of the proteins

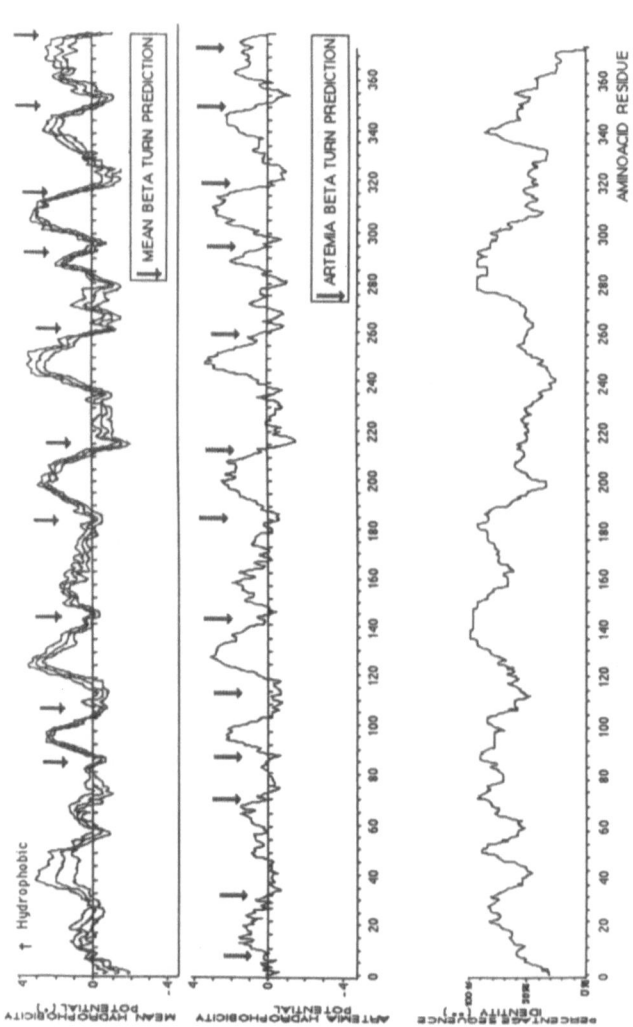

Fig. 7. Hydrophobicity prediction and primary structure comparisons for Artemia and other vertebrate and invertebrate mitochondrial apocytochrome b proteins.

*Mean hydrophobicities have been calculated comparing Artemia franciscana, Drosophila melanogaster, Drosophila yakuba, Xenopus laevis, Mus musculus and Homo sapiens. The accompanying curves indicate the behaviour of the standard deviations.

**Percentage of sequence identity has been calculated after alignment of the sequences of A: franciscana, D. melanogaster, D. yakuba, X. laevis, M. musculus and H. sapiens to maximize identity.

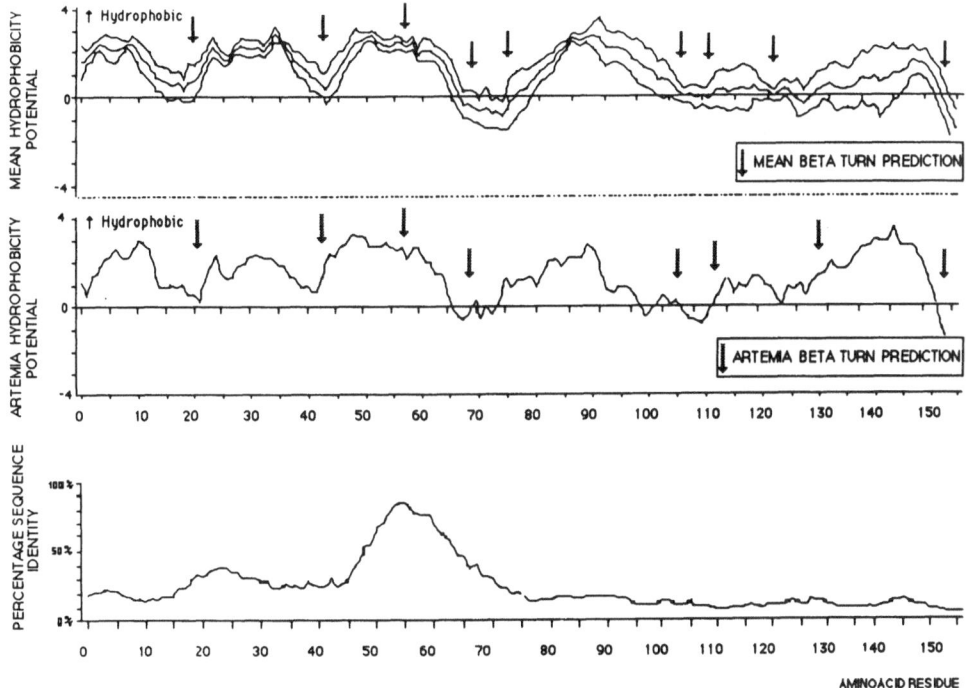

Fig. 8. Hydrophobicity and primary structure comparisons for <u>Artemia</u> and vertebrate and invertebrate mito-chondrial NAD dehydrogenase subunit 6. Symbols as in Fig. 7.

analyzed, apocytochrome b, shows considerable sequence conservation in several of its hydrophobic regions, probably those involved in forming the cytochrome b binding site. The ND proteins are part of a large complex, the electron transport chain complex I, NADH-ubiquinone oxidoreductase, in which many polypeptides take part, 6 encoded in the mt-DNA and the rest in the nuclear genome. The lower sequence conservation in the case of the two mt-ND genes studied here, ND-6 and ND-4L, suggests that they mostly contribute to the overall conformation of the complex. In fact, only one hydrophobic domain of the ND-6 gene (Figure 8) is strongly conserved at the level of the primary sequence, suggesting that it may be involved in the enzymatic function of the complex in a more direct way. Finally, ND-4L shows three well conserved hydrophobic domains (Figure 9), although the sequence conservation is quite low. Again, in the case of <u>Artemia</u>, there is some shifting in the location of the hydrophobic loops and beta turns, but these structural features are well conserved.

ACKNOWLEDGEMENTS

This work has been supported by grants from FIS and DGCYT of Spain. Dr. Fr. Amat kindly provided the samples of <u>Artemia</u> from Spain and other localities.

339

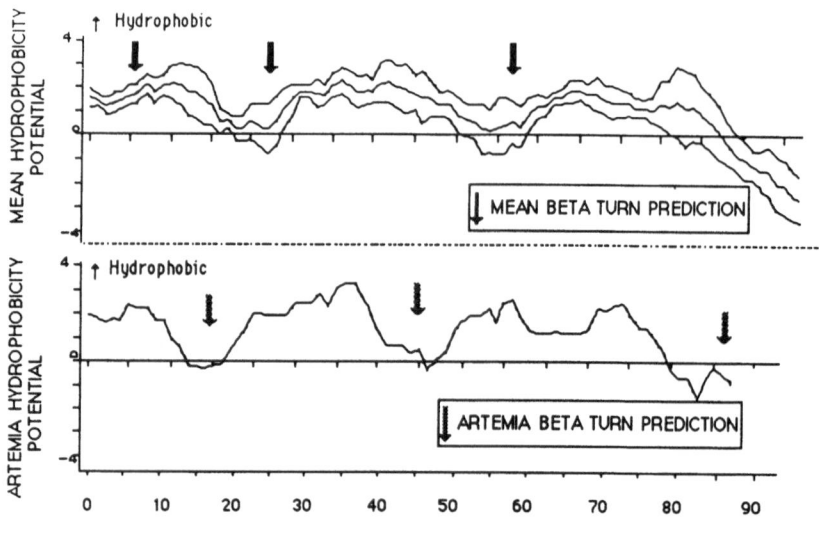

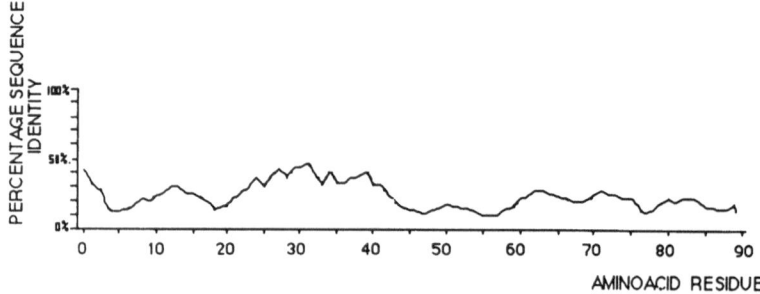

Fig. 9. Hydrophobicity and primary structure comparisons for
Artemia and vertebrate and invertebrate mitochondrial
NAD dehydrogenase subunit 4L. Symbols as in Fig. 7.

REFERENCES

1. R. Marco, A. Garesse and C. G. Vallejo, Mitochondrial unmasking and
   yolk platelets metabolization during early development in *Artemia
   salina*, in: "The Brine Shrimp *Artemia*," Vol. 2, G. Persoone, P.
   Sorgeloos, D. Roels and E. Jasper, eds., Universa Press, Wetteren
   (1980).
2. R. Marco, R. Garesse and C. G. Vallejo, Storage of mitochondria in the
   yolk platelets of *Artemia* dormant gastrulae, *Cell. Mol. Biol.* 27:
   515 (1981).
3. R. Marco, B. Batuecas, M. Calleja, M. Carratalá, M. Cervera, A. Domingo,
   C. Ferreiro, R. Garesse, C. Urquía and I. Vernós, Understanding the
   organization of cell metabolism in early embryonic systems. Developmental implications, in: "The Organization of Cell Metabolism,"
   R. G. Welch and J. G. Clegg, eds., Plenum, New York (1987).
4. H. Schmitt, J. S. Beckmann and U. Z. Littauer, Transcription of super-
   coiled mitochondrial DNA by bacterial RNA polymerase, *Eur. J. Biochem.* 47:227 (1974).
5. L. Pikó, A. Tyler and J. Vinograd, Amount, location, priming capacity,
   circularity and other properties of cytoplasmic DNA in sea urchin
   eggs, *Biol. Bull.* 132:68 (1967).
6. I. B. Dawid, Evolution of Mitochondrial DNA sequences in *Xenopus*,
   Develop. Biol. 29:139 (1972).

340

7. G. S. Michaels, W. W. Hauswirth and P. J. Laipis, Mitochondrial DNA copy number in bovine oocytes and somatic cells, Develop. Biol. 94: 246 (1982).

8. B. Batuecas, R. Marco, M. Calleja and R. Garesse, Molecular characterization of Artemia mitochondrial DNA: cloning, physical mapping and preliminary gene organization, in: "Artemia Research and its Applications," Vol. 2, W. Decleir, L. Moens, H. Slegers, E. Jasper and P. Sorgeloos, eds., Universa Press, Wetteren (1987).

9. D. R. Wolstenholme, J. L. MacFarlaine, R. Okimoto, D. O. Clary and J. A. Wahleithner, Bizarre tRNAs inferred from DNA sequences of mitochondrial genomes of nematode worms, Proc. Natl. Acad. Sci. U.S.A. 84:1324 (1987).

10. B. Batuecas, R. Garesse, M. Calleja, J. R. Valverde and R. Marco, Genome Organization of Artemia mitochondrial DNA, Nucleic Acid Res. 16:6515 (1988).

11. M. H. L. De Bruijn, Drosophila melanogaster mitochondrial DNA: a novel organization and genetic code, Nature 304:234 (1983).

12. R. Garesse, Drosophila melanogaster mitochondrial DNA: Gene organization and evolutionary considerations, Genetics 118:649 (1988).

13. D. R. Woltensholme and D. O. Clary, Sequence evolution of Drosophila mitochondrial DNA, Genetics 105:725 (1985).

14. J. M. Goddard and D. R. Woltensholme, Origin and direction of replication in mitochondrial DNA molecules from Drosophila melanogaster, Proc. Natl. Acad. Sci. 75:3886 (1978).

15. W. B. Upholt, Estimation of DNA sequence divergence from comparison of restriction endonuclease digests, Nucleic Acid Res. 4:1257 (1977).

16. C. Moritz, T. E. Dowling and W. M. Brown, Evolution of animal mitochondria: Relevance of population biology and systematics, Ann. Rev. Ecol. Syst. 18:262 (1987).

17. R. A. Raff and T. C. Kaufman, "Embryos, Genes and Evolution," MacMillan New York (1983).

18. F. A. Abreu-Grobois, A review of the genetics of Artemia, in: "Artemia Research and its Applications," Vol. 2, W. Decler, L. Moens, H. Slegers, E. Jaspers and P. Sorgeloos, eds., Universa Press, Wetteren (1987).

19. P. Cantatore, M. Roberti, G. Rainadi, C. Saccone and M. N. Gadaleta, Clustering of tRNA genes in Paracentrotus lividus mitochondrial DNA, Current Gen. 13:91 (1988).

20. H. T. Jacobs, D. J. Elliot, V. B. Math and A. Farquharson, Nucleotide sequence and gene organization of sea urchin mitochondrial DNA, J. Mol. Biol. 202:185 (1988).

21. B. A. Roe, D. -P. Ma, R. K. Wilson and J. F. -H. Wong, The complete nucleotide sequence of the Xenopus laevis mitochondrial genome, J. Biol. Chem. 260:9759 (1985).

22. M. J. Bibb, R. A. Van Etten, C. T. Wright, M. W. Walberg and D. A. Clayton, Sequence and gene organization of mouse mitochondrial genome, Cell 26:167 (1981).

23. S. M. Anderson, M. H. L. De Bruijn, A. R. Coulson, J. Drouin, I. C. Eperon, F. Sanger and I. G. Young, The complete sequence of bovine mitochondrial DNA: conserved features of the mammalian mitochondrial genome, J. Mol. Biol. 156:683 (1982).

24. S. M. Anderson, A. T. Bankier, B. G. Barrell, M. H. L. De Bruijn, A. R. Coulson, J. Drouin, I. C. Eperon, B. Nierlich, A. Roe, F. Sanger, P. H. Schreier, A. J. H. Smith, R. Staden and I. G. Young, Sequence and organization of human mitochondrial genome, Nature 290:457 (1981).

25. A. C. Wilson, V. M. Sarich and L. R. Maxson, The importance of gene rearrangement in evolution, Proc. Nat. Acad. Sci. 71:3028 (1974).

26. D. R. Wolstenholme, D. O. Clary, J. L. MacFarlane, S. A. Wahleithener and L. Nilcox, Organization and evolution of invertebrate mitochondrial genomes, in: "Artemia Research and its Applications," Vol. 2, W. Decleir, L. Moens, H. Slegers, E. Jaspers and P. Sorgeloos, eds., Universa Press, Wetteren (1987).

27. J. R. Palmer, Comparative organization of chloroplast genomes, Ann. Rev. Genet. 19:325 (1985).

28. R. M. Schwartz and M. O. Dayhoff, "Cytochromes Atlas of Protein Sequence and Structure," Vol. 5, Suppl 3, National Biomedical Foundation, Washington (1978).

29. M. S. Gardiner, "The Biology of Invertebrates," McGraw-Hill, New York, (1972).

30. K. G. Field, G. J. Olsen, D. J. Lane, S. J. Giovannoni, M. T. Ghiselin, E. C. Raff, N. R. Pace and R. A. Raff, Molecular phylogeny of the animal kingdom, Science 239:748 (1988).

31. P. Y. Chou and G. D. Fasman, Empirical predictions of protein conformation, Ann. Rev. Biochem. 47:251 (1978).

32. J. Kyte and R. F. Doolittle, A simple method for displaying the hydropathic character of a protein, J. Mol. Biol. 157:105 (1982).

IDENTIFICATION OF A PROTEIN WHICH SPECIFICALLY BINDS A HIGHLY REPETITIVE

HETEROCHROMATIC DNA FAMILY (Alu 1 FAMILY) OF <u>ARTEMIA FRANCISCANA</u>

Roberta Benfante, Claudio Barigozzi, Stefano Tenca and
Gianfranco Badaracco

Dipartimento di Genetica e
di Biologia dei Microrganismi Universita di Milano
Via Celoria 26
20133 Milano (Italy)

INTRODUCTION

Heterochromatin, the most highly condensed region of interphase
chromosomes and generally located in centromeric and telomeric regions,
contains nucleotide sequences of length from about 10 to more than 10^3
base pairs repeated thousands to millions of times per haploid genome and
arranged in long tandem arrays (satellite DNA) [1-5]. Even if the sequences
of the heterochromatic DNA are assumed to be the primary cause of differential condensation, other chromosomal constituents, such as proteins and
RNA, must mediate this folding [6,7]. Therefore, the understanding of the
molecular mechanisms involved in heterochromatin condensation requires
knowledge of heterochromatic DNA sequence organization as well as identification of proteins with the potential role to maintain higher order
heterochromatin structure.

A heterochromatin specific nucleosomal protein (D1 protein) was
isolated from <u>D</u>. <u>melanogaster</u> [8], and was demonstrated to be a stoichiometric component of <u>Drosophila</u> nucleosomes containing satellite DNA and to
bind <u>in vitro</u> preferentially to (A-T)-rich nucleic acids regions [9].
Analogous non-histone DNA-binding proteins, with undefined nucleosome-bind-
ing properties, have been detected in other eukaryotic cells [6-10]. More
recently, Strauss and Varshavsky [11,12] have demonstrated the existence
of a DNA-binding protein which recognizes a family of repeated nucleotide
sequences (alphoid DNA) and which probably acts as a nucleosome-positioning
factor in higher order chromatin structure.

In a recent paper, we have shown that in the brine shrimp <u>Artemia</u> a
DNA family of repeats (Alu 1 family) is positively correlated with the
chromocenters present in interphase nuclei and with the heterochromatic
blocks detectable in prophase chromosomes [13]. The Alu 1 repeated DNA
(about 5% of the <u>A. franciscana</u> haploid genome) is scarcely represented
in other sibling species and parthenogenetic populations belonging to the
same genus [14]. By using a gel electrophoresis retardation assay [15],
we now report the detection, in <u>A. franciscana</u> extract, of a binding
activity that specifically interacts with the Alu 1 DNA fragment. By
transblot renaturation experiments, we have also demonstrated that the
protein responsible for the specific DNA binding activity is a polypeptide
with a molecular weight of approximately 82 x 10^3 daltons (p82). Anti-p82

polyclonal antibodies have been used to demonstrate localization of p82 in the chromocenters of interphase nuclei.

MATERIALS AND METHODS

Artemia franciscana from S. Francisco Bay (USA) and _Artemia sp._ from Tsing-Tao (China) were provided as dry cysts, by _Artemia_ Reference Center of Ghent, Belgium. Cysts were developed in salt water to nauplii as previously described [13]. Nauplii were collected, washed with cold water, resuspended with 0.1 ml/gr of 50 mM Tris-HCl (pH 8.0), 5 mM $MgCl_2$, 1 mM PMSF (phenylmethylsulphonylfluoride), 1% DMSO (dimethylsulphoxide), 0.001 mg/ml pepstatin, 80% glycerol and quickly frozen in liquid nitrogen and stored to -80°C.

Preparation of Crude Extract and Partial Purification of Alu 1 DNA-Binding Protein

Frozen nauplii (40 to 50 gr) were resuspended in 100 ml of 50 mM Tris-HCl (pH 8.0), 100 mM NaCl, 5 mM $MgCl_2$, 1% DMSO, 0.001 mg/ml pepstatin and disrupted in a motor-driven Potter homogenizer. Ammonium sulphate was added to the cellular suspension (final concentration of 0.5 M) to obtain a protein solubilized extract. After 15 min. at 0°C, the extract was clarified by high speed centrifugation. Solid ammonium sulphate (0.361 g/ml) was added to the supernatant and the resulting precipitate was collected by centrifugation, and solubilized in 70 ml of 50 mM Tris-HCl (pH 8.0), 5 mM 2-mercaptoethanol (2-ME), 10% glycerol (buffer A), and dialyzed overnight against buffer A containing 50 mM NaCl (fraction I).

Fraction I was loaded onto a 40 ml heparin-sepharose column, previously equilibrated with dialysis buffer. The column was washed with buffer A containing 0.25 M NaCl and eluted with 0.5 M NaCl in buffer A. The 0.5 M NaCl eluate was dialyzed overnight against buffer A containing 50 mM NaCl (fraction II).

Fraction II was loaded onto a 10 ml phosphocellulose column, equilibrated with the dialysis buffer. The proteins, eluted with a single step at 0.5 M NaCl in buffer A (fraction III), were directly loaded onto a 3 ml hydroxylapatite column prequilibrated with buffer A containing 0.5 M NaCl. After the column was washed with the same buffer, proteins were eluted with a linear gradient of potassium phosphate (pH 7.2) from 50 to 400 mM made up in 5 mM 2-ME, 10% glycerol. The protein fractions containing DNA-binding activity were pooled and dialyzed against 50 mM NaCl in buffer A (fraction IV).

Fraction IV was then loaded onto a DEAE-sephadex A-25 column previously equilibrated with dialysis buffer. The column was washed with the same buffer and eluted with 0.5 M NaCl in buffer A. Fractions showing DNA binding activity were pooled and dialyzed against 50 mM Tris-HCl (pH 8.0), 50 mM NaCl, 5 mM 2-ME, 50% glycerol and stored at -20°C for several months without any loss of binding activity.

Preparation of Alu 1 DNA Probe

The 109 bp Alu 1 DNA monomer was used to detect DNA-binding proteins. The fragment was recovered by digestion of pUA-41 recombinant plasmid [13] with Bam Hl and Eco Rl, and isolated from a 2% agarose gel. The fragment was labeled with ^{32}P α-dCTP (1.0 mCi/µg DNA fragment) and separated from unincorporated nucleotides as described by Maniatis et al. [16]. The specific activity was of 2×10^5 cpm/µg of DNA.

Gel Retardation Assay

Reaction mixtures were obtained by mixing the labeled fragment and competitor DNAs before adding proteins in a final volume of 10 µl containing 50 mM Tris-HCl (pH 8.0), 2 mM 2-ME, 25 mM NaCl, 1 mM EDTA, 0.5 ng of labeled DNA fragment and 1-20 µg of proteins. The reactions were incubated at 35°C for 10 min. and then 4 µl containing 50 mM Tris-HCl (pH 8.0), 2 mM 2-ME, 25 mM NaCl, 1 mM EDTA, 50% glycerol, 0.5% bromophenolblue were added to the reaction mixture. The samples were electrophoresed in a low ionic strength 7% polyacrylamide gel made up in 1 mM EDTA, 3.3 mM Na-acetate, 6.7 mM Tris-HCl (pH 7.5). Electrophoresis was carried out as described by Strauss and Varshavsky [11].

Protein Electrophoresis and Western Blotting

SDS polyacrylamide gel electrophoresis (SDS-PAGE) was performed as described by Laemmli [17] except that protein samples were not boiled but incubated at 37°C for 5 min. Transfer of polypeptides from the gel to nitrocellulose filters and detection of immunoreactivity were as previously described [18].

Binding to DNA in situ

After transfer, nitrocellulose sheets were soaked with gentle agitation in renaturation buffer (10 mM Tris-HCl (pH 8.0), 10 mM magnesium acetate, 10 mM KCl, 10 mM 2-ME, 0.1 mM EDTA and 1x Denhardt's solution) for 60 min. at room temperature with one buffer change [16]. The filters were then incubated for 60 min. in binding buffer (see gel retardation assay) containing $0.1 - 0.2 \times 10^6$ cpm/ml of ^{32}P-DNA fragment as probe and E. coli competitor DNA. Filters were then washed twice with binding buffer before exposure for autoradiography.

Sucrose Gradient Fractionation of Partially Purified DNA Binding Proteins

Protein samples were layered onto a linear 5 to 20% (W/V) sucrose gradient made up in 50 mM Tris-HCl (pH 8.0), 0.5 M NaCl, 5 mM 2-ME, 10% glycerol and centrifuged at 45,000 rpm for 24 h at 4°C in a SW 55 Ti rotor in a Beckman ultracentrifuge. Fractions were collected from the top of the gradient with a Buchler Densi-flow apparatus, and aliquots were assayed for DNA binding protein activity. Protein fractions containing DNA binding activities were pooled, dialyzed against 50 mM Tris-HCl (pH 8.0), 50% glycerol and stored at -20°C. Ovalbumin and aldolase were separated on a parallel gradient and located by A^{280} measurements.

Immunological Reagents

Polyclonal antiserum against the p82 protein was obtained by separating 300 µg of the DEAE-Sephadex flow-through protein fraction on an SDS-PAGE gel. After gel staining, the protein band, corresponding to the 82 KDa position, was sliced out and homogenized in Tris-buffered saline (TBS) and the polyacrylamide suspension was used to immunize rabbits by multiple intradermal injections [20]. The specificity of the antiserum obtained was determined by immunoblotting and by the direct evaluation of its binding capacity versus a protein showing Alu 1-DNA binding activity. Briefly, goat antirabbit IgG was dotted onto a nitrocellulose filter and then the filter was saturated with bovine serum albumin (BSA). The filter was incubated with anti-p82 serum, followed by incubation with DEAE-Sephadex flow-through protein fraction and finally probed with the ^{32}P-Alu 1 DNA fragment.

Preparation of Nuclei for Immunolocalization

Cells were obtained from the living nauplii. The detailed procedure for the cytological preparation of nuclei was previously described [13]. Giemsa (C-banding) staining and immuno-peroxidase reaction for cytological localization of p82 protein were carried out as described [14,20].

RESULTS

Detection of Alu 1 DNA Binding Protein in Fractionated Cell Extract

Artemia nauplii homogenate was extracted with 0.5 M ammonium sulphate and the clarified crude extract was fractionated on heparine-Sepharose and phosphocellulose columns. The presence of DNA binding-proteins was detected by a gel retardation assay. Electrophoresis in low ionic strength 7% polyacrylamide gel showed that the Alu 1 DNA monomer (109 bp) migrated as a discrete band. On the contrary, in the presence of the partially purified Fraction III proteins, the DNA failed to enter the gel. When the phosphocellulose fraction was further purified by sucrose gradient sedimentation, several discrete Alu 1 DNA-protein complexes, migrating with lower mobility than that of naked DNA monomer, were observed. Competition experiments carried out on the different fractions of the gradient showed that in the high density region of the gradient a DNA binding activity with high affinity for the Alu 1 DNA can be found (data not shown). A band competition assay was then carried out to compare the binding specificity of the protein(s) contained in these fractions for the 109 bp Alu 1 DNA fragment (homologous reaction) and for a 110 bp pUC8 DNA fragment (heterologous reaction) (Fig. 1). Two DNA-protein complexes (I and II) were observed when Alu 1 DNA was used as probe (Fig. 1-panel A). Addition of increasing amounts of E.coli competitor DNA (lanes b to f) determines a lower complex II to complex I ratio, suggesting that the protein(s) involved in complex I formation has a higher affinity for the Alu 1 DNA than for E.coli DNA. Panel B show the band competition assay carried out with the heterologous DNA fragment (pUC8 110 bp). In this case only a DNA-protein complex, with the mobility of complex II, is formed. In both experiments (panel A and B), the disappearance of complex II is achieved by adding approximately the same amount of E.coli competitor DNA (500-to 1000-fold). These results suggest that complex 1 is formed by proteins that specifically interact with the Alu 1 DNA fragment. On the contrary, band II may represent a complex made up of proteins which bind homologous as well as heterologous DNA fragments with the same affinity.

A gel retardation assay was carried out to demonstrate that complex I is constituted by a protein that specifically recognizes the Alu 1 repeated DNA. Alu 1 repeated DNA was used as labeled probe in the presence of different competitors (Alu 1 repeated DNA 109 bp-fragment, pUC8 110 bp-fragment and a 110 bp A.franciscana genomic DNA fragment not related to Alu 1 heterochromatic DNA) (Fig. 2). The experiments shown in Fig. 2 (panel A), were performed in the presence of a 500-fold excess of E.coli DNA. This condition reproduces a pattern of complexes (I and II) similar to that shown in Fig. 1 e. By adding a 60-fold excess of cold homologous fragment (Alu 1 repeated DNA), only complex I shows a significant reduction (Fig. 2,d). On the contrary, complexes I and II were not, or only slightly, affected by the addition of a 60-fold excess of heterologous pUC8 DNA fragment (lanes e and f) and of unrelated A.franciscana DNA (lanes g and h). These results suggest that complex 1 is made up by a highly sequence-specific DNA binding-protein.

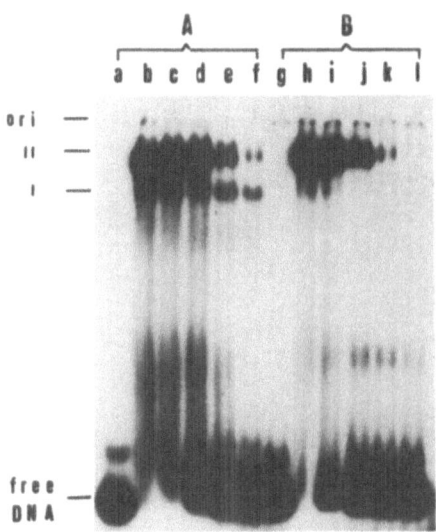

Fig. 1. Detection of Alu I repeated DNA binding protein by a competition
assay. A) The ^{32}P-labeled 109 bp Alu I monomer (0.05 ng; 25,000
cpm) was mixed with increasing amounts of unlabeled sonicated
E.coli competitor DNA, and with 3 μg of sucrose gradient active
protein fraction before electrophoresis in a low-ionic strength
7% polyacrylamide gel. Lane a contains Alu I DNA monomer in the
absence of added proteins. Lanes b-f contain 3 μg of proteins
and 0, 6.25, 12.5, 25 and 50 ng of E.coli DNA, respectively. B)
^{32}P-labeled 102 bp fragment excised from pUC8 plasmid with Hae III
(lanes g-l) was used instead of Alu I heterochromatic DNA fragment. Competition was set up as in panel A. I and II indicate
discrete Alu I-protein complex.

Purification and Characterization of Binding Activity Specific for Alu I Repeated DNA

To isolate the specific protein binding activity present in the
complex I, the phosphocellulose fraction was further purified by hydroxylapatite and DEAE-Sephadex chromatography (data not shown). By means of
these fractionation steps, the protein activity involved in the formation
of complex I was separated from other non-specific binding proteins. As
shown in Figure 2 (panel B), the DEAE flow-through protein fraction allows,
predominantly, the formation of complex I in the absence of any competitor
DNA (the ratio between complex I and II is estimated to be 20) (lane i).
The total elimination of complex II can be obtained by the addition of a
500-fold excess of E.coli competitor DNA (lane j). The binding specificity
of the protein present in the DEAE-Sephadex flow-through is also clearly
demonstrated by using different competitor DNAs (analogous to the experiment shown in Fig. 2, panel A in the absence of E.coli DNA Fig. 2, lanes
k-p). In particular, complex I is only affected by the addition of a 60-
fold excess of homologous heterochromatic DNA (lane l).

In order to determine the molecular weight of the protein that specifically interacts with the heterochromatic repetitive DNA, we have used a
DNA filter-binding assay after fractionation of the DEAE flow-through
protein on SDS-PAGE and renaturation [21]. As shown in Figure 3, only one

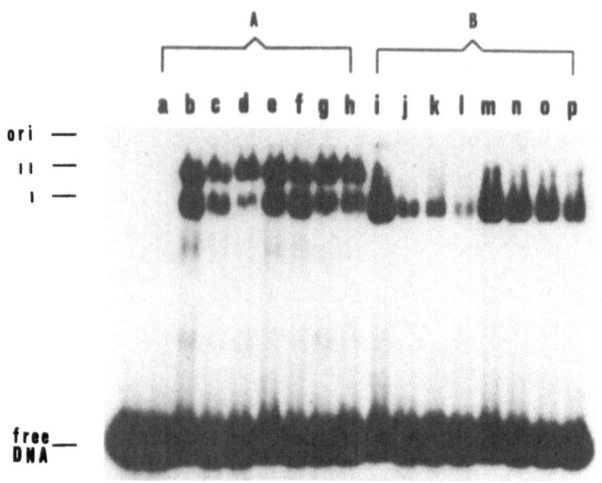

Fig. 2. Specificity of the binding activity evaluated by a competition
experiment. Alu I DNA-binding activity was detected using 3 μg
of protein from the sucrose gradient active fraction (panel A),
or 0.3 μg of DEAE-Sephadex flow-through protein fraction (panel
B). All samples contain 0.05 μg (25,000 cpm) of ^{32}P-Alu I DNA
fragment. Lane a has no protein added. Samples b and h contain
25 μg sonicated E.coli DNA as specific competitor (500-fold competition); samples c-h contain respectively 1.5 and 3 ng (30-and
60-fold competition) of cold homologous Alu I fragment, (c and d)
102 bp heterologous pUC8 fragment (e and f) 110 bp A.franciscana
genomic DNA fragment non-homologous to Alu I heterochromatic
DNA (g and h). Sample i does not contain any competitor DNA;
sample j contains 25 ng of sonicated E.coli DNA; samples k-p
contain competitors as well as in samples c-h.

band with a molecular weight of approximately 82 KDa shows a positive
reaction with the ^{32}P-labeled Alu I DNA monomer (lane 1). The specific
complex is unaffected by the addition of a large amount of unlabeled competitor DNA (100-fold E.coli DNA) (lane 2). In comparison, a very light
autoradiographic signal is detectable by using a labelled probe the heterologous 110 bp pUC8 fragment (lane 3). This result, together with the data
obtained in the gel retardation assay shown in Figure 2, suggests that the
DEAE-Sephadex flow-through protein fraction contains one protein with a
specific binding activity for the heterochromatic repetitive DNA. Using
the same experimental approach, we have also observed that the crude extract
contains several DNA-binding proteins, but a very low amount of specific
Alu I DNA-binding activity (data not shown).

Immunological Cytolocalization of the p82 Protein

To study the cytological location of p82 protein we have produced a
p82 antiserum in rabbits, as described in Materials and Methods. The
specific protein binding activity to heterochromatic DNA is visualized on
a nitrocellulose filter by using ^{32}P-Alu I DNA fragment as probe (Fig. 4A,
lane 2) after SDS-PAGE of the DEAE-Sephadex flow-through protein fraction,

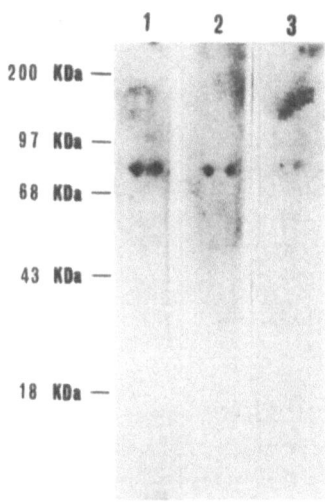

Fig. 3.   Detection of DNA binding activity in situ
after protein blotting. Partially purified
Alu I DNA binding proteins (20 μg of the DEAE-
Sephadex flow-through protein fraction) were
separated on a 10% polyacrylamide gel in the
presence of SDS [17], and electroblotted onto
nitrocellulose filter according to Plevani et
al. [18]. The renatured proteins (see methods)
were probed with 5 ml of reaction mixture containing 3 ng (1.5 x 10^6 cpm) of Alu I ^{32}P-DNA
fragment of the pUC8 plasmid (lane 3). Sonicated E.coli DNA (1200 ng) is present as non-
specific competitor in reaction mixture used
for probing lane 2. The markers used were:
blue dextran 2000, phosphorylase B, bovine
serum albumin, ovoalbumin and trypsin (200,
97, 68, 43 and 18 kDa, respectively).

electroblotting and protein renaturation. After autoradiography, the same
nitrocellulose filter was treated with the anti-p82 serum and the reaction
visualized with immunoperoxidase conjugated to anti-rabbit antibodies (Fig.
4A, lane 3). In both experiments one band, corresponding presumably to
the same polypeptide of 82 KDa, is clearly evident. Moreover, the same
nitrocellulose filter was stained with amido black to visualize the protein
pattern (Fig. 4A, lane 1). A control reaction, carried out with the preimmune rabbit serum, gives a negative result (Fig. 4A, lane 4).

The experiment shown in figure 4B was designed to establish the correlation between the p82 polypeptide and the protein responsible for the
formation of complex I. To this end, anti-rabbit IgG was dotted onto a
nitrocellulose filter followed by coating with BSA and the addition of
antibodies against p82 protein (dots 1 and 2) or the pre-immune rabbit serum
(dots 3 and 4). Dots 1 to 4 were treated with two different amounts of
DEAE-Sephadex flow-through fraction, and then with ^{32}P-Alu I DNA fragment.
Samples 1 and 2 bind a great amount of Alu I DNA, which is positively

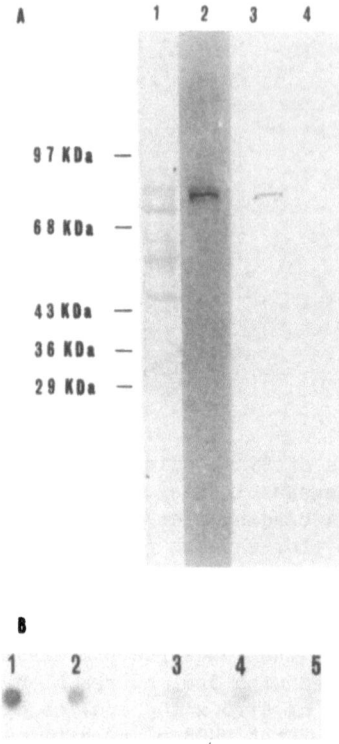

Fig. 4. Immunological correlation between p82 protein
and heterochromatic DNA binding activity.
Panel A (lane 2) shows the binding activity present in the DEAE flow-through protein fraction
(20 µg) detected as described in Materials and
Methods and in the legend of Figure 3. After
autoradiography the same nitrocellulose filter
was incubated with anti-p82 for detection of
immunoreactive polypeptides as described under
"Materials and Methods" (lane 3). Lane 1 represents the amido black staining of the sample in
lane 2 and 3. Lane 4 shows the immunoreaction of
20 µg of DEAE flow-through protein fraction with
preimmune serum. Marker proteins were as in
Figure 3. Panel B. Nitrocellulose filters (samples
1 to 5) were dotted with anti-rabbit IgG and saturated with BSA. Samples 1 and 2 were incubated
with anti- p82 serum, samples 3 and 4 with preimmune rabbit serum for 120 min. Samples 1, 3 and
2, 4 were respectively treated with 5 and 10 µg
of DEAE flow-through protein fraction for an additional 120 min. Finally, the filters (samples
1 to 5) were incubated in 0.2 ml of binding reaction mixture (see Materials and Methods) containing ^{32}P-DNA heterochromatic fragment (10^5 cpm)
for 30 min at 35°C. Filters were extensively
washed and autoradiographed.

correlated with the protein concentration, whereas the control samples (3 and 4) show only a weak retention of the labeled fragment similar to the background level in the absence of the protein fraction (dot 5). The use of anti-rabbit IgG was necessary to avoid non-specific retention of DNA by binding proteins in the serum. These results clearly suggest the specificity of the antiserum for the p82 protein which specifically recognizes the heterochromatic DNA fragment.

The p82 antiserum was then used to study the cytolocalization of the p82 protein in nuclei of <u>A.franciscana</u> (Fig. 5). <u>A.franciscana</u> interphase nuclei, stained with Giemsa (C-banding), show about 14-15 chromocenters (panel A) which were previously demonstrated by in situ hybridization to contain a considerable amount of repetitive Alu I DNA [13]. Panels B and C show the results obtained by using anti-p82 and preimmune sera, respectively. Nuclei treated with anti-p82 serum show, in positions corresponding to the chromocenters, a specific antibody reaction which is absent when using preimmune serum. The parthenogenetic population of <u>Artemia</u> from Tsing-Tao, almost completely devoided of chromocenters, does not show the formation of the specific complex I and also lacks the specific reaction in interphase nuclei (data not shown).

A

B

C

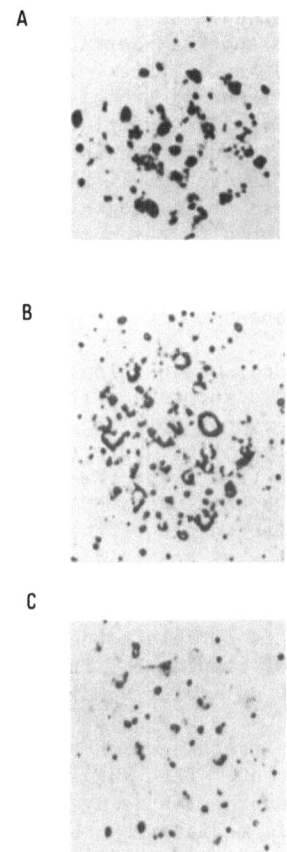

Fig. 5. Cytoimmunolocalization of p82 in interphasic nuclei. Chromocenters of interphase nuclei (Giemsa C-banding) of <u>A.franciscana</u> are shown in panel A. Panel B and C show interphase nuclei treated respectively with anti-p82 rabbit serum and with preimmune serum and subsequently with horseradish peroxidase conjugated with goat anti-rabbit IgG. Peroxidase was then visualized with benzidine:H_2O_2.

DISCUSSION

The brine shrimp Artemia seems to offer good possibilities to study
the mechanisms that determine the molecular structure of constitutive
heterochromatin and its role in the cell. The genus comprises sibling
species that are genetically isolated and also a large number of parthenogenetic populations that differ in their heterochromatic content. In the
species A.franciscana, we have demonstrated that part of the DNA organized
as heterochromatic masses is a repetitive Alu I DNA fragment of 109 bp in
length, with a reiteration frequency of approximately 6×10^5 copies per
haploid genome. This repetitive DNA is virtually absent in other populations such as Artemia of Tsing-Tao. The heterochromatic DNA fragment has
a nucleotide sequence which is particularly A + T rich (71%) [13, 14, 22].

By using a partially purified A.franciscana cell extract, we were able
to demonstrate the presence of a protein factor that specifically binds
the heterochromatic repetitive DNA fragment. In comparison, our attempts
to detect specific binding activities in crude extracts have been unsuc-
cessful. This failure is likely to be due to the very low amount of
specific protein compared to the non-specific ones that negatively interfere with the formation of complex I. The optimal visualization of the
nucleoprotein-specific complex requires, at least, two chromatographic
fractionation steps and a linear sucrose gradient. Furthermore, two additional columns, hydroxylapatite and DEAE-Sephadex, are required to eliminate
the undesirable residual binding activities. The specificity of the
complex formed between the Alu I heterochromatic DNA fragment and the DEAE-
Sephadex flow-through protein fraction (complex I) is deduced by the following observations: complex I is resistant to competitions carried out
with 1000-fold excess of heterologous E.coli DNA, but it is greatly reduced
by the addition of 60-fold excess of homologous cold DNA fragment. On the
contrary, the complex is unaffected by the same excess of heterologous
fragments, prepared from pUC8 plasmid or from A.franciscana genomic DNA.
These results exclude the possibility that complex I is formed by non-
specific proteins such as histones.

Renaturation experiments on nitrocellulose filter of the SDS-PAGE
separated DEAE-Sephadex flow-through protein fraction, allow us to determine that the binding activity responsible for the formation of complex I
is associated with a single polypeptide, with a molecular weight of 82 KDa
(p82). It is important to note that under these experimental conditions
the binding is not affected by heterologous DNA competition. The apparent
discrepancy between the molecular weight of the protein calculated by SDS-
PAGE and by the sedimentation of the specific binding activity on the
sucrose gradient in a position higher than aldolase (150 KDa; data not
shown) might be due to the aggregation of p82 protein molecules. An immunological approach was then used to correlate the p82 protein with the factor
responsible for the formation of complex I. The anti-p82 rabbit serum
recognizes, among the different molecular species present in the DEAE-
Sephadex flow-through, a polypeptide having a molecular weight of 82 kDa.
The anti-p82 serum also binds a protein which reacts specifically with the
Alu I heterochromatic DNA fragment. This suggests that the specific binding activity detected by the retardation assay and by protein blotting
after SDS-gel electrophoresis is mediated by the same protein. Moreover,
light-microscopic analysis of nuclei enzymatically labeled by the immunoperoxidase technique has shown the heterochromatic localization of the p82
protein.

Our results suggest the existence in A.franciscana of a DNA-binding
protein that specifically recognizes heterofchromatic DNA. The biological
function of this protein might be in generating the compact higher order
structure of constitutive heterochromatin mediating an arrangement of the

nucleosomes by formation of loops uniform in size. A similar mechanism of action was postulated for other non-histone heterochromatic proteins such as D1 protein [9], α-protein [11], BA protein [10]. Strauss and Varshavsky [11] suggest, for these proteins, a general role in temporal control at or near DNA sites required for initiation of transcription as a new genetic regulatory mechanism. Using the anti-p82 antibodies, we are currently attempting to isolate the gene coding for p82 protein. The variation of heterochromatin content in the different species and populations of Artemia may allow us to study the expression of the binding protein and consequently the mechanism of condensation of the constitutive heterochromatin and some of its functions.

ACKNOWLEDGEMENTS

We thank Prof. Enrico Ginelli for comments and discussion on the manuscript. We are also indebted to the Artemia Reference Center of the University of Ghent, Belgium which provided Artemia dry cysts. Roberta Benfante was supported by a fellowship from Fondazione Anna Villa Rusconi. This investigation was supported by the Consiglio Nazionale delle Ricerche, Italy, special grant IPRA.

REFERENCES

1. W. J. Peacock, A. R. Lohe, W. L. Gerlach, P. Dunsmuir, E. S. Dennis and R. Appels, "Fine structure and evolution of DNA in Heterochromatin," Cold Spring Harbor Symp. Quant. Biol. 42:1121 (1977).
2. B. John and G. L. C. Miklas, Functional aspects of satellite DNA and heterochromatin, Int. Rev. Cytol. 58:1 (1979).
3. D. L. Brutlag, Molecular arrangement and evolution of heterochromatic DNA, Ann. Rev. Genet. 14:121 (1980).
4. M. F. Singer, Highly repeated sequences in mammalian genomes, Int. Rev. Cytol. 76:67 (1982).
5. M. Gatty, D. A. Smith and B. S. Baker, A gene controlling condensation of heterochromatin in Drosophila melanagaster, Science 221:83 (1983).
6. T. Hsieh and D. L. Brutlag, A protein that preferentially binds Drosophila satellite DNA, Proc. Natl. Acad. Sci. U.S.A. 76:726 (1979).
7. C. R. Rodriguez-Alfageme, G. T. Rudkin and L. H. Cohen, Isolation, properties and cellular distribution of D1, a chromosomal protein of Drosophila, Chromosoma (Berl.) 78:1 (1980).
8. C. R. Rodriguez-Alfageme, G. T. Rudkin and L. H. Cohen, Locations of chromosomal proteins in polytene chromosomes, Proc. Natl. Acad. Sci. U.S.A. 73:2038 (1976).
9. L. Levinger and A. Varshavsky, Protein D1 preferentially binds A + T-rich DNA in vitro and is a component of Drosophila melanogaster nucleosomes containing A + T-rich satellite DNA, Proc. Natl. Acad. Sci. U.S.A. 79:7152 (1982).
10. F. C. Bennet, B. I. Rosenfeld, C. K. Huang and L. C. Yeoman, Evidence for two conformational forms of nonhistone protein BA which differ in their affinity for DNA, Biochem. Biophys. Res. Commun. 104:649 (1982).
11. F. Strauss and A. Varshavsky, A protein binds to a satellite DNA repeat at three specific sites that would be brought into mutual proximity by DNA folding in the nucleosome, Cell 37:889 (1984).
12. M. J. Solomon, F. Strauss and A. Varshavsky, A mammalian high mobility group protein recognizes any stretch of six A-T base pairs in duplex DNA, Proc. Natl. Acad. Sci. U.S.A. 83:1276 (1986).
13. C. Barigozzi, G. Badaracco, P. Plevani, L. Baratelli, S. Profeta, E.

Ginelli and R. Meneveri, Heterochromatin in the genus <u>Artemia</u>, <u>Chromosoma</u> (Berl.) 90:332 (1984).

14. G. Badaracco, L. Baratelli, E. Ginelli, R. Meneveri, P. Plevani, P. Valsasnini and C. Barigozzi, Variations in repetitive DNA and heterochromatin in the genes <u>Artemia</u>, <u>Chromosoma</u> (Berl.) 95:71 (1987).

15. M. M. Garner and A. Revzin, A gel electrophoresis method for quantifying the binding of proteins to specific DNA regions: Application to components of the <u>Escherichia coli</u> lactose operon regulatory system, <u>Nucl. Acids Res.</u> 9:3047 (1981).

16. R. Maniatis, E. F. Fritsch and J. Sambrook, "Molecular Cloning: A Laboratory Manual," Cold Spring Harbor Press, Cold Spring Harbor (1982).

17. U. K. Laemmli and M. Favre, Maturation of the head of bacteriophage T4. 1. DNA packaging events, <u>J. Mol. Biol.</u> 80:575 (1973).

18. P. Plevani, G. Badaracco, C. Angl and L. M. S. Chang, DNA polymerase I and DNA primase complex in yeast, <u>J. Biol. Chem.</u> 259:7532 (1984).

19. J. L. Vaitukatis, Production of antisera with small doses of immunogen: multiple intradermal injection, Methods in Enzymology 73:46 (1981).

20. S. N. Hsu, L. Daine and H. Fanger, Use of avidin-biotin-peroxidase complex (ABC) in immunoperoxidase techniques: a comparison between ABC and unlabeled antibody (PAP) procedures, <u>Jour. Histochem. Cytochem.</u> 29:577 (1981).

21. R. S. Jack, M. T. Brown and W. J. Gehsuing, Protein blotting as a means to detect sequence-specific DNA-binding proteins, <u>Cold Spring Harbor Symp. Quant. Biol.</u> XLVII:483 (1982).

22. J. Cruces, M. L. G. Wonenburger, M. Diaz-Guerra, J. Sebastian and J. Renart, Satellite DNA in the crustacean <u>Artemia</u>, Gene 44:341 (1986).

FUNCTIONAL AND STRUCTURAL STUDIES ON ARTEMIA POLYPEPTIDE CHAIN INITIATION

FACTOR 2. CLONING AND SEQUENCING OF eIF-2 α cDNA

J. N. Dholakia, N. S. Reddy and A. J. Wahba

Department of Biochemistry
The University of Mississippi Medical Center
Jackson, MS 39216-4505

INTRODUCTION

The eukaryotic polypeptide chain initiation factor 2 (eIF-2) plays a significant role in the regulation of protein synthesis [1,2]. The first step in polypeptide chain initiation is the formation of a ternary complex containing eIF-2, GTP and the initiator tRNA. This complex is transferred to a 40S ribosomal subunit [1-4] which later combines with a 60S ribosomal subunit and other initiation factors to form an 80S initiation complex, thereby setting the stage for polypeptide chain elongation. Upon formation of the 80S initiation complex, GTP is hydrolyzed and eIF-2 is released as the eIF-2·GDP binary complex [5,6]. In mammalian systems, this binary complex is stable in the presence of Mg^{2+} and is functionally inactive [7,8]. Regeneration of eIF-2·GTP·Met-tRNA$_f$ species require the guanine nucleotide exchange factor (GEF), which facilitates the exchange of eIF-2-bound GDP for GTP and facilitates the recycling of eIF-2 [9-12]. It is at this point in the eIF-2 cycle (Fig. 1) that regulation of polypeptide chain initiation occurs. Most of our understanding about this step in protein synthesis in eukaryotes is derived from studies with rabbit reticulocyte lysates. We have previously observed that dormant and developing embryos of Artemia contain equivalent amounts of eIF-2 activity [13]. Similar to reticulocyte eIF-2, the Artemia factor eIF-2 has 100 to 300-fold higher affinity for GDP than GTP [13] and its α-subunit gets phosphorylated by the rabbit reticulocyte heme-controlled repressor (HCR) [14]. Significant differences also exist between the reticulocyte and the Artemia factors. For example, the phosphorylation of the reticulocyte eIF-2 by HCR is associated with the cessation of protein synthesis and is due to the inability of GEF to catalyze the exchange of GTP for eIF-2-bound GDP [15,16]. In contrast to this, Artemia eIF-2 remains active in protein synthesis in reticulocyte lysates after phosphorylation by HCR [17]. We have so far been unable to demonstrate a requirement for a GDP/GTP exchange factor with Artemia eIF-2. In an attempt to understand some of these differences, we nave recently purified the factors to apparent homogeneity and compared their functional and structural properties. We have isolated and sequenced a cDNA clone for the α-subunit of Artemia eIF-2. This will permit the identification of the regulatory and catalytic sites and allow us to study the expression of the eIF-2α gene during development.

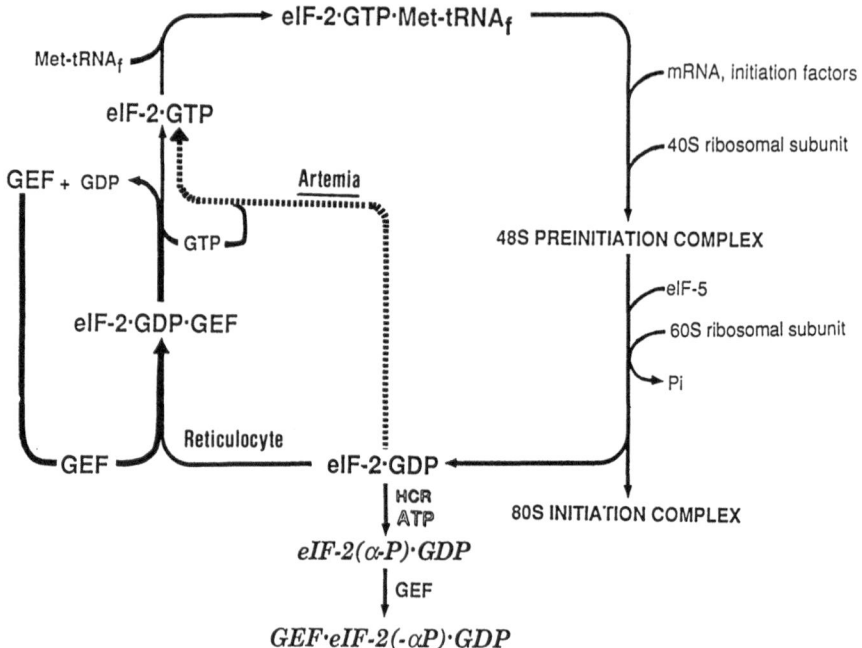

Fig. 1. eIF-2 cycle during polypeptide chain initiation

MATERIALS AND METHODS

Desiccated Artemia embryos (San Francisco Bay Brand, Newark, CA) and rabbit reticulocyte lysates (Green Hectares, Oregon, WI) were used as the source of initiation factors and kinases. Artemia and rabbit reticulocyte eIF-2 were purified from the high salt wash of ribosomes [17,18]. GEF and HCR were purified from the postribosomal supernatant of rabbit reticulocyte lysates as described by Dholakia and Wahba [19] and Kramer et al. [20], respectively. Casein kinase II was purified from Artemia embryos [21].

The preparation of eIF-2α-specific antibodies and the immunological methods used were as previously described [14]. Rabbit anti-Artemia eIF-2α and peroxidase conjugated goat anti-rabbit IgG were used as primary and secondary antibodies, respectively. Immunoreactive spots were visualized by using methanolic 4-chloro-1-napthol [18].

The activity of eIF-2 was determined by binary (eIF-2·GDP) and ternary (eIF-2·GTP·Met-tRNA$_f$) complex formation [15,18] and of GEF by monitoring the release of eIF-2-bound (^{3}H)GDP [11,19]. The phosphorylation of eIF-2 was carried out as previously described [14] and phosphorylated eIF-2 was reisolated by chromatography on a phosphocellulose column [17].

For the amino acid sequence determination of eIF-2α, the eIF-2 subunits were separated by sodium dodecylsulfate/polyacrylamide gel electrophoresis and visualized by brief staining. The gel slices containing the α-subunit were excised [14] and the α-subunit was electroeluted from the gel slices. The solution was then concentrated and applied to a polybrene-treated filter of the Applied Biosystems 470A gas-phase sequencer. Amino acids were identified with an Applied Biosystems PTH column attached to a Waters HPLC interfaced to the sequencer.

The procedures used for the synthesis of the oligodeoxyribonucleotide probes, purification and 5'-end labeling of the probes, nick translation of plasmid DNA, the preparation of poly(A)$^+$ RNA and agarose gel electrophoresis were previously described [22,23]. An Artemia eIF-2α cDNA clone was isolated by screening the Artemia cDNA library constructed in λgtll, using a mixed deoxyoligonucleotide probe, AIF-2α-51-mer (5'GACATTCACCATC-
                                      C   G   G
ACTACATCTTCCACCTCAGGAAATTTTCAGCATAATA3'). After subcloning into the bacteriophage M13mp18 and M13mp19, the DNA sequence of both strands was determined by the chain-termination method [24]. The EcoRI site at the 5'-end of the initiation codon and the HindIII site near the stop codon were created by using oligodeoxyribonucleotide directed mutagenesis [25]. The EcoRI/HindIII fragment was cloned into pSP65 to construct pSRI. In vitro transcription was carried out in a 100 μl reaction mixture containing 2.5 μg linearized pSRI plasmid DNA, 15 units of SP6 RNA polymerase, 80 units of RNasin, 40 mM Tris-HCl, pH 7.5, 6 mM MgCl$_2$, 2 mM spermidine, 10 mM NaCl, 10 mM dithiothreitol and 0.5 mM each of ATP, GTP, UTP and CTP. After incubation for 2 h at 37°C, 4 units of RQ1 DNAse was added and the incubation was continued for 30 additional min. The reaction was terminated by phenol extraction and RNA was precipitated with ethanol. This RNA was translated in a nuclease-treated rabbit reticulocyte lysate [17].

RESULTS AND DISCUSSION

Resumption of development of the encysted Artemia embryos is triggered by hydration [26], and growth of the freshly hydrated embryos is characterized by relatively synchronous development. Reinitiation of RNA and protein synthesis occurs during the 12 h of development, in the absence of DNA synthesis and cell division. Cell-free extracts from dormant embryos are inactive in the translation of natural mRNA [27] and this was attributed to a deficiency of chain initiation factors [28]. The eukaryotaic chain initiation factor 2 was initially reported to increase over 20-fold following resumption of embryo development [27,28]. However, a more thorough investigation indicated that approximately equal amounts of eIF-2 activity are present in dormant and developing Artemia embryos [13], and that unlike the reticulocyte eIF-2, the phosphorylation of Artemia eIF-2 by HCR does not regulate its activity [14,17]. Several significant differences between the Artemia and the reticulocyte system were also observed. In the presence of Mg^{2+}, ternary (eIF-2·GTP·Met-tRNA$_f$) and binary (eIF-2·GDP) complex formation with Artemia eIF-2 are stimulated and the Artemia eIF-2·GDP complex exchanges bound GDP readily with free GTP. In contrast, ternary and binary complex formation with reticulocyte eIF-2 are inhibited in the presence of 1 mM Mg^{2+} and a protein factor (GEF) isolated from reticulocytes is required for GDP/GTP exchange with reticulocyte eIF-2·GDP.

We have recently purified to apparent homogeneity Artemia and reticulocyte eIF-2. These preparations were phosphorylated either by HCR or casein kinase II and the reaction products analyzed by electrophoresis in SDS/polyacrylamide gels (Fig. 2). Casein kinase II phosphorylates the β-subunit of reticulocyte eIF-2, but specifically phosphorylates the α-subunit of Artemia eIF-2. HCR phosphorylates the α-subunit of both Artemia and reticulocyte eIF-2 (Fig. 2). Partial trypsin digestion suggested that the HCR phosphorylation sites on both eIF-2 preparations may reside in the NH$_2$-terminal region [14]. Upon extensive trypsin digestion, the two-dimensional phosphopeptide maps of the α-subunits of the reticulocyte and Artemia factors are indistinguishable [14]. Encouraged by these results, we determined the NH$_2$-terminal amino acid sequence of reticulocyte and Artemia eIF-2 (Fig. 3). Of the first 18 amino acids of the Artemia subunit, 14 amino acids in the reticulocyte factor are either the same or represent a conservative change. Based on this homology, we

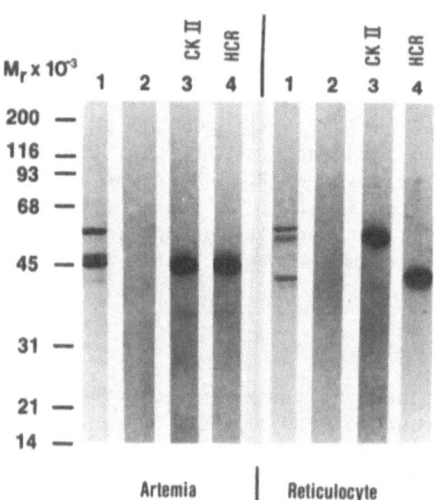

Fig. 2. Phosphorylation of <u>Artemia</u> and reticulocyte eIF-2 by
casein kinase II (CK II) and heme-controlled repressor
(HCR). Lane 1, Coomassie Brilliant Blue R-250 stained
gel of eIF-2 (2 µg), lanes 2 to 4, autoradiograph.
Phosphorylation was carried out either in the absence
of kinase (lane 2) or with casein kinase II (lane 3)
or HCR (lane 4).

Artemia eIF-2	:	Pro	–	Leu	Ser	Pro	Arg	Tyr	Tyr	Ala	Glu	Lys	Phe	Pro	Glu	Thr	Glu	Asp	Val	Val
Artemia cDNA	:	Pro	–	Leu	Ser	Cys	Arg	Tyr	Tyr	Ala	Glu	Lys	Phe	Pro	Glu	Thr	Glu	Asp	Val	Val
Retic. eIF-2	:	Pro	Gly	Leu	Ser	x	Arg	Phe	Tyr	Gln	His	Lys	Phe	Pro	Glu	Val	Glu	Asp	Val	Val

Fig. 3. NH_2-Terminal amino acid sequence of eIF-
2α. The gap (–) was inserted to increase the similarity between the sequences. x, not identified.

have synthesized a mixed 51-mer (AIF-2α-51-mer) oligodeoxyribonucleotide probe to screen the Artemia cDNA library.

The 5'-end labeled probe was hybridized to a northern blot of Artemia poly(A$^+$) RNA from 12 h developing embryos. The RNA was fractionated on a formaldehyde denaturing agarose gel prior to transfer to a nitrocellulose membrane. A single band of approximately 1600 bp was observed (Fig. 4).

We screened about 1.5 x 10^6 λ plaques and isolated two different positive recombinant phage plaques. One of these, phage AIF-2α1, contained a 4.25 kb KpnI/SstI fragment and the other, AIF-2α2, contained a 2.6 kb KpnI/SstI fragment. These fragments were subcloned into M13mp18 and MP13-mp19 and the nucleotide sequence was determined. The AIF-2α1 clone has an open reading frame coding for a protein of 335 amino acids with a calculated molecular weight of 38,000 daltons (Fig. 5). The NH$_2$-terminal amino acid sequence of the α-subunit of Artemia deduced from the cDNA sequence matches the NH$_2$-terminal amino acid sequence of Artemia and rabbit eIF-2 determined by protein sequencing (Fig. 3). The cDNA-derived amino acid sequence of Artemia eIF-2α was compared with that of human eIF-2α [29] and Ser-51 is identified as the putative phosphorylation site for the heme-

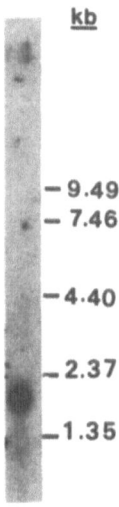

kb

− 9.49
− 7.46

− 4.40

− 2.37

− 1.35

Fig. 4.   Northern blot analysis. The poly(A)$^+$ RNA from 12
          h embryos of Artemia was electrophoresed into a
          formaldehyde 1% agarose gel, blotted on a nylon
          membrane and hybridized with a 5' end-labeled
          mixed probe (5'GACATTCACCATCACTACATCTTCCACCTCAGG-
             C   G G
          AAATTTTCAGCATAATA3'). The position of RNA molecular
          weight markers is indicated.

```
      -90         -80         -70         -60         -50         -40
       *           *           *           *           *           *
ACA TGG CTG AAT ATC GAC GGT TTC CAT ATG GGG ATT GGT GGC GAC GAC TCC TGG AGC CCG


  -30         -20         -10          1          10          20          30
   *           *           *           *           *           *           *
TCA GTA TCG GCG GGA TTT TAG CGG GTA AAG ATG CCT TTA TCA TGT CGA TAT TAT GCA GAG
                                    Met Pro Leu Ser Cys Arg Tyr Tyr Ala Glu


              40          50          60          70          80          90
               *           *           *           *           *           *
AAA TTC CCC GAG ACA GAG GAT GTA GTG GTA AAA GTC AGG TCA ATA GCT GAT ATG GGT
Lys Phe Pro Glu Thr Glu Asp Val Val Met Val Lys Val Arg Ser Ile Ala Asp Met Gly
                ----------


             100         110         120         130         140         150
               *           *           *           *           *           *
GCC TAT GTA AAC CTA TTA GAA TAT GAT GAT ATA GAA GGA ATG ATT TTA TTA TCA GAA TTG
Ala Tyr Val Asn Leu Leu Glu Tyr Asp Asp Ile Glu Gly Met Ile Leu Leu Ser Glu Leu
                                    ============


             160         170         180         190         200         210
               *           *           *           *           *           *
TCG AGG CGG AGA ATA AGA TCT ATT AAC AAA TTG ATT AGG ATT GGG AAA ACA GAA CCA GTG
Ser Arg Arg Arg Ile Arg Ser Ile Asn Lys Leu Ile Arg Ile Gly Lys Thr Glu Pro Val
==                          ==========


             220         230         240         250         260         270
               *           *           *           *           *           *
GTT GTT ATT AGG GTA GAT AAA GAC AAA GGC TAT ATC GAT TTG TCA AAA AGA AGA GTT TCT
Val Val Ile Arg Val Asp Lys Asp Lys Gly Tyr Ile Asp Leu Ser Lys Arg Arg Val Ser


             280         290         300         310         320         330
               *           *           *           *           *           *
GCA GAA GAC ATT GAC AAA TGT ACG GAG AAA TTT AGC AAA GCC AAA GCA GTA AAC TCA ATT
Ala Glu Asp Ile Asp Lys Cys Thr Glu Lys Phe Ser Lys Ala Lys Ala Val Asn Ser Ile


             340         350         360         370         380         390
               *           *           *           *           *           *
TTG AGA CAT GTT GCC TTT ATG TTG GAC TAT CAA ACA AAC GAG CAG CTT GAA GAG CTG TAC
Leu Arg His Val Ala Phe Met Leu Asp Tyr Gln Thr Asn Glu Gln Leu Glu Glu Leu Tyr


             400         410         420         430         440         450
               *           *           *           *           *           *
CAG AAA ACT GCG TGG TTA TTT GAT AAA AAA GCC AAA AGA CAA TCA GCT GCG TAT GAA GCC
Gln Lys Thr Ala Trp Leu Phe Asp Lys Lys Ala Lys Arg Gln Ser Ala Ala Tyr Glu Ala


             460         470         480         490         500         510
               *           *           *           *           *           *
TTT AAA CAA GCA GTC AAG GAT CCA TCA CTA CTA GAC GAA TGC GGA CTA GAT GAA AAG ACG
Phe Lys Gln Ala Val Lys Asp Pro Ser Leu Leu Asp Glu Cys Gly Leu Asp Glu Lys Thr


             520         530         540         550         560         570
               *           *           *           *           *           *
AAG GAG ACG CTT TTG AAT CAA ATT AAG CAG AAG TTA ACG CCA CAA GCA GTG AAA ATA AGA
Lys Glu Thr Leu Leu Asn Gln Ile Lys Gln Lys Leu Thr Pro Gln Ala Val Lys Ile Arg


             580         590         600         610         620         630
               *           *           *           *           *           *
GCC TAT GTT GAA TGC TCA TGT TAT GGA TAT GAA GGA ATT AAT GCA GTA AAG AGG GCC CTA
Ala Tyr Val Glu Cys Ser Cys Tyr Gly Tyr Glu Gly Ile Asn Ala Val Lys Arg Ala Leu


             640         650         660         670         680         690
               *           *           *           *           *           *
AGA GCT GGA ATG TCC GTC TCA ACC GAA GAA ATA CCT ATA AAA ATC AAT TTA GTA GCA TCT
Arg Ala Gly Met Ser Val Ser Thr Glu Glu Ile Pro Ile Lys Ile Asn Leu Val Ala Ser


             700         710         720         730         740         750
               *           *           *           *           *           *
CCC ATT TTT GTA ATT AAT ACT CAT ACA CAA GAG AAA GAA GAT GGC ATG AAG GCA TTG ACA
Pro Ile Phe Val Ile Asn Thr His Thr Gln Glu Lys Glu Asp Gly Met Lys Ala Leu Thr


             760         770         780         790         800         810
               *           *           *           *           *           *
TTG GCG CTA AAA GCC ATT GAA GAA GCA GCG GAA AAG GAA GGA GGG ACT TTT AGG ATT AAG
Leu Ala Leu Lys Ala Ile Glu Glu Ala Ala Glu Lys Glu Gly Gly Thr Phe Arg Ile Lys


             820         830         840         850         860         870
               *           *           *           *           *           *
TCA GCT CCA AAA GTC GTC ACA AAA GAA GAC GAA GCC GAA CAC GCT CGA CAG ATG GAA TTG
Ser Ala Pro Lys Val Val Thr Lys Glu Asp Glu Ala Glu His Ala Arg Gln Met Glu Leu


             880         890         900         910         920         930
               *           *           *           *           *           *
GCT GAG CAG GAA AGT CGA CAA GTT GCT GGT GAT GAT AGT GAA GAA GAA GAA TCC GAA GAG
Ala Glu Gln Glu Ser Arg Gln Val Ala Gly Asp Asp Ser Glu Glu Glu Glu Ser Glu Glu
                                                    --- --- ---     --- --- ---


             940         950         960         970         980         990
               *           *           *           *           *           *
GGC AGT GAC GAA GAA GAT GAC GAA AAG AAA GAA AAC GGA GTC AAA GAA TCC GAA GAA GAG
Gly Ser Asp Glu Glu Asp Asp Glu Lys Lys Glu Asn Gly Val Lys Glu Ser Glu Glu Glu
    --- --- ---                                             --- --- ---


            1000        1010        1020        1030        1040        1050
               *           *           *           *           *           *
GAA TGG GGC CTC GCG TAA TGC AAT ATT CCG CCT AAT TAT TAT ATT TTG GTT AAA TAG CAT
Glu Trp Gly Leu Ala ---


CGC TGT TTT TAG CTG AAC TAT ATT TCC GTT TAA ATT CTG AAA AGC TGA GTA CGA AAT GTT
TGG CAA TAC AAC TTT CAG GAA TAA CAT TCG CCA GCA CAA TAC TCA TTG CAC TAT TAC AAT
CAA TAT AAT TAT TTG AAT CAA TAA CAT TGA ACA TAC AGC CCA AAT AGG GGC AAG AAT CTA
GAA AAA GAC CAA TAC TCA ACG TTG GTA TTT GGG GGC TTT ACC GAG AGT CGG CAG AGG GAA
TGT ATA TTC GAT GTG ACC TTT TAC CAC ACT CAA AGT CTT AAA AAG GAA CGG AGT ATA AAT
AGA TGG CAC TAT TAA AAA ATA ACT ACT GGC AGC GTT GAC AGA AGA AGA AAA
AGA AAG GG
```

Fig. 5.   (legend on next page)

controlled repressor (HCR). The α subunit of <u>Artemia</u> eIF-2 also contains consensus phosphorylation sites for casein kinase II, one near the NH$_2$-terminal (Thr-15) and four sites near the carboxy terminal (Ser-303, Ser-308, Ser-312 and Ser-327). Based on the homology with the consensus sequences of GTP binding sites [30], a putative phosphoryl binding site (Asp Ile Glu Gly) and a putative guanine binding site (Asn Lys Leu Ile) are identified (Fig. 5). Recently, we have used the photoaffinity probe, 8-azido GTP [8-N$_3$GTP], to investigate the GTP binding domain of eIF-2. UV-irradiation resulted in covalent cross-linking of [γ-^{32}P]8N$_3$-GTP to the α-subunit of <u>Artemia</u> eIF-2 and the β-subunit of reticulocyte eIF-2 (data not presented). These results support the postulated GTP binding domains in the cDNA sequences of <u>Artemia</u> eIF-2α and the mammalian eIF-2β.

The EcoRI site at nucleotide position -11 to -16 and the HindIII site at 1024 to 1029 was created by using oligodeoxynucleotide directed mutagenesis [25]. The EcoRI and the HindIII fragment containing an entire coding sequence of eIF-2 was cloned into EcoRI/HindIII sites of SP65 and <u>in vitro</u> transcribed with SP6 RNA polymerase. The <u>in vitro</u> transcribed <u>Artemia</u> eIF-2 α mRNA was translated using [^{35}S]methionine in a nuclease-treated mRNA-dependent reticulocyte lysate. The <u>in vitro</u> transcribed mRNA directed the synthesis of a single polypeptide (Fig. 6) which was indistinguishable from <u>Artemia</u> eIF-2α on dodecylsulfate-/polyacrylamide gel electrophoresis and reacted specifically with eIF-2α antibodies (Fig. 6) thus confirming the specificity of the eIF-2 α cDNA clone. We plan to investigate the expression of the <u>Artemia</u> eIF-2α gene during development by using the eIF-2α clone and the eIF-2α-specific antibodies.

ACKNOWLEDGEMENT

This work was supported in part by USPH grant GM 25451.

Fig. 5. Nucleotide and derived amino acid sequence of cDNA for <u>Artemia</u> eIF-2α. Putative polyadenylation signal sequence and poly(A) addition sites are underlined. The putative HCR phosphorylation site is double underlined. Putative casein kinase II phosphorylation sites are indicated with dotted underlines. Putative phosphoryl binding sequences and GTP binding sites are indicated with double dotted underlines.

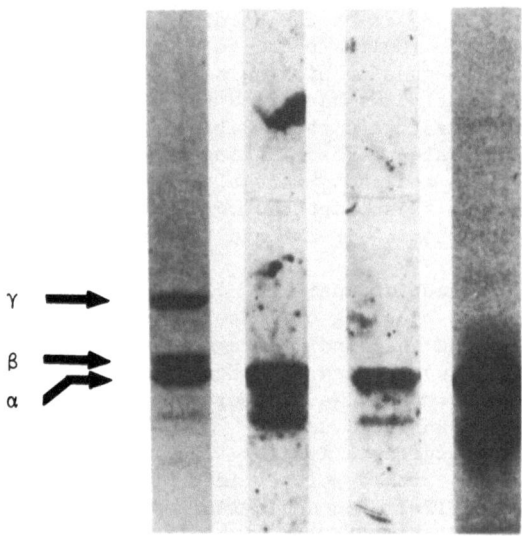

Fig. 6.   Western blot analysis. Artemia eIF-2 (lane 1
and 2) and in vitro translational products
synthesized by using Artemia eIF-2 α mRNA
derived from the cDNA clone (lane 3) were
electrophoresed in SDS/polyacrylamide gel and
transferred to nitrocellulose paper. The blot
containing lane 2 and 3 was treated with
rabbit anti-Artemia eIF-2α antibodies and
the immunoreactive spots were visualized as
described in Methods. Lane 1 is Coomassie
blue stained gel and lane 4 is the autoradiograph of lane 3.

REFERENCES

1. A. J. Wahba and C. L. Woodley, Molecular aspects of development in the
    brine shrimp Artemia, in: "Progress in Nucleic Acids Research and
    Molecular Biology," Vol. 31, W. Cohn and K. Moldave, eds., Academic
    Press, New York (1984).
2. V. M. Pain, Initiation of protein synthesis in mammalian cells, Biochem.
    J. 235:625 (1986).
3. K. Moldave, Eukaryotic protein synthesis, Ann. Rev. Biochem. 54:1109
    (1985).
4. S. Ochoa, Regulation of protein synthesis initiation in eukaryotes,
    Arch. Biochem. Biophys. 223:325 (1983).
5. H. Trachsel and T. Staehelin, Binding and release of eukaryotic initia-
    tion factor eIF-2 and GTP during protein synthesis initiation, Proc.
    Natl. Acad. Sci. U.S.A. 75:204 (1978).
6. M. S. Clemens, V. M. Pain, S. Wong and E. C. Henshaw, Phosphorylation
    inhibits guanine nucleotide exchange on eukaryotic initiation factor
    2, Nature (London) 297:93 (1982).
7. R. Panniers and E. C. Henshaw, A GDP/GTP exchange factor essential for
    eukaryotic initiation factor 2 cycling in Ehrlich ascites tumor
    cells and its regulation by eukaryotic initiation factor 2 phos-
    phorylation, J. Biol. Chem. 258:7928 (1983).

8. G. M. Walton and G. N. Gill, Nucleotide regulation of a eukaryotic protein synthesis initiation complex, <u>Biochim. Biophys. Acta</u> 390: 231 (1975).

9. J. Siekierka, L. Mauser and S. Ochoa, Mechanism of polypeptide chain initiation in eukaryotes and its control by phosphorylation of the α subunit of initiation factor 2, <u>Proc. Natl. Acad. Sci. U.S.A.</u> 79:2537 (1982)

10. M. Salimans, H. Goumans, H. Amesz, R. Benne and H. O. Voorma, Regulation of protein synthesis in eukaryotes. Mode of action of eRF, an eIF-2-recycling factor from rabbit reticulocytes involved in GDP/GTP exchange, <u>Eur. J. Biochem.</u> 145:91 (1984).

11. J. N. Dholakia and A. J. Wahba, The association of NADPH with the guanine nucleotide exchange factor from rabbit reticulocytes : A role of pyridine dinucleotides in eukaryotic polypeptide chain initiation, <u>Proc. Natl. Acad. Sci. U.S.A.</u> 83:6746 (1986).

12. A. Konieczny and B. Safer, Purification of the eukaryotic initiation factor 2-eukaryotic initiation factor 2/β complex and characterization of its guanine nucleotide exchange activity during protein synthesis initiation, <u>J. Biol. Chem.</u> 258:3402 (1983).

13. T. H. MacRae, M. Roychowdhury, K. J. Houston, C. L. Woodley and A. J. Wahba, Protein synthesis in brine shrimp embryos. Dormant and developing embryos of <u>Artemia salina</u> contain equivalent amounts of chain initiation factor 2, <u>Eur. J. Biochem.</u> 100:67 (1979).

14. H. B. Mehta, J. N. Dholakia, W. W. Roty, B. S. Parekh, R. C. Montelaro, C. L. Woodley and A. J. Wahba, Structural studies on the eukaryotic chain initiation factor 2 from rabbit reticulocytes and brine shrimp <u>Artemia</u> embryos. Phosphorylation by the heme-controlled repressor and casein kinase II, <u>J. Biol. Chem.</u> 261:6705 (1986).

15. H. B. Mehta, C. L. Woodley and A. J. Wahba, Protein synthesis in brine shrimp embryos and rabbit reticulocytes, <u>J. Biol. Chem.</u> 256: 3438 (1983).

16. D. J. Goss, L. J. Parkhurst, H. B. Mehta, C. L. Woodley and A. J. Wahba, Studies on the role of eukaryotic nucleotide exchange factor in polypeptide chain initiation, <u>J. Biol. Chem.</u> 259:7374 (1984).

17. C. L. Woodley, M. Roychowdhury, T. H. MacRae, K. W. Olsen and A. J. Wahba, Protein synthesis in brine shrimp embryos. Regulation of the formation of the ternary complex (Met-tRNA$_f$·eIF-2·GTP) by two purified protein factors and phosphorylation of <u>Artemia</u> eIF-2, <u>Eur. J. Biochem.</u> 117:543 (1981).

18. J. N. Dholakia and A. J. Wahba, The isolation and characterization from rabbit reticulocytes of two forms of eukaryotic initiation factor 2 having different β -polypeptides, <u>J. Biol. Chem.</u> 262:10164 (1987).

19. J. N. Dholakia and A. J. Wahba, Phosphorylation of the guanine nucleotide exchange factor from rabbit reticulocytes regulates its activity in polypeptide chain initiation, <u>Proc. Natl. Acad. Sci. U.S.A.</u> 85:51 (1988).

20. G. Kramer, J. M. Cimadevilla and B. Hardesty, <u>Proc. Natl. Acad., Sci. U.S.A.</u> 73:3078 (1976).

21. C. L. Woodley, J. N. Dholakia and A. J. Wahba, Polypeptide chain initiation in <u>Artemia</u> embryos: functional and structural studies on chain initiation factor in <u>Artemia</u>, in: "<u>Artemia</u> Research and its Applications," Vol. 2, W. Declair, L. Moens, H. Slegers, E. Jaspers and P. Sorgeloos, eds., Universa Press, Wetteren (1987).

22. H. A. Daum, III, P. W. Bragg, D. B. Sittman, J. N. Dholakia, C. L. Woodley and A. J. Wahba, The expression of a gene for eukaryotic elongation factor Tu in <u>Artemia</u> during development, <u>J. Biol. Chem.</u> 260:16347 (1985).

23. W. W. Roth, P. W. Bragg, M. V. Corrias, N. S. Reddy, J. N. Dholakia and A. J. Wahba, Expression of a gene for mouse eucaryotic elongation factor Tu during murine erythroleukemic cell differentiation, <u>Mol. Cell Biol.</u> 7:3929 (1987).

24. F. Sanger, S. Nicklen and A. R. Coulson, DNA sequencing with chain-terminating inhibitors, Proc. Natl. Acad. Sci. U.S.A. 74:5463 (1977).

25. Y. W. Hwang and D. L. Miller, A mutation that alters the nucleotide specificity of elongation factor Tu, a GTP regulatory protein, J. Biol. Chem. 262:13081 (1987).

26. J. S. Clegg, Interrelationships between water and cellular metabolism in Artemia cysts. V. $^{14}CO_2$ incorporation, J. Cell. Physiol. 89: 369 (1976).

27. W. Filipowicz, J. M. Sierra and S. Ochoa, Polypeptide chain initiation in eukaryotes : initiation factor MP in Artemia salina embryos, Proc. Natl. Acad. Sci. U.S.A. 72:3947 (1975).

28. W. Filipowicz, J. M. Sierra and S. Ochoa, Polypeptide chain initiation in eukaryotes : initiation factor requirements for translation of natural messengers, Proc. Natl. Acad. Sci. U.S.A. 73:44 (1976).

29. H. Ernst, R. F. Duncan and J. W. B. Hershey, Cloning and sequencing of complementary DNAs encoding the α-subunit of translational initiation factor eIF-2. Characterization of the protein and its messenger RNA, J. Biol. Chem. 262:1206 (1987).

30. T. E. Dever, M. J. Glynias and W. C. Merrick, GTP-binding domain: Three consensus sequence elements with distinct spacing, Proc. Natl. Acad. Sci. U.S.A. 84:1814 (1987).

STRUCTURE OF UBIQUITIN GENES IN <u>ARTEMIA</u>

E. Franco and J. Renart

Instituto de Investigagiones Biomédicas
Facultad de Medicina de la Universidad Autónoma
Arzobispo Morcillo, 4, 28029 Madrid, Spain

ABSTRACT

Ubiquitin is a small protein (76 aa) present in all eukaryotic organisms studied, that has been involved in several processes, all of which imply the covalent attachment of ubiquitin by its C-terminal glycine to the free amino groups of other proteins. Ubiquitin genes are unique in that they are organized as intronless tandem repetitions of the coding unit (228 bp). These genes are translated as polyproteins then cleaved specifically to ubiquitin units.

We have isolated and sequenced a truncated cDNA clone that contains three units (one of them incomplete) of ubiquitin plus 240 bp of 3' non-translated sequence. Using this cDNA as a probe, we have isolated three different genomic clones in lambda. One of these phages, λgU-5 has been studied in detail. It contains five ubiquitin units. The 5' upstream region has consensus promoter signals, as the TATA box (position -27) and the CAAT box (position -117). The first ATG is at position 53 after the putative transcription initiation site. In both the cDNA and the gene there is an extension of 13 (12) amino acid codons, very similar but not identical, between the last ubiquitin unit and the stop codons.

Analysis of mRNAs shows two bands of 2.7 and 3.5 kb. This finding suggests that the gene found in λgU-5 is not the major transcription unit except for the presence of long 3' non-translated regions.

ACKNOWLEDGEMENTS

This research was supported by grants from CAICYT and FISSS.

ISOLATION OF cDNA CLONES CODING FOR ARTEMIA TYPE P ATPASES

I. Palmero, J. Renart and L. Sastre

Instituto de Investigagiones Biomedicas
Facultad de Medicina de la Universidad Autonoma
Arzobispo Morcillo, 4, 28029 Madrid, Spain

ABSTRACT

Artemia cDNA libraries have been constructed in the lambda vector λgt11 using EcoRI linkers and in the plasmid vector pUC18, using BamHI linkers. Both libraries were screened for type P ATPases using an oligo-nucleotide probe for the autophosphorylation region, which is highly conserved between all the enzymes of this family. cDNA clones coding for two different genes have been isolated. Sequencing of these clones showed that some of them code for a protein very similar to rabbit muscle sarcoplasmic reticulum Ca^{2+}/Mg^{2+} ATPases while a second set of clones are similar to sheep kidney Na^+/K^+ ATPase. The clones coding for the Ca^{2+}/Mg^{2+} ATPase cover 3.1 kb and include the complete coding sequence of the protein, 79 bp of the 5' untranslated sequence and 200 bp of the 3' untranslated sequence. The deduced aminoacid sequence is 70% similar to both fast and slow twitch forms of the rabbit protein. Artemia Ca^{2+}/Mg^{2+} ATPase clones hybridize to a 4.5 Kb and a 5.2 Kb long mRNAs. The 4.5 Kb mRNA is present at all stages of development but the 5.2 Kb mRNA is only present in crypt-obiotic cysts and cysts developed for short periods of time and disappears after hatching. The clones coding for the Na^+/K^+ ATPase cover 1.5 Kb. Work is in progress to determine their complete nucleotide sequence as well as their levels of expression during Artemia development.

STRUCTURE AND EXPRESSION DURING DEVELOPMENT OF <u>ARTEMIA</u> c-<u>ras</u> GENES

M. Díaz-Guerra, A. Gandarillas, M. Quintanilla and J. Renart

Instituto de Investigagiones Biomédicas
Facultad de Medicina de la Universidad Autónoma
Arzobispo Morcillo, 4, 28029 Madrid, Spain

ABSTRACT

c-<u>ras</u> genes have been implicated in cellular growth control, and therefore they may play a role in the development of the <u>Artemia</u> embryo.

We have isolated a cDNA clone that contains the C-terminal half of the coding sequence p21 plus 300 bp of 3' non-translated sequences. The deduced p21 protein is 80% homologous to that of mammals and <u>Drosophila</u>.

Expression of c-<u>ras</u> was studied both at the mRNA and the protein levels. mRNA of 1.2 kb in length is present already in the dormant embryo, at 50% of the maximum level that is reached at 5-7 hr after initiation of development and prior to emergence. On the contrary, protein levels are barely detected prior to emergence and rise steadily after hatching. Protein was measured by immunoblotting and detection with a specific monoclonal antibody and also by measurement of GTP binding activity in immunocomplexes.

We have also isolated genomic clones that contain c-<u>ras</u> sequences. The phages obtained can be classified in two groups, those containing the N-terminal region and those containing the C-terminal region of p21. The c-<u>ras</u> gene studied has three introns, and the data suggest that these introns are larger than those found in <u>Drosophila</u> or the Harvay type in mammals.

ACKNOWLEDGEMENTS

This research was supported by grants from CAICYT and FISSS.

MOLECULAR BIOLOGY OF LARVAL OSMOREGULATION

Frank P. Conte

Department of Zoology
Oregon State University
Corvallis, Oregon 97331 U.S.A.

INTRODUCTION

Halophilic organisms have evolved several diverse strategies in solving problems of osmotic stress. For example, animal and plant halophiles find life can be sustained in large measure by controlling the intracellular osmotic pressure with small molecules. These organisms are referred to as osmoconformers. Somero and colleagues have cogently argued that halophilic osmoconformers utilize a family of organic solutes, termed osmolytes, to passively regulate the intracellular osmotic pressures[1]. This osmotic system is believed to be a more flexible and adaptive mechanism with simpler genetic controls than those proposed for the more "primitive" osmoregulators that are dependent upon genetic control of the mechanics of cell volume[2]. Similarly, halophilic microbes employ small ions as a simple osmolytes to passively control water fluxes across the bacterial cell wall. These bacteria utilize membrane mechanisms that pump into the cell high concentrations of K^+ ions. Again, it can be argued that the control of the osmotic system requires a simpler genetic mechanism and, together with the genetic selection for cytosolic proteins having a higher proportion of acidic amino acids, provides for vital subcellular organelles (ribosomes) and enzyme complexes to be functional at salt concentrations in excess of 1 molar. It is obvious that genetic evolution through natural selection has taken a long period of time for this osmotic system to ascertain which of the many changes in DNA sequences needed to code for these proteins was best in coping with widely fluctuating environmental salinites.

Likewise, I would like to propose that in some types of osmoregulators there has been an evolution of the osmotic system to adapt excretory glands as special organs of salt deposition[3]. For example, in the osmoregulating halophilic brine shrimp, the naupliar larval stage contains a gland that responds to changes in environmental salinities. The glandular epithelial cells, under an ionic stimulus, differentiate under gene expresssions that control the structure, organization and dynamics of the microdomains of the outer plasma membrane surfaces. At the present time, one of the gene expressions that we know most about is the one that controls the formation and distribution of the lipoprotein complex called the sodium pump (Na,K-ATPase). Like the halophilic bacterial cell, the genetic control of the epithelial Na,K-ATPase may be located in those genes which are most responsive to changing levels of intracellular K^+ ion. The net result is that by maintaining low rather than high ionic osmolytes (K^+) in the intracellular compartment of epithelial cells, halophilic osmoregulators can ingest large quantities of hypersaline fluids and couple the epithelial extrusion of surplus salts via salt glands to obtain sufficient amounts of water to combat osmotic stresses. In contrast to the dogma of diversity of osmotic strategies among halobionts, comparison between the two K-dependent systems may lead us into new insights on water balance. It is intriguing to speculate that in osmotic evolution one may have convergence of common genetic elements between osmoconformers with these types of osmoregulators.

To illustrate my point, I shall focus on the molecular events associated with the dynamics of the Na,K-ATPase as related to formation, assembly and growth in the plasma membrane of epithelial cells.

BIOGENESIS OF EPITHELIAL Na,K-ATPase

Experimental models available for the study of the biogenesis of epithelial Na,K-ATPase have been limited to embryonic tissues, such as the duck salt gland[4], brine shrimp nauplius[3] and the electroplax organ[5]. Those experimental systems using tissue culture have been limited to the use of isolated epithelial cell types, such as the Madin-Darby canine kidney (MDCK) cells[6] and toad urinary bladder cells[7].

Na,K-ATPase is an integral membrane protein, existing only in the plasma outer membrane, and consists of two dissimilar oligomers, a catalytic subunit polypeptide (α) and a glycopolypeptide (β). The molecular considerations relevant to the mechanism of active transport has been reviewed by Kyte[8]. The similarities of the chemical and kinetic characteristics between vertebrate and invertebrate epithelial Na,K-ATPase include molecular weight, phosphorylation site, ion cofactors, amino acid sequence for the α-subunits, proteolytic fragmentation pattern and metabolic inhibitors, such as ouabain and vanadate.

The complete purification of the Na,K-ATPase from larval brine shrimp has been a formidable challenge but it has been accomplished with details on the isolation, purification and characterization protocol already reported[3,9]. The amino acid sequence of the catalytic subunit has been completed.

However, there are some striking differences between vertebrate and brine shrimp Na,K-ATPase. The vertebrate epithelial Na,K-ATPase α-subunit is a non-glycosylated polypeptide which contains sialic acid and when separated on SDS-PAGE has usually demonstrated but a single epitopic form. Additionally, the biosynthesis of the α-subunit in cell free systems, either using mRNAα from polysomes or in pulse-chase experiments in whole MDCK cell cultures, indicates that the genes coding for the oligomer are translated on free ribosomes in the cytosol[10].

In contrast, the invertebrate epithelial Na,K-ATPase found in brine shrimp has both the α-subunit and β-subunit glycosylated but contains no sialic acid and when separated by SDS-PAGE yields two epitopic forms[11,12]. The biosynthesis of the α-subunit of the brine shrimp Na,K-ATPase in cell free systems indicates that mRNAα is translated on membrane-bound ribosomes whereas the β-subunit oligomer is translated on free ribosomes[13]. The rate of α-subunit synthesis and degradation together with temporal differentiation is quite different and appears to be non-hormonally regulated[11, 12,14]. In several species of vertebrate kidney epithelial cells, the Na,K-ATPase is under endocrine control and its rate of synthesis and degradation is markedly influenced by hormonal levels[7].

The functional importance regarding these post-translational differences in subunit polypeptides lies in fundamental questions relating to the subcellular construction and distribution of the "different" types of sodium pumps. For instance: 1). Are there two separate mRNAs coding for the two epitopes of the α-subunit or are both α-subunits derived from a single polycistronic message? 2). Are the two epitopes of the α-subunit assembled into a functional sodium pump at the cell surface or in different subcellular compartments and transported to the cell surface? 3). How does the non-glycosylated oligomer formed on the free ribosomes get inserted into and across the hydrophobic region of the membrane? 4). Are the two epitopes of the sodium pump, when found in a single cell, utilizing similar controls in regulating transcellular ion movement?

TRANSCRIPTION OF EPITHELIAL Na,K-ATPase SUBUNITS

The original finding of two forms of the phosphorylatable 100-kDa subunit ($\alpha 1$ and $\alpha 2$) in the brine shrimp Na,K-ATPase resolved on 8.75% polyacrylamide with the Laemmli system[15] was quite startling in light of the fact that phosphorylatable doublets had not been reported from other sources of purified Na,K-ATPases. Following this report, phosphorylatable doublets have been demonstrated to occur in brain tissue[16-18], in heart tissue[17] but only one form found in cardiac membranes[19], and in adipocytes and muscle cells[20]. Interestingly, kidney epithelial tissue demonstrated only one isoform of the phosphorylatable subunit for any species studied[21] until the investigation by Siegel et al.[22] where purified lamb kidney holoenzyme demonstrated doublets as did duck salt gland tissue and electroplax tissue.

Recently, Young and Lingrel[23] developed specific cDNA probes for the catalytic subunit isoforms of the rat Na,K-ATPase. Northern blot analyses were performed on mRNA isolated from kidney, brain, heart, adipose, muscle, stomach, liver, and lung tissues and showed the α-2 subunit probe, which is the lower mol. wt. isoform (α), hybridizes to a single mRNA of 3.7 kb in each tissue analyzed. In the α-1 subunit probe, which is the higher mol. wt. isoform (α), two species of mRNAs were found, exhibiting a 3.4 kb and 5.3 kb band size respectively, in every tissue except liver. Slot blot analyses were performed on these same tissues in order to establish the relative amounts of each isoform mRNAs. The α-1 (or α^+) mRNA was most abundant in muscle and brain or the excitable tissues whereas it was in much lower amounts in kidney, adipose, stomach and lung tissue. The α-2 (or α) mRNA is most abundant in kidney, an epithelial tissue, and virtually non-detectable in liver and barely detectable in stomach, lung and brain. Specific cDNA probes have been made for the brine shrimp Na,K-ATPase (Guo and Hokin, this volume), but specific cDNA probes for the two isoforms α-1 and α-2 remain to be developed. Recently, Urayama and Sweadner[24] reported that three isoform proteins for the catalytic subunit of rat brain Na,K-ATPase are products of different genes. Therefore, these data would support the hypothesis that two or more separate mRNAs are coding for the two isoforms of the brine shrimp phosphorylatable subunit.

However, the present results indicate that some discretion is needed concerning the interpretation as to whether these genes are being expressed in different cell types or whether individual cells are expressing only one or both of the genes in response to a pretranscriptional signal. Since the whole embryo of the brine shrimp has been used to prepare the Na,K-ATPase, it would be impossible at this time to determine the cellular origin of the α-1 and α-2 isoforms. What is needed is an in vitro cell culture of brine shrimp epithelial cells for studying this aspect of the problem but none is available. The development of an isolation protocol for the preparation of large quantities of salt glands (tissue which contains but a single type of epithelial cell) may offer a solution[25,26].

It is possible that the two molecular forms represent the growth and differentiation of two distinct types of epithelia, namely a salt gland secretory epithelium and a midgut absorptive epithelium[27].

PRETRANSLATIONAL REGULATION OF EPITHELIAL Na,K-ATPase

The signals which control the Na,K-ATPase genes appear to be changes in intracellular K^+ ion concentrations[28-30], whereas Boardman et al.[31] had proposed that the stimulation of Na,K-ATPase in Hela cells was signaled by an increase in intracellular Na+ concentrations. The most quantitative evidence which supports the concept of ionic signals regulating the synthesis, degradation and turnover of α- and β-subunit mRNA of the Na,K-ATPase enzyme come from the results of dot blot analysis of total RNA from MDCK cell cultures using α- and β-cDNA probes isolated from a λ-gt11 dog kidney cDNA library with synthetic oligonucleotide probes. The α-subunit cDNA probe contains 1,512 base pairs and corresponds to the region of the sheep kidney α-subunit between amino acids 242 and 746[28]. The results obtained from this investigation showed that α- and β-subunits mRNAs increase by 30 min after exposure of MDCK epithelial cells to a low K^+ incubation, peak at 1 h, then decrease over the next several hours. These results agree with the findings of Wolitzky and Fambrough[32] in chick skeletal muscle where veratridine stimulation, which increases intracellular Na^+, causes an increase in mRNA synthesis and is followed by a degradative mechanism. The rapid change in mRNA level indicates transcription is directly regulated, because it is unlikely that a regulatory protein could be synthesized and activate transcription within the short 30 min period.

The question of whether the K^+ or Na^+ or some other species of ion whose transport is linked to that of Na^+, such as H^+, is the factor controlling gene expression cannot be ruled out. There have been suggestions that hormonal control of Na,K-ATPase genes are mediated by changes in cell ion concentrations, but the evidence is contradictory. Recently, treatment of MDCK cell with the ionophore monensin or serum repletion protocol conducted in bicarbonate free medium to effect functional Cl/HCO3 antiporters have resulted in findings that are nearly identical to K^+ depletion. The modified MDCK cells evidenced a stimulation and elevation of Na,K-ATPase α- and β- mRNA levels[33,34].

The question of postranslational control of the two isoforms of the alpha subunit of brine shrimp Na,K-ATPase is dependent upon our understanding of the pretranslational control of alpha subunit mRNAs. At present, we have little data concerning the brine shrimp Na,K-ATPase gene(s). Should it be found that the mRNAs of the α- and β-subunit of the brine shrimp epithelial Na,K-ATPase respond in a similar fashion to changes of intracellular k^T, like the canine kidney epithelial Na,K-ATPase, then our understanding of the response of the larval salt gland to changes in environmental salinities could be much better.

In conclusion, investigations on the molecular genetic control of the larval salt gland has as an ultimate goal the elucidation of control mechanisms regulating electrolyte and water balance in animals.

ACKNOWLEDGEMENTS

This research was supported by National Science Foundation Grant PCM 76-14771 and a travel award from NATO and OSU Foundation Staff Travel Grant made to F.P. Conte.

REFERENCES

1. P. H. Yancey, M. E. Clark, S. C. Hand, R. D. Bowlus and G. M. Somero, Living with water stress: Evolution of osmolyte systems, Science 217:1214 (1982).
2. R. Gilles, "Mechanisms of Osmoregulation in Animals: Maintenance of Cell Volume," J. Wiley & Sons, New York (1979).
3. F. P. Conte, Structure and Function of the Crustacean Larval Salt Gland. in: "Inter. Rev. of Cytol," Danielli, J. ed., Academic Press, New York (1984).
4. R. J. Barnett, J. E. Mazurkiewicz and J. S. Addis, Avian salt gland: A model for the study of membrane biosis. in: "Methods in Enzymology," S. Fleischer and B. Fleischer, eds., Academic Press, New York (1983).
5. L. Churchill, G. L. Peterson and L. E. Hokin, The large subunit of (sodium+potassium-activated) adenosine triphosphatase from the electroplax of Electrophorus electricus in a glycoprotein, Biochem. Biophys. Res. Commun. 90:488 (1979).
6. A. McDonough and D. K. Mircheft, Synthesis and degradation of Na, K-ATPase alpha subunit in MDCK Cells, Fed. Proc. Abst. 42:1934 (1983).
7. B. C. Rossier, K. Geering and J. P. Kraeheubuhl, Regulation of the sodium pump: how and why? Trends in Biochem. Sci. 12:483 (1987).
8. J. Kyte, Molecular considerations relevant to the mechanism of active transport, Nature 292:201 (1981).
9. M. Morohashi and M. Kawamura, Solubilization and purification of Artemia salina Na, K-activated ATPase and NH_2-terminal amino acid sequence of its larger subunit, J. Biol. Chem. 259:14928 (1984).
10. A. Hiatt, A. A. McDonough and I. S. Edelman, Assembly of the (Na+K)-adenosine triphosphatase. Post-translational membrane integration of the α-subunit, J. Biol. Chem. 259:2629 (1984).
11. G. L. Peterson, L. Churchill, J. A. Fisher and L. E. Hokin, Structural and biosynthetic studies on the two molecular forms of the (Na + K+)-activated adenosine triphosphatase in Artemia salina nauplii, J. Exp. Zool. 221:295 (1982a).
12. G. L Peterson, L. Churchill, J. A. Fisher and L. E. Hokin, Structure and biosynthesis of (Na, K) -ATPase in developing brine shrimp nauplii, in "Ann. N.Y.. Acad. Sci: Transport ATPases," Vol. 402, E. Carafoili and A. Scarpa, eds., (1982b).
13. J. Fisher, L. Baxter-Lowe and L. Hokin, Site of synthesis of α and β-subunits of the Na, K-ATPase in brine shrimp nauplii, J. Biol. Chem. 259:14217 (1984).
14. J. Salon and I. S. Edelman, Developmental regulation of two catalytic forms of Na,K-ATPase in the brine shrimp, Fed. Proc. 45:A2884 (1986).
15. G. L. Peterson, R. D. Ewing, S. R. Hootman and F. P. Conte, Large scale partial purification and molecular and kinetic properties of the (Na+K)-activated adenosine triphosphatase from Artemia salina nauplii, J. Biol. Chem. 253:4762 (1978).
16. K. J. Sweadner, Two molecular forms of (Na^+K^+)-stimulated ATPase in brain, J. Biol. Chem. 254:6060 (1979).
17. T. Matsuda, H. Iwata and J. R. Cooper, Specific inactivation of alpha (+) molecular form of Na,K-ATPase by pyrithiamin, J. Biol. Chem. 259:3858 (1984).
18. G. D. Schellenberg, I. V. Pech and W. L. Stahl, Immunoreactivity of subunits of the Na,K-ATPase. Cross reactivity of the α, α+ and β forms in different organs and species, Biochim. Biophys. Acta 649:691 (1981).
19. A. McDonough and C. Schmitt, Comparison of subunits of cardia, brain, and kidney Na,K-ATPase, Am. J. Physiol. 248:C247 (1985).
20. J. Lytton, J. C. Lin and G. Guidotti, Identification of two molecular forms of Na,K-ATPase in rat adipocytes, J. Biol. Chem. 260:1177 (1985).

21. P. L. Jorgensen, Purification and characterization of (Na+ + K+)-ATPase. Estimation of the purity and of the molecular weight and polypeptide content per enzyme unit in preparations from the outer medulla of rabbit kidney, Biochim. Biophys. Acta 356:53 (1974).

22. G. J. Siegel, T. Desmond and S. A. Ernst, Immunoreactivity and Quabain-dependent phosphorylation of Na,K-ATPase catalytic subunit doublets, J. Biol. Chem. 261:13768 (1986).

23. R. M. Young and J. B. Lingrel, Tissue distribution of mRNAs encoding the alpha isoforms and beta subunit of rat Na,K-ATPase, Biochem. Biophys. Res. Comm. 145:52 (1987).

24. O. Urayama and K. J. Sweadner, Three isoform proteins of the catalytic subunit of rat brain Na,K-ATPase. Fed. Proc. 2:A5846 (1988).

25. R. J. Lowy and F. P. Conte, Isolation and functional characteristics of the larval brine shrimp salt gland (Artemia salina), Am. J. Physiol. 248:R702 (1984a).

26. R. J. Lowy and F. P. Conte, Morphology of isolated crustacean larval salt gland, Am. J. Physiol. 248:R709 (1984b).

27. J. Clegg and F. P. Conte, A review of the cellular and developmental biology of Artemia, in "The Brine Shrimp, Artemia," Vol. II, G. Persoone, P. Sorgeloos, O. Roels and E. Jaspers, eds., Universa Press, Wetteren (1980).

28. J. W. Bowen and A. McDonough, Pretranslational regulation of Na,K-ATPase in cultured canine kidney cells by low K+, Amer. J. Physiol. 252:C179 (1987).

29. L. R. Pollack, E. H. Tate and J. S. Cook, Na,K-ATPase in HeLa cells after prolonged growth in low K^+ or ouabain, J. Cell Physiol, 106:85 (1981).

30. J. P. T. Ward and I. R. Cameron, Adaptation of the cardiac muscle sodium pump to chronic potassium deficiency, Cardiovasc. Res. 18:257 (1984).

31. L. J. Boardman, J. M. Huett, J. F. Lamb and J. Polson, Effect of growth in lithium on ouabain binding, Na,K-ATPase and Na and K transport in HeLa cells, J. Physiol. Lond. 244:677 (1975).

32. B. A. Wolitzky and D. M. Fambrough, Regulation of the Na,K-ATPase in cultured chick skeletal muscle, J. Biol. Chem. 261:9990 (1986).

33. A. McDonough, J. W. Bowen, M. R. Quintero and D. S. Putnam, Ionic stimuli directly increase Na,K-ATPase α-subunit transcription, Fed. Proc. 2:A5854 (1988).

34. L. Lescale-Matys and A. McDonough, Serum repletion increases Na,K-ATPase synthesis pretranslationally in MDCK cells via sodium influx, Fed. Proc. 2:A5355 (1988).

DIFFERENTIAL GENE EXPRESSION OF Na,K-ATPase α- AND β-SUBUNITS IN THE

DEVELOPING BRINE SHRIMP, ARTEMIA

Jian Zhong Guo[a] and Lowell E. Hokin[b]

[b]Department of Pharmacology, University of Wisconsin Medical
School, 1300 University Avenue, Madison, Wisconsin 53706
[a]Present address: Department of Physiology, Faculty of Medicine
University of Manitoba, Winnipeg, Manitoba R3E 0W3

INTRODUCTION

Na,K-ATPase plays an important role in maintaining Na^+ and K^+ electro-chemical gradients across the plasma membrane [1]. It effects the coupled transports of three Na^+ ions out of the cell and two K^+ ions into the cell driven by the hydrolysis of one ATP. The ion gradients are essential for osmotic homeostasis and certain physiological processes such as the wave of excitation in nerve and muscle, ion transport in structures such as the kidney and intestine, and ion secretion in glands such as the avian salt gland and salivary glands. The enzyme has been purified from several tissues [2-4]. Numerous studies have shown that the enzyme molecule con-sists of two subunits: a catalytic α-subunit and a glycosylated β-subunit [2,3,5-7].

Recently, cDNAs encoding the α- and β-subunits from several tissues, e.g., the electric organ of Torpedo californica [8] and mammalian kidney [9], have been cloned and sequenced. We have now cloned in the brine shrimp cDNA fragments which cover the full length of the coding region of the α-subunit [10], and a cDNA fragment which shows homology to the se-quence of the β-subunit in other species. Some years ago we began studies on the structure and biosynthesis of Na,K-ATPase in developing brine shrimp [11]. Among other things, we reported two isoforms of the α-subunit of Na,K-ATPase in developing brine shrimp, which we termed α_1 and α_2 and which had molecular weights of 98,900 and 87,500, respectively [7]. More recently, we have been interested in gene expression in this model system. We have already reported studies [12] on the regulation of biosynthesis of the α- and β-subunits using radioimmunoassay and in vitro translation techniques.

RESULTS AND DISCUSSION

α- and β-mRNA Levels during Development

To study regulation of Na,K-ATPase gene expression, we have conducted Northern hybridization and nuclease protection assays to measure the mRNA levels of the two subunits in developing brine shrimp. The time that the desiccated cyst resumes development after suspension in sea water was set as zero time. Total cellular RNA was isolated from brine shrimp at: 0, 6, 12, 16, 24, and 36 hr after resumption of development. For the Northern study, equal amounts of RNA from each time point were resolved by electrophoresis on a denatured agarose gel, followed by transfering to filters. The restriction fragments of α - and β-subunit cDNA were subcloned into phagemid vectors. Different nick translated probes and labeled antisense RNA probes were prepared from the subclones for the hybridization analysis.

Under very stringent conditions, the labeled α-subunit cDNA probe hybridized to mRNAs of 5.1, 3.5 and 1.3 kilobases (Fig. 1). The 5.1 kb and 3.5 kb mRNAs in the brine shrimp are essentially identical in molecular weight to the 5.1 kb (α^{+}-form) and the 3.55 kb (α-III-form) mRNAs of the

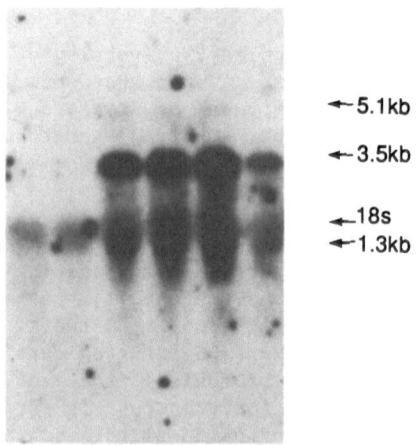

← 5.1kb

← 3.5kb

← 18s
← 1.3kb

0 6 12 16 24 36

Hours After Resumption
of Development

Fig. 1.   Northern blot analysis of Na,K-ATPase α-subunit mRNAs
in developing brine shrimp. Total cellular RNA (20 µg)
from different developmental stages of brine shrimp was
electrophoresed through a 1% agarose-formaldehyde gel.
The RNA was transferred to a Zeta bind filter and probed
with a ^{32}P-labeled α-subunit cDNA fragment. Hybridization was carried out at 42°C for 16 hr in a mixture
containing 50% deionized formamide, 5 x SET (1 x SET =
0.15 M NaCl, 0.03 M Tris-HCl, pH 8.0, 2 mM EDTA) 1 x
Denhardt's (0.01% polyvinylpyrolidone, 0.01% Ficoll,
and 0.01% bovine serum albumin), 0.02 M sodium phosphate
buffer, 0.1% SDS, and 50 µg/ml of denatured herring
sperm DNA. The filter was washed once at 55°C for 30
min with 2 x SSC (1 x SSC = 0.15 M NaCl, 0.015 M sodium
citrate, pH 7.0), 0.1% SDS, 0.1% sodium pyrophosphate,
and 0.025 M sodium phosphate buffer, then three times
at 68°C for 45 min each with 0.1 x SSC in the same
buffer. The position of 18s RNA marker is indicated.
DNA molecular weight markers were used as references
and converted to the molecular weight of the RNA
species. The molecular weight sizes of the RNAs are
shown at the right.

rat brain, respectively [13]. It is possible that our 3.5 kb form is the ∝-form of the rat brain mRNA, which is 3.65 kb. A fragment around 1.3 kb has also been seen on Northern hybridization of mRNA in other species, but its significance is unknown. In rat brain, the 5.1 kb form is more abundant than the 3.55 or 3.65 kb ∝-form [13], while in the brine shrimp the 3.5 kb form is the more abundant isoform. Figure 1 shows that there was no detectable ∝-subunit mRNA at 0 time and 6 hr after resumption of development. After 6 hr, there was a significant increase in the mRNA level, which peaked at 16 hr and then declined at 36 hr. Different probes prepared from different cDNA fragments were used in the Northern hybridization studies and gave consistent results.

To verify that the mRNA species which appeared on the Northern blots were indeed ∝-subunit messengers, we conducted ribonuclease protection assays [14] to study ∝-subunit mRNA levels. ^{32}P-Labeled antisense RNA was synthesized from phagemid vectors by in vitro transcription with phage RNA polymerases. An excess of labeled probe was used to hybridize the total cellular RNA, followed by RNase digestion. Only the 100% matched hybrids are protected from the ribonuclease digestion. After electrophoresis on urea-polyacrylamide gel, the autoradiograph should show the exact length and amount of the protected mRNA fragments. Figure 2 is an autoradiograph of the assay which revealed a similar developmental pattern for the 3.5 kb mRNA as we observed in Northern blotting studies.

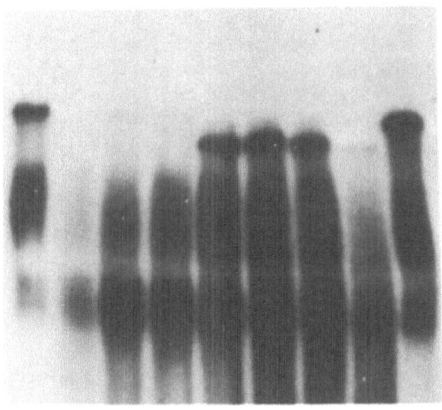

S   C   0   6   12 16 24 36 S

**Hours After Resumption
of Development**

Fig. 2. Ribonuclease protection assay of Na,K-ATPase ∝-subunit mRNA in developing brine shrimp. The labeled antisense RNA were synthesized from the cDNA subclone in phagemid vector by in vitro transcription with phage RNA polymerase. An excess of the RNA probe was hybridized with total cellular RNA prepared from brine shrimp at different developmental stages (0-36 hr) or yeast tRNA (lane C) subsequently digested with ribonucleases A and T1. Lane S is an undigested RNA probe. The sizes of labeled RNA and protected RNA fragment are indicated.

By using the same procedures, we have studied levels of the Na,K-ATPase β-subunit mRNA in developing brine shrimp (Fig. 3). The Northern study for the β-subunit mRNA revealed that, contrary to the α-subunit mRNA, the β-subunit mRNA was present in the desiccated cysts. Its level increased continuously after resumption of development and plateaued after 16 hr. The nuclease protection assay (Fig. 4) confirmed this pattern.

To determine whether ATPase gene expression might be different from that of another "house-keeping" gene, actin mRNA levels were measured by Northern hybridization. A chicken β-actin probe was used to detect mRNA species. Because of the homology between the sequences of β-actin and α-actin, the labeled β-actin probe hybridized with both the α- and β-actin mRNAs (Fig. 5). The actin mRNA levels were developmentally regulated. There was no detectable actin mRNA until after 8 hr, at which time both α- and β-actin mRNA levels increased dramatically. However, the β-actin mRNA level started to decrease soon after reaching a maximum level, and by 24 hr it was almost zero. The α-actin mRNA remained high at 24 and 36 hr.

To obtain a relatively quantitative estimate of the kinetics of the mRNA levels for the genes we have studied, the autoradiograms from Northern blots were measured densitometrically. Figures 6, 7, and 8 show the changes in relative abundance of different mRNAs in developing brine shrimp. The density of each band was standardized to that of 18s ribosomal RNA at each time point. Every point was an average of three measurements from different blots. It is clear that the kinetics of accumulation of different mRNA's showed differential regulation during development.

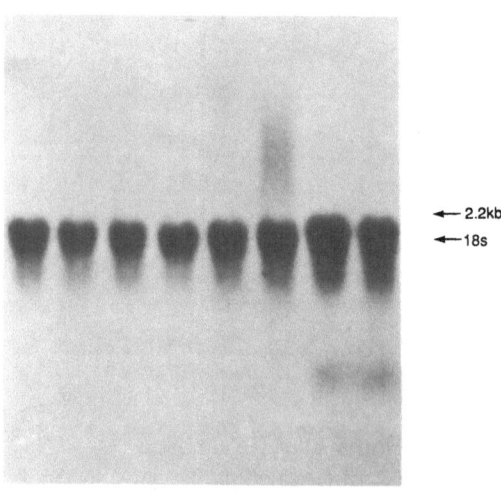

-4   0   4   8   12 16 24 36

Hours After Resumption
of Development

Fig. 3. Northern blot analysis of Na,K-ATPase
β-subunit mRNA in developing brine
shrimp. Hybridization was performed
with a labeled β-subunit cDNA probe,
as described in Figure 1.

No Regulation of Gene Expression at the Genomic Level

The regulation of gene expression could occur at one or more levels: gene amplification, gene rearrangement, DNA methylation, transcription, or post-transcriptional steps. To determine whether the developmental regulation of Na,K-ATPase gene expression occurs at the genomic level, we we conducted Southern hybridization of genomic DNA from different stages of brine shrimp development with labeled cDNA probes. High molecular weight DNA isolated from brine shrimp at different developmental stages was digested with several restriction enzymes and analyzed by Southern blotting. For the Na,K-ATPase α-subunit (Fig. 9), three bands were identified in EcoRI digests, one band in BamHI digests, and two bands in MspI and HpaII digests. The banding patterns of each restriction digest at different

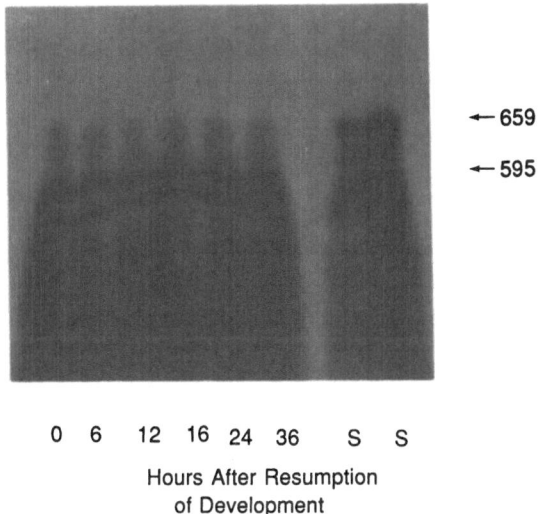

0 6 12 16 24 36 S S

Hours After Resumption
of Development

Fig. 4.   Ribonuclease protection assay of Na,K-ATPase β-subunit mRNA in developing brine shrimp. The assay was performed with labeled β-subunit cDNA probe, as described
in Figure 2.

times were identical. There was some variation in amounts of DNA due to errors in aliquoting DNA samples. The restriction enzyme MspI recognizes both sequences, CCGG and metCCGG, while the restriction enzyme HpaII only recognizes CCGG. There were no differences between the digestion patterns of MspI and HpaII.

The Southern blots, hybridized with the β-subunit cDNA probe showed the same banding patterns for each restriction digest at different time points (Fig. 10). The data indicate that gene amplification, rearrangement, and DNA methylation did not occur within the Na,K-ATPase coding region during development, which implies that the regulation of Na,K-ATPase gene expression was not at the level of DNA modification.

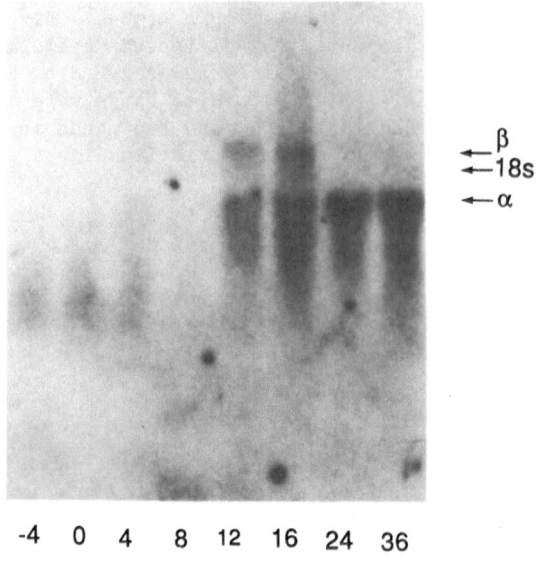

-4 0 4 8 12 16 24 36

Hours After Resumption
of Development

Fig. 5. Northern blot analysis is actin mRNAs in developing brine shrimp.
Hybridization was performed with labeled probe which was prepared
from a chicken β -actin gene fragment [10], except the temperature
of hybridization was 42°C and that of washing, 50°C.

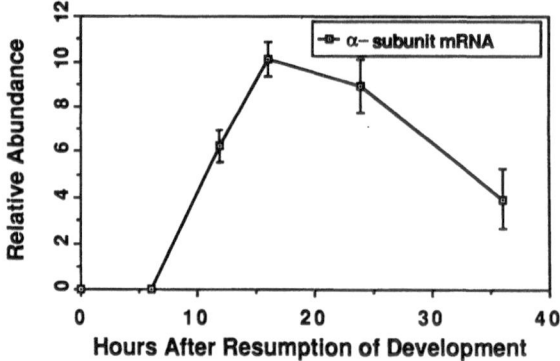

Fig. 6. Densitometric analysis of a Northern blot of the Na,K-ATPase α-
subunit mRNA in developing brine shrimp. The density of each
mRNA band was standardized to that of an 18s RNA for that time
point and expressed as relative abundance of the mRNA. Each
determination was an average of three measurements of different
blots. The error bars represent ± 5.0.

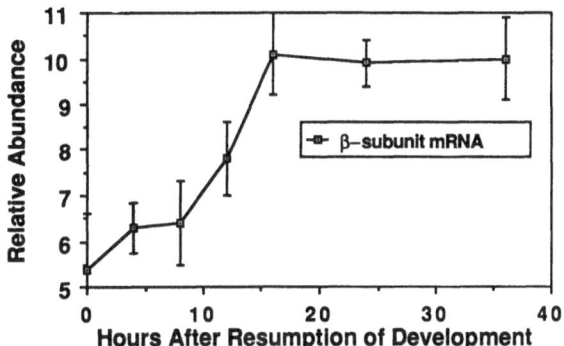

Fig. 7. Densitometric analysis of the Northern blot
of the Na,K-ATPase β-subunit mRNA in developing brine shrimp. The analysis was carried
out as described in Figure 6.

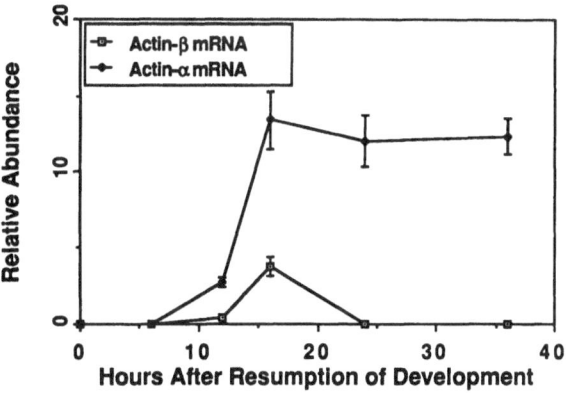

Fig. 8. Densitometric analysis of the Northern blot
of α- and β-actin mRNAs in developing
brine shrimp. The analysis was carried out
as described in Figure 6.

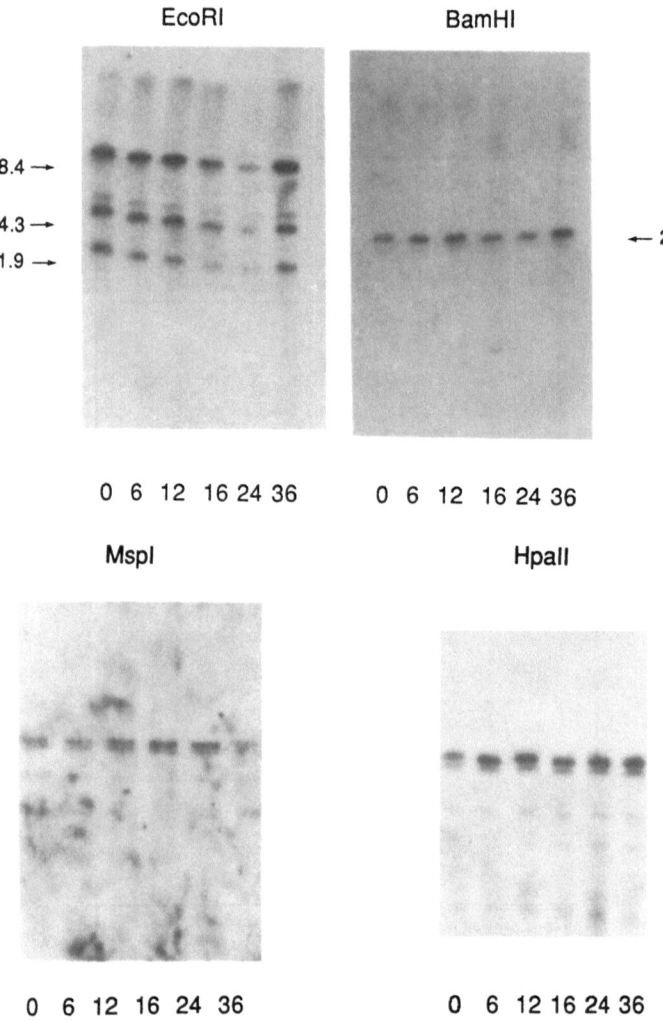

Hours After Resumption of Development

Fig. 9.   Southern blot analysis for Na,K-ATPase α-
subunit genomic DNA of developing brine
shrimp. High molecular weight DNA (10 μg)
was prepared from brine shrimp at different
stages of development and digested with
restriction enzymes EcoRI, BamHI, MspI and
HpaII, separately as shown. The DNA fragments
were electrophoresed through 1.0% agarose gel
and transferred to a Zeta bond filter. The
labeled cDNA fragment of the Na,K-ATPase α-
subunit was used as the hybridization probe.
The sizes of the restriction bands are indicated.

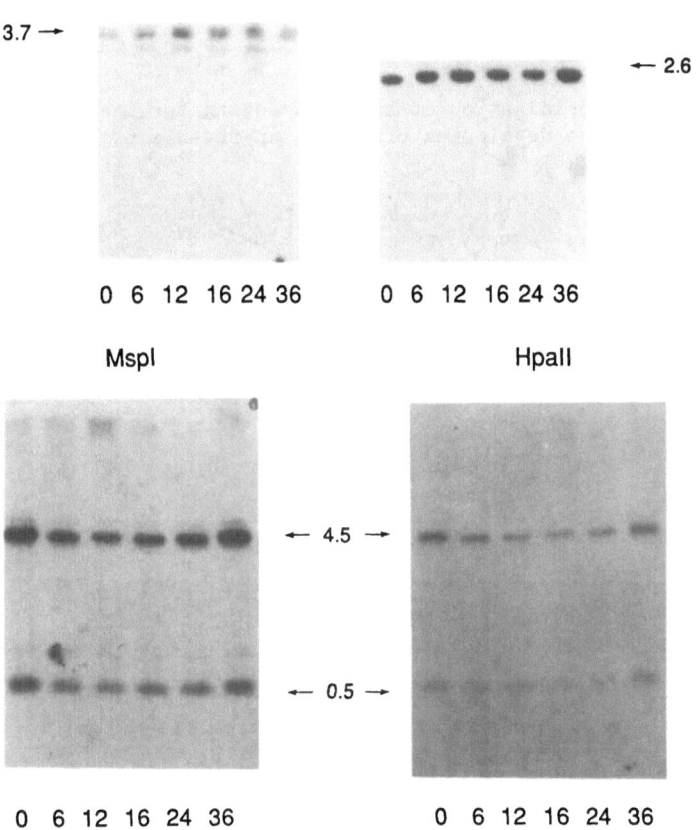

EcoRI   BamHI

3.7 →

← 2.6

0 6 12 16 24 36   0 6 12 16 24 36

MspI   HpaII

← 4.5 →

← 0.5 →

0 6 12 16 24 36   0 6 12 16 24 36

Hours After Resumption of Development

Fig. 10. Southern blot analysis of the Na,
K-ATPase β-subunit genomic DNA
in developing brine shrimp. Analysis was carried out as described
in Figure 9, except that a labeled
β-subunit cDNA fragment was used
as the probe. The restricition
enzymes were EcoRI, BamHI, MspI
and HpaII, as shown.

Correlation between mRNA Levels of the α- and β-Subunits and the Biosynthetic Rates of the Protein Products

When we compared the relative levels of Na,K-ATPase α- and β-subunit
mRNAs with the data for biosynthesis of the enzyme subunits in developing
brine shrimp (Fig. 11), the profiles matched very well. It is reasonable
to conclude that the regulation of Na,K-ATPase gene expression is very
likely to be at the transcriptional and/or posttranscriptional level. The
differential expression of the Na,K-ATPase α- and β-subunit genes in
developing brine shrimp embryos is interesting and requires further investigation.

SUMMARY

1. Northern hybridization of the mRNA coding for the α-subunit of
the Na,K-ATPase in the developing brine shrimp reveals two mRNAs of 5.1

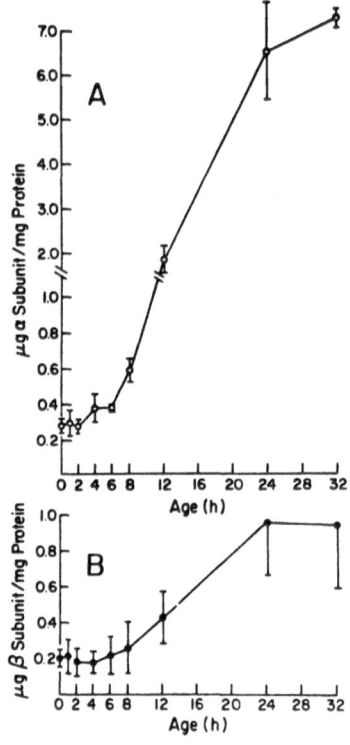

Fig. 11.   Radioimmunoassay of α- and β-subunits.
           Homogenates of brine shrimp nauplii were
           prepared at the indicated times, and the
           amounts of α- or β-subunit in 80 µl
           aliquots were determined by RIA as des-
           cribed [16]. The total amount of protein
           in each aliquot was determined by Peterson's
           modification [7] of the Lowry assay [17].
           The amount of α-subunit/mg of protein is
           shown in A, and the amount of β-subunit/
           mg of protein is shown in B. Reproduced
           by permission of the Journal of Biological
           Chemistry.

and 3.5 kb, which are essentially identical in molecular weight to the 5.1 kb (α-form) and the 3.55 kb (α-III-form) or 3.65 kb (α-form) mRNAs of the rat brain, respectively.

2. The mRNA levels of Na,K-ATPase α- and β-subunits are differentially regulated in developing brine shrimp. There is little or no α-subunit mRNA at the point of resumption of development. The α-subunit mRNA level increases significantly after 6 hr, reaches a peak at 16 hr, and then gradually decreases. The mRNA of the β-subunit is present in the desiccated cysts, and it accumulates continuously after 0 time. The β-subunit mRNA level plateaus after 16 hr.

3. The genes encoding the Na,K-ATPase α- and β-subunits do not amplify and rearrange and are not methylated during development of the brine shrimp embryo.

4. Thus, the regulation of gene expression of Na,K-ATPase does not occur at the DNA level but probably at the transcriptional and/or post-transcriptional level.

ACKNOWLEDGEMENT

This work was supported by the National Institutes of Health, grant GM33850.

REFERENCES

1. B. C. Rossier, K. Greering and J. P. Kraehenbuhl, Regulation of the sodium pump: How and why? Trends Biochem. Sci. 12:483 (1987).
2. J. Kyte, Properties of the two polypeptides of sodium- and potassium-dependent adenosine triphosphatase, J. Biol. Chem. 247:7642 (1972).
3. L. E. Hokin, J. L. Dahl, J. D. Deupree, J. F. Dixon, J. F. Hackney and J. F. Perdue, Studies on the characterization of the sodium-potassium transport adenosine triphosphatase. X. Purification of the enzyme from the rectal gland of Squalus acanthias, J. Biol. Chem. 248:2593 (1973).
4. P. L. Jorgenson, Purification and characterization of Na,K-ATPase. III. Purification from the outer medulla of a mammalian kidney after selective removal of membrane components by sodium dodecyl-sulphate, Biochim. Biophys. Acta 356:36 (1974).
5. J. R. Perrone, J. F. Hackney, J. F. Dixon and L. E. Hokin, Molecular properties of purified (sodium + potassium)-activated adenosine triphosphatases and their subunits from the rectal gland of Squalus acanthias and the electric organ of Electrophorus electricus, J. Biol. Chem. 250:4178 (1975).
6. I. M. Glynn, in: "The Enzymes of Biological Membranes", Vol. 3, A. Martonosi, ed., Plenum Publishing, New York (1985).
7. G. L. Peterson and L. E. Hokin, Molecular weight and stochiometry of the sodium- and potassium-activated adenosine triphosphatase subunit, J. Biol. Chem. 256:3751 (1981).
8. K. Kawakami, S. Noguchi, M. Noda, H. Takahashi, T. Ohta, M. Kawamura, H. Nojima, K. Nagano, T. Hirose, S. Inayama, H. Hayashida, T. Miyata and S. Numa, Primary structure of the α-subunit of Torpedo californica Na,K-ATPase deduced from cDNA sequence, Nature 316:733 (1985).
9. G. E. Shull, A. Schwartz and J. B. Lingrel, Amino-acid sequence of the catalytic subunit of the Na,K-ATPase deduced from a complementary DNA, Nature 316:691 (1985).
10. L. A. Baxter-Lowe, J. Z. Guo, E. Bergstrom and L. E. Hokin, Molecular

cloning and cDNA-derived amino acid sequence of the Na,K-ATPase in developing brine shrimp, Submitted (1988).

11. G. J. Peterson, L. Churchill, J. A. Fisher and L. E. Hokin, Structure and biosynthesis of Na,K-ATPase in developing brine shrimp nauplii, Ann. N.Y. Acad. Sci. 402:185 (1982).

12. J. A. Fisher, L. A. Baxter-Lowe and L. E. Hokin, Regulation of Na,K-ATPase biosynthesis in developing Artemia salina, J. Biol. Chem. 261:515 (1986).

13. G. E. Shull, J. Greeb and J. B. Lingrel, Molecular cloning of three distinct forms of the Na,K-ATPase α-subunit from rat brain, Biochemistry 25:8125 (1986).

14. M. Gilman, in: "Current Protocols in Molecular Biology," F. M. Ausubel, R. Brent, R. E. Kingston, D. D. Moore, L. G. Seidman, J. A. Smith and K. Struhl, eds., John Wiley and Son, New York (1987).

15. T. A. Kost, N. Theodorakis and S. H. Hughs, The nucleotide sequence of the chick cytoplasmic β-actin gene, Nucleic Acids Res. 11:8287 (1983).

16. G. L. Peterson, A simplification of the protein assay method of Lowry et al. which is more generally applicable, Anal. Biochem. 83:346 (1977).

17. O. H. Lowry, N. J. Rosebrough, A. L. Farr and R. J. Randall, Protein measurement with the Folin phenol reagent, J. Biol. Chem. 193:265 (1951).

TUBULIN SYNTHESIS AND MICROTUBULE ORGANIZATION IN <u>ARTEMIA</u>

Thomas H. MacRae, Elizabeth J. Campbell and Carrie M. Langdon

Department of Biology
Dalhousie University
Halifax, N. S., Canada

OVERVIEW

Many aspects of eukaryotic cell function and structural organization depend on the cytoskeleton, an interconnected array of filamentous elements including microtubules, microfilaments and intermediate filaments. The main structural protein of microtubules is tubulin, a heterodimer of α- and β- tubulin. The tubulins are in most cases the products of small gene families [1-3], giving rise to collections of isotubulins. The best characterized tubulin gene family occurs in the chicken where several, if not all, of the α- and β- genes have been cloned and sequenced. At least seven β-tubulin genes encoding six β-tubulin isotypes are present, and except for the β-tubulin gene expressed in erythrocytes, they are relatively highly conserved [2]. Most of the diversity within the chicken β-tubulins exists in a region near the carboxy terminus, with a smaller amount nearer the amino end of the protein. The β- isotypes, based on the variable region sequences and their tissue - specific expression, are divided into families. Each member of the family in the chicken shares an identical or almost identical sequence, even within the variable region, with a β-tubulin gene from human or other vertebrates. The genes of corresponding sequence are expressed in similar tissues from one species to another. The chicken α-tubulin gene family has at least five functional genes, and the five polypeptides they encode are more heterogeneous than those encoded by the β-tubulin genes [3]. The α-tubulins do not readily segregate into conserved classes within the vertebrates, although their major region of heterogeneity is at the carboxy terminus of the protein. In organisms other than the vertebrates, the number and arrangement of the tubulin genes is widely divergent but, like the vertebrate genes, they are subject to developmental regulation and yield isotubulin families of differing complexity [1].

Superimposed on the tubulin diversity generated by the expression of several members of a gene family, is that due to post-translational tubulin modification. The three major types of post-translational modifications are detyrosination/tyrosination [4-8], acetylation [9-14] and phosphorylation [15-19]. For the first type of post-translational modification, a gene-encoded tyrosine located at the carboxy terminus of α-tubulin is removed by a tubulin carboxypeptidase and subsequently replaced by a tubulin tyrosine ligase (reviewed in 4,5). The carboxypeptidase prefers polymeric tubulin as a substrate, whereas the ligase prefers tubulin monomers [5,6]. Not all α-tubulins contain a carboxy terminal tyrosine upon synthesis and they are not subject to this particular post-translational modification [3,7]. No specific function has been demonstrated for detyrosinated vs tyrosinated tubulin, although the former tends to predominate in stable microtubules whereas the latter predominates in dynamic microtubules. The stability of microtubules enriched in detyrosinated tubulin is not, however, a function of the level of detyrosination [4]. Of particular interest, both α-tubulins of <u>Typanosoma brucei brucei</u> undergo detyrosination/tyrosination [8]. The process is related to cell cycle and shape changes in the organism. Results suggest the tyrosination state may influence α-tubulin's role in the cell during development, morphogenesis and progression through the cell cycle.

Acetylation of α-tubulin is catalyzed by α-tubulin acetyltransferase and reversed by tubulin deacetylase [9]. The functional significance of acetylated tubulin is unknown but its presence does not seem to be absolutely required for viability as at least one cell type, PtK2, lacks detectable acetylated tubulin [10]. Like detyrosinated tubulin, acetylated tubulin tends to occur in stable microtubules [11,12] and in some cases both post-translational modifications occur on the same microtubule [13]. It seems again, as for detyrosinated tubulin, that α-tubulin acetylation is as likely to be a result of stabilization as it is to be a cause. A possible role for acetylated tubulin in formation and function of the cell cytoskeleton is suggested by its location in the stable subpellicular microtubules of the trypanosome. Further, Wolf et al. [12] in their study of <u>Drosophila</u> embryogenesis, suggested acetylation of α-tubulin may allow interaction of microtubules with membranes, microtubule-associated proteins (MAP) or other microtubules. As such, α-tubulin acetylation may be both an effector and an indicator of morphological and behavioral changes undergone by cells during development.

Phosphorylation as a tubulin post-translational modification has been studied the least, perhaps because of the highly variable nature of the change [15-19] and the lack of well-defined antibodies specific for phosphorylated tubulin, an advantage enjoyed by people who study tubulin detyrosination/tyrosination and acetylation. Depending on the type of kinase used and the assembly state of the tubulin, the phosphate group(s) is added to different amino acid residues on either α- or β-tubulin [15-18] and it may be preferentially added to only one of the β-tubulin isotypes in the cell, at least in vertebrates [19]. Phosphorylation of α-tubulin at the carboxy terminal tyrosine prevented its assembly into microtubules whereas phosphorylation of α-tubulin lacking the carboxy terminal tyrosine did not prevent its polymerization [15]. Moreover, when insulin-dependent kinase was used [15], the phosphorylation of the carboxy terminal tyrosine seemed to prevent the incorporation of phosphate into other residues within the same α-tubulin monomer, and tyrosine phosphorylation occurred most readily when dimeric tubulin, rather than microtubules, was used as enzyme substrate. Phosphorylation of tubulin affects binding of MAP to tubulin [18] and may, under some conditions [17] influence the incorporation of tubulin into membranes. It is proposed that phosphorylation induces a conformational change in tubulin, resulting in the exposure of previously hidden hydrophobic groups and subsequent modulation of tubulin function. Whatever the case, phosphorylation presents novel potential mechanisms for the utilization and subcellular compartmentation of tubulins.

When the biochemical complexity of tubulin within each species is realized, one is forced to ask why multiple isoforms exist. Do organisms possess multiple tubulin genes to allow selective developmental synthesis of tubulins which can be used interchangeably to assemble microtubules of different functions? Or, as stated in the multitubulin hypothesis [20,21],

390

are the isotubulins functionally specific? Based on work by Cleveland [22, 23] and Bulinski [13], where microtubules were shown to be copolymers of isotubulins, one can probably eliminate the presence, at least in cells of metazoan animals, of microtubules formed from single α- and β-tubulins. From this, the simplest possibility for the division of tubulin labor, one α- and β-tubulin forming a microtubule with one function, seems unlikely. This is probably also the case for structurally simpler eukaryotes since in the trypanosomes all post-translationally modified isotubulins share all of the subcellular tubulin compartments within the cell [18]. Moreover, since heterologous [24] or hybrid [25] tubulins are incorporated into functional microtubules and in some organisms a single isotubulin apparently carries out several functions [26], there is not an absolute requirement for a unique tubulin in order for the cell to carry out a particular activity. Thus, if an isotubulin is to exert a specific influence on microtubule function, it must do so in the company of its fellow isotubulins, and probably in association with nontubulin proteins termed MAPs. The characterization of MAPs thus constitutes an important component of current microtubule research. The assembly of the cytoskeleton depends on the cross-linking of its various components, and within a more general picture, the MAPs in all eukaryotic cells undoubtedly have a major influence on tubulin/microtubule function.

The MAPs were first found in mammalian brain tissue as high molecular weight proteins now termed MAP 1A, 1B, 1C, 2A and 2B and as tau proteins in the 65,000 - 70,000 molecular weight range [27]. Neural MAPs were identified by their ability to copurify with tubulin during assembly/disassembly and to stimulate tubulin assembly in vitro but, as a functional definition, this is now too narrow (see subsequent discussion). The synthesis [28] and cell/tissue distribution [29-30] of the neural MAPs are subject to regulation, possibly reflecting their specific functions within cells. Neural MAPs or their structural/functional analogues have been reported in several nonneural cell types and in nonmammalian species [31-33]. Other MAPs, some of which were first seen in neural cell extracts, include: the chartins [34], STOP [35], MAP 3 [36] and MAP 4 [37]. Some of these proteins are found in a wide range of cell types and MAP 4 may actually be a species-specific analogue of a group of proteins in the 200,000 MW range which were initially observed in HeLa cells [38]. Nonneural cells have yeilded MAPs which are not present or have yet to be found in brain [39-41].

Because the number of MAPs and their proposed functions are becoming so large, it is profitable to divide the MAPs into two major groups, the energy transducing and architectural. The energy transducing MAPs are exemplified by kinesin [42] and cytoplasmic dynein [43]. These MAPs are primarily involved in moving cellular constituents along microtubules which act as rails to guide the movement. The energy transducing MAPs are most often recognized by the ATP-dependent nature of their binding to microtubules.

The architectural MAPs are concerned with cell shape and microtubule organization within the cell. These MAPs are recognized in vitro by their ability to cross-link or bundle microtubules [39,40] (Campbell et al., manuscript submitted), although in vivo they probably have an important role in binding microtubules to the cytoplasmic membrane [39,44-46], or to other cytoskeletal elements [32,50]. One example of an architectural MAP-microtubule complex is the marginal band of chicken erythrocytes [33] in which microtubules composed mostly of a highly divergent tubulin are presumably cross-linked to one another by a tau-like MAP. The erythrocyte situation is, however, poorly understood. For example, the need for a specific tubulin isoform has been undermined by results showing neural tubulin can substitute for erythrocyte tubulin during reconstruction of marginal bands in vitro [24].

Of importance when considering the role of architectural MAPs in binding of microtubules to one another or to membranes is the actual linkage mechanism or the chemical basis of cross-link-microtubule interaction. Neural MAPs interact, in an apparent phosphate-dependent manner [18], with the acidic carboxy terminus of tubulin [21,51], a region previously described in this chapter as being highly variable in sequence and subject to post-translational modification by detyrosination/tyrosination. Tau also binds to a region on the amino terminus of α-tubulin [52]. Neural MAP binding stimulates tubulin assembly and results in formation of projections on the surface of the microtubule, possibly with the ability to bind other cytoskeletal elements. It is not known if microtubule cross-linking MAPs interact with the same sites on tubulin as do other MAPs although in at least two examples [39,40], the cross-linking MAPs join microtubules with a periodicity equivalent to the length of a tubulin dimer, suggesting the presence of one binding site per dimer. The MAP binding site may be a homologous region of primary sequence or higher structural level or an area of the tubulin molecule, such as the carboxy terminus, which shares similar charge properties.

Do the microtubule cross-linking proteins mediate membrane attachments, or is there a more complex system of proteins analogous to the red blood cell spectrin network [53]? The mechanism requiring the smallest number of proteins is one in which tubulin, as a membrane protein, is linked to microtubules by the same protein that cross-links the microtubules to one another. Tubulin is a stable component of some types of membrane [54], and tubulin can interact with artificial membranes [55] and mild detergents [56] in a reversible manner possibly influenced by the extent of tubulin phosphorylation [17]. Tubulin is, then, an amphitropic protein, that is, a protein existing as a soluble cytoplasmic molecule or as a membrane component with the ability to interact with lipids and the hydrophobic part of membranes [57]. Whether tubulin, as a membrane protein, is involved in organization of microtubule-membrane complexes remains to be seen. What is clear is that the architectural MAPs play an important role in organization of the cell cytoplasm and the interaction of the cytoskeleton with cellular membranes. How this is accomplished is a fundamental problem in cell biology.

ARTEMIA TUBULIN

There are several reasons, other than ease of experimental manipulation, why Artemia is a suitable model organism for the study of tubulin and MAPs. The brine shrimp has an unusual developmental cycle characterized by a period of development in the virtual absence of mitosis and cell division [58]. As such, their use offers one of the few possibilities to study the role of microtubule proteins in morphogenesis and development independently of mitosis. Studying microtubule proteins during an unusual type of development may reveal fundamental information about similar processes in organisms undergoing so-called normal development, a situation analogous to the study of mutants.

Brine shrimp tubulin has been purified to apparent homogeneity [59,60] and shown by two dimensional gel electrophoresis to consist of at least three α- and two β-tubulins (Fig. 1). The isotubulin composition does not change during the first twenty-four hours of post-gastrula development, that is, from initial incubation of cysts to formation of the first instar free-swimming nauplius. An absence of change in the isotubulin composition during development was somewhat of an unexpected result in light of the

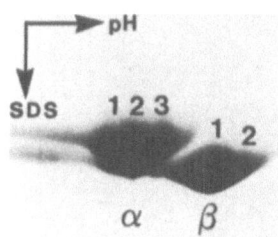

Fig. 1. Two-dimensional gel of purified <u>Artemia</u>
tubulin from organisms allowed to develop
15 h. Migration was from left (basic)
to right (acidic) during isoelectric
focusing and from top to bottom for SDS-
polyacrylamide gel electrophoresis as
indicated in the figure. The gel was
stained with Coomassie blue and only the
tubulin region is shown.

strict tissue-specific and/or developmental regulation of tubulin synthesis
common to most species [1-3,21,26], but it provided an opportunity for
interesting speculation [21,60]. For example, the results indicate the brine
shrimp are able to increase in complexity and accumulate cells with a greater
range of activities, without the need to modify their tubulin in order to
carry out new functions. The organisms are either storing tubulin for rather
long periods of time in anticipation of upcoming events, or they are using
tubulins interchangeably, a proposal contradictory to the multitubulin
hypothesis [20,21].

It is possible that none of the tubulins has a function during the
prenauplius development of the organism, rendering its isotubulin composition of little consequence until later in the animal's life cycle. The
actual function of cytoplasmic microtubules and the need for specific isotubulins during interphase of actively growing and dividing cells in culture
have been difficult to demonstrate. Tubulin may be required only to construct the mitotic apparatus in dividing cells and to form specialized
organelles such as flagella or centrioles [21]. In differentiating cells,
which are no longer undergoing division and are committed to a developmental
pathway, a major consequence is the specific functional deployment of
tubulin, whether it be to allow directed transport of cellular components
or formation and maintenance of a desired morphology. At this time, the
isotubulin composition of the cell may be important, but there is little
proof, other than the coordinate appearance of isotubulins and functions.

<u>Artemia</u> tubulin may be modified during development but we are unable
to detect the difference(s) by isoelectric focusing or two-dimensional gel
electrophoresis. Resolution of a change by either of the electrophoretic
methods requires a relatively large shift in the molecular weight of the
proteins or a charge difference occasioned by substitution of amino acid
residues. To circumvent limitations of the electrophoretic analyses, we
began an immunological characterization of <u>Artemia</u> tubulin on western blots
using isotype-specific monoclonal and polyclonal antibodies. Our preliminary analysis of partially purified <u>Artemia</u> tubulin revealed acetylated
and tyrosinated tubulins within the organism after fifteen hours of development (Fig. 2) and they remain relatively constant in amount up to formation
of the first instar. Detyrosinated tubulin, on the other hand, could not
be detected in cell-free supernatants of the organism at any stage of

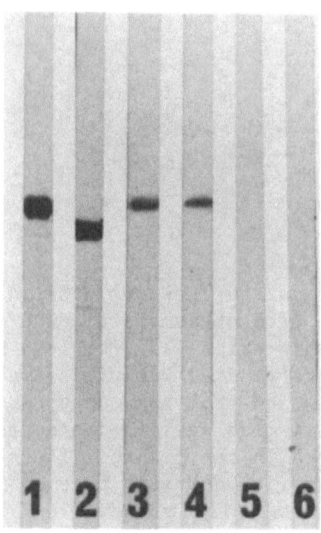

Fig. 2. Immunoperoxidase staining of western blots.
Partially purified Artemia tubulin was resolved on one-dimensional SDS-polyacrylamide
gels, transferred to nitrocellulose and
stained by the immunoperoxidase method. The
antibodies used were: 1) DMIA, a general
anti α-tubulin; 2) DMIB, a general anti β-
tubulin; 3) YLI/2, specific for tyrosinated
tubulin; 4) 6-11B-1, specific for acetylated
tubulin; 5+6) controls for the anti-rat and
anti-mouse IgG secondary antibodies. DMIA
and DMIB were obtained from Dr. S. H. Blose,
YL1/2 was from Dr. J. V. Kilmartin, 6-11B-1
was from Dr. G. Piperno and secondary antibodies were from Sigma Chemical Co., St.
Louis, Mo.

development examined, although it is observed in purified tubulin, presumably because more tubulin could be applied to the gel under these conditions.
No new spots are generated on two-dimensional gels stained with Coomassie
blue by the detyrosinated tubulin, but we are able to resolve the post-
translationally modified tubulins on western blots by staining with an antibody that specifically recognizes this type of tubulin. It seems the Artemia
isotubulin family is more complex than previously realized and it undergoes
some change during development. We are not, however, much closer to resolution of the problem of isotubulin function.

The demonstration of post-translationally modified α-tubulins reveals
that at least part of the brine shrimp tubulin microheterogeneity is generated after the tubulins are synthesized. To better understand the origins
and interrelationships of the Artemia isotubulins, we used cloned Drosophila
α- and β-tubulin genes as molecular probes. The Drosophila tubulin genes
hybridize very efficiently to Southern blots of restriction digested Artemia
DNA under stringent conditions (Fig. 3). The blots shown in Fig. 3 were
hybridized to the labelled probes and washed to 1 X SET at 65°C. A further
wash with 3 mM Tris at room temperature caused only a slight decrease in

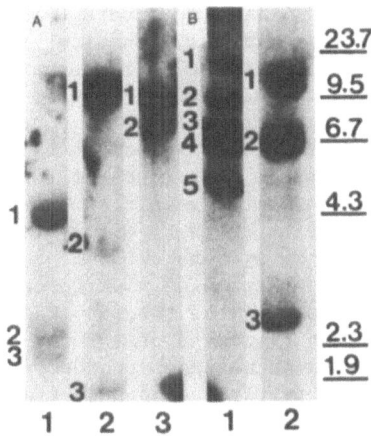

Fig. 3. Hybridization of cloned <u>Drosophila</u> tubulin
genes to <u>Artemia</u> DNA. <u>Artemia</u> DNA was
digested with restriction enzymes, electrophoresed on 1% agarose gels, transferred
to nitrocellulose by the method of Southern
and hybridized at 65°C to nick translated
<u>Drosophila</u> tubulin genes. The blots were
washed to 1 x SET containing 1% SDS and
0.1% sodium pyrophosphate at 65°C. SET
contained 0.15 M NaCl, 0.03 M Tris-HCl
and 2 mM EDTA adjusted to pH 8.0. The
probes were pDmT α1 obtained from Dr. P.
C. Wensink in panel A and DTB2 from Dr.
J. E. Natzle in panel B. The enzymes in
panel A were Kpn I, Bgl II and Bam H I in
lanes 1-3 respectively and Sal I and Bam
H I respectively in lanes 1 and 2 of panel
B. The bands of hybridization are numbered
and size markers in base pairs (x 10^{-3})
are on the right side of the blot.

the intensity of the bands, demonstrating excellent homology between the
<u>Artemia</u> and <u>Drosophila</u> tubulin genes. The result is particularly interesting
when viewed in context of our failure to achieve efficient hybridization of
cloned tubulin genes from chicken and <u>Chlamydomonas</u> to <u>Artemia</u> DNA. Hybridization was very weak or nonexistent after several attempts at different
stringencies even though parallel experiments with the <u>Drosophila</u> tubulin
genes gave strong bands of blots. There is apparently much less homology
between <u>Artemia</u> tubulin genes and those from chicken and <u>Chlamydomonas</u> than
there is between <u>Artemia</u> and <u>Drosophila</u> tubulin genes. The experimental
results reflect the phylogenetic interrelationships of the organisms tested,
with <u>Artemia</u> being much more closely related to <u>Drosophila</u> than to either
chicken or <u>Chlamydomonas</u>.

The cloned <u>Drosophila</u> genes and parts thereof have been used to quant-
itate <u>Artemia</u> tubulin genes and mRNA. For <u>Artemia</u> there is a good indication
of only one α - but two or more β -tubulin genes. No other metazoan animal
possesses only one α-tubulin gene, rendering the situation in <u>Artemia</u>
very interesting, especially from the developmental perspective. The

suggestion is that all of the α-tubulin diversity in Artemia must arise by post-transcriptional/post-translational mechanisms rather than transcription of different genes. We know this to be at least partially true since we have shown post-translationally modified α-tubulins in Artemia but we cannot yet account for generation of all the α-tubulins by this process. Analysis of α-tubulin mRNA on Northern blots reveals only one size class of message (Fig. 4). The possible occurrence of two or more different α-tubulin mRNAs has not been disproved by this experiment, since messengers coding for similar sized proteins are very likely to be closely related in size and not resolved on the gel systems we used. We have, however, eliminated the possibility of messengers widely different in size and are currently analyzing the products of the tubulin mRNA after in vitro translation. The occurrence of two or more spots on two-dimensional gels will indicate two different α-tubulin messages and demonstrate that the α-tubulin microheterogeneity is partly due to post-transcriptional (but not post-translational) mechanisms or to the presence of additional transcribed α-tubulin genes not revealed on our Southern blots. The observation of one spot on a two-dimensional gel, although not definitively eliminating the transcription of genes yielding identical electrophoretic products, will strongly indicate that the Artemia α-tubulin diversity results solely from post-translational processes. Formation of the Artemia β-tubulins is most simply explained by the expression of different β-tubulin genes in the animal, since our analyses of Southern blots indicates two or more β-tubulin genes. If true, messages of similar size must be produced, as we see only one band on Northern blots of Artemia poly (A$^+$) mRNA when they are probed with the Drosophila β-tubulin gene (Fig. 4). Again, we are investigating this possibility by in vitro translation of Artemia mRNA and analysis of products on two-dimensional gels.

The assembly properties of Artemia tubulin are being studied in our laboratory and compared to those of neural tubulin, which, due to its abundance and ease of purification, is the most widely studied. Artemia tubulin, in contrast to neural tubulin, assembles readily in the absence of MAPs but its assembly is stimulated by MAPs, from bovine brain (Fig. 5) with a resultant drop in its critical concentration from 0.6 mg/ml to 0.2 mg/ml [60,61]. The critical concentration is the concentration of tubulin below which no net assembly of tubulin into microtubules occurs; stated another way, it is the amount of tubulin remaining in solution as a dimer when the maximum amount of assembly has occurred. Not only do neural MAPs stimulate Artemia tubulin assembly, but they bind to the microtubules to become an integral part of the structure [60].

The results demonstrate that Artemia tubulin has a MAP binding domain, probably sharing at least some of the properties of the analogous site on neural tubulin. The MAP binding site of neural tubulin is the highly variable, negatively charged carboxy terminal region of the molecule [21,51] which, if removed, permits the assembly of neural tubulin at low concentration in the absence of MAPs. The resulting microtubules have an abnormal morphology. The proposed role of the neural MAPs in stimulation of assembly is to bind to and shield the negatively charged domains from one another. If the proposal is true, then the enhanced, MAP-independent assembly of Artemia tubulin can be explained by a reduced overall negative charge in its carboxy terminal region. In support of this, we showed that the isoelectric point of Artemia tubulin is higher than that of neural tubulin [59], perhaps reflecting a less acidic carboxy terminus.

We used vinblastine, a microtubule-reactive drug, to study the assembly properties of Artemia tubulin [61,62]. Vinblastine, at appropriate concentrations, induces the formation of coiled polymeric structures when mixed with neural microtubules formed from tubulin and MAPs, but causes complete disassembly of the microtubules if MAPs are absent [61-63]. It is thought

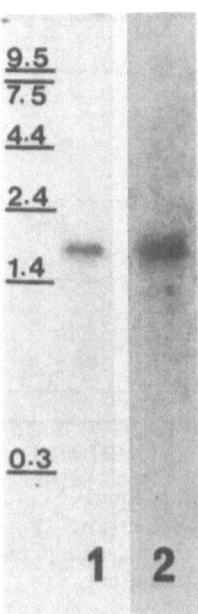

Fig. 4. Hybridization of cloned <u>Drosophila</u> tubulin
genes to <u>Artemia</u> mRNA. <u>Artemia</u> poly (A)$^+$ mRNA
was purified by chromatography on oligo-dT-
cellulose, electrophoresed on 1.5% agarose gels
and transferred to nitrocellulose by the
Northern method. Hybridization was at 42°C to
nick translated probes in a solution containing 50% deionized formamide, 50 mM sodium
phosphate, 5 x SSC, 5X Denhardt's medium, 0.1%
SDS, 100 μg/ml sheared salmon sperm DNA and
10 μg/ml polyriboadenylate. After hybridization, the filters were washed at room temperature in 2 x SSC containing 0.1% SDS followed
by 0.1 x SSC containing 0.1% SDS at 50°C. SSC
contained 0.15 M NaCl and 0.015 M sodium
citrate at pH 7.0. The probes were pDmT α1
(lane 1) and DTB2 (lane 2). Markers in bases
(x 10^{-3}) are on the left side of the picture.

that the drug-induced coil formation does not involve complete microtubule
disassembly. Rather, the protofilaments of the microtubule peel apart as
lateral tubulin-to-tubulin interactions used in formation of the microtubule
wall are disrupted. The protofilaments remain intact due to the MAP-induced
stabilization of the longitudinal tubulin-to-tubulin interactions used in
their construction. If MAPs are absent, the protofilaments fall apart
because of the inherent instability caused by the unshielded negative charges
of the tubulin carboxy terminus. For <u>Artemia</u> tubulin, vinblastine induced
coil formation either in the presence or absence of neural MAPs, although
the coils were not identical in both cases. These results imply that the
vinblastine-induced protofilaments of <u>Artemia</u> tubulin are sufficiently
stable, even in the absence of MAPs to resist vinblastine-induced depoly-
merization. The most probable cause for the enhanced stability is a decrease
in the negative charge of the carboxy terminus of <u>Artemia</u> tubulin.

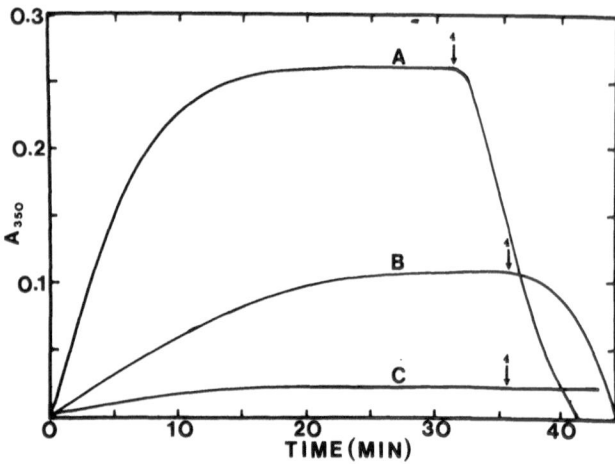

Fig. 5. Effect of neural microtubule-associated proteins on the assembly of Artemia tubulin. Purified Artemia tubulin was incubated at 37°C in the presence (A) or absence (B) of phosphocellulose-purified and heated bovine neural MAPs. For curve C, no tubulin was added. Tubulin assembly was monitored by turbidity measurement at 350 nm and the formation of morphologically normal microtubules was verified by examination of negatively stained samples with the electron microscope. The temperature was reduced to 4°C at the times indicated by the arrows, causing disassembly of the microtubules. Adapted with permission from Biochimica et Biophysica Acta, 882, 419-426, copyright C, 1986.

Our analyses of MAP and vinblastine effects on Artemia tubulin prompted us to ask if Artemia contains MAPs, and if so, what is their function in the organism. Do the MAPs stimulate tubulin assembly or are they involved in cell architecture [39,40] or movement [42,43]? We used the drug, taxol, to stimulate assembly of Artemia tubulin in cell free extracts of the organism [60] (manuscript submitted for publication). Taxol lowers the critical concentration of tubulin, leading to microtubule formation under conditions normally preventing assembly. At the same time, taxol does not interfere with the interaction of tubulin and MAPs [64]. Taxol-induced Artemia microtubules were pelleted through 15% sucrose cushions and either examined in the electron microscope (Fig. 6) or on SDS-polyacrylamide gels (Fig. 7). Microtubules in the negatively stained pellets were cross-linked by round or somewhat elongated structures. The pellets, when solubilized and electrophoresed on gels, revealed α- and β-tubulin and many non-tubulin proteins.

The non-tubulin proteins were purified by extraction from the taxol-induced microtubules with buffers containing 1.0 M NaCl, centrifugation through Sephadex columns and subsequent co-assembly with purified Artemia tubulin (Fig. 7). The co-assembly with tubulin resulted in reconstitution of the cross-linking (Fig. 6) and resolution on gels of four potential candidates for the cross-linking role. We believe, due to its efficient co-assembly with tubulin, that the 49kD protein is the cross-linking agent, but we are not yet able to eliminate the possibility that one or more of the other polypeptides appearing on the gel is required for the cross-linking.

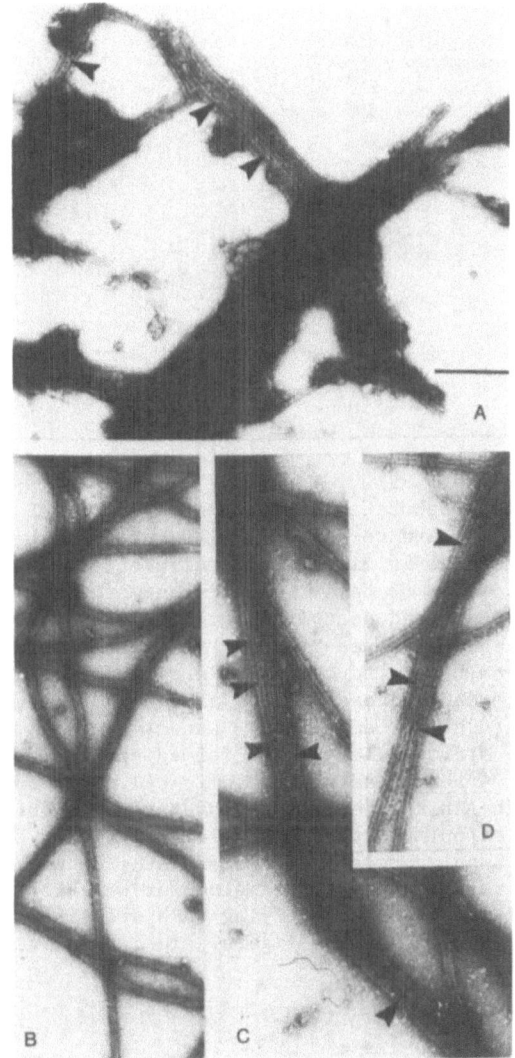

Fig. 6. Electron micrographs of cross-linked <u>Artemia</u> microtubules. A, microtubules induced to assemble in cell-
free extracts from <u>Artemia</u> by the addition of 10 μM
taxol were collected by centrifugation through sucrose
cushions, stained and examined in the electron microscope. B, microtubules formed by the assembly of
purified <u>Artemia</u> tubulin in the absence of MAPs. C
and D, microtubules formed by the assembly of purified <u>Artemia</u> (C) and neural (D) tubulin in the presence of <u>Artemia</u> MAPs. The arrowheads indicate
cross-linking structures between microtubules. The
samples were negatively stained with 1% aqueous
uranyl acetate. The bar in A represents 0.2 μm and
all pictures are the same magnification.

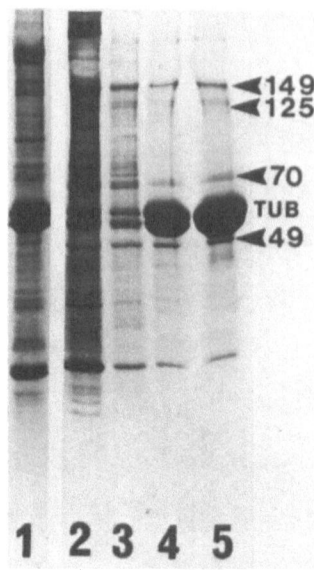

1 2 3 4 5

Fig. 7. Electrophoretic analysis of Artemia MAP purification and co-assembly with tubulin. Lane 1,
proteins in microtubule pellets formed upon
addition of taxol to cell-free Artemia supernatants and centrifugation through a sucrose
cushion. Lanes 2 and 3, proteins extracted
from taxol-induced Artemia microtubules with
1 M NaCl before (2) and after (3) passage
through Sephadex G-25. Lanes 4 and 5, proteins
in microtubule pellets formed upon co-assembly
of Artemia MAP with Artemia (4) or bovine (5)
tubulin and centrifugation through a sucrose
cushion. Samples were electrophoresed on 10%
SDS-polyacrylamide gels and stained with
Coomassie blue. The molecular weights (x 10^{-3})
of the major copurifying MAPs are listed on the
right side of the figure. TUB, tubulin.

The Artemia MAPs bind to tubulin in the presence of ATP, they survive at
least one cycle of assembly/disassembly of tubulin, they do not stablize
microtubules to cold and they bind to neural tubulin causing cross-linking
of the microtubules. The combined analyses indicate Artemia contain a unique
group of architectural MAPs, but the MAPs share a similar binding site on
phylogenetically disparate tubulins. Artemia MAP function is uncertain.
They may be required for organization of the mitotic apparatus or the
establishment and maintenance of cell morphology. We are currently investigating the possibilities by preparing antibody to the MAPs so as to identify the actual cross-linker polypeptide(s) and determine its intracellular
location.

On a more speculative note, MAPs may not be required in vivo to stimulate assembly of tubulin from any source, even though they have the function
in vitro. The critical concentration of most non-neural and of some neural
[65] tubulins is lower than for mammalian neural tubulins, indicating the

400

need for assembly-stimulating MAPs is less than one might imagine if only
MAPs from mammalian brain are considered. The possibility exists, even for
brain, that stimulation of tubulin assembly by MAPs in vitro reflects only
a secondary or nonphysiological function of MAPs. Their true function
may be stabilization of microtubules formed in cellular extensions, cross-
linking of microtubules with one another or with other cell structures [39,
40,44-51], or as motors for movement of subcellular components along microtubules [42,43].

The integrated biochemical/molecular approach to the study of Artemia
tubulin is providing a great deal of useful information. To complement the
work and to extend it to the cellular level, we have begun to apply, in
collaboration with Dr. John Freeman of the University of South Alabama (see
chapter elsewhere in this book), immunofluorescence technologies to the
study of microtubule arrangement in brine shrimp. Using several polyclonal
and monoclonal antibodies to tubulin, we have successfully stained cells in
whole organisms (Fig. 8). Cloned α- and β-tubulin genes are being used
to localize Artemia tubulin mRNA in situ. Our long range goal is to deter-

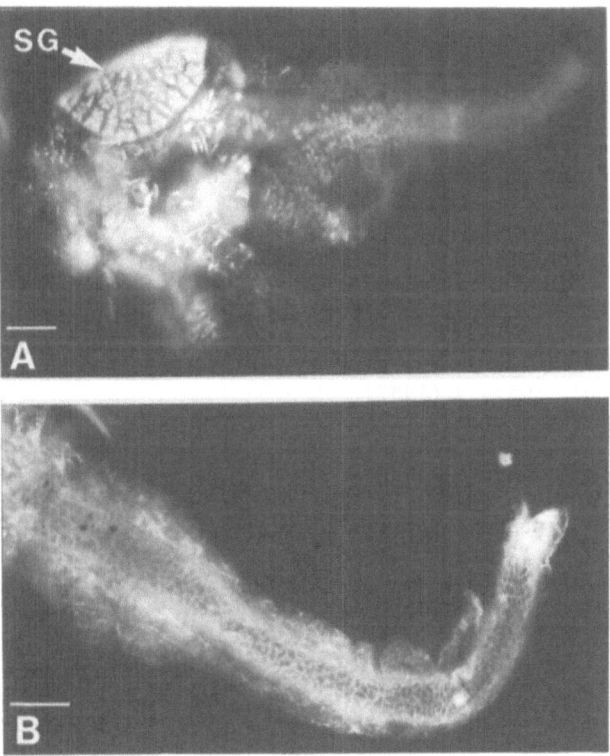

Fig. 8.   Immunofluorescence staining of Artemia nauplii. Nauplii, fixed
          in methanol containing 2.4 μM taxol followed by 4% paraformalde-
          hyde, were stained with the anti-tubulin antibodies 6-11B-1 (A)
          and DMIA (B) followed by an FITC-conjugated secondary antibody
          obtained from Sigma Chemical Co., St. Louis, Mo. The salt gland
          (SG) and other regions of intense staining are seen in A whereas
          individual cells consisting of a ring of fluorescence surrounding
          an unstained nucleus are visible in the mid-to posterior region
          of the tail in B. The bars represent 100 μM in each figure. The
          stained nauplii were prepared by Dr. John Freeman, University of
          South Alabama.

mine whether there are periodic molt-related or other developmental changes in Artemia tubulin mRNA, tubulin and MAPs. Localization of specific tubulins or MAPs within cells may also provide clues to their functions. Artemia is proving to be a suitable organism for the integation of cell/molecular/biochemical techniques in the study of the cytoskeleton. Its use promises some answers to important questions regarding the role of microtubules in the organization and function of intracellular cytoplasmic components.

REFERENCES

1. D. W. Cleveland and K. F. Sullivan, Molecular biology and genetics of tubulin, Ann. Rev. Biochem. 54:331 (1985).
2. M. J. Monteiro and D. W. Cleveland, Sequence of chicken cβ 7 tubulin. Analysis of a complete set of vertebrate β-tubulin isotypes, J. Mol. Biol. 199:439 (1988).
3. L. F. Pratt and D. W. Cleveland, A survey of the α-tubulin gene family in chicken: unexpected sequence heterogeneity in the polypeptides encoded by five expressed genes, EMBO J. 7:931 (1988).
4. S. Khawaja, G. G. Gundersen and J. C. Bulinski, Enhanced stability of microtubules enriched in detyrosinated tubulin is not a direct function of detyrosination level, J. Cell Biol. 106:141 (1988).
5. J. Wehland and K. Weber, Turnover of the carboxy-terminal tyrosine of α-tubulin and means of reaching elevated levels of detyrosination in living cells, J. Cell Sci. 88:185 (1987).
6. D. M. Beltramo, C. A. Arce and H. S. Barra, Tyrosination of microtubules and non-assembled tubulin in brain slices, Eur. J. Biochem. 162:137 (1987).
7. L. F. Pratt, S. Okamura and D. W. Cleveland, A divergent testis-specific α-tubulin isotype that does not contain a coded C-terminal tyrosine, Mol. Cell. Biol. 7:552 (1987).
8. T. Sherwin, A. Schneider, R. Sasse, T. Seebeck and K. Gull, Distinct localization and cell cycle dependence of COOH terminally tyrosinated α-tubulin in the microtubules of Trypanosoma brucei brucei, J. Cell Biol. 104:439 (1987).
9. H. Maruta, K. Greer and J. L. Rosenbaum, The acetylation of α-tubulin and its relationship to the assembly and disassembly of microtubules, J. Cell Biol. 103:571 (1986).
10. G. Piperno, M. LeDizet and X. Chang, Microtubules containing acetylated α-tubulin in mammalian cells in culture, J. Cell Biol. 104:289 (1987).
11. W. S. Sale, J. C. Besharse and G. Piperno, Distribution of aceylated α-tubulin in retina and in in-vitro-assembled microtubules, Cell Motil. Cytoskel. 9:243 (1988).
12. N. Wolf, C. L. Regan and M. T. Fuller, Temporal and spatial pattern of differences in microtubule behavior during Drosophila embryogenesis revealed by distribution of a tubulin isoform, Development 102:311 (1988).
13. J. C. Bulinski, J. E. Richards and G. Piperno, Posttranslational modifications of α-tubulin: detyrosination and acetylation differentiate populations of interphase microtubules in cultured cells, J. Cell Biol. 106:1213 (1988).
14. A. Schneider, T. Sherwin, R. Sasse, D. G. Russell, K. Gull and T. Seebeck, Subpellicular and flagellar microtubules of Trypanosoma brucei brucei contain the same α-tubulin isoforms, J. Cell Biol. 104:431 (1987).
15. F. Wandosell, L. Serrano and J. Avila, Phosphorylation of α-tubulin carboxyl-terminal tyrosine prevents its incorporation into microtubules, J. Biol. Chem. 262:8268 (1987).
16. L. Serrano, J. Diaz-Nido, F. Wandosell and J. Avila, Tubulin phosphory-

lation by casein kinase II is similar to that found in vivo, J. Cell Biol. 105:1731 (1987).

17. A. J. Hargreaves, F. Wandosell and J. Avila, Phosphorylation of tubulin enhances its interaction with membranes, Nature 323:827 (1986).

18. F. Wandosell, L. Serrano, M. A. Hernandez and J. Avila, Phosphorylation of tubulin by a calmodulin-dependent protein kinase, J. Biol. Chem. 261:10332 (1986).

19. R. F. Luduena, H. -P. Zimmerman and M. Little, Identification of the phosphorylated β-tubulin isotype in differentiated neuroblastoma cells, FEBS Lett. 230:142 (1988).

20. D. W. Cleveland, The multitubulin hypothesis revisited, in: "What have we learned?" J. Cell. Biol. 104:381 (1987).

21. T. H. MacRae, Nonneural microtubule proteins: structure and function, Bio Essays 6:128 (1987).

22. H. C. Joshi, T. Y. Yen and D. W. Cleveland, In vivo co-assembly of a divergent β-tubulin subunit (cB6) into microtubules of different function, J. Cell Biol. 105:2179 (1987).

23. M. A. Lopata and D. W. Cleveland, In vivo microtubules are copolymers of available β-tubulin isotypes: localization of each of six vertebrate β-tubulin isotypes using polyclonal antibodies elicited by synthetic peptide antigens, J. Cell Biol. 105:1707 (1987).

24. J. A. Swan and F. Solomon, Reformation of the marginal band of avian erythrocytes in vitro using calf brain tubulin: peripheral determinants of microtubule form, J. Cell Biol. 99:2108 (1984).

25. J. R. Bond, J. L. Fridovich-Keil, L. Pillus, R. C. Mulligan and F. Solomon, A chicken-yeast chimeric β-tubulin protein is incorporated into mouse microtubules in vivo, Cell 44:461 (1986).

26. E. C. Raff, Genetics of microtubule systems, J. Cell Biol. 99:1 (1984).

27. J. B. Olmsted, Microtubule-associated proteins, Ann. Rev. Biol. 2:421 (1986).

28. D. G. Drubin, D. Caput and M. W. Kirschner, Studies on the expression of the microtubule-associated protein, tau, during mouse brain development, with newly isolated complimentary DNA probes, J. Cell Biol. 98:1090 (1984).

29. L. I. Binder, A. Frankfurter and L. I. Rebhun, The distribution of tau in the mammalian central nervous system, J. Cell Biol. 101:1371 (1985).

30. G. S. Bloom, T. A. Schoenfeld and R. B. Vallee, Widespread distribution of the major polypeptide component of MAP 1 (microtubule-associated protein 1) in the nervous system, J. Cell Biol. 98:320 (1984).

31. F. Wandosell and J. Avila, Microtubule-associated proteins present in different developmental stages of Drosophila melanogaster, J. Cell Biochem. 35:83 (1987).

32. N. Hirokawa, 270k microtubule-associated protein cross-reacting with anti-MAP 2 IgG in the crayfish peripheral nerve axon, J. Cell Biol. 103:33 (1986).

33. S. Kim, M. Magendantz, W. Katz and F. Solomon, Development of a differentiated microtubule structure: formation of the chicken erythrocyte marginal band in vivo, J. Cell Biol. 104:51 (1987).

34. J. M. Aletta and L. A. Greene, Sequential phosphorylation of chartin microtubule-associated proteins is regulated by the presence of microtubules, J. Cell Biol. 105:277 (1987).

35. R. L. Margolis, C. T. Rauch and D. Job, Purification and assay of a 145-kDa protein (STOP 145) with microtubule-stabilizing and motility behavior, Proc. Natl. Acad. Sci. U.S.A. 83:639 (1986).

36. G. Huber, G. Pehling and A. Matus, The novel microtubule-associated protein MAP 3 contributes to the in vitro assembly of brain microtubules, J. Biol. Chem. 261:2270 (1986).

37. L. M. Parysek, J. J. Wolosewick and J. B. Olmsted, MAP 4: a microtubule-associated protein specific for a subset of tissue microtubules, J. Cell Biol. 99:2287 (1984).

38. S. Kotani, H. Murofushi, S. Maekawa, H. Aizawa and H. Sakai, Isolation of rat liver microtubule-associated proteins with molecular mass of around 200,000 which distribute widely among mammalian cells, J. Biol. Chem. 263:5385 (1988).

39. G. T. Bramblett, S. Chang and M. Flavin, Periodic crosslinking of microtubules by cytoplasmic microtubule-associated and microtubule-corset proteins from a trypanosomatid, Proc. Natl. Acad. Sci. U.S.A. 84:3259 (1987).

40. E. J. Aamodt and J. G. Culatti, Microtubules and microtubule-associated proteins from the nematode Caenorhabditis elegans: Periodic cross-links connect microtubules in vitro, J. Cell Biol. 103:23 (1986).

41. L. S. B. Goldstein, R. A. Laymon and J. R. McIntosh, A microtubule-associated protein in Drosophila melanogaster: identification, characterization and isolation of coding sequences, J. Cell Biol. 102:2076 (1986).

42. B. W. Neighbors, R. C. Williams, Jr. and J. R. McIntosh, Localization of kinesin in cultured cells, J. Cell Biol. 106:1193 (1988).

43. R. B. Vallee, J. S. Wall, B. M. Paschal and H. S. Shpetner, Microtubule-associated protein IC from brain is a two-headed cytosolic dynein, Nature 332:561 (1988).

44. T. Seebeck, V. Kung, T. Wyler and M. Muller, A 60-kDa cytoskeletal protein from Trypanosoma brucei brucei can inteact with membranes and with microtubules, Proc. Natl. Acad. Sci. U.S.A. 85:1101 (1988).

45. A. Schneider, W. Eichenberger and T. Seebeck, A microtubule-binding protein of Trypanosoma brucei which contains covalently bound fatty acid, J. Biol. Chem. 263:6472 (1988).

46. J. M. Murray, Disassembly and reconstitution of a membrane-microtubule complex, J. Cell Biol. 98:1481 (1984).

47. G. Mithieux, C. Audebet and B. Rousset, Association of purified thyroid lysosomes to reconstituted microtubules, Biochem. Biophys. Acta 969:121 (1988).

48. M. M. Pratt, Stable complexes of axoplasmic vesicles and microtubules: protein composition and ATPase activity, J. Cell Biol. 103:957 (1986).

49. V. J. Allan and T. E. Kreis, A microtubule-binding protein associated with membranes of the Golgi apparatus, J. Cell Biol. 103:2229 (1986).

50. V. I. Rodionov, E. S. Nadezhdina, E. V. Leonova, E. A. Vaisberg, S. A. Kuznetsov and V. I. Gelfand, Identification of a 100kD protein associated with microtubules, intermediate filaments and coated vesicles in cultured cells, Exp. Cell Res. 159:377 (1985).

51. R. B. Maccioni, L. Serrano and J. Avila, Structural and functional domains of tubulin, BioEssays 2:105 (1985).

52. U. Z. Littauer, D. Giveon, M. Thierauf, I. Ginsberg and H. Ponstingl, Common and distinct tubulin binding sites for microtubule-associated proteins, Proc. Natl. Acad. Sci. U.S.A. 83:7162 (1986).

53. V. Bennett, The membrane skeleton of human erythrocytes and its implications for more complex cells, Ann. Rev. Biochem. 54:273 (1985).

54. R. E. Stephens, S. Oleszko-Szuts and M. J. Good, Evidence that tubulin forms an integral membrane skeleton in molluscan gill cilia, J. Cell Sci. 88:527 (1987).

55. J. M. Caron and R. D. Berlin, Dynamic interactions between microtubules and artificial membranes, Biochemistry 26:3681 (1987).

56. J. M. Andreu and J. A. Munoz, Interaction of tubulin with octyl glucoside and deoxycholate. 1. Binding and hydrodynamic studies, Biochemistry 25:5220 (1986).

57. P. Burn, Amphitropic proteins: a new class of membrane proteins, Trend. Biochem. Sci. 13:79 (1988).

58. P. Rafiee, C. O. Matthews, J. C. Bagshaw and T. H. MacRae, Reversible arrest of Artemia development by cadmium, Can. J. Zool. 64:1633 (1986).

59. T. H. MacRae and R. F. Luduena, Developmental and comparative aspects

of brine shrimp tubulin, <u>Biochem. J.</u> 219:137 (1984).

60. P. Rafiee, S. A. MacKinlay and T. H. MacRae, Taxol-induced assembly and characterization of microtubule proteins from developing brine shrimp (<u>Artemia</u>), <u>Biochem. Cell Biol.</u> 64:238 (1986).

61. S. A. MacKinlay, R. F. Luduena and T. H. MacRae, Vinblastine-induced aggregation of brine shrimp (<u>Artemia</u>) tubulin, <u>Biochem. Biophys. Acta</u> 882:419 (1986).

62. S. A. MacKinlay, R. F. Luduena and T. H. MacRae, Temperature effects on solutions of vinblastine-induced polymers assembled from brine shrimp (<u>Artemia</u>) tubulin, <u>FEBS Lett.</u> 203:301 (1986).

63. R. F. Luduena, A. Fellows, L. McManus, M. A. Jordan and J. Nunez, Contrasting roles of tau and microtubule-associated protein 2 in the vinblastine-induced aggregation of brain tubulin, <u>J. Biol. Chem.</u> 259:12890 (1984).

64. R. B. Vallee, A taxol-dependent procedure for the isolation of microtubules and microtubule-associated proteins (MAPs), <u>J. Cell Biol.</u> 92:435 (1982).

65. H. W. Detrich, III, V. Prasad and R. F. Luduena, Cold-stable microtubules from Antarctic fishes contain unique α-tubulin, <u>J. Biol. Chem.</u> 262:8360 (1987).

REPEATED SEQUENCES FLANK THE TUBULIN GENES OF <u>ARTEMIA</u>

M. C. Lemaire, M. C. Sanders, L. J. Grzyb, G. R. Farley and
J. C. Bagshaw

Department of Biology and Biotechnology
Worcester Polytechnic Institute
Worcester, Massacheusettes 01609, U.S.A.

ABSTRACT

We have isolated separate genomic DNA clones containing one α- and two β- tubulin genes of the brine shrimp, <u>Artemia</u>, which we will refer to here as $\alpha 2$, $\beta 3$ and $\beta 7$. We have mapped the coding region of each isolate, and we have discovered that the non-coding regions of these isolates contain at least three different kinds of repeated DNA sequence elements. (1) All three isolates contain sequence elements repeated several hundredfold and scattered throughout the <u>Artemia</u> genome (LINES and/or SINES). There is no homology between the elements flanking the α-tubulin gene and those within 2kb of the α-tubulin genes. (2) At least one repeated sequence element in $\alpha 2$ is also present in $\beta 7$ at a distance of more than 2kb from the tubulin coding region. (3) An inverted repeat element, detected as a "snap-back" structure by S1 nuclease treatment, is adjacent to the coding region in $\beta 7$. This sequence element is about 2kb in length and is not present in either $\alpha 2$ or $\beta 3$. Direct DNA sequencing of these repeated elements is in progress.

ACKNOWLEDGEMENT

This research was sponsored by NSF Grant DCB85-10471.

MICROTUBULE CROSS-LINKING PROTEINS FROM ARTEMIA

E. J. Campbell and T. H. MacRae

Biology Department
Dalhousie University
Halifax, N. S., Canada B3H 4J1

ABSTRACT

As determined by examination of negatively stained samples with the electron microscope, microtubules assembled in cell-free Artemia extracts upon addition of taxol are cross-linked. Taxol-induced microtubules, collected by centrifugation through sucrose cushions and extracted with 1 M NaCl, yielded several non-tubulin proteins, herein termed microtubule-associated proteins (MAP). Although their dependence on microtubule assembly for sedimentation through sucrose cushions and their extraction from taxol-stabilized pellets with salt-containing buffers are characteristics shared by MAP from other organisms, the Artemia MAP do not stimulate tubulin assembly. Assembly of the Artemia MAP with purified Artemia or neural tubulin does, however, allow reconstitution of the microtubule cross-links. Electrophoretic analysis of cross-linked microtubules demonstrated that the same non-tubulin proteins co-sedimented with both types of tubulins. Formation of cross-links between the microtubules does not appear to be influenced by ATP. Probing of western blots with polyclonal and monoclonal antibodies indicates that Artemia MAP are different from neural MAP, although some epitopes may be shared by MAP from both sources. Neural MAP1 and MAP2 cDNA clones do not hybridize to Southern blots of restriction-digested Artemia DNA. The results clearly demonstrate that Artemia contain novel protein(s) with the ability to cross-link microtubules from phylogenetically disparate organisms.

ACKNOWLEDGEMENT

This research was supported by NSERC of Canada Grant A7661.

THE DIVERSITY OF α-TUBULIN DURING <u>ARTEMIA</u> PRELARVAL DEVELOPMENT

C. M. Langdon and T. H. MacRae

Department of Biology
Dalhousie University
Halifax, N. S., Canada   B3H 4J1

ABSTRACT

<u>Artemia</u> possess three major α-tubulins, as defined by migration
patterns on two-dimensional gels, but appears to contain only one α-
tubulin gene. To determine how the isotubulin diversity is generated, we
have characterized <u>Artemia</u> tubulin mRNA and immunologically analyzed the
α-tubulins during development of the organisms. Hybridization of a cloned
<u>Drosophila</u> α-tubulin gene to northern blots reveals only one size class
of tubulin mRNA in <u>Artemia</u> at all developmental stages examined. Quantitation on dot blots with the same tubulin gene shows that the amount of
α-tubulin message remains relatively constant during development. <u>In vitro</u>
translation of poly (A)$^+$ mRNA yields radioactively labelled tubulin which
can be recovered by taxol-driven coassembly with purified <u>Artemia</u> tubulin
and subsequent centrifugation through sucrose. Preliminary results from
two-dimensional gel electrophoresis and fluorography indicate the synthesis
of two α-tubulin isoforms <u>in vitro</u>. Anti-tubulin antibodies, when used
to stain western blots of cell-free extracts or purified tubulin from
<u>Artemia</u>, reveal the presence of acetylated and tyrosinated α-tubulin.
Acetylated tubulin has been shown by others to result from post-translational mechanisms whereas the tyrosinated tubulin may arise by post-translational modification or as a primary gene product. The results indicate
that the α-tubulin diversity in <u>Artemia</u> is due to post-transcriptional
mechanisms and support our earlier conclusion that the α-tubulin composition does not change during prelarval development.

ACKNOWLEDGEMENT

This research was supported by NSERC of Canada, Grant A7661.

THE MOLECULAR BIOLOGY OF ARTEMIA HAEMOGLOBINS

Anthony M. Manning, Craig J. Marshall, Robert J. Powell,
Clive N. Trotman and Warren P. Tate

Department of Biochemistry
University of Otago
Dunedin, New Zealand

INTRODUCTION

Haemoglobin is a major protein induced during post-gastrula development of Artemia. Three haemoglobins are formed from the association of two
different subunits, and they are expressed differentially during post-
gastrula development. These haemoglobins are unusual in that they are of
high molecular weight (Mr 260,000) and are composed of two globin chains
(Mr 130,000) each of which is a polymer of eight covalently linked
myoglobin-like domains [1]. Each polypeptide is encoded by a single large
mRNA which may have originated as a result of successive tandem duplication
and fusion of a single ancestral globin gene [2].

We wish to study the organization of the globin genes and how their
expression is controlled during development. We report here the relationship between the developmental appearance of the globin mRNA and the
protein it encodes, and the isolation of cDNAs for at least two of the
globin genes.

MATERIALS AND METHODS

Western Analysis of Proteins

Artemia cysts collected from the shores of salterns around Lake
Grassmere, New Zealand, were cultured in natural seawater as previously
described [3]. Extracts of partially purified haemoglobin were prepared
by homogenising cultures in a CO-saturated buffer (50mM Tris-Cl, pH 7.5;
125mM NaCl; 0.3mg/ml Soyabean Trypsin Inhibitor) containing 8% (w/v)
polyethylene glycol 6000 and 0.05% (v/v) Tween 20 and then by centrifuging
to remove precipitated material. Contaminating lipid was removed from the
supernatant by centrifuging for 10 minutes at 28psi in a Beckman Airfuge.
The supernatants were analysed by either Western blotting after gel electrophoresis or by a dot blot assay using antibodies raised to the major
haemoglobin, Hb II, the heterodimer.

Isolation of RNA

RNA was isolated from cultures of Artemia by the guanidinium thiocyanate/CsCl centrifugation method and poly(A+) RNA was selected by oligo dT cellulose chromatography as previously described [3].

Northern Analysis of mRNA

Poly(A+) RNA was fractionated by denaturing agarose gel electrophoresis in the presence of formaldehyde [4] and blotted in 1M ammonium acetate to nylon membranes (Amersham Hybond-N). Hybridization with labelled oligo-nucleotide (17mer, labelled greater than 1×10^8 cpm/μg) was performed in 6xSSC, 10x Denhardts solution [5] 0.5% (w/v) sodium dodecyl sulfate and 0.1mg/ml salmon sperm DNA at 40°C for 18-24 hours. Hybridized membranes were washed twice at T_H in 6xSSC, 0.5% sodium dodecyl sulfate. Autoradiographs were analysed by scanning laser densitometry.

In vitro Translation of mRNA

Poly(A+) RNA was translated in a micrococcal nuclease-treated rabbit reticulocyte lysate available commercially. Translations of agarose plugs from denaturing methylmercury gels containing fractionated mRNA were performed in 50 μl containing 110mM K^+, 2.1mM Mg^{2+} and ^{35}S-methionine (0.5 mCi/ml) at 37°C for 60 minutes [6]. Translation products were separated by SDS polyacrylamide gel electrophoresis and visualised by fluorography.

Poly(A+) RNA was microinjected into batches of 10 Xenopus oocytes and incubated at room temperature for 24 hours [7]. ^{35}S-Methionine was then added at 0.5mCi/ml and incubation continued for another 24 hours. The oocytes were then homogenised and the translation products analysed by immunoprecipitation.

cDNA Synthesis and Cloning

cDNA was synthesised from 44 hour poly(A+) RNA by a one-tube reaction utilising AMV Reverse Transcriptase for synthesis of first strand cDNA and RNase H and E.coli DNA Polymerase for second strand synthesis [8]. The double stranded cDNA was efficiently blunt-ended with T4 DNA Polymerase. cDNA was synthesised using either oligo dT_{15} as the primer of first strand synthesis or by using a globin-specific oligonucleotide followed by primer extension.

cDNA synthesised by primer extension from the globin-specific oligo-nucleotide was cloned into the SmaI cloning site of pUC9 and transformed into E.coli strain JM83 [9]. cDNA synthesised with oligo dT as the primer was cloned by Eco R1 linker addition into λgt10. Phage DNA was packaged in vitro and propagated on E.coli strains L87 and NM514 [10].

DNA Sequence Analysis

Fragments of cDNAs to be sequenced were generated by restriction endonuclease digestion with either EcoR1, Rsal or Alul. In the latter cases the fragments were shotgun cloned into the Smal site of the sequencing vector M13mp8 [9] and sequenced by the dideoxynucleotide chain termination method [11]. DNA and protein sequences were compared using Stadens programs [12], FASTP [13] and the haemoglobin template of Bashford et al. [14].

Southern Analysis of Artemia Genomic DNA

Artemia DNA was digested with restriction endonucleases, including Eco RI, Kpn I, Bam HI, Pst I and Hind III before electrophoresis on 0.6% agarose gels. The DNA was transferred to Zetaprobe membranes by alkali transfer [15] and hybridized with globin cDNA fragments labelled by random

oligomer labelling [16] in 2 x SSPE, 0.5% Trim milk, 0.5% Sodium dodecyl
sulfate and 0.2mg/ml salmon sperm DNA at 65°C for 18-24 hours. Hybridized
membranes were washed at 65°C and autoradiographed.

RESULTS AND DISCUSSION

Developmental Appearance of Haemoglobin Protein

    We first set out to follow the early developmental expression of the
haemoglobin genes, by comparing the appearance of the globin mRNA and
protein following reinitiation of development of desiccated cysts up until
several days of age. Extracts of partially purified haemoglobin were
prepared from cultures grown to different developmental stages up to 44
hours of development. The presence of haemoglobin protein was assayed by
Western blotting after polyacrylamide gel electrophoresis using antibodies
to Hb II (Fig. 1), which recognises all three haemoglobin species strongly
[17]. An alternative, simpler dot blot assay was established and by this
method we could detect as little as 1 nanogram (4 fmoles) of haemoglobin.
No globin protein was detectable until between 20 and 22 hours of development. The level of haemoglobin rose rapidly to reach an apparently steady
state level by 44 hours. While some slight variation was observed in different cultures as to the time of first detection of haemoglobin, in all
cases the appearance of haemoglobin protein was closely linked to the
appearance of nauplii.

Developmental Appearance of Haemoglobin mRNA

    We first investigated the possibility of assaying for the appearance
of the globin mRNA by in vitro translation of poly(A+) RNA isolated throughout early development. We have previously reported the inability of the
globin mRNA to be translated efficiently either in translation systems
derived from wheat germ or in rabbit reticulocyte lysate [2]. It was possible, however, by methylmercury hydroxide gel electrophoresis to purify
a fraction of mRNA containing almost pure globin mRNA (fraction 3, Fig. 2A),
which could direct the efficient translation of the full globin polypeptide
(Fig. 2B, lane 3). The identity of the polypeptide was confirmed by immunoprecipitation with an antibody to the haemoglobin. The inefficient translation of the globin mRNA in unfractionated poly(A+) RNA (Fig. 2B, lane 10)
may be due to either an inability to compete with other mRNAs for available
translational components, or to a possible complex secondary structure
which would lead to the premature termination of polypeptide synthesis.
Only in the presence of the powerful denaturant, methylmercury hydroxide,
is there significant synthesis of an intact globin polypeptide in unfractionated mRNA. The other gel plugs (fractions 1,2,4-9) direct the synthesis
of other products of a size reflecting the size of the mRNAs. The purification of a single fraction of mRNA containing almost pure globin mRNA
confirmed that the large globin polypeptide is encoded by a single mRNA
species and that this mRNA is of high molecular weight, being larger than
23S.

    Although the reticulocyte lysate system can translate the globin mRNA
when the mRNA has been fractionated, it is not particularly suitable as a
routine assay for developmental studies. We have therefore used the Xenopus
oocyte translation system as an alternative assay for the appearance of
the globin mRNA. Unfractionated poly(A+) RNA was able to direct the synthesis of the globin polypeptide of the expected molecular weight and the
identity of this polypeptide could be confirmed by immunoprecipitation with
an antibody against the haemoglobin (Fig. 3A). The oocyte system was used
as an assay for the appearance of the globin mRNA during development, with
globin mRNA detectable by translation at 20 hours of development and the

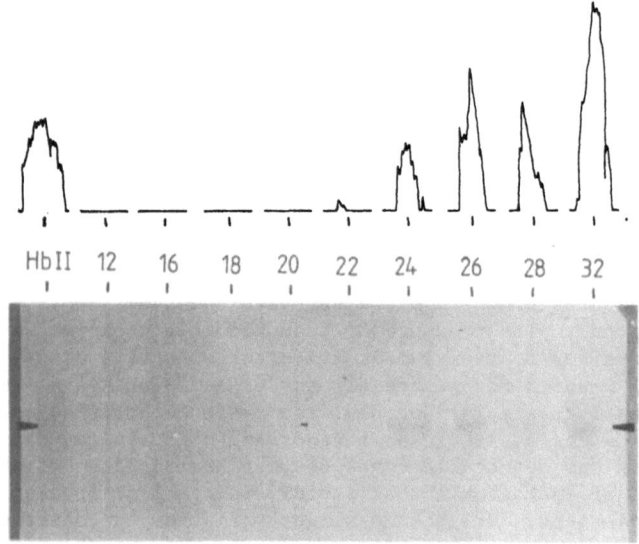

Fig. 1. Western blotting of protein extracts using an anti-HbII antibody. Samples of crude proteins extracted from 12-32 hour cultures of <u>Artemia</u> were electrophoresed on polyacrylamide gels and electro-blotted to nitrocellulose membranes. Haemoglobin was detected by immunoreaction with an anti-HbII antibody. The filter was then scanned by laser densitometry.

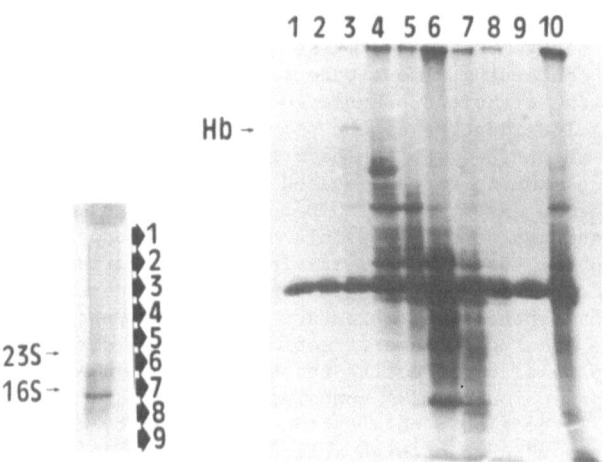

Fig. 2. Left. Fractionation of <u>Artemia</u> poly (A+) RNA (100h) by methyl-mercury hydroxide gel electrophoresis. The gel track from which plugs were taken for translation is represented. The positions of <u>E. coli</u> rRNA markers are shown. Right. Translation of frac-tionated poly (A+) RNA rabbit reticulocyte lysate. The gel plugs as shown in Panel A were translated in the lysate (tracks 1-9 re-presenting plugs 1-9). The products of translation were separ-ated by electrophoresis and fluorographed. Track 10 contains the products of translation of unfractionated poly (A+) RNA.

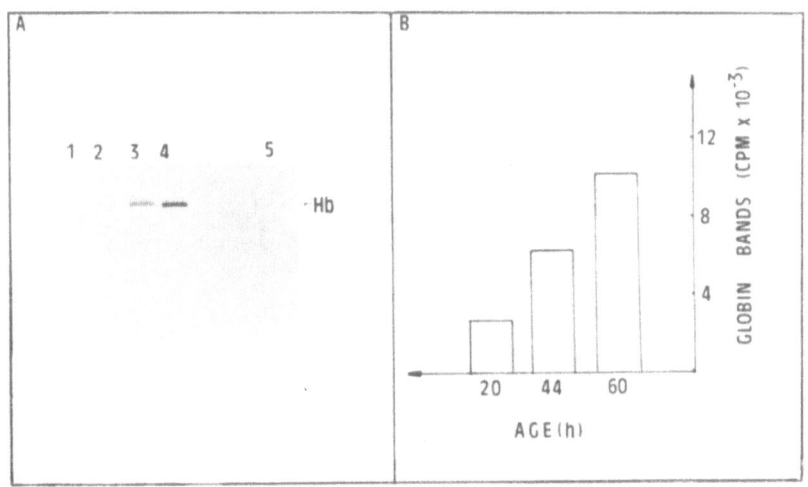

Fig. 3. Translation of <u>Artemia</u> poly (A+) RNA from various
stages of development in <u>Xenopus</u> oocytes. Panel A,
lane 1, RNA (50ng) from the developmental times
shown were translated and the globin products immun-
oprecipitated with an antibody to Hb II. Water-
injected oocytes; 2,20h RNA; 3,44h RNA; 4,60h RNA;
5, ^{14}C-labelled haemoglobin. Panel B. The gel corresponding to the globin bands was excised and the
radioactivity determined by scintillation counting.

level increasing until at least 60 hours. The radioactivity in the excised
bands is shown in Fig. 3B. Assaying for the appearance of the globin mRNA
by <u>in vitro</u> translation, however, may not be indicative of the levels of
globin mRNA actually present at any given time during development. An
increase or decrease in either the stability or specific translation activity of the globin mRNA will be seen as a change in the abundance of the
translated product, without there being any change in the actual level of
the globin mRNA. We wished, therefore, to establish an assay for the
globin mRNA itself. From available protein sequence (L. Moens, person.
comm.) several oligonucleotides were constructed and used in Northern
analysis of poly(A+) RNA isolated during development. By this method it
was possible to identify a single large globin mRNA species, of 4-5 kilobases in length, as was expected from the identification of the globin mRNA
by size fractionation and <u>in vitro</u> translation (Fig. 4). This mRNA was
not detectable until 20 hours of development and its level increased till
at least 44 hours. The results of scanning the autoradiograph (Fig. 4,
left panel) are shown in the right panel. The globin mRNA is not stored
in a latent form and the appearance of haemoglobin protein follows closely
the appearance of this mRNA. The appearance of the haemoglobin mRNA and
the protein it encodes correlates ontogenetically with the appearance of
the free swimming nauplius larvae (Fig. 5), and so we conclude that the
process of hatching of the prenauplius to the nauplius larvae acts as a
physiological switch, initiating the transcription of the globin mRNA and
its subsequent translation to produce functional haemoglobin. The mechanism by which this physiological switch is mediated remains unknown.

417

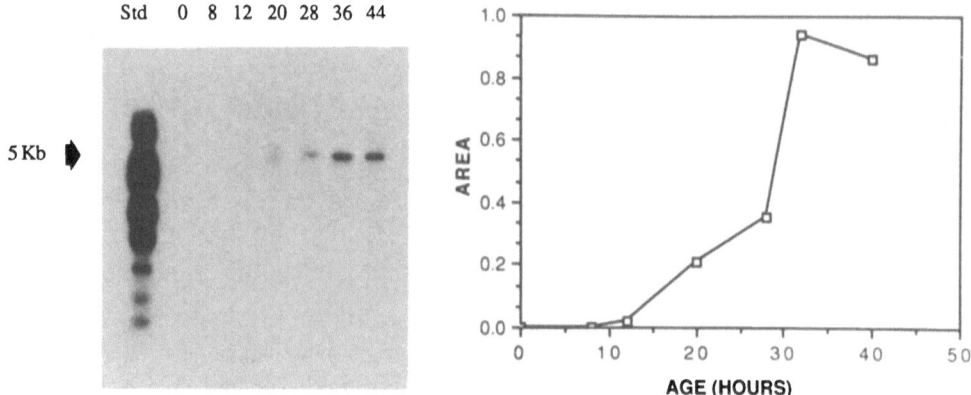

Fig. 4. Developmental appearance of globin mRNA. Left. Autoradiograph of <u>Artemia</u> poly(A+) RNA from different stages
of development, separated on formaldehyde-agarose gels
and capillary-blotted to Nylon membranes, was probed
with a globin-specific oligonucleotide prepared by end-
labelling with T4 polynucleotide kinase. Right. The
autoradiograph was analysed by scanning laser densitometry.

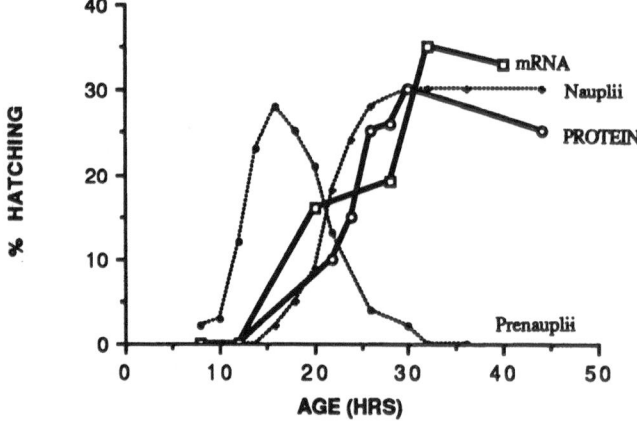

Fig. 5. Developmental appearance of haemoglobin protein and
mRNA. Expression of the globin mRNA and protein is
linked to a physiological switch. Schematic representation of the relative levels of globin mRNA and
protein, estimated by scanning laser densitometry of
Northern and Western blots, respectively, is plotted
along with the data for the appearance of the prenauplius and nauplius larvae during early development
of <u>Artemia</u>.

One strategy used in cDNA cloning involves the synthesis of oligonucleotide probes to known protein sequence. However, the degeneracy of
the genetic code requires synthesis of oligonucleotide mixtures for use
as hybridization probes [18]. Developing the conditions to distinguish
an authentic signal from spurious hybridization is difficult and time consuming if degenerate probes are used. To overcome these limitations, we
have attempted to generate globin-specific cDNA probes by primer extension
of an oligonucleotide constructed to available protein sequence data [19],
(Fig. 6). This olignucleotide (17mer) corresponds to the coding region
for the putative C helix of the El polypeptide, which corresponds to approximately one domain of the haemoglobin. The oligonucleotide hybridized
to only the globin mRNA on Northern blots and so was suitable as a primer
for the synthesis and cloning of globin-specific cDNA.

Preparations of mRNA identified as being enriched with globin mRNA
were used as the template for cDNA synthesis. The oligonucleotide, after
purification by denaturing polyacrylamide gel electrophoresis, was used
as the primer for cDNA synthesis and the resulting cDNA was cloned into
the SmaI restriction site of the cloning vector pUC9. 750 recombinants
were screened for globin sequences using the oligonucleotide. Approximately

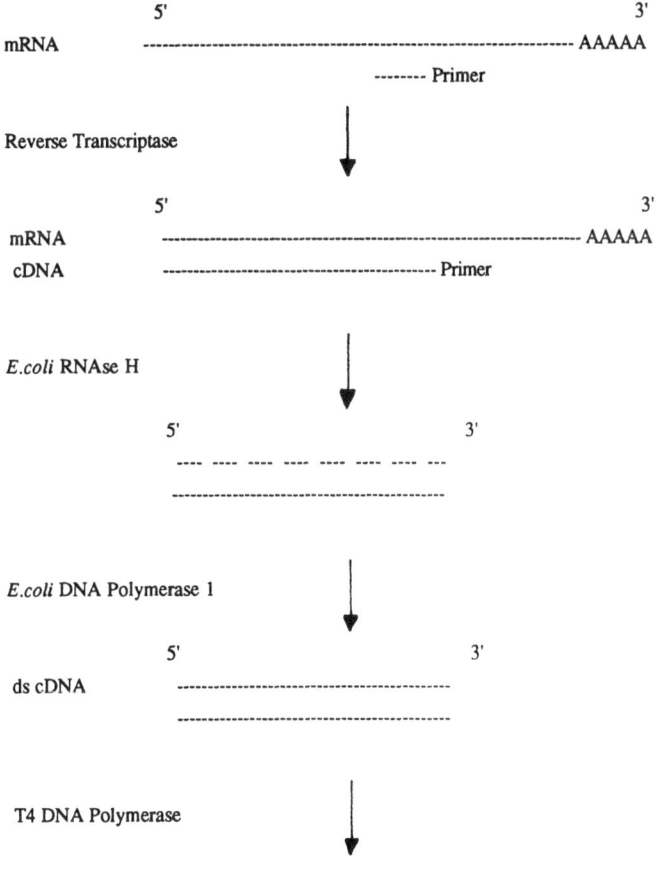

Fig. 6. Strategy for cDNA Cloning by Primer Extension

120 positive clones were identified and 72 of these were combined into 12 pools of 6 clones each for further characterization. Three individual clones from the pool containing the largest cDNA inserts were sequenced to confirm their identity. Only one of the three clones analysed, pAGp27, was found to be a globin-specific clone. It was possible that non-specific binding of the oligonucleotide had led to primer extension on mRNA species other than globin mRNA.

DNA sequencing revealed that the globin-specific clone had, in fact, lost the first two bases of the oligonucleotide during the cloning procedure. However, the amino acid sequence derived from the cDNA sequence extends from the oligonucleotide site in the C helix of the E1 domain (91-95, Fig. 7) out to and extending past the N-terminal region of E1 (Fig. 7), across a putative linker region connecting two domains (49-55) and into the C-terminal region (H,GH,G helices, 1-54) of the preceding domain. These sequences have provided us with new information as to how the multiple haem-binding domains are linked, an important consideration in understanding how the genes for separate domains may have duplicated and fused during evolution. The amino acid sequence of pAGp27 also differed in several residues from those found in E1, and these differences may reflect either species differences or polymorphism at the DNA level which affects the amino acid sequence (Fig. 7). These differences could, however, suggest that Artemia contains a globin multigene family and that the mRNA upon which the oligonucleotide has primed may not be the same mRNA encoding the globin chain from which the E1 domain was purified. The cDNA clone obtained was small (280 base pairs) in comparison to the globin mRNA (4-5 kilobases), but it was a suitable probe for the isolation of larger cDNA clones.

Isolation of Globin cDNAs from λgt10 Libraries

cDNA was synthesised using oligo dT as the primer (complementary to the 3' end of the mRNA) and this cDNA was size-fractionated by Bio-Gel A50m column chromatography. cDNA of greater than 2 kilobasepairs was

```
         1                11               21               31

p27   D I S H F Q N F R V T L L E Y L T E N G M N G A Q K A S W N K A F D A F E K Y I

E1
         41               51               61               71

      I M G L S S L E R V D P I T G L S G L E K N A I L│S│T W G K V R G N L Q E V G K

              E R V D P I T G L S G L E K N A I L│D│T W G K V R G N L Q E V G K


         81               91

      A T F G K L F│T│A H P E Y Q Q

      A T F G K L F│A│A H P E Y Q Q
```

Fig. 7. Comparison of the derived amino acid sequence of the partial cDNA clone pAGp27, isolated by the primer extension strategy, with the amino acid sequence of the domain E1. The numbering of residues, from 1-95, bears no relationship to their position in the myoglobin structural motif.

cloned by EcoRl linker addition into λgtl0 and 25,000 recombinants were screened with the partial cDNA clone described above. Eleven positive clones were purified and the cDNA they contained was analysed (Fig. 8). These clones all appeared to be greater than 4 kilobases in length and they fell into two classes by restriction endonuclease analysis, suggesting they may represent the two globin gene families.

EcoRl fragments from representatives of these two classes of cDNAs were subcloned into both pUC9 for amplification and into M13 for DNA sequencing with dideoxynucleotide chain terminators. DNA sequencing revealed that the two cDNAs coded for two similar but distinct mRNAs. The homology between the two cDNAs was high and explains how the partial cDNA clone (pAGp27) could hybridize to cDNAs representing two different mRNAs. This is further confirmed by the fact that the El-specific oligonucleotide hybridizes with equal strength to the two cDNAs. This also confirmed that the El domain is fortuitously near the N-terminus of the polypeptide and so aided the identification of only large cDNAs primed with oligo dT. It is possible these two classes of cDNA clones represent mRNAs for the two different polypeptides (α and β) which combine to form the three haemoglobins. However, it is also possible that they may represent two different mRNAs for only one of these polypeptides. Further characterization of these cDNAs and the genes which encode them may help resolve these possibilities.

Analysis of the Nature of Domain Linkage

From DNA sequence data obtained from the globin cDNA clones and from the available protein sequence data we have addressed the question of how the multiple domains of Artemia haemoglobin are connected. Using the Bashford template of globin sequences [14] we have been able to determine the optimal alignment of amino acid sequences for several different regions linking two myoglobin-like domains together (Fig. 9). The partial cDNA clone p27 spans a connection between the El domain and the preceding domain in the polypeptide as described above. We have also identified a region of cDNA towards the carboxy-terminus of the polypeptide which also spans a connection between two different domains (the large EcoRl fragment in λAG 3.2, Fig. 8). This is referred to as AG1.E2 in Fig. 9. The derived amino acid sequence for the H helical region of this cDNA shows high homology to the same region of the El domain and so has allowed us to align

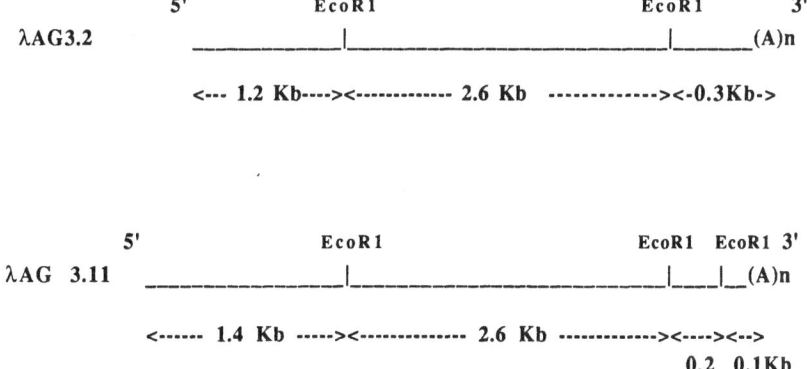

Fig. 8. Artemia globin cDNA clones. EcoRl restriction map of two cDNAs representative of the two classes of globin cDNA clones isolated from an Artemia λgtl0 cDNA library.

421

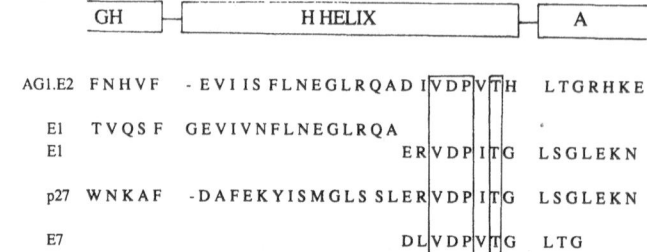

GH		H HELIX		A

```
AG1.E2   FNHVF   - EVI IS FLNEGLRQAD I VDP V T H   LTGRHKE

E1       TVQS F   GEVIVNFLNEGLRQA
E1                                    ER VDP I T G   LSGLEKN

p27      WNKAF   -DAFEKYISMGLS SLER VDP I T G   LSGLEKN

E7                                   DL VDP V T G   LTG
```

Fig. 9. Alignment of amino acid sequences around the linker
        region between two myoglobin-like domains. The
        Bashford template has been used to align amino
        acid sequences derived from cDNA sequences (p27
        and AG1.E2) and from available protein sequence
        (E1 and E7), spanning three different linker
        regions between domains of Artemia haemoglobin.
        Highly conserved amino acids are identified by
        boxes.

these two regions. This alignment suggests that the first eight amino
acids of the E1 protein sequence can indeed be identified as the linker
region between the E1 domain and the preceding domain, as has been suggested [20]. The consensus template obtained using the Bashford template
(-VDP-T) suggests that the end of the H region of the domain has been
adapted to form a linker between two myoglobin-like domains rather than
there having been the insertion of a specialised linker during the evolution of the globin polypeptide. The fact that this region is highly conserved between the three linker sequences suggests that there may be tight
structural constraints on the amino acid sequence of this region, as would
be expected for such a linker. It is also interesting to note that the
two classes of cDNAs do not differ at the DNA sequence level in the carboxy-
terminal region and that the carboxy terminus of E7, another domain for
which amino acid sequence data has been obtained [20] contains a region
displaying the same consensus linker sequence of -VDP-T-.

Southern Analysis of Artemia Genomic DNA

     Genomic DNA was probed with a 1.2 kilobase EcoR1 fragment derived from
the 5' end of one of the cDNAs. This fragment is large enough to contain
up to three domains and is known to contain the equivalent of the E1 domain,
as it is to this fragment that the E1-specific oligonucleotide and the
partial cDNA clone hybridize. This probe identified several distinct high
molecular weight fragments as having high homology on Southern blots, with
these hybridizing fragments spanning nearly 100 kilobases of DNA (fig. 10).
The molecular size of these fragments precludes their having originated
from only one gene and this finding, along with the high DNA sequence
homology found between the two classes of cDNAs, suggests that Artemia
contains a globin multigene family.

CONCLUSION

     We have identified the globin mRNA encoding the haemoglobins of Artemia

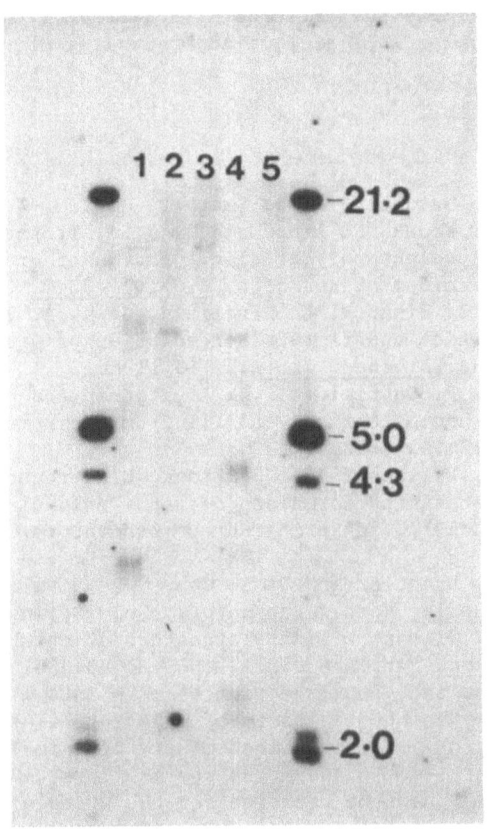

Fig. 10. Southern analysis of <u>Artemia</u> genomic DNA using
the cDNAs as a probe. Genomic DNA was digest-
ed with restriction endonucleases as below,
electrophoresed and capillary blotted to Zeta
Probe membrane and hybridized at high string-
ency with the 5' EcoRl fragment of the cDNA
clone λAG 3.2, prepared by random oligo labelling. Lane 1, EcoRl; 2, Kpnl; 3, Pstl; 4,
HindIII and 5, BamHl. Molecular weight markers
(λ DNA digested with HindIII and EcoRl) are
as indicated.

as being a single, high molecular weight species. There is no latent store
of this mRNA in early development and the appearance of the mRNA is followed
closely by the appearance of globin protein. Appearance of the globin
mRNA and protein is linked closely to the appearance of free-swimming
nauplii, suggesting the process of hatching acts as a physiological switch
inducing expression of the haemoglobin genes. We have isolated two classes
of cDNA clones which may represent the mRNAs for the α and β polypeptides.
DNA sequence obtained from these clones, along with available amino acid
sequences, have allowed us to postulate that the multiple domains of the
haemoglobin are joined by an adapted H polypeptide and that this structure
has been highly conserved. The complete cDNA sequence will aid the analysis
of the three dimensional structure of these proteins. Southern analysis
of genomic DNA using the cDNAs as probes suggests that <u>Artemia</u>, in fact,

contains a globin multigene family. The elucidation of the genetic structure of these haemoglobins will aid our understanding of the processes of
globin gene evolution.

REFERENCES

1. L. Moens, M-L. Van Hauwaert, K. De Smet, D. Geelen, G. Verpooten, J.
   Van Beeumen, S. Wodak, P. Alard and C. N. A. Trotman, A structural
   domain of the covalent polymer globin chains of Artemia, Interpretation of amino acid sequence data, J. Biol. Chem. 263:4679 (1988).
2. A. M. Manning, G. S. Ting, B. C. Mansfield, C. N. A. Trotman and W. P.
   Tate, The isolation and faithful translation of Artemia naupliar
   haemoglobin mRNA, Biochem. Int. 12:715 (1986).
3. C. J. Marshall, J. F. Cutfield, C. N. A. Trotman and W. P. Tate, Purification of the haemoglobins I and III from the brine shrimp, Artemia,
   Biochem. Int. 12:693 (1986).
4. T. Maniatis, E. F. Fritsch and J. Sambrook, Electrophoresis of RNA
   through Gels containing Formaldehyde, in "Molecular Cloning: A
   Laboratory Manual," Cold Spring Harbor Laboratory, Cold Spring
   Harbor (1982).
5. D. J. Denhardt, D. Dressler and D. S. Ray, "The Single Stranded DNA
   Phages," Cold Spring Harbor Laboratory, Cold Spring Harbor (1978).
6. N. Lonsberg and W. Gilbert, Primary structure of chicken muscle pyruvate
   kinase mRNA, Proc. Nat. Acad. Sci. (U.S.A) 80:3661 (1983).
7. C. Lane and J. Knowland, The injection of mRNA into living cells: The
   use of frog oocytes for the assay of mRNA in the study of the control
   of gene expression, in: "The Biochemistry of Animal Development,"
   Vol. 3; R. Weber, ed., Academic Press, New York (1975).
8. U. Gubler, A simple one-tube reaction for the synthesis of double-
   stranded cDNA, Nucleic Acids Res. 16:2726 (1988).
9. J. Messing, R. Crea and P. H. Seeburg, A system for shotgun DNA sequencing, Nucleic Acids Res. 9:309 (1981).
10. T. V. Huynh, R. A. Young and R. W. Davis, Constructing and screening
    cDNA libraries in λgt10 and λgt11, in "DNA Cloning: A Practical
    Approach," Vol. 1, D. Glover, ed., IRL Press, Oxford (1985).
11. F. Sanger, S. Nicklen and A. R. Coulson, DNA sequencing with chain-
    terminating inhibitors, Proc. Nat. Acad. Sci. (U.S.A) 74:5463 (1977).
12. R. Staden, A software system for DNA sequence manipulation by computer,
    Nucleic Acids Res., 10:4731 (1982).
13. D. J. Lipmann and W. R. Pearson, Rapid and sensitive protein similarity
    searches, Science 227:1435 (1985).
14. D. Bashford, C. Chothia and A. M. Lesk, Determinants of a protein fold.
    Unique features of the globin amino acid sequences, J. Mol. Biol.
    196:199 (1987).
15. K. C. Reed and D. A. Mann, Rapid transfer of DNA from agarose gels to
    Nylon membranes, Nucleic Acids Res. 13:7207 (1985).
16. A. P. Feinberg and B. Vogelstein, A technique for radiolabelling DNA
    restriction fragments to high specific activity, Anal. Biochem.
    132:6 (1983).
17. C. J. Marshall J. F. Cutfield, C. N. A. Trotman and W. P. Tate, An
    analysis of Artemia haemoglobins by comparison of immunoreactivities,
    Biochem. Int. 15:925 (1987).
18. A. A. Reyes and R. B. Wallace, in "Gentic Engineering: Principles and
    Methods," Vol. 6., Plenum, New York (1984).
19. A. M. Manning, C. N. A. Trotman and W. P. Tate, Isolation of cDNA
    clones for the haemoglobins of the brine shrimp, Artemia, Proc. Univ.
    Otago. Med. Sch. 66:9 (1988).
20. K. De Smet, M-L. Van Hauwaert, L. Moens and J. Van Beeumen, The
    structure of Artemia sp. haemoglobins II. A comparison of the
    structural units composing the Artemia sp. globin chains, in

"_Artemia_ Research and its Applications," Vol. 2, W. Decleir, L. Moens, H. Slegers, E. Jaspers and P. Sorgeloos, eds., Universa Press, Wetteren (1987).

LOCALIZATION OF PUTATIVE HAEMOGLOBIN BY IMMUNOGOLD STAINING IN <u>ARTEMIA</u>

C. N. A. Trotman[1], A. Kean[1], W. P. Tate[1], L. Moens[1] and
W. Jacob[2]

[1]Department of Biochemistry, University of Otago
Dunedin, New Zealand
[2]Universitaire Instelling Antwerpen
Antwerp, Belgium

ABSTRACT

The distribution of immunoreactive haemoglobin in <u>Artemia</u> was sought
by means of the immunogold technique and electron microscopy with primary
antibodies raised against HbII.

Before affinity purification of the primary antibody, gold deposits
were found in a number of regions that would not have been regarded as
primary candidates for haemoglobin localization. Most conspicuously staining was the cuticular region of the cyst wall between the inner and outer
cuticular membranes. Other positive regions were the developing exoskeletal
region of the encysted embryo; the naupliar exoskeleton; and the microvillous surface of cells lining the digestive tract.

Primary antibody purified by elution from immobilised haemoglobin
bound less to the above structures but more to intercellular spaces interpreted as haemocoel. This would be compatible with a purification of the
antibody from a contaminant fraction. However the localization in the unexpected structures was supported by positive and negative controls and
may indicate the incorporation of antigenic fragments recognized by one
fraction of the antibody. Since significant synthesis of globin does not
occur in the encysted embryo, the cyst wall cuticular globin immunoreactivity may be of maternal origin.

THE STRUCTURE OF <u>ARTEMIA</u> HEMOGLOBIN AND HEMOGLOBIN DOMAINS

L. Moens[1], K. Ver Donck[1], K. De Smet[1], M. L. Van Hauwaert[1],
J. Van Beeumen[2], P. Allard[3], S. Wodak[3] and C. N. A. Trotman[4]

[1]Department of Biochemistry, Universitaire Instelling
Antwerpen, Universiteitsplein 1. B-2610 Antwerpen-Wilrijk,
Belgium
[2]Laboratory of Microbiology, State University of Ghent
B-9000 Gent, Belgium
[3]Biological Macromolecular Conformation Unit. Free University
of Brussels (ULB) and Plant Genetic Systems. B-1050
Brussels, Belgium
[4]Department of Biochemistry, University of Otago, Dunedin
New Zealand

INTRODUCTION

Unlike the hemoglobins of the vertebrates, which are almost invariably
intracellular and tetrameric, the intra- and extracellular hemoglobins of
the invertebrates show a wide variety in their molecular size (Mr 16,000
to $\sim 1.7.10^6$) and architecture [1-4]. Intracellular hemoglobins usually
have low Mr's whereas extracellular hemoglobins have high Mr's which are
advantageous in minimizing excretion and avoiding excessive osmotic pressure. A high Mr can be achieved either by aggregation of many low Mr chains
into a functional hemoglobin, as in annelids, or by concatenation of the
low Mr chains into polymeric globins, as in molluscs and arthropods [4].
Despite this heterogeneity, Svedberg & Hedenius [5] suggested that all
these pigments are built up from myoglobin-like polypeptide chains of Mr
16,000 containing one heme group and able to bind oxygen reversibly. Polypeptide chains, or fragments of much longer chains having these characteristics (Mr 16,000; one heme), were defined by Vinogradov [4] as "heme-
binding domains". Based on the number of domains and subunits in the
native molecule, the invertebrate extracellular hemoglobins can be classified into four groups:
a) Single-domain, single-subunit hemoglobins, consisting of a single polypeptide chain, containing one heme group and having a Mr $\sim$ 16,000
   (<u>Chironomus</u>)
b) Single-domain, multi-subunit hemoglobins consisting of aggregates of
   monomeric subunits, some of which are connected by disulfide bonds
   (<u>Annelida</u>)
c) Two-domain, multi-subunit hemoglobins consisting of aggregates of dimeric
   polypeptide chains (Mr 30,000-40,000), each containing two heme-binding
   domains (<u>Arthropoda</u>)
d) Multi-domain, multi-subunit hemoglobins, consisting of two or more polypeptide chains each comprising eight to twenty heme binding domains
   (<u>Arthropoda</u>).

Artemia has three extracellular hemoglobins in its hemolymph. All
three, hemoglobins have a native Mr of about 260,000, and each is a dimer
of two globin chains (α or β) of similar size. They bind oxygen with low
cooperativity. The β_2 form, which has the highest oxygen affinity, is
specifically and reversibly inducible by low environmental oxygen tension.
The individual chains have Mr 130,000 with a minimum Mr calculated from
amino acid composition and heme content of 16,000, suggesting the presence
of eight structural and functional domains. Cooperativity is not expressed
between different domains within the isolated polymeric subunits; it only
exists between oxygen binding sites from different globin chains within
the native dimer. Limited proteolysis of the native hemoglobin with subtilisin resulted in a cleavage pattern compatible with a random cleavage
of a polymer containing eight domains of Mr $\sim$16,000 [6-12]. The Artemia
hemoglobins are thus representatives of the multi-domain, multi-subunit
hemoglobin type. No structural information is available on the multi-domain
globin chains. In order to establish their structural and evolutional
relationship with the classical globin family, we have studied the molecular
structure of the Artemia hemoglobins in detail.

RESULTS AND DISCUSSION

The Aminoterminal Sequence of the α and β Globin Chain and some of their
Domains

The multidomain globin chains of Artemia might be built up as homo-
or heteropolymers of Mr 16,000 domains. A comparison of the tryptic map
of the isolated domains (E_1 to E_8) suggests that they are heteropolymers,
although the differences are minor [13]. Limited proteolysis is expected
to cleave the intact globin chains between the compact tertiary structures.
However cleavage within domains can not be excluded. A comparison of the
aminoterminal sequences of the purified domains and the intact globin
chains will provide more information on these points.

The separation of the intact globin chains from Hb II by reversed
phase chromatography and of the domains by isoelectric focusing is illustrated respectively in Fig. 1 and Fig. 2. Only domains E_2 and E_5 were not
pure enough for sequencing. Aminoterminal sequencing of the polypeptides
was performed by vapor phase sequencing [14,15]. The sequences obtained
were compared with the aminoterminal sequence of sperm whale myoglobin.
Figure 3 clearly shows that the aminoterminal sequence of both intact
chains and of the domains E_1, E_4, E_6, E_7 and E_8 are compatible with the
aminoterminal sequence of the classical globin type. In contrast the
sequence obtained for domain E_3 seems to be more homologous to the GH-H
region of myoglobin. The sequence differences are considerable, validating
the conclusion that the globin chains are heteropolymers. Subtilisin seems
to cleave preferentially between the domains. However, as the sequence of
E_3 shows, cleavage within domains also occurs. This may explain the high
number of domains seen in Fig. 2.

Alignment of the Amino Acid Sequences of Domain E_1 and E_7 with other Globins

The amino acid sequences of two domains, E_1 and E_7, were determined
by a combination of vapor phase sequencing and manual techniques [12,13].
The sequences obtained were aligned with globins of which the tertiary
structure is known. This was done by constructing a computer alignment
using existing programs [27]. The alignment obtained was manually adjusted,
taking into account the X-ray crystallographic data [16]. Finally the
alignment was checked against the Bashford templates [17]. The consensus

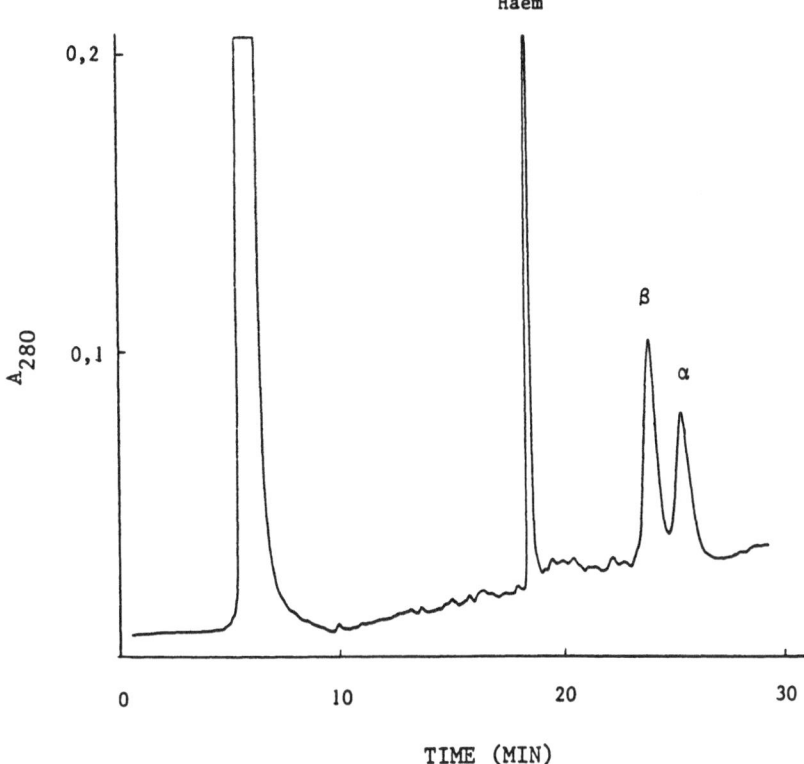

Fig. 1.   Separation of the α and β globin chain by reversed
          phase chromatography. 100 μg hemoglobin II was de-
          natured in 6 M guanidinium hydrochloride, 50 mM Tris-
          HCl pH 7 by heating at 100°C for 5 min. After de-
          naturation the sample was immediately loaded onto a
          C_4 reverse phase column (Bakerbond-wide pore) equili-
          brated in 0.1% trifluoroacetic acid. The column was
          developed with a gradient of 0-75% acetonitrile, 0.1%
          trifluoroacetic acid at a flow rate of 1 ml/min over
          60 min. Detection was at 280 nm and the α and β
          chain were identified by SDS-polyacrylamide gel
          electrophoresis.

alignment is shown in Fig. 4. This alignment clearly shows that the E_1
and E_7 sequences are most closely related, respectively, to the myoglobin
and the β chain type (22 and 24% homology). All helices A to H could be
unequivocally identified. The major length discrepancies are localized
at the amino- and carboxy termini and in the EF, FG and GH bends. The
majority of residues determining the secondary and tertiary structure in
the key globins are identical or functionally conserved in domains E_1 and
E_7 (Fig. 4, Table 1). The length discrepancy at the amino and carboxy
termini of both domains is undoubtedly the result of the proteolytic
cleavage of the domains from the intact molecules, which is also illustrated
by the amino terminal sequence of domain E_3 (Fig. 3). Interpretation of
the discrepancies in the EF/FG and GH regions and non-conservative substitutions such as Tyr at C_4 are more difficult to interpret from the
alignment alone. Despite the low homology, domains E_1 and E_7 can thus be

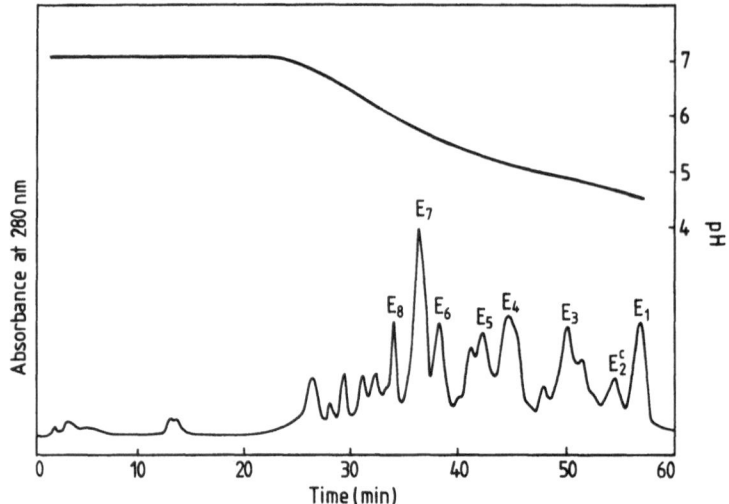

Fig. 2. Chromatofocusing of the domains composing the <u>Artemia</u>
hemoglobins. Total hemoglobin was subjected to limited
proteolysis and the resulting fragment mixture sized by
gel filtration [9]. The single domain fraction (Fraction E - Mr ~16,000) was separated by chromatofocusing
on a Mono P HR 5/20 column equilibrated in 0.025 M Bis-
Tris/iminodiacetic acid pH 7.1. The individual domains
were eluted with 9% Polybuffer (7-4) iminodiacetic acid
pH 4.0 at a flow rate of 1 ml/min [13]. Detection was
at 280 nm and the separated domains were collected manually.

```
            NA1 5          A1   5    10   15   AB1  B1   5    10   15   C1   5    CD1 5
            :   :          :    :    :    :    :    :    :    :    :    :    :    :   :

MYO         -------VL      SEGEWOLVLHVWAKVE   A--  DVAGHGODILIRLFKS   HPETLEK   FDRFKHLK

ALPHA AR    --AE-VSGI      LVSDKATIKRTWSIVN   ---  DLPSFGRNVFLSNFAA   KPIYDXX   --------
BETA AR     --AE-IRGI      LXDSKMTIKLTXAQVT   ---  DLPSFGNNVDASVRAA   KXEYDIG   F-------

E1AR        ERVDPITGL      SGLEKNAILDTWGKVR   G--  NLQEVGKATFGKLFAA   HPEYQQM   FRFFQGVQ
E4AR        -------L       VALVKXAXEVXWV---   ---  ----------------   -------   --------
E6AR        --VDVITGL      SLXEKNAILHTWGKVK   ---  PNLXXGGNKFEK----   -------   --------
E7AR        -------AL      TALEKQSIQDIWTILK   AV-  GLEFLQVKMFGKLFAD   HPEYKAH   FDNFLTAI
E8AR        -DL-PITV-      KTYLKKKVVAA------  ---  ----------------   -------   --------
```

```
            NA  A1   5    10   15   AB   B1   5    10   15   C1   5    CD1 5    D1   5
            :   :    :    :    :    :    :    :    :    :    :    :    :   :    :    :

MYO         VL  SEGEWOLVLHVWAKVE   A    DVAGHGQDILIRLFKS   HPETLEK   FDRFKHLK   TEAEMKA

E3AR        --  ----------------   -    ---------FLKKFLT   QKGQAKQ   VKLATKFT   ---FIL-
```

Fig. 3. Aminoterminal sequence of the intact <u>Artemia</u> globin chains
and some of their domains. X : unidentified residues.

```
           NA1 5       A1  5   10  15  AB1  B1  5   10  15   C1  5      CD1 5
           :   :       :   :   :   :   :    :   :   :   :    :   :      :   :
MYWHP      -------VL   SEGEWOLVLHVWAKVE   A--  DVAGHGODILIRLFKS   HPETLEK   FDRFKHLK
HAHU       -------VL   SPADKTNVKAAWGKVG   A--  HAGEYGAEALERMFLS   FPTTKTY   FPHF-DLS
HBHU       ------VHL   TPEEKSAVTALWGKV-   ---  NVDEVGGEALGRLLVV   YPWTQRF   FESFGDLS
E1AR       ERVDPITGL   SGLEKNAILDTWGKVR   G--  NLQEVGKATFGKLFAA   HPEYQQM   FRFFQGVQ
E7AR       -------AL   TALEKQSIQDIWTILK   AV-  GLEFLQVKMFGKLFAD   HPEYKAH   FDNFLTAI
GGICE3     --------L   SADQISTVQASFDKVK   G--  ----DPVGILYAVFKA   DPSIMAK   FTQFAG-F


           D1  5       E1  5   10  15  20    EF1 5   10      F1  5   10   FG1
           :   :       :   :   :   :   :     :   :   :       :   :   :    :
MYWHP      TEAEMKA     SEDLKKHGVTVLTALGAILK   K--K--GHHEAE   LKPLAQSHAT   KHKI
HAHU       -----HG     SAQVKGHGKKVADALTNAVA   HV-D---DMPNA   LSALSDLHAH   KLRV
HBHU       TPDAVMG     NPKVKAHGKKVLGAFSDGLA   HL-D---NLKGT   FATLSELHCD   KLHV
E1AR       LAFLVQ-     SPKFAAHTQRVVSALDQTLL   ALNRPSDQFVYM   IKELGLDHIN   RGT-
E7AR       FSVAEDL     VPKLRAHLHRVIDAFDLVIF   AL--GRESLRGS   LKDLGIFHTG   RDIV
GGICE3     DLESIKG     TAPFETHANRIVGFFSKIIG   ELP----NIEAD   VNTFVASHKP   RG-V


           G1  5   10  15      GH1 5   H1  5   10  15  20  25   HC1
           :   :   :   :       :   :   :   :   :   :   :   :    :
MYWHP      PIKYLEFYSEAIIHVLHSR   HPGDF   GADAQGAMNKALELFRKDIAAKYKEL   GYQG
HAHU       DPVNFKLLSHCLLVTLAAH   LPAEF   TPAVHASLDKFLASVSTVLTSKYR--   ----
HBHU       DPENFRLLGNVLVCVLAHH   FGKEF   TPPVQAAYQKVVAGVANALAHKYH--   ----
E1AR       DRSFVEYLKESLGDSVDEF   TVQSF   GEVIVNFLNEGLRQA-----------   ----
E7AR       DPVESLTGFKLMVAVIEEG   LDT-F   --RAVPEYSKGLEGRFGNVDNINENA   PFR-
GGICE3     THDQLNNFRAGFVSYMKAH   T--DF   A-GAEAAWGATLDTFFGMIFSKM---   ----
```

MYWHP = sperm whale myoglobin   E1AR = Artemia E1 hemoglobin
HAHU = human alpha hemoglobin   E7AR = Artemia E7 hemoglobin
HBHU = human beta hemoglobin   GGICE3 = Chironimus III hemoglobin

Fig. 4. Alignment of the E_1 and E_7 domains with selected globin sequences.

considered as homologous with the classical globin family. Furthermore, the hydrophobicity profiles [18] of myoglobin and domain E_1 and of the chain and domain E_7 show with the exception of the C/CD/D region, a good overall similarity (results not shown).

Model Building of the E_1 and E_7 Domains

Homologous proteins, having the same pattern of residue hydrophobicity along their amino acid sequence tend to share the same type of folding [22]. This suggests that the domains E_1 and E_7 may have a tertiary structure to the globin fold. Although it is impossible to predict protein tertiary structure from sequence alone, it is possible to test whether a sequence could conform to a predefined structure or not. Computer model building, using the Brugel software [23] was applied on the discrepant regions of the Artemia domain E_1 [12] and E_7. This was done by searching the Brookhaven Database for structures with similar coordinates in main chain flanking regions and similar in length [24]. Such fragments were mainly found as homologous regions in Chironomus globin IIIA, which is evolutionarily the most closely related globin sequence [25]. The fragment was then substituted into the template chain, after which the individual residues were "mutated" to the Artemia ones. The compatibility of the structure obtained was then studied in terms of bond distances, energies and character. This approach is illustrated for the GH region of domain E_7 in Fig. 5. Similarly, the steric compatibility of non-conservative substitutions, such as the replacement of Thr at C_4 by Tyr (Fig. 4), can be studied by modeling (Fig. 6). From this modeling approach we can conclude that

Table 1. Comparison of the Amino Acids Identical in Myoglobin and Haemoglobin with Artemia E_1 and E_7 and some Globin Chains with Known Atomic Structure

Position	Function	Myo/Hb	Artemia E_1	Artemia E_7	CTT
NA2	H-helix contact	L	L	L	L
A8	H-helix contact : hydrophobic cluster at bottom of heme pocket	V	I	I	V
A12	Hydrogen bond to E helix	W	W	W	F
A14	External salt bridge	K	K	I	K
A15	Hydrophobic cluster at bottom of heme pocket spacer between B and G-helices	V	V	L	V
·B6	Close contact with G at E_8 where helices cross	G	G	Q	P
B10	In hydrophobic cluster at right of heme	L	F	F	L
B12	In Hb : hydrogen bond to F at GH5 and E at B8 in myo : hydrogen bond to water	R	K	K	A
C2	Sharp turn between B and C helices	P	P	P	P
C4	Near heme	T	Y	Y	K
CD1	Packed against heme	F	F	F	F
CD4	In hydrophobic cluster at right of heme	F	F	F	F
CD7	In hydrophobic cluster at right of heme	L	V	A	-
E5	External salt bridge	K	A	R	E
E7	Interacts with sixth heme ligand	H	H	H	H
E8	Close contact with G at B6 where helices cross	G	T	L	A
E11	Packed against heme in right hydrophobic cluster	V	V	V	I
F4	Packed against heme in right hydrophobic cluster	L	L	L	F
F8	Fifth ligand to heme	H	H	H	H
FG2	Unknown	K	R	R	R
G16	Packed against A-helix	L	V	I	M
GH5	In hydrophobic cluster at bottom of heme pocket as a spacer between G and H helices	F	F	F	F
H10	External salt bridges	K	E	K	A
H22	External salt bridges	K	-	I	K
H23	Hydrogen bond to FG5 to stabilize end of H-helix	Y	-	N	-

Myo : sperm whale myoglobin
Hb : human haemoglobin
CTT : Chironomus thumni III

References : [19,20,21]

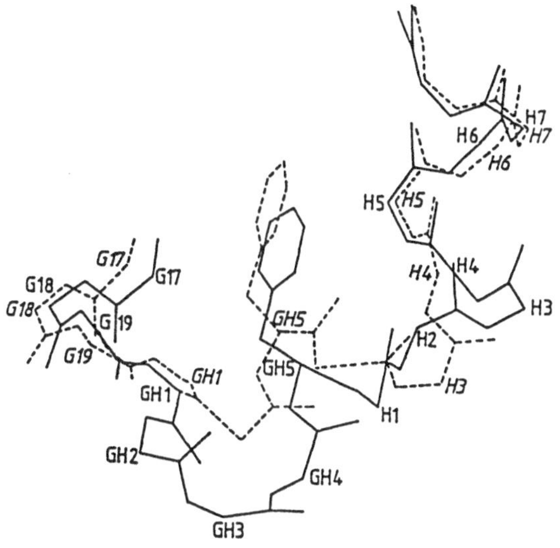

```
                G15      GH1 5  H1  5    10
                 :        :  :   :  :    :
        HBHU    VCVLAHH   FGKEF  TPPVQAAYQK
        E7AR    VAVIEEG   LDT-F  --RAVPEYSK
        GGICE3  VSYMKAH   T--DF  A-GAEAAWGA
```

Fig. 5. The GH region of domain E_7. Compared to the human
         β chain, three deletions occur in the GH corner
         of E_7 : one before the conserved GH5 and two after
         it. The database provided a fragment from <u>Chironomus</u>
         globin IIIA. This fragment, however, showed two
         deletions before GH5 and one after it. Neverthe-
         less, the fit on the anchoring regions is good and
         brings the Phe side chain of GH5 in approximately
         the same place and orientation. It therefore shows
         that the E_7 GH corner, despite the deletions, is
         comparable with that of the human β chain.
         Human β chain ——; <u>Artemia</u> E_7 ----.

the E_7 domain, in the discrepant regions as well as in the total molecule,
is compatible with the globin fold.

Hypothetical Model for the Artemia Hemoglobin

    Assuming that the domains have dimensions comparable to a globin
folded structure, an elipsoid of 45 Å x 35 Å x 25 Å, a speculative model
for the <u>Artemia</u> hemoglobin can be proposed (Fig. 7). Each globin forms a
disk with a diameter of 120 A and a height of 35 Å. Two disks stacked
on each other form the native molecule. Within a globin chain the eight
domains can be orientated in several ways, two of which will be considered
here. One way is asymmetric where four pairs of domains are formed in
each globin chain. Each pair has an upright and an inverted domain packed
with their E helices in contact. The heme groups are facing each other.
Such asymmetric linking of the domains implies that there must be two types
of linker regions; a long one between the domains in a pair and a short
one between pairs. A second way is symmetric with the domains in the same
direction and having the E and F helices coplanar and the heme oriented
towards the center of the molecule. Covalent linking is achieved by a
larger linker formed by a pre-A and post-H segment (Fig. 7B) [26]. Low
angle X-ray diffraction and a study by computer modeling of the docking
of the domains inside the globin chains will help to resolve these models.

                                                                    435

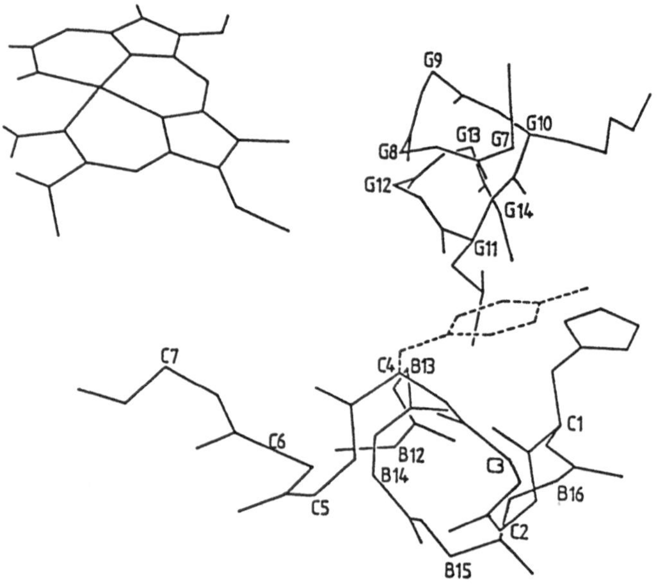

	B10 15	C1 5	CD1 5
	: :	: :	: :
MYWHP	LIRLFKS	HPETLEK	FDRFKHLK
HBHU	LGRLLVV	YPWTQRF	FESFGDLS
E1AR	FGKLFAA	HPEYQQM	FRFFQGVQ
E7AR	FGKLFAD	HPEYKAH	FDNFLTAI

Fig. 6. The BC and C region of domain E_7. In myoglobin, the C and G helices cross in close proximity, thus creating a problem of accomodating
the bulky Tyr at C_4. In domain E_1 this could
only be solved by a concomitant adjustment of
the neighbouring residues [12]. In the
chain, however, the C and G helices are more
spaced so that the Tyr residue at position C_4
of E_7 has enough space to turn the phenolic
group to the surface, whereas the hydrophobic
contact between the C and G helices is provided
by the aromatic ring.

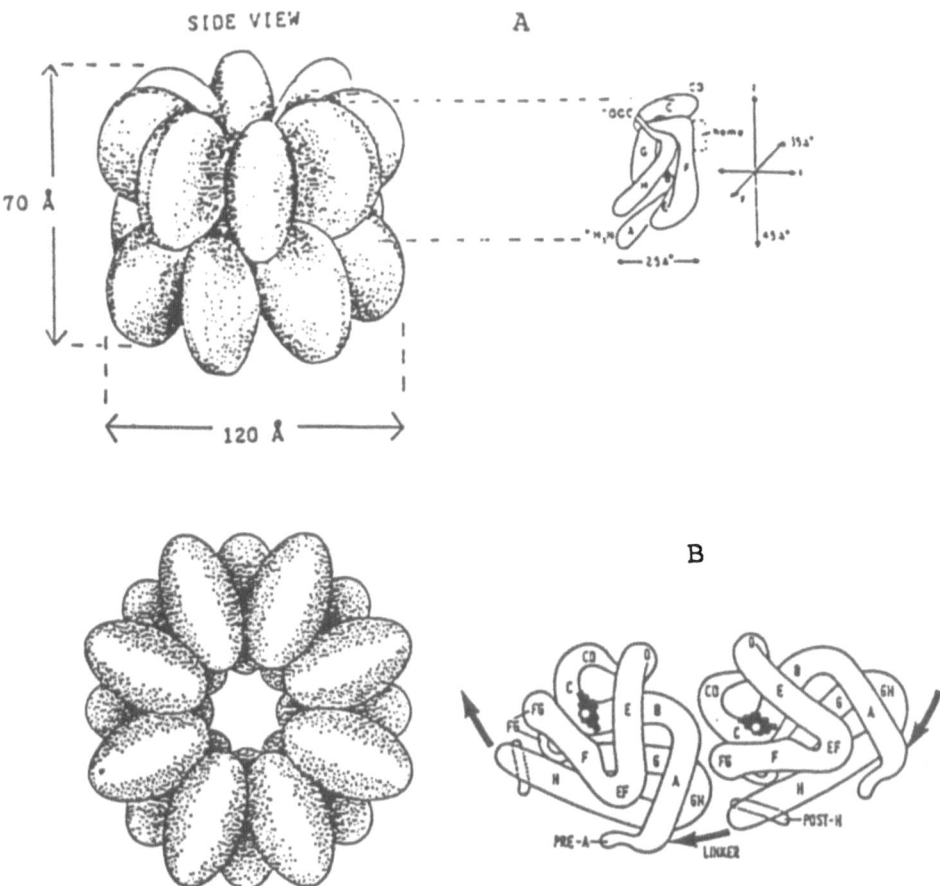

SIDE VIEW   A

70 Å

120 Å

TOP VIEW

Fig. 7. Hypothetical models for the _Artemia_ hemoglobin.
A. Top and side view of the native hemoglobin.
B. A symmetric way of linking two domains.

REFERENCES

1. M. C. H. Chung and M. D. Ellerton, The physico-chemical and functional properties of extracellular respiratory haemoglobins and chlorocruorins, _Prog. Biophys. Mol. Biol._ 35:53 (1979).

2. E. J. Wood, The oxygen transport and storage of proteins of invertebrates, _Essays Biochem._ 16:1 (1980).

3. R. E. Dickerson and I. Geis, "Hemoglobin : Structure, Function, Evolution and Pathology," Benjamin/Cummings, Menlo Park (1983).

4. S. Vinogradov, The structure of invertebrate extracellular hemoglobins (erythrocruorins and chlorocruorins), _Comp. Biochem. Physiol._ 82B: 1 (1985).

5. T. Svedberg and A. Hedenius, Sedimentation constants of the respiratory proteins, _Biol. Bull._ 66:191 (1934).

6. L. Moens and M. Kondo, Evidence for a dimeric form of _Artemia salina_ extracellular hemoglobins with high molecular weight subunits, Eur. _J. Biochem._ 82:65 (1978).

7. J. D'Hondt, L. Moens, J. Heip, A. D'Hondt and M. Kondo, Oxygen-binding characteristics of three extracellular haemoglobins of _Artemia salina_, _Biochem. J._ 171:705 (1978).

8. E. J. Wood, C. Barker, L. Moens, W. Jacob, J. Heip and M. Kondo, Bio-
    physical characterization of <u>Artemia</u> <u>salina</u> extracellular haemo-
    globins, <u>Biochem.</u> <u>J.</u> 193:353 (1981).
9. L. Moens, D. Geelen, M. L. Van Hauwaert, G. Wolf, R. Blust, R. Witters
    and R. Lontie, The structure of <u>Artemia</u> <u>sp.</u> hemoglobin. Cleavage
    of the native molecule into functional units by limited subtilisin
    digestion, <u>Biochem.</u> <u>J.</u> 223:801 (1984).
10. L. Moens, M. L. Van Hauwaert and G. Wolf, The structure of <u>Artemia</u> <u>sp.</u>
    haemoglobins. III. Purification of a structural unit to homogeneity,
    <u>Biochem.</u> <u>J.</u> 227:917 (1985).
11. L. Moens, M. L. Van Hauwaert, D. Geelen, G. Verpooten and J. Van
    Beeumen, in: "<u>Artemia</u> Research an its Applications," Vol. 2, W.
    Decleir, L. Moens, H. Slegers, E. Jaspers and P. Sorgeloos, eds.,
    Universa Press, Wetteren (1987).
12. L. Moens, M. L. Van Hauwaert, K. De Smet, D. Geelen, G. Verpooten, J.
    Van Beeumen, S. Wodak, P. Allard and C. Trotman, A structural domain
    of the covalent polymer globin chains of <u>Artemia</u>, <u>J.</u> <u>Biol.</u> <u>Chem.</u>
    263:4679 (1988).
13. K. De Smet, M. L. Van Hauwaert, L. Moens and J. Van Beeumen, The
    structure of <u>Artemia</u> <u>sp.</u> haemoglobins. II. A comparison of the
    structural units composing the <u>Artemia</u> <u>sp.</u> globin chains, in:
    "<u>Artemia</u> Research and its Applications", Vol. 2, W. Decleir, L.
    Moens, H. Slegers, E. Jaspers and P. Sorgeloos, eds., Universa Press,
    Wetteren (1987).
14. R. M. Hewick, M. W. Hunkapiller, L. E. Hood and W. J. Dreyer, A gas-
    liquid solid phase peptide and protein sequenator, <u>J.</u> <u>Biol.</u> <u>Chem.</u>
    256:7990 (1981).
15. M. W. Hunkapiller, R. Hewick, R. M. Dreyer and L. E. Hood, High sen-
    sitivity sequencing with a gas-phase sequenator, <u>Methods Enzymol.</u>
    91:393 (1983).
16. A. M. Lesk and C. Chothia, How different amino acid sequences determine
    similar protein structures : the structure and evolutionary dynamics
    of the globins, <u>J.</u> <u>Mol.</u> <u>Biol.</u> 136:225 (1980).
17. D. Bashford, C. Chothia and A. M. Lesk, Determinants of a protein fold.
    Unique features of the globin amino acid sequences, <u>J.</u> <u>Mol.</u> <u>Biol.</u>
    196:199 (1987).
18. J. Kyte and R. F. Doolittle, A simple method for displaying the hydro-
    pathic character of a protein, <u>J.</u> <u>Mol.</u> <u>Biol.</u> 157:105 (1982).
19. G. Fermi, Three-dimensional fourier synthesis of human deoxy haemoglobin
    at 2.5 Å resolution. Refinement of the atomic model, <u>J.</u> <u>Mol.</u> <u>Biol.</u>
    97:237 (1976).
20. R. Huber, O. Epp, W. Steigemann and H. Formanek, The atomic structure
    of erythrocruorin in the light of the chemical sequence and its
    comparison with myoglobin. <u>J.</u> <u>Mol.</u> <u>Biol.</u> 52:349 (1971).
21. T. Takano, Structure of myoglobin refined at 2.0 Å resolution, <u>J.</u> <u>Mol.</u>
    <u>Biol.</u> 110:537 (1988).
22. D. Eisenberg, Three-dimensional structure of membrane and surface pro-
    teins, <u>Ann.</u> <u>Rev.</u> <u>Biochem.</u> 53:595 (1984).
23. P. Delhaise, M. Van Belle, M. Bardiaux, P. Allard, P. Hamers, E. Van
    Cutsem and S. Wodak, Analysis of data from computer simulations on
    macromolecules using the CERAN package, <u>J.</u> <u>Mol.</u> <u>Graphics</u> 3:116
    (1985).
24. A. T. Jones and S. Thirup, Using known substructures in protein model
    building and crystallography, <u>EMBO</u> <u>J.</u> 5:819 (1986).
25. M. Goodman, J. Pedwaydon, J. Czelusniak, T. Suzuki, T. Gotoh, L. Moens,
    F. Shishikura, D. Walz and S. Vinogradov, An evolutionary tree for
    invertebrate globin sequences, <u>J.</u> <u>Mol.</u> <u>Evol.</u> 27:236 (1988).
26. C. N. A. Trotman, W. P. Tate, L. Moens and S. Wodak, <u>Artemia</u> haemoglobin
    compared with mammalian globins, <u>Proc.</u> <u>Univ.</u> <u>Otago</u> <u>Med.</u> <u>Sch.</u> 66:19
    (1988).
27. P. A. Stockwell and G. B. Petersen, Homed: a homologous sequence editor,
    <u>CABIOS</u> 3:37 (1987).

COMPUTER UTILIZATION IN <u>ARTEMIA</u> RESEARCH

Robert P. McCourt

Artemia Bibliographic Center (ABC)
Department of Biology
Hofstra University
Hempstead, New York (USA) 11550

INTRODUCTION

In addition to specific computer applications, several general applications of computers are available for persons performing research in any aspect of <u>Artemia</u> biology. A general description of each of these and information regarding sources of further information are included below.

THE <u>ARTEMIA</u> BIBLIOGRAPHY

Extensive bibliographies of literature pertinent to <u>Artemia</u> research have been published [1-4]. The two most recent examples have made use of computer-based storage, retrieval, and search capabilities. The current version is now available using several different microcomputers (Apple II+, e, c and some Apple compatibles; Apple MacIntosh; IBM-PC and most IBM compatibles) and a commercial program (BOOKENDS; Sensible Software, Inc., 335 E. Big Beaver, Troy, Michigan 48083 (USA)) which is available at a minimal cost (List: $100.00-$150.00 US). The reference data bases (approximately 5200 entries) are available by mail from the Artemia Bibliographic Center (ABC) at cost (five 5 ¼ floppy disks and postage) or by a computer network (see below). Also, updates will be available periodically via mail or computer network.

The ABC would appreciate information regarding errors, omissions of earlier references, and <u>particularly</u> new references. Please send the following information and a copy of the reference if possible:

1. Author(s) with Full Names.
2. Date of Publication.
3. Complete Title in Original Language and English translation.
4. Complete Reference (Journal; Volume; Page Numbers)
   Complete Book or Monograph Title; Editor(s); Publisher; Location of Publication; Page Numbers).
5. Suggested keywords.

439

COMMUNICATION VIA COMPUTER NETWORKS

Several national and international computer networks are available for use by members of educational and research institutions. A recent publication [5] describes most of the commonly available networks and discusses some of the possible internetwork connections. Communication on a single network is relatively simple and communication between different networks, although somewhat more complicated initially, is simple once the correct sequences have been determined. It is suggested that all persons performing research with Artemia investigate the availability of these networks at their own institution. If one of these networks is or becomes available, attempt to communicate with the ABC (BITNET: BIORPM@HOFSTRA) and the ARC (BITNET: ARTEMIA@BGERUG51). The publication mentioned above and the assistance of your computer specialists may be necessary depending upon the network you have available. A confirmation that your attempt has been successful will be returned as soon as possible. Both the ABC and the ARC will collect a list of users and their network addresses and these will be available to other users. Also, other information about the networks and their use, information about the bibliography, the bibliography files, periodic updates of the bibliography, and other information about Artemia research can then be communicated to other users.

LIBRARY OF DNA, RNA, AND AMINO ACID SEQUENCES

Some of the participants in the workshop have suggested that since many investigators are beginning to determine DNA, RNA and amino acid sequences for parts of the Artemia genome, that a computer-based library be available for the exchange of this information. If there appears to be an interest in such a project, this might be incorporated into the network discussed above. Those interested in establishing such a library should communicate with the ABC and/or the ARC. Further information about this project will then be made available.

REFERENCES

1. J. L. Littlepage and M. N. McGinley, A bibliography of the Genus Artemia (Artemia salina) 1812-1962, Special Publication Number 1, San Francisco Aquarium Society, San Francisco, California (1965).
2. R. P. McCourt and P. Lavens, in "The Brine Shrimp Artemia Bibliography", Artemia Reference Center, State University of Ghent, Ghent, Belgium (1985).
3. P. Sorgeloos, A bibliography on the brine shrimp Artemia salina L, in "Special Publication No.1. European Mariculture Society", E. Jaspers, ed., Bredene, Belgium (1976).
4. P. Sorgeloos, R. P. McCourt, P. Lavens, L. Spectorova and A. N. Khalaf, Updated bibliography of the brine shrimp Artemia, Special Publication No. 5. European Mariculture Society, Bredene, Belgium (1980).
5. J. S. Quarterman and J. C. Hoskins, Notable computer networks, Communications of the ACM 29(10):932 (1986).

PARTICIPANTS

R. A. Acey
Department of Chemistry
California State University
Long Beach, Cal. 90840, U.S.A.

G. Badaracco
Dipartimento di Genetica e di
Biologia die Microrganismi
Universita di Milano
20133 Milano, ITALY

J. C. Bagshaw
Department of Biology and Biotech.
Worcester Polytechnic Institute
Worcester, MA. 01609, U.S.A.

J. E. Breckenridge
Department of Biology
Dalhousie University
Halifax, Nova Scotia B3H 4J1 CANADA

M. G. Cacace
Istituto di Biochimica delle
Proteine ed Enzimologia, CNR
Via Toiano 6, 80072 Arco Felice
ITALY

E. J. Campbell
Department of Biology
Dalhousie University
Halifax, Nova Scotia B3H 4J1 CANADA

J. S. Clegg
Bodega Marine Laboratory
Bodega Bay, Cal. 94923, U.S.A.

F. P. Conte
Department of Zoology
Oregon State University
Corvallis, Oregon 97331-2914, U.S.A.

G. R. J. Criel
Department of Anatomy
University of Gent
Ledeganckstraat, 35, B-9000 Gent
BELGIUM

J. Cruces
Instituto de Investigaciones
Biomedicas Facultad de Medicina
de la Universidad Autonoma,
Arzobispo Morcillo, 4, 28029 Madrid
SPAIN

W. Decleir
Laboratorium voor Biochemie
en Algemene Dierkunde,
University of Antwerp
Gronenborgerlaan 171, B-2020 Antwerp
BELGIUM

J. A. Freeman
Department of Biology
University of South Alabama
Mobile, Alabama 36680, U.S.A.

M. Gorski
Department of Biological Sciences
University of Windsor
Windsor, Ontario N9B 3P4, CANADA

D. J. Goss
Department of Chemistry
Hunter College of CUNY
695 Park Ave., New York, N.Y. 10021
U.S.A.

J.-Z. Guo
Department of Physiology
Faculty of Medicine
University of Manitoba
Winnipeg, Man. R3E 0W3 CANADA

A. Hernandorena
Laboratoire du Museum National
d'Historie Naturelle
Plateau de l'Atalaye
64202 Biarritz, FRANCE

L. E. Hokin
Department of Pharmacology
Medical Sciences Center
University of Wisconsin
Madison, Wis. 53706, U.S.A.

M. N. Horst
Division of Basic Sciences (Biochem.)
School of Medicine
Mercer University, 1550 College St.,
Macon, Georgia 31207, U.S.A.

C. Langdon
Department of Biology
Dalhousie University
Halifax, Nova Scotia B3H 4J1, CANADA

M. C. Lemaire
Department of Biology and Biotech.
Worcester Polytechnic Institute
Worcester, MA. 01609, U.S.A.

T. H. MacRae
Department of Biology
Dalhousie University
Halifax, Nova Scotia B3H 4J1 CANADA

A. G. Manning
Department of Biochemistry
University of Otago, P.O. Box 56,
Dunedin, NEW ZEALAND

R. Marco
Department of Biochemistry
Universidad Autonoma de Madrid
Arzobispo Morcillo, 4, 28029 Madrid,
SPAIN

R. P. McCourt
Department of Biology
Hofstra University
Hempstead, New York 11550 U.S.A.

A. G. McLennan
Department of Biochemistry
The University of Liverpool
Liverpool, U.K. L69 3BX

D. S. Miller
Department of Chemistry
California State University
Long Beach, Cal. 90840 U.S.A.

L. Moens
Department of Biochemistry
University of Antwerp (U.I.A.),
Universiteitsplein 1, 2610 Wilrijk,
BELGIUM

N. Munuswamy
Department of Zoology
University of Madras, Guindy Campus
Madras-600 025, INDIA

M. Prescott
Department of Biochemistry
University of Liverpool
Liverpool, ENGLAND L69 3BX

M. Raineri
Institute of Comparative Anatomy
University of Genoa, v. Balbi 5,
16136 Genova, ITALY

L. M. V. Raja
Centre for Biotechnology
Anna University
Madras 600 025, INDIA

J. Renart
Instituto de Investigaciones
Biomedicas Facultad de Medicina
de la Universidad Autonoma
Arzobispo Morcillo, 4, 28029 Madrid
SPAIN

J. F. Samain
IFREMER DRV/PA, BP 70,
29263 Plouzane, FRANCE

L. Sastre
Instituto de Investigaciones
Biomedicas Facultad de Medicina
de la Universidad Autonoma
Arzobispo Morcillo, 4, 28029 Madrid
SPAIN

H. Slegers
Department of Biochemistry
University of Antwerp (U.I.A.)
Universiteitsplein 1, B-2610
Wilrijk, BELGIUM

M. Sonnenfeld
Department of Biological Sciences
University of Windsor
Windsor, Ontario N9B 3P4, CANADA

P. Sorgeloos
Artemia Reference Centre
State University of Ghent
Rozier 44, B-9000 Ghent, BELGIUM

S. L. Squires
Department of Chemistry
California State University
Long Beach, Cal. 90840, U.S.A.

K.-D. Spindler
Institute of Zoology
University of Dusseldorf
D-4000 Dusseldorf, F.R.G.

C. N. A. Trotman
Department of Biochemistry
University of Otago, P.O. Box 56,
Dunedin, NEW ZEALAND

C. G. Vallejo
Instituto de Investigaciones
Biomedicas CSIC, Facultad de Medicina
UAM, Arzobispo Morcillo, 4, 28029
Madrid, SPAIN

A. Van der Linden
Laboratory of Biochemistry and
General Zoology,
University of Antwerp,
Groenenborgerlaan 171, B-2020
Antwerpen, BELGIUM

J. C. Vaughn
Department of Zoology
Miami University,
Oxford, Ohio 45056 U.S.A.

A. H. Warner
Department of Biological Sciences
University of Windsor
Windsor, Ontario N9B 3P4, CANADA

G. Wolf
Laboratorium voor Biochemie en
Algemene Dierkunde,
University of Antwerp,
Groenenborgerlaan 171, B-2020
Antwerp, BELGIUM

WORKSHOP OBSERVERS

S. Bailey
Department of Chemistry
California State University
Long Beach, Cal. 90840, U.S.A.

C. Bedolla
Department of Chemistry
California State University
Long Beach, Cal. 90840, U.S.A.

D. Guttridge
Department of Chemistry
California State University
Long Beach, Cal. 90840, U.S.A.

M. Mustillo
Department of Chemistry
California State University
Long Beach, Cal. 90840, U.S.A.

V. Paranjape
School of Life Sciences
Jawaharlal Nehru University
New Delhi-110 067, INDIA

S. Rivera
Department of Chemistry
California State University
Long Beach, Cal. 90840 U.S.A.

E. N. Waindi
Department of Zoology
University of Nairobi
Nairobi, KEYENA

OTHER CONTRIBUTORS

M. Aerden
Department of Biochemistry
Universitaire Instelling Antwerpen
Universiteitsplein 1 - B 2610
Antwerpen-Wilrijk, BELGIUM

C. Barigozzi
Dipartimento di Genetica e
di Biologia dei Microrganismi
Universita di Milano - Via Celoria 26,
20133 Milano, ITALY

B. Batuecas
Departmento de Bioquímica
Instituto de Investigaciones Biomedicas
Facultad de Medicina, Universidad
Autónoma, Arzobispo Morcillo,
4 Madrid 28029 SPAIN

T. C. Becker
Department of Chemistry
Hunter College of the City University
of New York   New York, N.Y. U.S.A.

R. Benfante
Dipartimento di Genetica e
di Biologia dei Microrganismi
Universita di Milano - Via Celoria 26,
20133 Milano, ITALY

G. M. Blackburn
Department of Chemistry
University of Sheffield
Sheffield S3 7HF, U. K.

R. Blust
Laboratory of Biochemistry and Gen'l
Zoology, Groenenborgerlaan, 171,
2020 Antwerp, University of Antwerp,
RUCA, BELGIUM

M. Calleja
Departmento de Bioquímica
Instituto de Investigaciones Biomedicas
Facultad de Medicina, Universidad
Autónoma, Arzobispo Morcillo,
4 Madrid 28029 SPAIN

V. Carratora
Istituto di Biochimica delle Proteine
ed Enzimologia CNR, Via Toiano 6
80072 Arco Felice, ITALY

J. Y. Daniel
IFREMER DRV/PA
BP 70
29263 Plouzane, FRANCE

B. De Wachter
Laboratory of Biochemistry and
General Zoology
University of Antwerp (R.U.C.A.)
Groenenborgerlaan 171, B-2020
Antwerp, BELGIUM

J. N. Dholakia
Department of Biochemistry
University of Mississippi
Medical Center
Jackson, Miss. 39216-4506 U.S.A.

M. Diaz-Guerra
Instituto de Investigaciones Bioméd.
Departmento de Bioquímica de la
Autónoma, Arzobispo Morcillo,
4. 28029-Madrid, SPAIN

F. Diaz-Otero
Departmento de Bioquímica
Instituto de Investigaciones
Biomédicas, Facultad de Medicina,
Universidad Autónoma
Arzobispo Morcillo,
4 Madrid 28029 SPAIN

J. Diez-Sebastian
Departmento de Bioquímica
Instituto de Investigaciones
Biomédicas, Facultad de Medicina,
Universidad Autónoma
Arzobispo Morcillo,
4 Madrid 28029 SPAIN

E. Dominguez
Instituto de Investigaciones
Biomédicas, Dept. de Bioquimica
de la Universidad Autónoma,
Arzobispo Morcillo,
4. 28029-Madrid, SPAIN

R. Dommisse
Laboratory of Organic Chemistry
Groenenborgerlaan, 171, 2020 Antwerp
University of Antwerp
RUCA, BELGIUM

M. E. Edep
University of California
School of Medicine
San Francisco, Cal. 94143 U.S.A.

G. R. Farley
Department of Biology and Biotech.
Worcester Polytechnic Institute
Worcester, Mass. 01609, U.S.A.

E. Franco
Instituto de Investigaciones Biomédica
Departmento de Bioquimica de la
Universidad Autónoma,
Arzobispo Morcillo, 4. 28029-Madrid,
SPAIN

K. A. Frondorf
Department of Zoology
Miami University
Oxford, Ohio 45056 U.S.A.

B. Funke
Institut fur Zoologie, Lehrstuhl fur
Hormon- und Entwick lungsphysiologie,
Universitaet Duesseldorf, Universit-
aetsstr. 1, D-4000 Duesseldorf, F.R.G.

A. Gandarillas
Instituto de Investigaciones Bioméd.
Facultad de Medicina de la
Universidad Autónoma
Arzobispo Morcillo, 4. 28029 Madrid,
SPAIN

R. Garesse
Departmento de Bioquímica
Instituto de Investigaciones Bioméd.
Facultad de Medicina,
Universidad Autónoma
Arzobispo Morcillo, 4 Madrid 28029
SPAIN

S. P. Gieseg
Department of Biochemistry
University of Otago
Dunedin, NEW ZEALAND

I. Gil
Instituto de Investigaciones Bioméd.
Departmento de Bioquímica de la
Universidad Autónoma,
Arzobispo Morcillo, 4. 28029 Madrid,
SPAIN

R. Gomez-Rioja
Departmento de Bioquímica
Instituto de Investigaciones Biomed.
Facultad de Medicina,
 Universidad Autónoma
Arzobispo Morcillo, 4 Madrid 28029
SPAIN

L. J. Grzyb
Department of Biology and Biotech.
Worcester Polytechnic Institute
Worcester, Mass. 01609, U.S.A.

P. A. Healy
Department of Chemistry
California State University
Long Beach, Cal. U.S.A 90840

F. Hens
Laboratory of Biochemistry and
General Zoology (RUCA)
Groenenborgerlaan, 171 B-2020
Antwerpen, BELGIUM

S. A. Jackson
University of California
Bodega Marine Laboratory
Bodega Bay, Cal. 94923 U.S.A.

W. Jacob
Universitaire Instelling Antwerpen,
Antwerp, BELGIUM

A. Kean
Department of Biochemistry
University of Otago
Dunedin, NEW ZEALAND

H. T. Koller
Department of Zoology
Miami University
Oxford, Ohio 45056 U.S.A.

J. R. Le Coz
IFREMER DRV/PA
BP 70
29263 Plouzane, FRANCE

M. T. Macias
Instituto de Investigaciones Biomed.
Departmento de Bioquímica de la
Universidad Autónoma,
Arzobispo Morcillo, 4. 28029-Madrid,
SPAIN

C. J. Marshall
Department of Biochemistry
University of Otago
Dunedin, NEW ZEALAND

P. D. Maschner
Department of Zoology
Miami University
Oxford, Ohio 45056 U.S.A.

A. D. Milne
Department of Biochemistry
University of Liverpool
Liverpool L69 3BX, U. K.

J. Moal
IFREMER DRV/PA
BP 70
29263 Plouzane, FRANCE

C. Moratilla
Departmento de Bioquímica
Instituto de Investigaciones Biomed.
Facultad de Medicina,
Universidad Autónoma
Arzobispo Morcillo, 4. Madrid 28029
SPAIN

I. Palmero
Instituto de Investigaciones Biomed.
Departmento de Bioquímica de la
Universidad Autónoma,
Arzobispo Morcillo, 4. 28029-Madrid,
SPAIN

A. S. Pandey
Department of Biology
Dalhousie University
Halifax, N.S. B3H 4J1, CANADA

R. S. Pirie
Department of Biochemistry
University of Otago
Dunedin, NEW ZEALAND

R. J. Powell
Department of Biochemistry
University of Otago
Dunedin, NEW Zealand

M. Quintanilla
Instituto de Investigaciones Biomed.
Departmento de Bioquímica de la
Universidad Autónoma de Madrid
Arzobispo Morcillo, 4. 28029 Madrid,
SPAIN

N. S. Reddy
Department of Biochemistry
University of Mississippi
Medical Center
Jackson, Miss. 39216-4506 U.S.A.

D. J. Rounds
Department of Chemistry
Hunter College of the City University
of New York
New York, N.Y. U.S.A.

A. Sada
Istituto di Biochimica delle Proteine
ed Enzimologia CNR, Via Toiano 6
80072 Arco Felice, ITALY

M. C. Sanders
Department of Biology and Biotech.
Worcester Polytechnic Institute
Worcester, Mass. 01609, U.S.A.

W. H. Swanson
Department of Zoology
Miami University
Oxford, Ohio 45056 U.S.A.

W. P. Tate
Department of Biochemistry
University of Otago
Dunedin, NEW ZEALAND

G. E. Taylor
Department of Chemistry
University of Sheffield,
Sheffield S3 7HF, U. K.

S. Tenca
Dipartimento di Genetica e
di Biologia dei Microrganismi
Universita di Milano - Via Celoria 26,
20133 Milano ITALY

A. Valencia
Departmento de Bioquímica
Instituto de Investigaciones Bioméd.
Facultad de Medicina, Universidad
Autónoma, Arzobispo Morcillo,
4 Madrid 28029 SPAIN

J. R. Valverde
Departmento de Bioquímica
Instituto de Investigaciones Bioméd.
Facultad de Medicina,
Universidad Autónoma
Arzobispo Morcillo,
4 Madrid 28029 SPAIN

A. J. Wahba
Department of Biochemistry
University of Mississippi
Medical Center
Jackson, Miss. 39216-4506 U.S.A.

B. N. Yoshida
Department of Biology
California Institute of Technology
Pasadena, Cal. 91125 U.S.A.

Tubulin carboxypeptidase, 389
Tubulin tyrosine ligase, 389

Ubiquitin
  gene expression, 323
  gene structure, 365
UDP-agarose, 62, 64, 72
UDP-hexanolamine, 60

Vas deferens, 95
Villi, 105
Vitelline, 91
  granule, 147
  membrane, 120
Vitellogenesis, 100, 106, 108, 111,
    117
Vitellolysis, 144

WGA agarose, 63, 66

Yolk, 141
  bodies, 101, 103, 144, 147, 151
  degradation of, 131, 137, 145, 148,
    149
    regulation, 187
  formation, 106, 114, 115
  lipid droplets, 103, 106, 110, 120
  nucleus, 101, 106
    formation of, 120
  platelets, 101, 103, 106, 122, 144, 221
  vesicular bodies, 106, 110, 120
Y-organ, 87

Zinc binding proteins, 207